HANDBOOK OF SEMICONDUCTOR WAFER CLEANING TECHNOLOGY

HANDBOOK OF SEMICONDUCTOR WAFER CLEANING TECHNOLOGY

Science, Technology, and Applications

Edited by

Werner Kern

Werner Kern Associates
East Windsor, New Jersey

np

NOYES PUBLICATIONS
Park Ridge, New Jersey, U.S.A.

Library of Congress Catalog Card Number: 93-4078
ISBN: 0-8155-1331-3
Printed in the United States

Published in the United States of America by
Noyes Publications
Mill Road, Park Ridge, New Jersey 07656

10 9 8 7 6 5 4 3 2

Library of Congress Cataloging-in-Publication Data

Handbook of semiconductor wafer cleaning technology : science, technology, and applications / edited by Werner Kern.
p. cm.
Includes bibliographical references and index.
ISBN 0-8155-1331-3
1. Semiconductor wafers--Cleaning. I. Kern, Werner. 1925-
TK7871.85.H335 1993 93-4078
621.3815'2--dc20 CIP

MATERIALS SCIENCE AND PROCESS TECHNOLOGY SERIES

Electronic Materials and Process Technology

HANDBOOK OF DEPOSITION TECHNOLOGIES FOR FILMS AND COATINGS, Second Edition: edited by Rointan F. Bunshah

CHEMICAL VAPOR DEPOSITION FOR MICROELECTRONICS: by Arthur Sherman

SEMICONDUCTOR MATERIALS AND PROCESS TECHNOLOGY HANDBOOK: edited by Gary E. McGuire

HYBRID MICROCIRCUIT TECHNOLOGY HANDBOOK: by James J. Licari and Leonard R. Enlow

HANDBOOK OF THIN FILM DEPOSITION PROCESSES AND TECHNIQUES: edited by Klaus K. Schuegraf

IONIZED-CLUSTER BEAM DEPOSITION AND EPITAXY: by Toshinori Takagi

DIFFUSION PHENOMENA IN THIN FILMS AND MICROELECTRONIC MATERIALS: edited by Devendra Gupta and Paul S. Ho

HANDBOOK OF CONTAMINATION CONTROL IN MICROELECTRONICS: edited by Donald L. Tolliver

HANDBOOK OF ION BEAM PROCESSING TECHNOLOGY: edited by Jerome J. Cuomo, Stephen M. Rossnagel, and Harold R. Kaufman

CHARACTERIZATION OF SEMICONDUCTOR MATERIALS, Volume 1: edited by Gary E. McGuire

HANDBOOK OF PLASMA PROCESSING TECHNOLOGY: edited by Stephen M. Rossnagel, Jerome J. Cuomo, and William D. Westwood

HANDBOOK OF SEMICONDUCTOR SILICON TECHNOLOGY: edited by William C. O'Mara, Robert B. Herring, and Lee P. Hunt

HANDBOOK OF POLYMER COATINGS FOR ELECTRONICS, Second Edition: by James Licari and Laura A. Hughes

HANDBOOK OF SPUTTER DEPOSITION TECHNOLOGY: by Kiyotaka Wasa and Shigeru Hayakawa

HANDBOOK OF VLSI MICROLITHOGRAPHY: edited by William B. Glendinning and John N. Helbert

CHEMISTRY OF SUPERCONDUCTOR MATERIALS: edited by Terrell A. Vanderah

CHEMICAL VAPOR DEPOSITION OF TUNGSTEN AND TUNGSTEN SILICIDES: by John E. J. Schmitz

ELECTROCHEMISTRY OF SEMICONDUCTORS AND ELECTRONICS: edited by John McHardy and Frank Ludwig

HANDBOOK OF CHEMICAL VAPOR DEPOSITION: by Hugh O. Pierson

DIAMOND FILMS AND COATINGS: edited by Robert F. Davis

ELECTRODEPOSITION: by Jack W. Dini

HANDBOOK OF SEMICONDUCTOR WAFER CLEANING TECHNOLOGY: edited by Werner Kern

CONTACTS TO SEMICONDUCTORS: edited by Leonard J. Brillson

HANDBOOK OF MULTILEVEL METALLIZATION FOR INTEGRATED CIRCUITS: edited by Syd R. Wilson, Clarence J. Tracy, and John L. Freeman, Jr.

HANDBOOK OF CARBON, GRAPHITE, DIAMONDS AND FULLERENES: by Hugh O. Pierson

Ceramic and Other Materials—Processing and Technology

SOL-GEL TECHNOLOGY FOR THIN FILMS, FIBERS, PREFORMS, ELECTRONICS AND SPECIALTY SHAPES: edited by Lisa C. Klein

FIBER REINFORCED CERAMIC COMPOSITES: edited by K. S. Mazdiyasni

ADVANCED CERAMIC PROCESSING AND TECHNOLOGY, Volume 1: edited by Jon G. P. Binner

FRICTION AND WEAR TRANSITIONS OF MATERIALS: by Peter J. Blau

SHOCK WAVES FOR INDUSTRIAL APPLICATIONS: edited by Lawrence E. Murr

SPECIAL MELTING AND PROCESSING TECHNOLOGIES: edited by G. K. Bhat

CORROSION OF GLASS, CERAMICS AND CERAMIC SUPERCONDUCTORS: edited by David E. Clark and Bruce K. Zoitos

HANDBOOK OF INDUSTRIAL REFRACTORIES TECHNOLOGY: by Stephen C. Carniglia and Gordon L. Barna

CERAMIC FILMS AND COATINGS: edited by John B. Wachtman and Richard A. Haber

Related Titles

ADHESIVES TECHNOLOGY HANDBOOK: by Arthur H. Landrock

HANDBOOK OF THERMOSET PLASTICS: edited by Sidney H. Goodman

SURFACE PREPARATION TECHNIQUES FOR ADHESIVE BONDING: by Raymond F. Wegman

FORMULATING PLASTICS AND ELASTOMERS BY COMPUTER: by Ralph D. Hermansen

HANDBOOK OF ADHESIVE BONDED STRUCTURAL REPAIR: by Raymond F. Wegman and Thomas R. Tullos

CARBON–CARBON MATERIALS AND COMPOSITES: edited by John D. Buckley and Dan D. Edie

CODE COMPLIANCE FOR ADVANCED TECHNOLOGY FACILITIES: by William R. Acorn

Preface

The cleaning of semiconductor wafers has become one of the most critical operations in the fabrication of semiconductor devices, especially advanced ULSI silicon circuits. A considerable body of technical and scientific literature has been published on this important subject; however, it is widely dispersed in numerous journals and symposia proceedings. It is the objective of this book to bring together in one volume all pertinent knowledge on semiconductor wafer cleaning and the scientific and technical disciplines associated directly or indirectly with this subject. The book provides the first comprehensive and up-to-date coverage of this rapidly evolving field. Its thirteen chapters were written by nineteen scientists who are recognized experts in each topic.

The scope of this book is very broad, covering all aspects of wafer cleaning. Emphasis is on practical applications in the fab combined with authoritative scientific background information to provide a solid scientific basis for the understanding of the chemical and physical processes involved in cleaning and in the analytical methods of testing and evaluation.

This user-friendly handbook has been based on lectures presented in an intensive two-day course on wafer cleaning technology that was organized by the editor and held in San Francisco and Princeton with the participation of several of the chapter authors. The enthusiastic response by the course attendees convinced us that a book treating this material in greater depth in the format of topical overviews is indeed highly desirable, if not urgently needed.

In setting out to create a comprehensive handbook that would fulfill this need, each chapter author or group of co-authors contributed specialized and complementary expert knowledge in covering this multidisciplinary subject.

The book comprises the following five parts, with one to four chapters each:

I. Introduction and Overview

II. Wet-Chemical Processes

III. Dry Cleaning Processes

IV. Analytical and Control Aspects

V. Conclusions and Future Directions

The reviews presented on these chapters were completed by mid-1992 and include the literature up to this time. Each chapter was closely edited to eliminate excessive duplication.

The depth and breadth of the material should appeal to those new in the field as well as to experienced professionals. The volume is intended to serve as a handbook for practitioners and professionals in the field, including fab engineers, scientists and technicians working in the production or development of semiconductor microelectronic devices. It should also prove useful to manufacturers of processing equipment, persons concerned with contamination control and analysis, and students attending advanced or specialized technical courses.

Finally, I wish to thank the contributing authors for their considerable efforts in summarizing the voluminous literature of their specialties and for preparing chapters of outstanding quality. I am also indebted to George Narita, Vice President and Executive Editor of Noyes Publications, for his helpful cooperation in this exciting venture.

East Windsor, New Jersey
October, 1992

Werner Kern

Contributors

Donald C. Burkman
Unit Instruments
Yorba Linda, CA

Yves J. Chabal
AT&T
Bell Laboratories
Murray Hill, NJ

Bruce E. Deal
Department of
Electrical Engineering
Stanford University
Stanford, CA

Donald Deal
FSI International, Inc.
Chaska, MN

Robert P. Donovan
Research Triangle Institute
Research Triangle Park, NC

G. (John) Foggiato
Quester Technology, Inc.
Fremont, CA

Donald C. Grant
FSI International, Inc.
Chaska, MN

C. Robert Helms
Department of
Electrical Engineering
Stanford University
Stanford, CA

Gregg S. Higashi
AT&T
Bell Laboratories
Murry Hill, NJ

Richard S. Hockett
Charles Evans & Associates
Redwood City, CA

Emil Kamieniecki
SemiTest, Inc.
Billerica, MA

Werner Kern
Werner Kern Associates
East Windsor, NJ

Venu B. Menon
SEMATECH, Inc.
Austin, TX

Robert J. Nemanich
Department of Physics
North Carolina State University
Raleigh, NC

Charlie A. Peterson
Empak
Chanhassen, MN

Ronald A. Rudder
Research Triangle Institute
Research Triangle Park, NC

Jerzy Ruzyllo
Department of Electrical
and Computer Engineering
The Pennsylvania State University
University Park, PA

Raymond E. Thomas
Research Triangle Institute
Research Triangle Park, NC

Donald L. Tolliver
Motorola/SEMATECH, Inc.
University of Arizona
Tucson, AZ

John R. Vig
U.S. Army Research Laboratory
Department of the Army
Fort Monmouth, NJ

Contents

Part I. Introduction and Overview

Part II. Wet-Chemical Processes

Part III. Dry Cleaning Processes

Part I.

Introduction and Overview

1

Overview and Evolution of Semiconductor Wafer Contamination and Cleaning Technology

Werner Kern

1.0 INTRODUCTION

1.1 Importance of Clean Wafer Surfaces

The importance of clean substrate surfaces in the fabrication of semiconductor microelectronic devices has been recognized since the dawn of solid-state device technology in the 1950s. It is well known that the device performance, reliability, and product yield of silicon circuits are critically affected by the presence of chemical contaminants and particulate impurities on the wafer or device surface. Effective techniques for cleaning silicon wafers initially and after oxidation and patterning are now more important than ever before because of the extreme sensitivity of the semiconductor surface and the submicron sizes of the device features. As a consequence, the preparation of *ultraclean* silicon wafers has become one of the key technologies in the fabrication of ULSI silicon circuits, such as 64- and 256-megabit DRAM devices. The term "ultraclean" may be defined in terms of the concentration of both chemical contaminants and particles on the silicon surface. Specifically, total metallic impurities should be, conservatively speaking, less than 10^{10} atoms per cm^2. Particles larger than 0.1 µm in size should be fewer than approximately 0.1 per cm^2, which translates to fewer than 30 particles per 200 mm wafer. These extremely low numbers are impressive indeed! The reason for these stringent specifications is the fact that the overall device quality, as noted, is critically affected by trace impurities. Each of the hundreds of processing steps in the fabrication of advanced silicon circuits can contribute to contamination.

1.2 Wafer Cleaning Technology

The objective of wafer cleaning is the removal of particulate and chemical impurities from the semiconductor surface without damaging or deleteriously altering the substrate surface. Dry-physical, wet-chemical, and vapor-phase methods can be used to achieve these objectives. An array of equipment is available for implementing the various processes for industrial applications.

Nearly all wafer cleaning processes have been developed specifically for silicon since silicon semiconductor devices are by far the most important industrially for the fabrication of integrated circuits (ICs). Special procedures must be used for germanium and compound semiconductor materials.

The traditional approach of wafer cleaning is based on wet-chemical processes, which use mostly hydrogen peroxide solutions. Successful results have been achieved by this approach for the past twenty-five years. However, the relatively large consumption of chemicals required by these processes, the disposal of chemical waste, and the incompatibility with advanced concepts in integrated processing (such as cluster tooling) are the main reasons why methods based on gas-phase cleaning are now being developed that are less affected by these limitations.

The development of wafer cleaning technology had a slow start in the early period of 1950 to 1970, but then accelerated with the refinements of semiconductor device architecture and the increasing criticalness of contaminant-free surfaces. The greatly increased level of research and development of improved and new cleaning processes, coupled with advances in analytical methodology and instrumentation for the detection and characterization of impurities and surface structures, has been especially pronounced for the past three years. This increased level of activity is exemplified by recent international conferences and symposia on wafer cleaning and related science and technology (1)-(8), and the appearance of several volumes on particle contamination (9)-(11).

1.3 Scope and Organization of This Chapter

This chapter is intended as a general introduction to and overview of semiconductor wafer cleaning, as discussed in this book. There are many important aspects and topics associated with this subject, both scientific and technological, that should be addressed to gain a well-rounded understanding of this important field of semiconductor processing. Details are kept to a minimum in this overview, except for some topics that merit special emphasis or that are not discussed elsewhere in this volume, since the various chapters treat the material in depth.

It is informative and interesting to review the development of wafer cleaning science and technology from the beginning to the present time. I have attempted to trace who has done what, when, and how, documenting the contributions in the literature references and stressing major achievements for proper perspective.

The chapter is divided into three main sections:

- Overview of wafer contamination aspects
- Overview of wafer cleaning technology
- Evolution of wafer cleaning science and technology

2.0 OVERVIEW OF WAFER CONTAMINATION ASPECTS

2.1 Types and Origins of Contaminants

Contaminants on semiconductor wafer surfaces exist as contaminant films, discrete particles or particulates (groups of particles) and adsorbed gases, as summarized in Table 1. Surface contaminant films and particles can be classified as molecular compounds, ionic materials, and atomic species. Molecular compounds are mostly particles or films of condensed organic vapors from lubricants, greases, photoresists, solvent residues, organic components from DI water or plastic storage containers, and metal oxides or hydroxides. Ionic materials comprise cations and anions, mostly from inorganic compounds that may be physically adsorbed or chemically bonded (chemisorbed), such as ions of sodium, fluorine and chlorine. Atomic or elemental species comprise metals, such as gold and copper, that may be chemically or electrochemically plated out on the semiconductor surface from hydrofluoric acid (HF)-containing solutions, or they may consist of silicon particles or metal debris from equipment.

The sources of contaminants are listed in Table 2 and are seen to be manifold. Particles can originate from equipment, processing chemicals, factory operators, gas piping, etc. Mechanical (moving) equipment and containers for wafers and liquids are especially serious sources, whereas materials, liquids, gaseous chemicals, and ambient air tend to cause less particle contamination, but all contribute significantly to the generation of contaminant films. Static charge built-up on wafers and carriers is a powerful mechanism of particle deposition that is often overlooked. The origins of particles in advanced integrated circuits (DRAMs or Dynamic Random Access Memory devices) are shown in Table 3 in terms of percent distribution of the total. As the complexities of the devices increase to the present 16-Mbit types and to the future 64-Mbit types, the distribution shifts as indicated.

Table 1. Forms and Types of Contaminants

▲ **Equipment:**	Mechanical	Gas piping
	Deposition systems	Metal tweezers
	Ion implanters	Liquid-containers
▲ **Humans:**	Factory operators	
	Process engineers	
▲ **Materials:**	Liquid chemicals	Photoresists
	Etchants	Air
	D.I. water	Gasses
▲ **Processes:**	Combination of all above sources	

Table 2. Sources of Particles and Contaminant Films

▲ **Forms:**	Films, discrete particles, particulates, micro-droplets, vapors and gases
▲ **Types:**	Molecular, ionic, atomic, gaseous
▲ **Ionic:**	Physisorbed and chemisorbed cations and anions from inorganics; e.g. Na^+, Cl^-, SO_4^{-2}, fluoride species
▲ **Atomic:**	Elemental metal films and particles; e.g. electrochemically plated Au, Ag, Cu films; particles of Si, Fe, Ni
▲ **Gaseous:**	Adsorbed gases and vapors; generally of little practical consequence

Table 3. Distribution of Particle Defects (>0.5 μm) in DRAMs. *(Source: Lam Research Corp. 1990.)*

Particle Origin	*% Distribution in DRAMs*			
	1-Mbit	*4-Mbit*	*16-Mbit*	*64-Mbit*
▲ Tools	40	40	35	25
▲ Process	25	25	40	60
▲ Environment	25	25	15	10
▲ Handling	10	10	10	5

2.2 Types of Semiconductor Wafers

The vast majority of semiconductor wafers used in the fabrication of solid-state microelectronic devices are silicon. However, a small fraction of all wafers are compound semiconductors, such as gallium arsenide, gallium phosphide, and numerous complex alloys. The importance of these materials has grown steadily for unique applications in opto-electronics. The great differences in chemical properties between these semiconductors and those of silicon require different and specialized cleaning treatments. Published information on cleaning of compound semiconductor wafers is scarce and, therefore, has to be restricted in this chapter (and in the book as a whole) to limited comments whenever possible.

Semiconductor wafers can be in the form of mechanically lapped and chemo-mechanically polished slices cut from single-crystal ingots. They may be coated with an epitaxial layer of the semiconductor with different dopant type and concentration, or they may be coated with a film of uniform or patterned silicon dioxide. Up to this point in the processing, wafer cleaning operations can utilize highly reactive chemicals that do not attack these corrosion-resistant materials. The situation changes drastically once layers or patterns of deposited metals are present on the wafers. The cleaning chemistry must then be confined to mild and noncorrosive treatments, such as rinses with dilute acids, de-ionized (DI) water, and selected organic solvents.

2.3 Effects of Contaminants on Semiconductor Devices

The effects of contaminants on semiconductor materials and dielectrics during wafer processing, and the effects on the finished semiconductor devices, are complex and depend on the nature and quantity of a specific type of contaminant. The importance of their control and minimization is obvious from the fact that over fifty percent of the yield losses in IC manufacturing are caused by microcontamination. A selection of representative papers published from 1987 to 1992 on this important subject is included as Refs. 12 - 31. A brief summary of contamination effects is presented in Table 4.

Table 4. Effects of Contaminants

▲ **Molecular types:**	Block and mask operations Impair adhesion Form deleterious decomposition products Nucleate defects in films
▲ **Ionic types:**	Diffuse on surface, in bulk, at interfaces Cause electrical device defects Degrade device performance and yield Cause crystal defects Lower oxide breakdown field
▲ **Atomic types:**	Can diffuse readily Cause surface conduction Decrease minority-carrier lifetime Degrade electrical device performance Lower product yield Nucleate crystal defects Particles short-out conductor lines

Molecular contaminant films on wafer surfaces can prevent effective cleaning or rinsing, impair good adhesion of deposited films, and form deleterious decomposition products. For example, organic residues, if heated to high temperatures in a non-oxidizing atmosphere, can carbonize and, in the case of silicon wafers, form silicon carbide that can nucleate polycrystalline regions in an epitaxial deposit.

Ionic contaminants cause a host of problems in semiconductor devices. During high-temperature processing, or on application of an electric field, they may diffuse into the bulk of the semiconductor structure or spread on the surface, leading to electrical defects, device degradation, and yield losses. For example, highly mobile alkali ions in amorphous SiO_2 films on Si may cause drift currents and unstable surface potential, shifts in threshold and flat-band voltages, surface current leakage, lowering of the oxide breakdown field of thermally grown films of SiO_2, etc. In the growth of epitaxial silicon layers, sufficiently high concentrations of ions can give rise to twinning dislocations, stacking faults, and other crystal defects.

Certain metallic contaminants are especially detrimental to the performance of semiconductors devices, as indicated in Table 5. Since silicon is above hydrogen in the electromotive series of the elements, heavy metals tend to deposit from solution on its surface by galvanic action, actually plating out with high efficiency, especially from HF-containing etchants. If not removed, these impurities may diffuse into the silicon substrate during later heat treatments and introduce energy levels into the forbidden band to act as traps or generation/recombination centers, cause uncontrolled drifts in the semiconductor surface potential, affect the surface minority-carrier lifetime and the surface recombination velocity, lead to inversion or accumulation layers, cause excessive leakage currents, and give rise to various other device degradation and reliability problems. Metal contaminants in or on semiconductor wafers can lead to structural defects in vapor-grown epitaxial layers and degrade the breakdown voltage of gate oxides.

Table 5. Impurity Elements

The following common impurity elements from chemicals and processing can be deleterious to silicon devices:

▲ **Heavy metals** (most critical)
Fe, Cu, Ni, Zn, Cr, Au, Hg, Ag

▲ **Alkali metals** (critical)
Na, K, Li

▲ **Light elements** (less serious)
Al, Mg, Ca, C, S, Cl, F

Particles can cause blocking or masking of wafer processing operations, such as photolithography, etching, deposition, and rinsing. They may obstinately adhere to surfaces by electrostatic adsorption and may become embedded during film formation. Deposition and removal of particles becomes exacerbated as the size decreases because of the extremely strong adhesion forces. Furthermore, particles constitute a potential source of chemical contamination, depending on their composition. Particles that are present during film growth or deposition can lead to pinholes, microvoids, microcracks, and the generation of defects as noted above, depending on their chemical composition. In later stages of device fabrication, particles can cause shorts between conductor lines if they are sufficiently large, conductive, and located adjacently between conductor lines. They are considered potential device killers if their size is larger than one tenth the size of the minimum size feature of the particular integrated circuit.

Adsorbed gases and moisture have much less serious effects, but they can cause problems by outgassing in vacuum systems and affecting the quality of deposited films.

2.4 Prevention of Contamination from Equipment and Processing

Semiconductor surface microcontamination and its control (9), prevention, detection, and measurement are topics that may seem beyond the scope of wafer cleaning. However, these aspects are intimately interconnected with wafer cleaning and are, therefore, included briefly in this chapter and more extensively in several chapters of the present volume.

The key notion of this topic is *prevention.* If we can prevent contamination during the entire semiconductor device manufacturing process by creating and maintaining super-clean conditions in equipment, materials, and environment, there would be little need for wafer cleaning. Furthermore, it is generally easier (or less difficult!) to prevent contamination rather than to remove it once it has taken place. Therefore, avoiding contamination must be the first priority, and strict contamination control should be exercised to the fullest extent throughout device manufacturing (32). Changes in contamination control requirements have been reassessed recently (17)(33).

Processing equipment has been the major source of particle contamination (34) and must be controlled effectively by eliminating dust particles through scheduled maintenance and by electrostatic charge removal (34). Particle generation can be further minimized by eliminating friction of moving equipment parts, avoidance of turbulent gas flows, reduction of operator handling through automation, and by exercising periodic cleanup actions (35).

Next to equipment, semiconductor wafer processing in the fab must be controlled. Carefully optimized processing conditions for film deposition, plasma etching, ion implantation, thermal treatments, and other critical processing steps are effective measures for preventing particle deposition. For example, in chemical vapor deposition, particle-generating homogeneous gas-phase nucleation must be minimized by optimizing the reaction parameters. Recirculation of the partially depleted reaction gases must be avoided by improved equipment design. Sudden bursts in the introduction of gases into a system should be prevented by using "soft starts" to gradually increase the gas flow rates, and so forth (36).

2.5 Purity of Chemicals (37)

Many chemicals are used in the fabrication of semiconductor devices and impurities in those chemicals can critically affect the quality of the devices. Stringent control must be imposed to minimize contamination by transfer of impurities from these sources. Most chemicals used in wafer processing are either gases or liquids. Gases are obtainable at very high purity and can be ultra-filtered relatively easily to remove particles quite effectively. Liquid processing chemicals are numerous, as shown in the compilation of Table 6.

Impurity levels in chemicals can be quite variable. A list of trace metals found in several liquid chemicals of typical electronic purity grade is presented in Table 7 as an example. It can be seen that the impurity concentrations vary considerably for any one chemical, depending on its source. Great efforts have been made in recent years by the producers of chemicals to provide the electronics industry with *ultrapure* materials (7)(38)-(40). "Ultrapure" can be defined as meaning total impurity concentrations in the low ppb (parts per billion) range. A more restricted alternative for producing ultrapure acids is the point-of-use generation with reprocessors (41)(42).

The method of dispensing can affect the particle concentration very significantly, as exemplified by the results in Table 8 where bulk-distributed chemicals are compared with their bottle-dispensed counterparts. The latter exhibit very high particle counts from recontamination during the dispensing step. This secondary type of contamination of originally ultrapure chemicals that is introduced from containers, pipes, valves, etc. has been a serious and often neglected problem that demands stringent control (38)(39).

Table 6. Liquid Chemicals in the Semiconductor Industry

▲ Acids	▲ Alkalis	▲ Resist chemicals
HF	NH_4OH	Resist preparations
H_2SO_4	KOH	Developers
HCl	NaOH	Solvents
H_3PO_4	Choline	Strippers
HNO_3	Tertiary amines	Adhesion promoters
CH_3CO_2H		Coupling agents

▲ Solvents	▲ Reactants	▲ Organometallics of
Isopropanol	H_2O_2	Si, Ge
Ethanol	NH_4F	B, P, As, Sb
Trichloroethylene	$SiCl_4$	Al, Cu, Ga, In
Acetone	$SiHCl_3$	Ta, Nb
Toluene	$Si(C_2H_5O)_4$	▲ Universal
Ethylacetate	Br	D.I. Water
Methylene chloride	EDTA	
Freons	Surfactants	

Table 7. Trace Metallic Impurities in some Liquid Chemicals. *(Source: Balazs Analytical Labs.)*

Chemical	*No. Brands*	*No. Metals*	*ppb Range*
H_2O_2	5	30	6.2-160
HF	5	30	25-133
IPA	2	26	44, 184
MeOH	2	26	100, 184
H_2SO_4	6	30	28-246
HCl	5	30	81-470
Acetone	2	26	30, 1,043
Resists	5	16	430-1,400
Strippers	7	17	264-4,945

Table 8. Bulk-Distributed vs. Bottled Chemicals. *(Source: Grant and Schmidt, FSI)*

▲ *Bulk,** *op. hrs*	*>0.5μm Particles per mL* H_2SO_4	NH_4OH	H_2O_2	HCl
0.1	27	28	0.06	0.02
1.0	7.0	1.9	0.03	0.01
8.0	2.0	0.17	0.02	<0.01
▲ *Bottled***	60	500	100	40

* FSI Chemfill™
** Cleanroom™ Low-particle, 1 gallon bottle, Ashland Chemical Company

An interesting approach to circumvent these difficulties is in situ generation of aqueous chemicals by use of high-purity DI water and reactant gases, such as ozone (43)-(45), HCl, and NH_3 (45).

De-ionized water, which could be considered the primary chemical, is used widely as a diluent, cleaning chemical, and rinsing agent in many wafer cleaning process steps. Chemical impurities, organics, and particles in DI water must also be carefully controlled to prevent contamination of the semiconductor wafers (38)(46)(47). Typical impurity levels of ultrapure DI water are compared with those of high-purity sulfuric acid and hydrofluoric acid in Table 9.

In summary, it can be said that ultrapure, low-particulate chemicals, organic solvents, and DI water, should be used for all critical wafer processing operations. High-purity gas, water, and chemical bulk-distributed delivery subsystems with point-of-use ultra-filtration should be installed in the fabrication areas to eliminate particle-generating dispensing from individual source containers. In situ generation of aqueous chemicals should be implemented whenever possible.

Table 9. Impurities in Chemicals vs. Water in 1991. *(Source: Millipore Corp.)*

	98% H_2SO_4	*49% HF*	*Ultrapure Water*
▲ Particles (≥0.2μm)	100-500 per mL	10-50 per mL	6 per mL
▲ Individual metal ions	1-5 ppb	1 ppb	<0.1 ppb
▲ Total metal ions	≤70 ppb	≤40 ppb	<1 ppb

2.6 Analytical Methods

Another important aspect of contamination control that can only be briefly discussed is the analytical methodology for measuring and monitoring trace and ultra-trace levels of impurities and particles in processing reactants, DI water, gases, and the ambient air in processing areas on the one hand, and on the semiconductor wafer surface on the other. Analysis and measurement of contaminants in source materials and on the wafer surface are needed in establishing process conditions to identify what specific impurities are transferred during processing to the critical semiconductor surface and at what concentration. This information is the first step to be taken in improving a process to attain ultraclean wafers. On-line, in situ analytical testing is desirable in many critical operations to continuously monitor the concentration of specific contaminants such as particles, gaseous species, moisture, or metal ions so as to allow real-time process control (48).

Impressive progress has been made in recent years in increasing the sensitivity and speed of non-destructive instrumental analysis methods for ultra-trace impurities on surfaces. Outstanding examples are secondary ion mass spectroscopy (SIMS) and total reflection x-ray fluorescence (TXRF) analysis, both combined with vapor-phase decomposition techniques to concentrate the impurities. These advanced methods of analysis now offer detection capabilities down to an astonishing 10^7 to 10^8 metal atoms per cm^2. Volume-sensitive methods for bulk contamination analysis of semiconductors have also been refined or developed, such as surface photovoltage (SPV) for measuring minority-carrier diffusion length, and deep-level transient spectroscopy (DLTS) for lifetime measurements (49)-(56).

The atomic structure and morphology of semiconductor surfaces after various chemical treatments has been studied by diffraction, imaging, and vibrational methods. These sophisticated techniques of analysis are discussed in Ch. 10 on surface chemical composition and morphology. An excellent example of applications of some of these methods (including x-ray photoelectron spectroscopy, high resolution electron energy-loss spectroscopy, and angle resolved light scattering) has been published in Ref. 57.

3.0 OVERVIEW OF WAFER CLEANING TECHNOLOGY

3.1 Approaches for Attaining Clean Semiconductor Wafers

A combination of several approaches should be used for most effectively achieving the required high level of cleanliness of the semiconductor surface:

1. Control and prevention of contamination from processing equipment and chemicals
2. Removal of wafer contaminants by liquid cleaning methods
3. Removal of wafer contaminants by gas-phase cleaning methods

The first approach is applicable to all processing steps in the manufacture of semiconductor devices and has been discussed in Sec. 2.4. The contaminant removal methods noted in items 2 and 3 refer specifically to wafer cleaning steps prior to metal deposition and are reviewed in this section. References are representative rather than comprehensive, and the most recent advances will be discussed in more detail in Sec. 4.4.

3.2 Liquid Cleaning Methods

Liquid cleaning refers to processes that use liquid cleaning reagents or mixtures. They are usually, but not always, chemical in nature and are based on water as the liquid component, hence the more popular but incorrect term *wet-chemical*. Cleaning with organic solvents is neither chemical nor wet. Cleaning with aqueous oxidizing agents, on the other hand, is true wet-chemical.

The mechanism of liquid cleaning can be purely physical dissolution and/or chemical reaction dissolution. Chemical etching occurs when materials are removed by a chemical transformation to soluble species. Traditionally, chemical etching is expected to remove substantial quantities of a material, such as a deposited film on a substrate. Some chemical dissolution reactions may result in the removal of only a few atomic layers of material, and by above definition should be considered chemical etching processes; perhaps the term *microetching* should be used for such processes, which actually occur in most wet-chemical wafer cleaning processes.

Conventional wet-chemical etching processes for the removal of bulk quantities of electronic materials are beyond the scope of this book. Several comprehensive reviews have been published that cover this topic thoroughly (58)-(61).

Liquid cleaning methods for semiconductor wafers are based on the application of mineral acids, aqueous solutions including DI water, hydrogen peroxide containing mixtures, and organic solvents. The requirements for a typical wafer cleaning process are outlined in Table 10. Different process combinations and sequences are used for specific applications. A variety of technical equipment is available commercially for efficiently implementing cleaning processes for high-volume fabrication of ICs. Rinsing, drying, and storing of cleaned wafers is intimately connected with cleaning operations and is also addressed in the section that follows.

Organic solvents are rarely used for cleaning pre-metallized silicon wafers where much more effective cleaning agents can be used. Compound semiconductor wafers, however, are frequently treated with organic solvents to attain some degree of cleaning, since suitable and safe wet-chemical cleans cannot always be used or may not be available. Chlorofluorocarbon compounds, acetone, methanol, ethanol, and isopropyl alcohol (IPA) are solvents that are frequently used to remove organic impurities. IPA is generally the purest organic solvent available, as seen from Table 7, and is used extensively for vapor drying of water-rinsed wafers. Chlorofluorocarbon solvents are being phased out rapidly because of ecological problems related to ozone destruction.

Table 10. Requirements for a Silicon Wafer Cleaning Process

1. Effective removal of all types of surface contaminants
2. Not etching or damaging Si and SiO_2
3. Use of contamination-free and volatilizable chemicals
4. Relatively safe, simple, and economical for production applications
5. Ecologically acceptable, free of toxic waste products
6. Implementable by a variety of techniques

3.3 Wet-Chemical Cleaning Processes

Hydrofluoric Acid Solutions. Mixtures of concentrated hydrofluoric acid (49 wt% HF) and DI water have been widely used for removal by etching of silicon dioxide (SiO_2) films and silicate glasses (e.g., phosphosilicates, borophosphosilicates) that were grown or vapor deposited on semiconductor substrate wafers (58)(59). The chemical dissolution reactions have been identified and described in the literature and are discussed in Chs. 7 and 10.

The thin layer of native oxide on silicon, typically 1.0 to 1.5 nm thick, is removed by a brief immersion of the wafers in diluted (typically 1:50 or 1:100) ultrapure filtered HF solution at room temperature. The change of the wetting characteristics of the initially hydrophilic surface to a hydrophobic surface, which strongly repels aqueous solutions, indicates when the oxide dissolution is complete. The effect is due to the hydrogen-passivated silicon surface that results from exposure to HF solutions. The resulting hydrogenated silicon surface is highly sensitive to oppositely charged particles in solution, and also to organic impurities from DI water and ambient air that are strongly attracted to the surface. Treatments with HF solutions that leave the semiconductor surface bare must, therefore, be carried out with ultrapure and ultra-filtered HF solution in a very clean atmosphere. In addition to being an etchant for oxides and silicates, HF solutions desorb certain metallic impurities from the surface. On the other hand, they tend to contaminate silicon with iron deposits. The primary purpose of HF treatments, however, is to expose the silicon surface to subsequent attack by other cleaning agents, such as in an H_2O_2-based cleaning sequence (62)(63).

Mixtures of hydrofluoric acid and ammonium fluoride (NH_4F) are known as buffered oxide etch (BHF) and are used for pattern delineation etching of dielectric films to avoid loss of the photoresist polymer pattern that would not withstand the strongly acidic HF solution without a buffering agent. Whereas the free acid is the major etching species in aqueous HF solutions, the ionized fluoride associate HF_2^- is the major etchant species in buffered HF solutions. Addition of NH_4F increases the pH to 3 - 5, maintains the concentration of fluoride ions, stabilizes the etching rate, and produces the highly reactive HF_2^- ions. A commonly used BHF composition of 7:1 NH_4F (40%)-HF (49%) has a pH of about 4.5 and appears to contain only HF_2^- and F^-, with very little free HF acid. The SiO_2 etching rate of HF_2^- is four to five times as fast as that for the HF species in aqueous hydrofluoric acid (59). Additional details of the chemistry involved in these reactions (and of wet-chemical etching in general) can be found in Refs. 58 and 59. The subtle differences in the silicon surface morphology resulting from these two types of etchants are discussed in detail in Ch. 10.

Very recent references on acid etchants (H_2O-H_2O_2-HF, HNO_3-HF) and very dilute acids (1:10^3 - 1:10^6) for silicon cleaning are described in Sec. 4.4.

Sulfuric-Acid/Hydrogen-Peroxide Mixtures. Removal of heavy organic materials from silicon wafers, such as photoresist patterns and other visible gross contaminants of organic nature, can be accomplished with mixtures of 98% H_2SO_4 and 30% H_2O_2. Volume ratios of 2 - 4:1 are used at temperatures of 100°C and above. A treatment of 10 - 15 min at 130°C is most effective, followed by vigorous DI water rinsing to eliminate all of the viscous liquid. Organics are removed by wet-chemical oxidation, but inorganic contaminants, such as metals, are not desorbed. These mixtures, which are also known as "piranha etch" (because of their voracious ability to eradicate organics) or, incorrectly, "Caros acid", are extremely dangerous to handle in the fab; goggles, face shields, and gloves are needed to protect the operators. Finally, it is advantageous after the rinsing to strip the impurity-containing oxide film on silicon or on the thermal SiO_2 layer by dipping the wafers for 15 sec in 1% HF-H_2O (1:50), followed by a DI water rinse (62)(63).

Conventional RCA-Type Hydrogen Peroxide Mixtures (62)-(64). These are the most widely used and best established cleaning solutions for silicon wafers. They are made up of ultra-filtered, high-purity DI water, high-purity "not stabilized" hydrogen peroxide, and either electronic-grade ammonium hydroxide or electronic-grade hydrochloric acid. The hydrogen peroxide must be very low in aluminum and stabilizer additives (sodium phosphate, sodium stannate, or amino derivatives) to prevent wafer recontamination. These mixtures, used in two process steps, have become known as RCA standard cleans (SC-1 and SC-2). The treatment is usually preceded by the preliminary cleaning described in the previous section. The development of the *original* SC-1 and SC-2 mixtures is described in Section 4.2. Tables 11 and 12 summarize the salient features of these two solutions.

The first step uses a mixture (SC-1) of 5:1:1 vol. DI water, H_2O_2 (30%,"not stabilized"), and NH_4OH (29 w/w% as NH_3) at 70°C for 5 min, followed by quench and rinse with cold ultra-filtered DI water. This deceptively simple procedure removes any remaining organics by oxidative dissolution. Many metal contaminants (group IB, group IIB, Au, Ag, Cu, Ni, Cd, Co, and Cr) are dissolved, complexed, and removed from the surface.

Table 11. Cleaning Processes Based on Diluted Mixtures of H_2O_2 and NH_4OH or HCl

- ▲ First systematically developed wafer cleaning process for bare and oxidized Si. Introduced to RCA device fabrication in 1965, published in 1970.
- ▲ RCA cleaning process is based on a 2-step wet-oxidation and complexing treatment in aqueous H_2O_2- NH_4OH and H_2O_2 - HCl mixtures at 75-80°C for 10 min.
- ▲ Chemical Principles:
  1. H_2O_2 at high pH is a powerful oxidant,decomposing to $H_2O + O_2$
  2. NH_4OH is a strong complexant for many metals
  3. HCl in H_2O_2 forms soluble alkali and metal salts by dissolution and/or complexing
  4. Mixtures formulated not to attack Si or SiO_2

Table 12. Original RCA Cleaning Solutions

Step 1

- ▲ **"Standard Clean 1 or SC-1"**, alkaline peroxide mixture consisting of 5 vol H_2O + 1 vol H_2O_2 30% + 1 vol NH_4OH 29%, followed by D.I. water rinse
- ▲ Effects wet oxidation removal of organic surface films and exposes the surface for desorption of trace metals (Au, Ag, Cu, Ni, Cd, Zn, Co, Cr, etc.)
- ▲ Keeps forming and dissolving hydrous oxide film

Step 2

- ▲ **"Standard Clean 2 or SC-2"**, acidic peroxide mixture consisting of 6 vol H_2O + 1 vol H_2O_2 30% + 1 vol HCl 37%, followed by D.I. water rinse
- ▲ Dissolves alkali ions and hydroxides of Al^{+3}, Fe^{+3}, Mg^{+2}
- ▲ Desorbs by complexing residual metals
- ▲ Leaves protective passivation hydrated oxide film

The solution temperature should be 70°C for sufficient thermal activation, but must not exceed 80°C to avoid excessively fast decomposition of the H_2O_2 and loss of NH_3. The treatment is terminated by, ideally, an overflow quench with cold DI water to displace the surface layer of the liquid and to reduce the temperature to prevent any drying of the wafers on withdrawal from the bath.

SC-1 slowly dissolves the thin native oxide layer on silicon and forms a new one by oxidation of the surface. This oxide regeneration has a self-cleaning effect and aids the removal of particles by dislodging them. These effects account undoubtedly for some of the beneficial results achieved by the treatment.

SC-1 also etches silicon at a very low rate. The standard 5:1:1 composition can have a surface roughening effect due to non-uniform local micro-etching. Lower fractions of NH_4OH have been proposed to avoid micro-roughening of the silicon, as further discussed in Sec. 4.4.

The second step in the conventional RCA cleaning procedure uses a mixture (SC-2) consisting of 6:1:1 vol. DI water, H_2O_2 (30%,"not stabilized"), and HCl (37 w/w%). A solution temperature of 70°C for 5 - 10 minutes is used, followed by quenching and rinsing as in the SC-1 treatment. SC-2 removes alkali ions, NH_4OH-insoluble hydroxides such as $Al(OH)_3$, $Fe(OH)_3$, $Mg(OH)_2$, and $Zn(OH)_2$, and any residual trace metals (such as Cu and Au) that were not completely desorbed by SC-1.

SC-2 does not etch oxide or silicon, and does not have the beneficial surfactant activity of SC-1. Redeposited particles are, therefore, not removed by this mixture. The exact composition of SC-2 is much less critical than that of SC-1. The solution has better thermal stability than SC-1, and thus the treatment temperature need not be controlled as closely.

An optional etching step with dilute HF solution can be used between the SC-1 and SC-2 treatments of bare silicon wafers. Since the hydrous oxide film from the SC-1 treatment may entrap trace impurities, its removal before the SC-2 step should be beneficial. A 15-second immersion in 1% HF-H_2O (1:50) solution is sufficient to remove this film, as evidenced by the change from the hydrophilic oxidized surface to hydrophobic after stripping. However, unless high-purity and point-of-use ultra-filtered and particle-free HF solution is used under controlled conditions, recontamination will result. A silicon surface that was exposed to HF solution immediately attracts particles and organic contaminants from solutions, DI water, and the ambient air, as noted before. In contrast to SC-1, the subsequent SC-2 solution does not release these contaminants. It may therefore be preferable to rely on the dissolution and regrowth action of SC-1. If a pre-clean is used, then the 1% HF step *prior* to SC-1 is recommended, since SC-1 will remove most particles and other contaminants. Exposure of bare silicon wafers to HF solution after SC-2 is generally not advisable since it would cause loss of the protective oxide film that passivates the silicon surface.

Choline Solutions. Choline is trimethyl-2-hydroxyethyl ammonium hydroxide and can be used as a replacement of inorganic bases for etching and cleaning. It is a strong and corrosive base without alkali elements and etches silicon like other bases. A formulation of the chemical is available commercially (Summa-Clean SC-15M, Mallinckrodt) which is a diluted choline solution containing a surfactant and methanol. Bulk etching of silicon can be prevented by adding H_2O_2 as an oxidant (65). This mixture, which has excellent wetting characteristics for silicon wafers (66), is similar to SC-1 in its effect. In fact, some researchers have found it more efficient than SC-1 and other pre-oxidation cleans (67). Despite the fact that choline has been examined for many years, there is relatively little published information available, most data being contained in proprietary technical reports, sometimes with contradictory results. An automatic dual-cassette spray machine that uses a warm choline-H_2O_2-H_2O mixture and a DI water spray rinse is available commercially (68). In some procedures the mixture replaces only SC-1 in the RCA cleaning procedure.

3.4 Implementation of Wet-Chemical Cleaning Processes

The processes described in the preceding sections can be implemented in the fab by the following techniques (63):

1. Immersion tank technique (Table 13)
2. Centrifugal spraying (Table 14)
3. Megasonic processing (Table 15)
4. Closed system method (Table 16)
5. Brush scrubbing for hydrodynamically removing large particles from wafers with special brushes and DI water or isopropyl alcohol.
6. High-pressure fluid jet cleaning with a potentially harmful high-velocity jet of liquid sweeping over the surface at pressures of up to 4000 psi (68).

Detailed descriptions of these techniques are included in Chs. 3 and 4.

Table 13. Immersion Tank Technique

- ▲ Original technique for RCA cleaning process
- ▲ Vessels of fused silica used with SC-1, SC-2 to prevent leaching of Al, B, Na from Pyrex. (Polypropylene vessel for optional HF-H_2O).
- ▲ Batch of wafers immersed in SC-1 followed by water rinse and SC-2 for each 10 min at 75-80°C
- ▲ Reactions terminated by overflow quenching with cold, ultrafiltered D.I. water followed by drying
- ▲ Refined wet bench production systems are commercially available

Table 14. Centrifugal Spraying

- ▲ First automatic centrifugal spray cleaning machine for corrosives introduced in 1975 by FSI
- ▲ Wafers rotate past stationary spray column
- ▲ Filtered solutions, including HF-H_2O, hot SC-1 and SC-2, are dispensed as spray onto spinning wafers
- ▲ Solutions are freshly mixed just before spraying; reduced volumes are adequate
- ▲ Faster than immersion techniques. Wafersremain enclosed for cleaning, rinsing, and spin drying.
- ▲ Cleaning efficiency comparable with immersion technique, but particles are removed more efficiently
- ▲ Centrifugal spray machines are maintenance extensive

Table 15. Megasonic Processing

- ▲ **Developed at RCA** for removing particles on wafers to complement peroxide immersion cleaning process.
- ▲ **Megasonics:** A highly effective non-contact scrubbing method with high-pressure waves in a cleaning solution.
- ▲ **Achieved by ultrahigh frequency** sonic energy at 0.85 - 0.90 MHz, generated by array of piezoelectric transducers.
- ▲ **Particles** of 0.1 μm to several microns are removed from front and backside of wafers with input power densities of 5-10 W/cm^2; (ultrasonics at 20-80 kHz requires up to 50% higher power densities and is less effective for small particles).
- ▲ **Megasonics with SC-1** at 35-42°C simultaneously removes particles and contaminant films. Chemisorbed inorganics require higher temperatures (70°C) for complete desorption with SC-1/SC-2.
- ▲ **Commercial megasonic systems** available from Verteq and Semiconductor Technologies.

Table 16. Closed System Method by Liquid Displacement

- ▲ System developed in 1986 by CFM Technology ("Full Flow"™)
- ▲ Keeps wafers enclosed and stationary during entire cleaning, rinsing, and drying cycle
- ▲ Vessel containing wafer batch is hydraulically controlled to remain filled with hot or cold process fluids that flow sequentially and continuously over wafers in cassette
- ▲ Repeated crossing through gas/liquid phase boundaries is avoided, thus eliminating recontamination of wafers that occurs on conventional pullout from immersion tank

3.5 Wafer Rinsing, Drying, and Storing

The last steps in wet-chemical wafer cleaning are rinsing and drying (69)(70); both are extremely critical because clean wafers become recontaminated very easily. Rinsing should be done with flowing high-purity and ultra-filtered high-resistivity DI water, usually at room temperature. The results of several recent studies have been published (28)(71)-(73). Megasonic rinsing is advantageous (72) and is the most effective technique for reducing the critical boundary layer between the wafer surface and the rinse water (73). Centrifugal spray rinsing (74) and rinsing in a closed system (75) have the advantage that the wafers are not removed between cleaning, rinsing, and drying.

Wafer drying after rinsing should be done by physical removal of the water rather than by allowing it to evaporate. Spin drying accomplishes this and has been the most widely used technique, although recontamination occurs frequently. Filtered hot forced air or nitrogen drying is a preferred technique with less chance for particle recontamination (76)(77). Capillary drying is based on capillary action and surface tension to remove the water. Individual wafers are pulled out of DI water at 80 - 85°C; less than 1% of the water is said to evaporate, leaving a particle-free surface (69).

In solvent vapor drying, wet wafers are moved into the hot vapor of a high-purity water-miscible solvent, usually IPA (isopropyl alcohol), which condenses and displaces the water. The wafers dry particle-free when the cassette is withdrawn above the vapor zone. Commercial drying systems for IPA and for non-flammable solvent mixtures are available (69)(75). The purity of the solvent is extremely important, and the water content during processing must be closely controlled so as not to exceed a critical concentration to achieve an ultraclean surface (69)(78)-(81). A comparative evaluation of spin rinse/drying and IPA vapor drying has been published recently (82). IPA on silicon surfaces could not be detected by SIMS, but the growth rate of native oxide films was depressed, indicating the presence of a thin IPA film; the electrical properties of the oxide improved substantially (83).

A recently developed technique of drying is known as "Marangoni drying". During removal of the wafer from the DI rinse water, the air-water-silicon surface is exposed to a stream of water-miscible organic solvent vapor. Surface tension effects cause the water to sheet off a planar wafer surface (84), leaving a very dry hydrophilic silicon surface. Finally, a novel low-pressure cleaning technique has been described recently (85).

Extreme care must be taken to avoid recontamination of clean device wafers during storage if immediate continuation of processing is not possible. Wafers should be placed, preferably, in a chemically-cleaned closed glass container or in a stainless steel container flushed with high-purity filtered nitrogen and stored in a clean room. Metal tweezers must never be used to handle semiconductor wafers since this will invariably cause contamination by traces of metals. The final criterion of the success of all wafer cleaning operations is the purity of the wafer surface after the last treatment. No matter how effective the various cleaning steps may be, improper rinsing, drying, and storage can ruin the best results (63).

3.6 Vapor-Phase Cleaning Methods

General Considerations. Vapor-phase cleaning is often called *dry cleaning* in contrast with *wet cleaning*. However, just like liquid cleaning is a more accurate term than the more restricted "wet cleaning" term, vapor-phase cleaning is a more correct expression than "dry-cleaning" since these processes are carried out in the gas or vapor phase, which are not necessarily dry. As noted in Sec. 1.1, vapor-phase cleaning has many actual and potential advantages over liquid cleaning methods in the fabrication of advanced semiconductor devices. Its development for commercial application is, as a consequence, being pursued vigorously. In this section we present a broad classification of vapor-phase cleaning processes and discuss some basic approaches for the removal of various groups of contaminant types. Referenced details have been included in Sec. 4.

Classification of Vapor-Phase Cleaning Methods. First, the impurities and contaminants on semiconductor wafers that must be removed by any method can be listed as follows:

1. Gross and trace organics
2. Native and chemical thin films of oxides
3. Physically and chemically adsorbed ions
4. Deposited and adsorbed metals
5. Particles and particulates
6. Impurities absorbed or entrapped by oxides.

Second, the methods of vapor-phase cleaning can be classified into the following major categories and specific types of processes and techniques:

1. Physical interactions:
   thermal and low-pressure techniques
   sublimation
   evaporation and volatilization
   argon ion oxide sputter etching
2. Physically-enhanced chemical reactions:
   reactive ion bombardment
   glow discharge plasma reactions
   remote RF and microwave discharge hydrogen plasma reactions
   electron cyclotron resonance hydrogen plasma reactions
   ultraviolet-ozone-atomic oxygen organics oxidation
   photochemically-enhanced metal reactions
   ultraviolet-activated halogen metal reactions
3. Chemical thermal reactions:
   thermal oxide decomposition in vacuum
   volatile metal organics formation
   anhydrous HF gas-phase oxide etching
   wet HF vapor-phase oxide etching
   HF-alcohols vapor-phase oxide etching
4. Mechanical techniques:
   high-velocity dry-ice (CO_2) jet scrubbing
   cryogenic argon-aerosol jet impingement
   pulsed laser radiation particle dislodgment
   supercritical/subcritical fluid (CO_2) pressure cycling

Removal of Contaminants by Vapor-Phase Cleaning Methods. The methods and techniques listed in the preceding section are not just hypothetical possibilities. Specific examples can be cited that demonstrate their practical feasibility and implementation (see Sec. 4.0). The summary below indicates in general terms which of these methods is most appropriate for removing the various types of contaminants from semiconductor wafer surfaces.

Organic contaminants can be removed, depending on their composition, by one of the following methods: volatilization in vacuum at elevated temperature, oxidative degradation by the UV/O_3 reaction, remote or downstream oxygen plasma treatment, and plasma glow discharge reactions. These methods are applicable to all types of semiconductor wafers.

Native and chemical thin films of oxides and silicate glasses require chemical etching or physical sputter etching for their removal. The latter can lead to erosion of the semiconductor surface. Techniques for gas-phase etching with anhydrous HF, or vapor-phase etching with HF-H_2O, have been well established (see Ch. 7). Alternative techniques for oxide removal are reduction annealing in H_2 under UHV (ultra-high vacuum) conditions at high temperature, low-energy ECR plasma etching in Ar or with NF_3, thermal desorption at high temperature in vacuum or by plasma enhancement for pre-epitaxial cleaning of silicon wafers, and remote hydrogen plasma techniques.

Physisorbed and chemisorbed ions and deposited elemental metals require chemical processes to remove them from the semiconductor or oxide surfaces. Techniques of physical enhancement are frequently used to decrease the thermal energy requirement by supplying electromagnetic radiation or energy from plasma environments. However, these activated chemical reactions can have adverse radiation effects and cause ion-induced damage of the substrates. Typical examples are the removal of metallic impurities by remote microwave plasma or by photo-induced reactions, such as UV-generated chlorine radicals. Strictly chemical approaches are based on the formation of volatilizable species, such as metal chelates or nitrosyl compounds. The key requirement for removing chemical impurities is the formation of volatilizable species by reaction at low temperature, followed by their elimination at an elevated temperature and at low pressure.

Special attention must be paid to the state in which contaminants occur. Elemental metals and other impurities are often present as absorbates or inclusions in the native oxide film rather than being exposed in pure form on the semiconductor surface. Etching in HF gas may be necessary to first remove the oxide cover and make the impurities accessible for chemical reactions.

The removal of particles and particulates in the gas or vapor phase from wafer surfaces is another difficult problem. Chemical reactions as described could be used to transform them into some volatile species, depending on their composition and size. Vapor etching of an oxide film on which particles are located or in which they are imbedded could be effective if the wafers are positioned vertically and if a vigorous flow of inert gas would sweep the freed particles away from the wafer.

The mechanical gas-phase techniques listed above have shown promise for the removal of both inert and reactive particles, but there may be problems of substrate damage and technical implementation that must be overcome.

4.0 EVOLUTION OF WAFER CLEANING SCIENCE AND TECHNOLOGY

In this section we will trace the evolution of semiconductor wafer cleaning science and technology from the beginning; that is, from the advent of semiconductor device fabrication, to the present time. Four periods of this development can be discerned: *(i)* the early days from 1950 to 1960; *(ii)* the development period of wet-chemical cleans from 1961 to 1971; *(iii)* the widespread evaluation, application, and refinement of wet cleans from 1972 to September 1989, and finally *(iv)* the era of explosive research and development activity in both wet and dry cleaning methods from October 1989 to mid-1992.

The introduction section of each chapter in this book usually includes historical remarks pertaining to the chapter topic. The present systematic overall survey of the entire field should serve in establishing a comprehensive perspective of this evolution.

4.1 Period from 1950 to 1960

Harmful effects of impurities on the performance of simple transistors were recognized already in the early days of germanium devices; these problems became more apparent with the advent of silicon transistor fabrication in the later 1950s. Some sort of wafer cleaning was deemed necessary as part of the device manufacturing process.

Early cleaning techniques consisted of mechanical and chemical treatments. Particulate impurities were removed by ultrasonic treatment in detergent solutions or by brush scrubbing. The first caused frequent wafer breakage and the second often deposited more debris from the bristles than it removed from the wafer surfaces. Organic solvents were used to dissolve wax residues and other gross organic impurities.

Chemical treatments consisted of immersion of the wafers in aqua regia, concentrated hydrofluoric acid, boiling nitric acid, and hot acid mixtures as cleaning chemicals. Mixtures of sulfuric-acid/chromic-acid led to chromium contamination and caused ecological problems of disposal. Mixtures of sulfuric acid and hydrogen peroxide caused sulfur contamination. Nitric acid and hydrofluoric acid were impure and led to redeposition of impurities. In general, impurity levels and particles in process chemicals were high and in themselves tended to lead to surface contamination.

Aqueous solutions containing hydrogen peroxide had long been used for cleaning electron tube components (86)(87), but had never been investigated for possible applications to semiconductor wafer cleaning.

Plasma ashing was the first dry process applied to wafer cleaning for removing organic photoresist patterns, but left high concentrations of metallic compounds and other inorganic impurity residues.

4.2 Period from 1961 to 1971

This period can be considered one of research into semiconductor surface contamination and the systematic development of wafer cleaning procedures.

Radiochemical Contamination Studies. Radioactive isotopes offered a unique opportunity for the study of surface contamination with an unprecedented degree of sensitivity. Although it had been applied to several other areas of semiconductor research, very few papers had been published prior to 1963 on the use of radioactive tracers for surface contamination studies, notably papers by Wolsky et al. for germanium (88), Sotnikov and Belanovskii for silicon (89), and Larrabee for GaAs and InSb (90).

In a series of intensive contamination studies, Kern applied radioactive tracer methods to investigate the concentrations of contaminant elements that were transferred onto electronic materials during manufacturing operations. It may be of interest to look at some of these still useful early results, which were published in 1963 (91)(92).

The adsorption of sodium ions on the assembly parts of germanium transistors (RCA 2N217, pnp) during the germanium anodic etching step in alkali electrolyte solution was investigated with sodium-22 radioactive tracer. The desorption of sodium ions was tested after preliminary water rinses. The efficiency of various treatments was then assessed by measuring the radioactivity as a function of treatment time. Germanium pellets that has been cascade-rinsed with DI water had a concentration of 6.2 x 10^{14} sodium ions per cm^2. The plots in Fig. 1 demonstrate several interesting results. For example, EDTA chelating solution (as the free acid) was 280 times more effective than counter-current cascade rinsing with cold DI water.

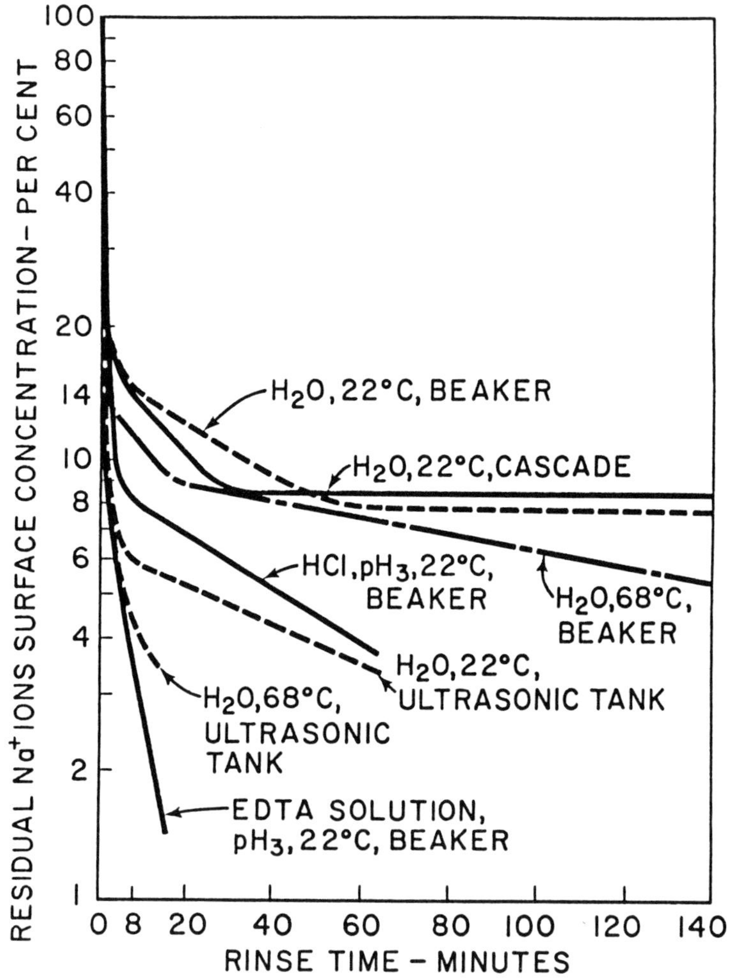

Figure 1. Efficiencies of various desorption treatments for removing sodium ions from germanium transistors electrolytically etched in radioactive $Na^{22}OH$ solution (90)(91).

A similar investigation was conducted with the components of a silicon power transistor (RCA 2N1482, npn). In this work, sodium-24, the isotope of sodium with a half-life of 15 hours, was created by neutron activation to attain several thousand times greater radioactivity levels than were available for the work with the germanium transistors, so as to achieve a much greater analytical sensitivity. Etching of various transistor parts in this highly radioactive NaOH solution and preliminary rinsing were performed entirely by remote-control manipulation in a radiation hot-cell. The efficiencies of several different treatments for desorbing sodium ions from various transistor parts were then determined by measuring the radiation intensity. The final concentration on silicon wafers after a dip in HCl solution and rinsing in DI water was $<8 \times 10^{11}$ Na^+/cm^2.

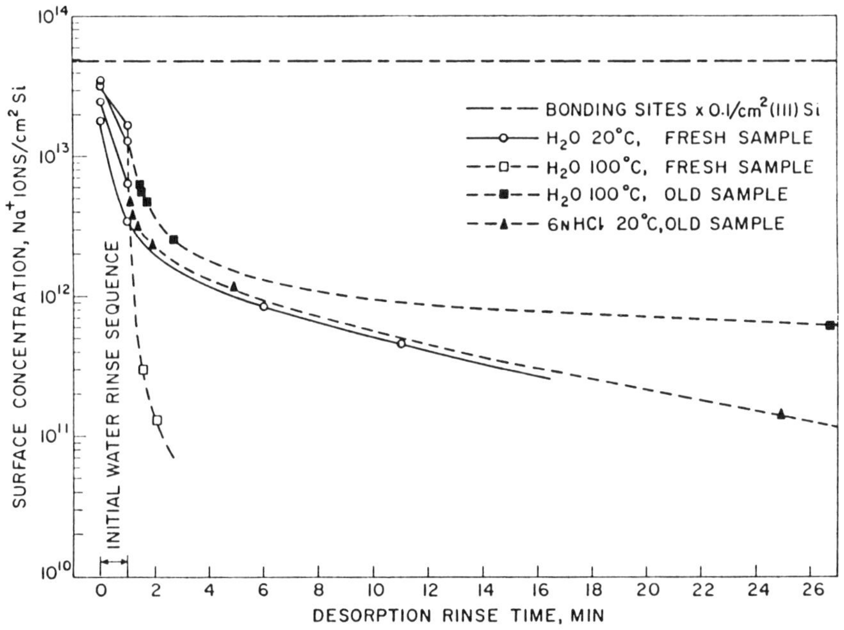

Figure 2. Inhibition effects of aging on desorption of Na^+ ions from silicon wafers with DI H_2O and 6N (19%) HCl. Wafers were immersed in 0.025N $Na^{22}OH$ followed immediately by initial rinsing for 60 sec. Subsequent desorption treatments within 24 hrs *(open symbols)* are compared with those after several weeks of storage *(solid symbols)* (93 l).

Contamination of silicon transistors by metallic impurities from solutions was assessed by use of the radioactive isotopes chromium-51, iron-59, copper-64, and gold-198. Techniques were devised to minimize residual concentrations, leading to a product yield increase for a silicon power transistor type (RCA 2N2102) of over two hundred percent in production.

The contamination of silicon, germanium, and gallium arsenide wafers during wet-chemical etching and processing was also investigated with radiotracers. Known quantities of radioactively tagged trace metals were added to the processing liquids used for treating the wafers. The spatial distribution of residual metals after rinsing with DI water was examined by autoradiographic film techniques. The surface concentrations were determined by correlation with radiation intensity measurements. The effectiveness of various rinsing and cleaning treatments was also measured quantitatively.

The adsorption of constituents of etchant solutions on semiconductor wafers was investigated by use of the appropriate radionuclides: sodium-22 and -24 for NaOH, fluorine-18 for HF-containing etchants, chlorine-38 for HCl, iodine-131 for polishing etchants containing iodine, and carbon-14 labeled acetic acid for isotopic etchants containing this chemical. Analytical techniques similar to those described for trace contaminants in solutions were applied to measure adsorption and desorption phenomena.

These radioactive tracer investigations were later extended to include additional metallic contaminants, substrates etchants, solutions, and cleaning agents. The radionuclides used as contaminant tracers included Mn^{54}, Zn^{65}, Mo^{99}, Sb^{122}, and Sb^{124} in addition to the previously used metal tracers Cr^{51}, Fe^{59}, Cu^{64}, and Au^{198}, and the reagent component tracers C^{14}, F^{18}, Na^{22}, Na^{24}, Cl^{38}, and I^{131}. Discs of fused quartz were used as substrates in addition to Si, Ge, and GaAs wafers. Contaminant adsorption was measured for many typical etchants and reagents containing the radioactively marked ions, and their desorption was investigated for many reagents, chelating agents, and cleaning solutions. For example, acidic H_2O_2 solutions were most effective for desorbing Au, Cu, and Cr from Si and Ge; HCl for Fe on Si; EDTA and other chelates suppressed deposition of Cu on Ge from solutions by 10^2 to 10^3. The results for silicon and quartz were reported in three extensive papers published in 1970 (64)(93 I)(93 II); the work with germanium and gallium arsenide was published in 1971 (93 III). A few typical metal desorption data plots of current interest are reproduced in Figs. 3 - 8 for silicon and in Fig. 9 for gallium arsenide.

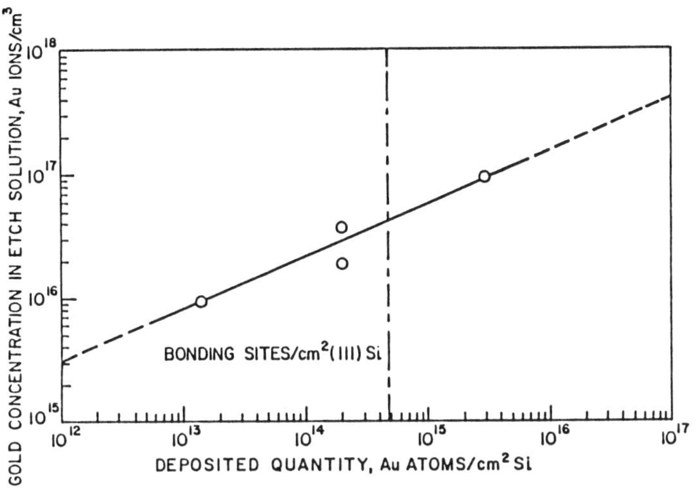

Figure 3. Numbers of gold atoms deposited on silicon wafers from etchant as a function of gold ions in solution. Au^{198} was added as a radioactive tracer to the HF-HNO_3-CH_3CO_2H-I_2 etchant. Vertical dashes indicate number of bonding sites on (111)-Si (as a reference only) (93 II).

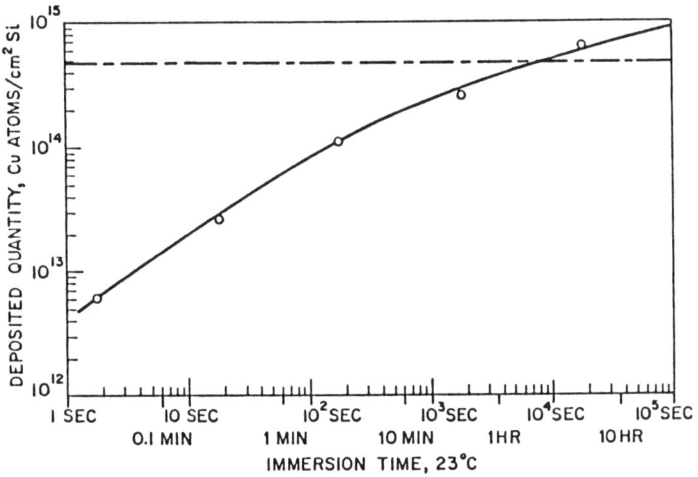

Figure 4. Number of copper atoms deposited on silicon wafers from 49% HF solution as a function of immersion time at 23°C. The acid contained 0.1 ppm copper tagged with Cu^{64} (93 II).

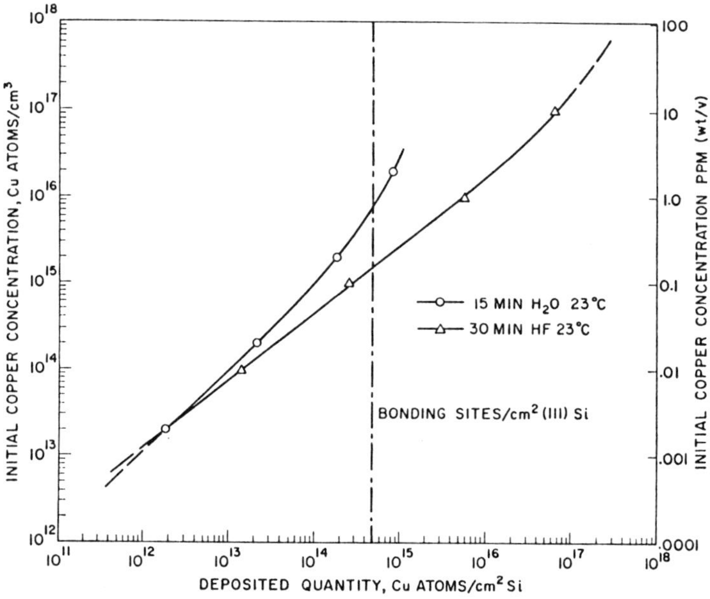

Figure 5. Quantity of copper deposited on silicon wafers from 49% HF solution and DI-distilled water as a function of copper concentration in solution. Cu^{64} was used as the radioactive tracer. Prior to immersion in the radioactive water the wafers had been dipped in non-radioactive HF and rinsed in non-radioactive water (93 II).

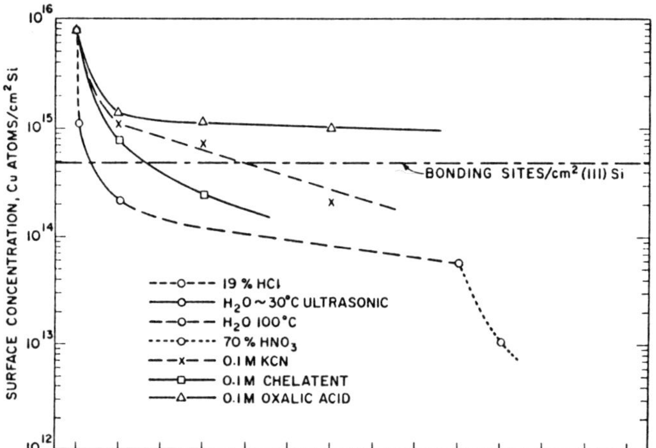

Figure 6. Effectiveness of various cleaning agents for desorbing heavy copper deposits from hot 5% NaOH solution with Cu^{64}-labeled Cu. Desorbing treatments were conducted at 23°C except for water rinses. Chelating agent is pentasodium diethylenetriamine pentacetate (93 II).

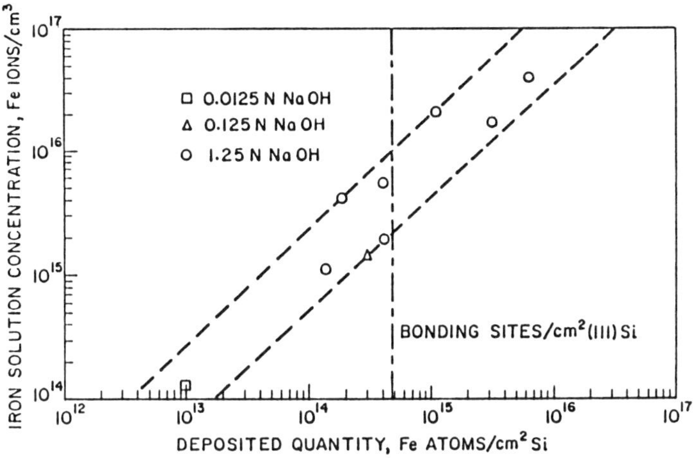

Figure 7. Quantity of iron deposited on silicon wafers from NaOH solutions as a function of iron solution concentration. Fe^{59} was used as the radioactive tracer in NaOH solutions of the concentrations indicated. Wafers were immersed at 100°C for 1 min (93 II).

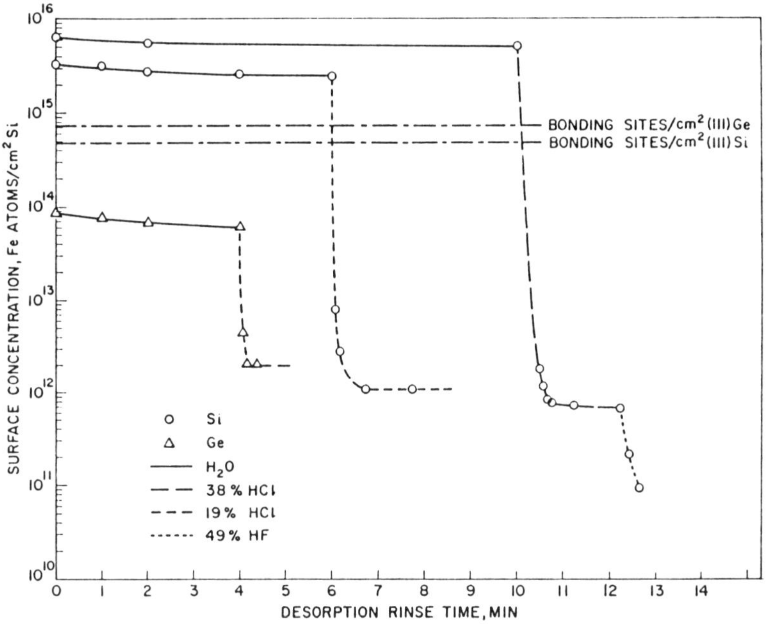

Figure 8. Efficiency of water and acid solutions at 23°C for desorbing iron adsorbates from silicon and germanium wafers. Fe^{59}-containing deposits from hot NaOH were used. Number of bonding sites for Si and Ge are indicated for reference (93 II).

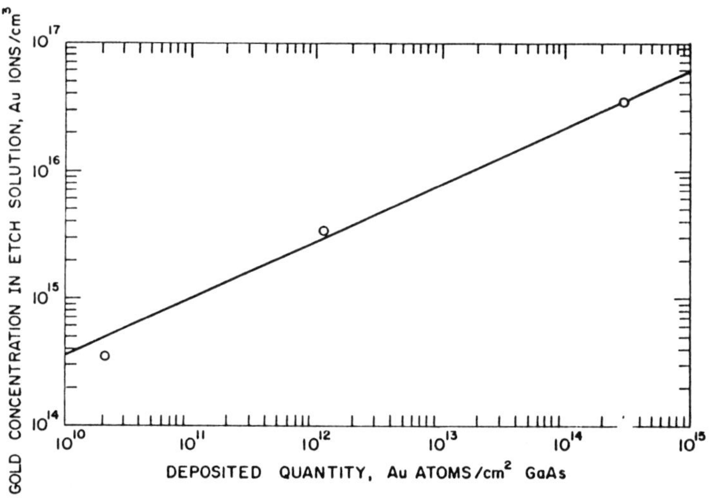

Figure 9. Number of gold atoms deposited on (100)-GaAs wafers from etchant as a function of gold ions in solution. A polishing etchant of 5 vol H_2SO_4 (98%) - 1 vol H_2O_2 (30%) - 1 vol H_2O containing Au^{198} as radioactive tracer was used. Each data point is the average of radioactivity measurements from three wafers (93 III).

The strong concentrating efficiency of silicon for gold from HF solution is shown in Fig. 10a, and its possible utilization for purifying HF solution is demonstrated in Fig. 10b. The dashed gamma radiation spectrum in Fig. 10a shows that of a 1N HF solution containing antimony-122, molybdenum-99 with its associated technetium-99, and gold-198 showing as a minor peak. The solid curve, obtained from silicon wafers that had been immersed in this solution, was normalized with the TC^{99m} peak maximum of the solution spectrum. The extremely high degree of selective deposition of metallic gold on the silicon is dramatically evident by comparing the Au^{198} peak intensities. The utilization of this effect is exemplified in Fig 10b: percolating a 49% HF solution containing Au^{198} through a column of high-purity silicon crystal pieces resulted in a retention of more than 95% of the gold in the first sixth of the column, and more than 98.8% removal in a single pass. Similar results were obtained with copper and other heavy metals in HF, BHF, and H_2O_2 solutions.

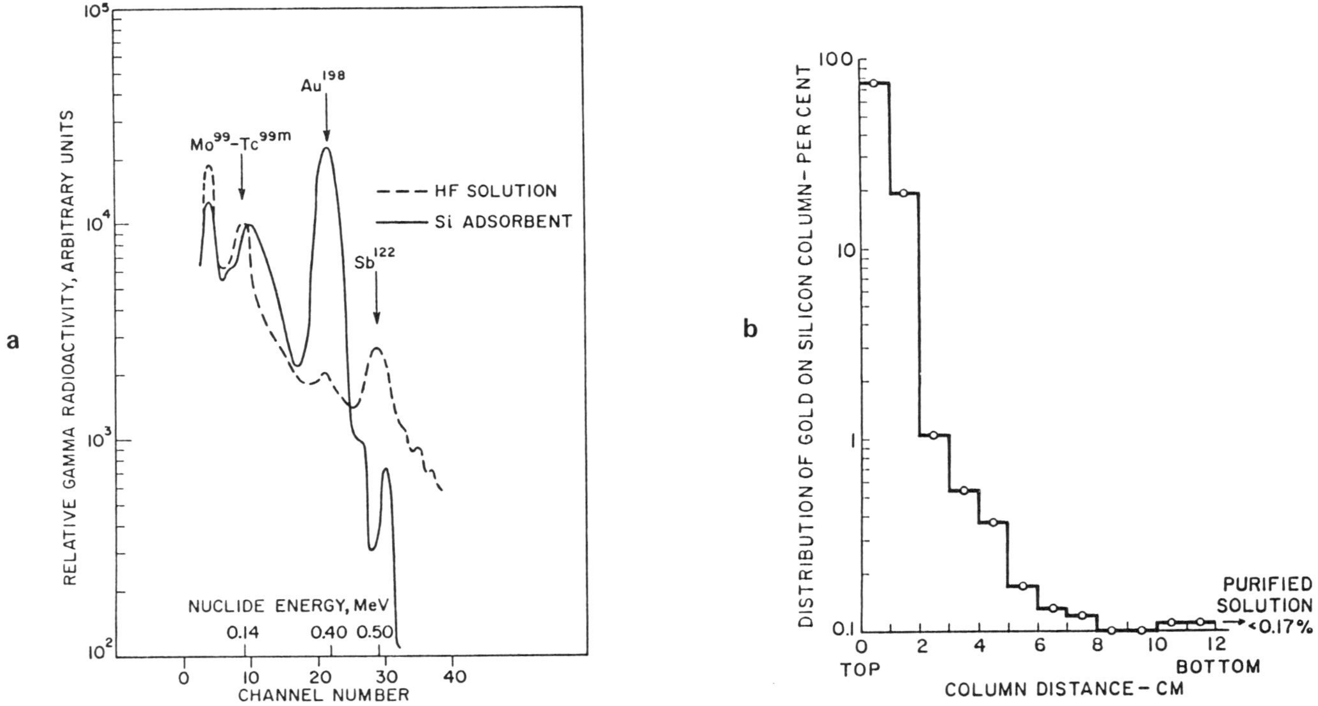

Figure 10. *(a)* Gamma radiation spectra of an HF solution and its adsorbate on silicon, demonstrating the strong accumulation efficiency of gold on silicon (93 II). *(b)* Distribution of radioactively marked gold from 49% HF solution on a chromatographic column of silicon crystal particles (91).

An unconventional approach to contaminant transfer measurements by radiochemical techniques should be mentioned before we leave this topic. This study, which had also been conducted and published by the author (95), concerned the transfer of impurity elements from crucibles. In this study the crucibles were radioactivated by bombardment with thermal neutrons in a nuclear reactor to produce radionuclides from the quartz impurity elements and the silicon. The GaAs crystals were then synthesized in these highly radioactive crucibles. Gamma ray scintillation spectrometry of the ingot sections allowed identification and quantitative measurements of the transferred contaminants. The results led to improved processing techniques for the crystal growth synthesis of GaAs.

Development of the Original RCA Wafer Cleaning Procedure. In this section we present a brief commentary on the development of the original RCA standard cleans by Kern and Puotinen, as published in 1970 (64).

The development of an optimized wet-cleaning procedure for silicon wafers proceeded concurrently with the contamination studies described in the previous section, which provided important information on adsorption and desorption characteristics of many contaminants.

It was realized early on that the first step should remove organic contaminants to expose the silicon surface; some kind of wet oxidant was needed to achieve this. Adsorbed ions and metals would then have to be removed by some solubilizing oxidant reagent. To prevent redeposition of the dissolved ionic contaminants, some complexing agent would be required. In addition, a set of important technological considerations, listed in Table 10, would have to be fulfilled in formulating an ideal procedure.

Thermodynamic reasoning based on the oxidation potentials of several possible candidates were an important consideration in reactant selection. Figure 11, reproduced from the original paper in 1970, shows the oxidation potentials for several common reactions as a function of pH; note that oxidizing power increases with decreasing (more negative) electrode potential. For equivalent concentrations, the peroxide oxidation reaction is the most powerful oxidizing agent shown. The selection of hydrogen peroxide as the primary reagent in the procedure was, therefore, the obvious choice, especially since it also met the criteria in Table 10 and had been used in cleaning mixtures for electron tube components. The reactive additives selected were ammonium hydroxide for the first solution and hydrochloric acid for the second. Both chemicals have the desired chemical reactivity, are volatile, are compatible with the criteria in Table 10, and were readily accessible in relatively pure form like the hydrogen peroxide. Salient features are summarized in Tables 11 and 12, which we presented in Sec. 3.3.

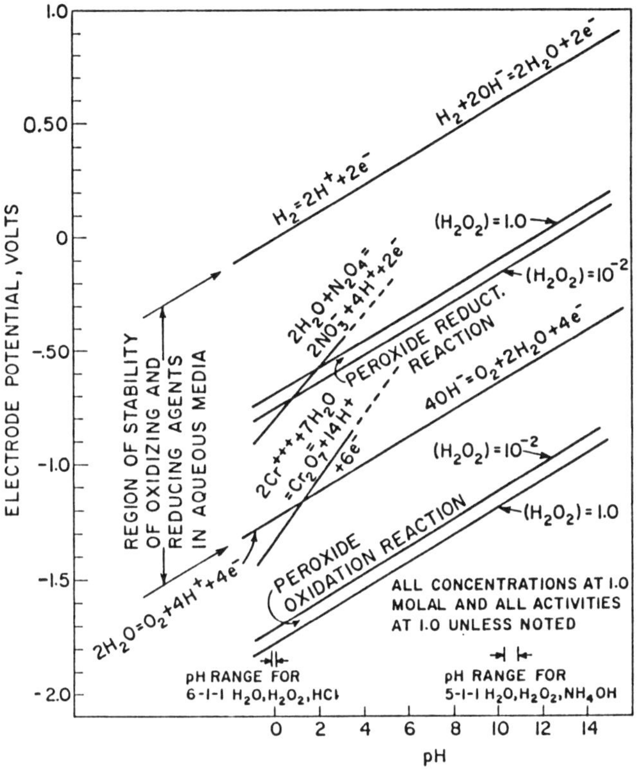

Figure 11. Electrode potentials vs. pH for various redox systems at 25°C (64).

The first solution (SC-1) was designed to remove organic surface impurities by the solvating action of the NH_4OH and the powerful oxidation capability of the H_2O_2. It was realized that the NH_4OH would also serve as a complexant for many metallic contaminants; copper, for example, forms the $Cu(NH_3)_4^{+2}$ amino-complex.

The volume ratios for H_2O-H_2O_2-NH_4OH that were recommended originally are in the range of 5:1:1 to 7:2:1 (or 5:1.4:0.7). Most people have used the 5:1:1 ratio since there seemed to be no obvious difference within this range.

The second solution (SC-2) was formulated to remove residual heavy-metals and prevent electrochemical displacement replating from solution by forming soluble complexes with the resulting ions. This strongly acidic mixture also dissolves alkali ions and metal hydroxides. The volume ratios recommended for H_2O-H_2O_2-HCl ranged from 6:1:1 to 8:2:1 (or 6:1.5:0.75). The first ratio has been used most often; sometimes the same ratio as for SC-1 (5:1:1) is used for simplicity, without adverse effects.

The volume ratios and processing conditions were determined empirically by performance tests. The originally-used conditions were 10 - 20 min at 75 - 85°C (these were later reduced to 10 min at 75 - 80°C, and finally to 10 min at 70°C, to minimize etching effects and the decomposition rate of the H_2O_2). Additional details have been given in Sec. 3.3.

The cleaning procedure was carried out by simple immersion of the wafer batches in quartz carriers into SC-1 and SC-2 baths. Vessels of fused quartz were specified to prevent leaching of aluminum, boron, and alkalies from glass if Pyrex had been used. A condenser minimizes volatilization of NH_3 from SC-1 and HCl from SC-2. The graph in Fig. 12 shows the worst-case situation (open vessel, excessively high temperature, long use time) for the critical SC-1 mixture in which the H_2O_2 decomposes much more rapidly than in the acidic SC-2. We recommended that the reactions be terminated by overflow quenching with cold DI water before transferring the wafer batches to a flow rinse station with ultra-filtered DI water.

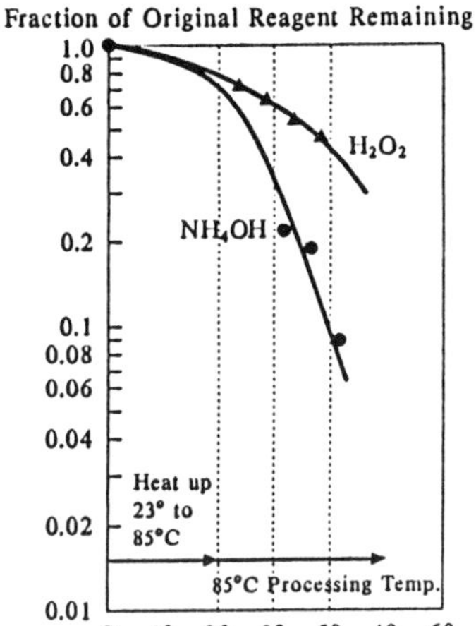

Figure 12. Decrease of the H_2O_2 and NH_4OH concentration in 5:1:1 SC-1 as a function of use time at excessively high temperature and long time periods in an open container. The decomposing H_2O_2 emits O_2, and the NH_4OH gives off NH_3. (After Ref. 62.)

Stripping of the native or chemically formed hydrous oxide films before and after SC-1 with very dilute (1:50 - 1:100) HF-H_2O was also investigated but considered optional because no conclusive experimental results were available to clearly justify its use. These aspects are further discussed in Sec. 3.3.

The effectiveness of the RCA cleaning procedure for silicon wafers, as described in the original publication (64), had been assessed with sensitive water spray wetting tests for organic contaminants, capacitance-voltage bias-temperature measurements on metal-oxide-semiconductor capacitors for ionic contaminants, and radioactive tracer analysis for metal desorption, as shown in the reproduced Figs. 13 - 15.

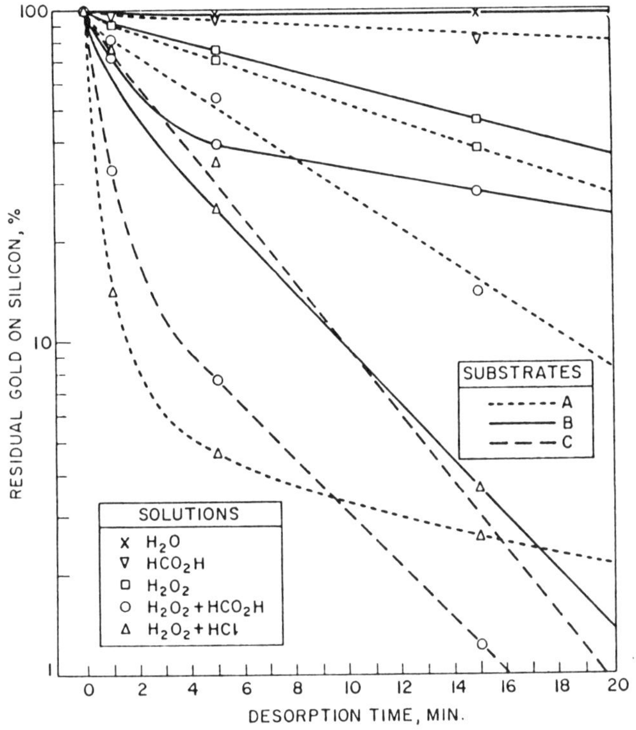

Figure 13. Efficiency of various cleaning agents at 90°C for desorbing heavy gold deposits from silicon. The wafers were etched in Au^{198}-containing solutions and had the following surface concentrations (Au/cm^2): curve A (dotted): 7×10^{13}, from HNO_3-HF, curve B (solid): 2×10^{15}, from HF, curve C (dashed): 3×10^{15}, from HF-HNO_3-I_2-CH_3CO_2H. Desorption solutions: **x** = DI-distilled H_2O, **∇** = H_2O-HCO_2H (90%) (9:1), **□** = H_2O-H_2O_2 (30%) (9:1), **O** = H_2O-H_2O_2-HCO_2H (8:1:1), **Δ** = H_2O-H_2O_2-HCl (1N) (8:1:1) (64).

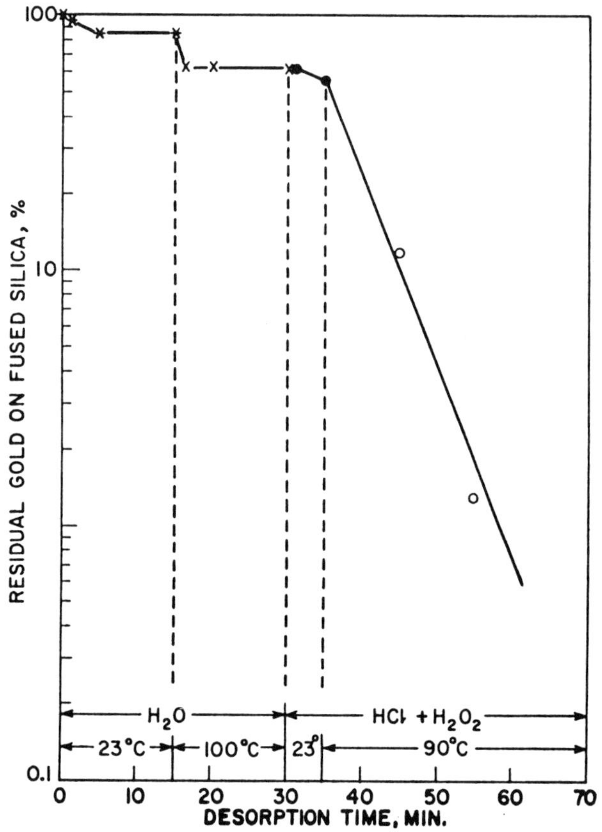

Figure 14. Desorption efficiency for gold (marked with Au^{198}) from fused quartz surfaces under various conditions. Quartz plates were etched in 49% HF containing Au^{198}; 100% = 1.6 x 10^{12} Au atoms/cm^2. The HCl + H_2O_2 mixture consisted of 8 vol H_2O - 1 vol diluted HCl (1N) - 1 vol H_2O_2 (30%) (64).

Silicon etching effects for various solution compositions and the addition of fluoride were also studied. No silicon etching of the 5:1:1 SC-1 solution at that time could be observed even if substantially lower H_2O_2 concentrations were used. These results demonstrated a wide margin of process safety when used in the fab (see Fig. 16).

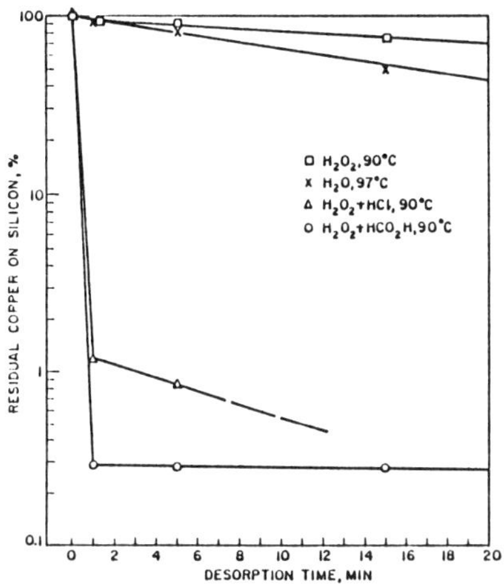

Figure 15. Desorption efficiency of various agents for removing Cu^{64}-labeled heavy copper deposits from silicon wafers. The initial Cu concentration from HF solution was five times higher for the H_2O_2 - HCl test than for the others (64).

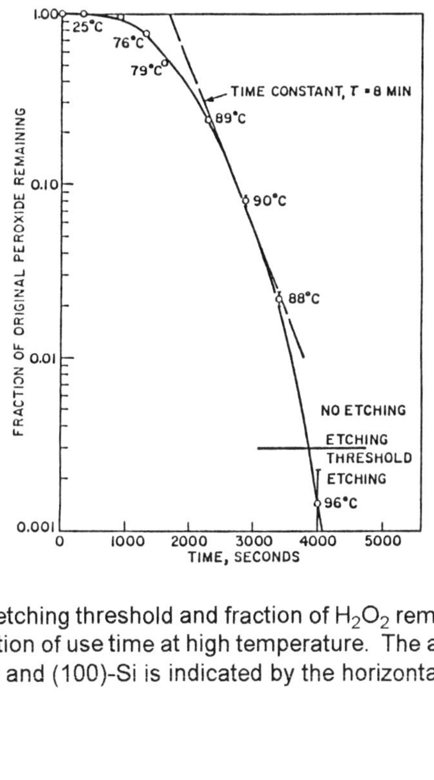

Figure 16. Silicon etching threshold and fraction of H_2O_2 remaining for 5:1:1 SC-1 solution as a function of use time at high temperature. The approximate etching threshold for (111)- and (100)-Si is indicated by the horizontal bar (64).

4.3 Period from 1972 to 1989

Chronological Survey of the Literature on H_2O_2-based Cleans*. Beginning in 1972, independent investigators examined and verified by various analytical methods the effectiveness of the RCA cleaning method published in 1970 (64). In this section we review chronologically references on silicon wafer cleaning pertaining primarily to hydrogen peroxide solutions. The period covered extends from 1972 up to the First International Symposium on Cleaning Technology in Semiconductor Device Manufacturing in October 1989 (1).

In 1972, Henderson published results of the evaluation of SC-1/SC-2 cleaning, using high-energy electron diffraction and Auger electron spectroscopy as analytical methods (96). He concluded that the process is well suited for wafer cleaning prior to high-temperature treatments, as long as quartzware is used for processing, as specified by us (64). An additional final etch in HF solution after SC-1/SC-2 caused carbon contamination and surface roughening during vacuum heating at 1100°C due to loss of the protective 1.5 nm thick C-free oxide film remaining after SC-2. Meek et al. (1973) investigated the removal of inorganic contaminants, including Cu and heavy metals, from silica-sol polished wafers by several reagent solutions (97). Using Rutherford backscattering, they concluded that SC-1/SC-2 pre-oxidation cleaning removes all elements heavier than Cl. Sulfur and chlorine remained after either SC-1, SC-2, or other cleaning procedures at $10^{13}/cm^2$. SC-1/SC-2 cleaning eliminated Ca and Cu much more reliably than did HF-HNO_3.

Amick (1976) reported the presence of Cl on Si after SC-2 and S after H_2SO_4-H_2O_2; he used spark-source mass spectrometric analysis (98). In 1976 Kern and Deckert published a brief review of surface contamination and semiconductor cleaning as part of a book chapter on etching (58). Murarka et al. (1977) studied methods for oxidizing Si without generating stacking faults and concluded that SC-1/SC-2 prior to oxidation is essential for this purpose (99). Gluck (1978) discussed removal of gold from Si by a variety of solutions. The desorption efficiency of SC-1 was more effective than that of SC-2, but the recommended sequential treatment of SC-1 followed by SC-2 was found the most effective method at high gold surface concentrations ($10^{14}/cm^2$ range) (100).

Peters and Deckert (1979) investigated photoresist stripping by

---

* This section is reprinted by permission of the publisher; the paper was originally presented at the 1989 fall meeting of The Electrochemical Society, held in Hollywood, Florida (63).

solvents, chemical agents, and plasma ashing, The SC-1 procedure was the only acceptable technique by which the residues could be removed completely (101). Burkman (1981) reported on desorption of gold with several reagent solutions by centrifugal spraying. SC-1 type solution was much more effective than H_2SO_4-H_2O_2, while a SC-2 type alone showed poor efficiency (74).

Phillips et al. (1983) applied SIMS (secondary ion mass spectroscopy) to determine the relative quantities of contaminants on Si. Cleaned wafers were purposely contaminated with gross quantities of numerous inorganic materials and then cleaned by immersion or spray techniques with various aggressive reagents, including aqua regia, hot fuming HNO_3, and H_2SO_4-H_2O_2. The lowest residual concentrations for most impurity elements were obtained by spray cleaning with H_2SO_4-H_2O_2 followed by the SC-1/HF/SC-2 type cleaning sequence (102). Goodman et al. (1983) demonstrated by minority-carrier diffusion-length measurements the effectiveness of SC-1/SC-2 for desorbing trace metals on Si (103). The author (1983) published a review of the subject on the occasion of the Citation Classic declaration of the original 1970 paper (104).

In 1983 Watanabe et al. (105) reported dissolution rates of SiO_2 and Si_3N_4 films in SC-1. The dissolution rate of thermally grown SiO_2 in SC-1 during 20 min at 80°C was a constant 0.4 nm/min, a significant rate for structures with thin oxide layers. The etch rate of CVD Si_3N_4 was 0.2 nm/min under the same conditions. Measurements by the author in 1981 (and published in 1984), however, indicated much lower oxide dissolution rates under nearly identical conditions (62). Film thicknesses were measured by ellipsometry after each of four consecutive treatments in fresh 5:1:1 SC-1 at 85°C and totaled only 7.0 nm/80 min, or 0.09 nm/min. Under the same conditions, 6:1:1 SC-2 showed no loss. Similar results averaging 0.13 nm/min were obtained with thermal SiO_2 films grown on lightly or heavily doped Si. Wafers from the same sets were used to determine the etch rates of exposed Si in SC-1 solutions with decreasingly lower H_2O_2 content. Little etching or attack of Si occurred (less than 0.8 nm/min), even when the H_2O_2 concentration was reduced by 90% (62).

Bansal (1984, 1985) reported extensive results on particle removal from Si wafers by spray cleaning with SC-1/SC-2, H_2SO_4-H_2O_2, and HF solutions of various purity grades. He found the RCA cleaning solutions the most effective (106)(107). Shwartzman et al. (1985) described simultaneous removal of particles and contaminant films by megasonic cleaning with SC-1 solutions (77). Ishizaka and Shiraki (1986) showed that atomically clean Si surfaces for MBE can be prepared below 800°C in UHV by thermal desorption (volatilization) of a thin (0.5 - 0.8 nm) passivating oxide layer that protects from carbon contamination (108). The layer was formed in a series of wet oxidation (HNO_3, SC-1) and HF-stripping steps, terminating with an SC-2 type treatment.

Wong and Klepner (1986) used XPS analysis to examine Si after wet chemical treatments. RCA cleaning without buffered HF stripping resulted in about 30% of the Si atoms in the top 1.0 nm being oxidized, whereas with a final BHF step, less than one monolayer of suboxide coverage resulted (109). Grundner and Jacob (1986) conducted extensive studies of Si surfaces after treatment with SC-1/SC-2 or 5% HF solutions, using x-ray photoelectron and high-resolution electron energy-loss spectroscopy. Oxidizing solutions produced hydrophilic surfaces, whereas HF solution led to hydrophobic surfaces consisting mainly of Si-H with some Si-CH_x and Si-F (110).

In 1986 Becker et al. (111) reported on decontamination by different reagent sequences. SIMS analysis was used to test for the removal of Na, K, Ca, Mg, Cr, Cu, Al, and particle impurities. The best cleaning sequence for metallics was H_2SO_4-H_2O_2/SC-1/HF/SC-2. Reversing the order of SC-1 and HF was most effective for particle removal and slightly less so for metal ions. Kawado et al. (1986) found by SIMS that Al on Si wafers originated from impure H_2O_2 used in SC-2. Very high concentrations resulted if Pyrex vessels were used in the processing instead of fused quartz (112). In 1986 McGillivray et al. (113) investigated effects of reagent contaminants on MOS capacitors. Low-field breakdown was more prevalent if pre-oxidation cleaning with SC-2 was terminated with HF solution instead of omitting it. No other significant differences in electrical properties resulted from these two treatments.

Lampert (1987) examined growth and properties of oxide films on Si in various aqueous solutions, including SC-1 and SC-2 (114). Gould and Irene (1987) studied the influence of pre-oxidation cleaning on Si oxidation kinetics (115). They found significant rate variations depending on treatment (SC-1/SC-2/HF, SC-1, SC-2, HF, no clean). Ruzyllo (1987) reported on similar experiments and found that various pre-oxidation cleans seem to affect structure and/or composition of the subsequently grown oxide rather than the reactivity of the Si surface (116). Slusser and MacDowell (1987) found that sub-ppm levels of Al in H_2O_2 used for SC-1/SC-2 cause a substantial shift (up to 0.2 V) in the flat-band voltage levels of a dual dielectric. Aluminum concentrates on the wafer surface, and basic media such as SC-1 lead to five times higher levels than acidic (SC-2) solutions (12). In 1987 Kern and Schnable reviewed wafer cleaning in a book chapter on wet-chemical etching (59).

Probst et al. (1988) stated that for achieving predictable diffusion from implanted doped poly-Si into single-crystal Si, an SC-1/SC-2 treatment of the substrate prior to poly-Si deposition is imperative (117). Khilnani (1988) discussed various aspects of semiconductor cleaning, including the RCA process (14). Peterson (1988) showed that the sequencing of cleaning solutions (H_2SO_4-H_2O_2, SC-1, SC-2, HF) can have dramatic effects on particle levels (118).

In 1989 Morita et al. (119) reported on the contamination of SC-1/SC-2 cleaned wafers by Na, K, Al, Cr, Fe, Ni, and Cu from solutions, showing that the absence or presence of an SiO_x layer on the Si surface strongly affects adsorption. Desorption of Al and Fe was most effective with HF-H_2O, and that of Cu and Cr with SC-2. The same authors (120) postulated that metals of high enthalpy of oxide formation adsorb on the oxidized Si surface by oxide formation, whereas metals of low ionization tendency deposit galvanically on the bare Si. Gould and Irene (1989) studied the etching of native SiO_x and Si in NH_4OH-H_2O, BHF, and SC-1 by ellipsometry. Severe Si surface roughness resulted from NH_4OH, less with BHF, and none with SC-1 (121).

Ohmi et al. (1989) compared particle removal efficiency of several cleaning solutions. They found that 5:1:1 SC-1 efficiently removes particles larger than 0.5 μm, but increased those smaller than 0.5 μm (haze) unless the NH_4OH ratio was decreased to one half or less, in which case both types of particles were reduced efficiently. However, no processing conditions and effects of low-NH_4OH SC-1 on removal of chemical contaminant films were mentioned (78). Menon et al. (1989) evaluated effects of solution chemistry (5:1:1, SC-1, DI water) and particle composition on megasonic cleaning efficiency at various power levels. They concluded that cleaning efficiency depends on several factors and that megasonics can provide wafer cleanliness levels not previously attainable (72).

Other Important Advances. The previous section, being concerned with H_2O_2-based cleans, has not included other important developments that had taken place within the period of 1972 to September 1989. Since most of this work will be discussed in other chapters of this book, we will cover this topic in broad terms with only a few key references.

Significant advances in the science and technology of particles have led to a better understanding of the forces of adhesion, the nature of submicron particles in liquids and gases, and the mechanisms of transfer to solid surfaces. Most of this work, such as that by Bowling (122), Mishima et al. (80) and Menon et al. (72)(123), was published between 1985 and 1989 and has resulted in improved high-purity processing and effective removal of particles from wafer surfaces, especially by the extended application of megasonic cleaning techniques.

Important advances were made in the area of dry cleaning of semiconductor wafers. The work by Vig (124), Kaneko et al. (125) and others extended the use of the UV/ozone method for removal of organic contaminants on silicon and III-V compound semiconductor surfaces. Mishima et al. (79)(80) and Ohmi et al. (78) conducted and published much research on wafer drying techniques, especially the subtleties of IPA vapor drying.

The introduction of anhydrous HF gas etching and HF-H_2O vapor etching technology for removing oxide films on wafers to avoid particle contamination was pioneered by Claevelin, Duranko, and Novak (126), Clements et al. (127), and Duranko et al. (128). This major advance set the stage for the development of vapor-phase cleaning science and technology in general. Early investigations were conducted in remote plasma cleaning, an entirely different approach to dry cleaning, as exemplified by the work of Fountain et al. (129).

Another area associated with wafer cleaning where significant progress was made is the refinement of microanalytical chemical and instrumental methods for detecting and quantifying trace contaminants and for exploring the atomic structure and morphology of semiconductor surfaces. Much insight has been gained by elucidating the chemical surface reactions with HF reactants and silicon, for example, and the nature of the resulting passivated, hydrogenated silicon surface. Example references for papers on this subject are by Burrows et al. (130), Hahn et al. (131), Zazzera and Moulder (132), and Chabal et al. (133).

4.4 Period of October 1989 to Mid-1992

As already mentioned, this is the period of explosive growth in the field of wafer cleaning science and technology, and the rate of progress is still accelerating. Rather than attempting a comprehensive coverage of all results reported, which is beyond the scope of this chapter, we will consider highlights of the progress achieved in *(i)* wet-chemical cleaning processes and *(ii)* gas-phase cleaning methods, as they pertain to the removal of specific types of contaminants. Most of the information is contained in the voluminous literature (1)-(11) cited in the Introduction Sec. 1.2.

Wet-Chemical Cleaning Processes. New observations on the performance and effects of H_2O_2-based cleaning solutions have led to some modifications of the original RCA SC-1 cleaning procedure. In addition, high-purity chemicals have become available, including Al-free H_2O_2, low-particulate HF, and low-metal HCl and NH_4OH solutions that have led to much lower trace metal contamination levels from these sources than was previously possible.

Van den Meerakker and Van der Straaten (134) elucidated the mechanism of silicon etching inhibition by H_2O_2 in SC-1 and studied the kinetics of etching. They reported a half-life for the standard 5:1:1 SC-1 of 16 min at 70°C and 9.3 min at 80°C. The authors stated that no etching occurs at 70°C as long as there is a H_2O_2 concentration of at least 3 x 10^{-3} M present to passivate the silicon surface, which represents 0.2% of the H_2O_2 concentration in the 5:1:1 SC-1. In other words, no etching was found as long as at least 1/500 of the original H_2O_2 was present. The result agrees with the etching threshold of 0.4% indicated in Fig. 16 for a higher solution temperature (64).

Tanaka et al. (135) reported silicon etch rates for 5:1:1 SC-1 of 0.5 nm/min at 75°C and 0.8 nm/min at 85°C, with higher values for decreasing H_2O_2 or increasing NH_4OH concentrations. These values are higher than those determined by the writer, which averaged 0.05 nm/min at 80 - 85°C (62). The discrepancy could be due to the long etch time used in the treatments by Tanaka et al. that could lead to a loss of the etch-protective oxide layer (134); variations in the silicon wafer properties could also be a cause (62). Although these etch rates are relatively low, it would be prudent not to exceed the SC-1 cleaning temperature of 70°C and the time of 10 min, as recommended in Sec. 3.3.

Microroughening of the silicon surface as a result of nonuniform microetching by 5:1:1 SC-1 solution has been investigated by many researchers, including Miyashita et al. (136), Ohmi et al. (137)(138), and Heyns et al. (139). This effect has become detectable on an atomic level only recently with the advent of atomic force microscopy. microroughening has detrimental consequences on the quality and breakdown voltage characteristics of thin, thermally grown gate oxide films, as reported by Meuris et al. (26), Verhaverbeke et al. (27)(31), Ohmi et al. (138), and Heyns et al. (139). Miyashita et al. (136) reported that a reduction of the NH_4OH concentration in the 5:1:1 H_2O-H_2O_2-NH_4OH SC-1 mixture down to 5:1:0.1 to 5:1:0.01 not only eliminated the roughening but also enhanced* the removal of particles. A reduction of the NH_4OH concentration to 5 - 10%

---

* This statement is opposite to that demonstrated clearly by Meuris et al. (26).

of that used in the conventional 5:1:1 SC-1 did not impair the desorption of Fe, Cu, Zn, and Ni from the Si surface. Meuris et al. (26) proposed a ratio of 5:1:0.25 as the best compromise for a modified SC-1 in terms of particle removal efficiency and avoidance of microroughening of the silicon. The authors who are cited in this paragraph have also examined the correlation between these effects, metal contamination, and electrical properties of grown oxide films (26)(27)(31)(138)-(140).

Sakurai et al. (141) reported that the thickness of the chemically grown oxide films on silicon during SC-1 cleaning does not depend on temperature, time, and solution composition (except for very low NH_4OH concentrations). The thickness of the films was 0.5 - 0.6 nm as determined by XPS analysis, or 1.2 nm, if measured by ellipsometry.

New results on wet-chemical contamination and on desorption cleaning by various processes and techniques were reported during this period by a number of researchers (19)-(31)(73)(139)(140)(143)-(153) as follows:

Heyns et al. (139) found that a dip of the wafers in dilute HF solution after SC-1/SC-2 removes any metal contaminants that may still be present, apparently without introducing new impurities (139). However, one has to be aware that post-cleaning exposure of the wafers to HF solutions will give rise to recontamination by organics, particles, and possibly trace metals. According to Rubloff (142), it should be applied only in the case of subsequent low-temperature epitaxial vapor growth where the absence of an oxide layer is crucial. Treatment by HF is not appropriate as a pre-oxidation clean where the formation of a passivating oxide layer is essential to prevent thermal surface etching and roughening of the silicon, with consequent degradation of the oxide qualities in MOS devices.

Verhaverbeke et al. (140) investigated the characteristics of HF-treated silicon surfaces as a function of immersion time in dilute HF solutions. They demonstrated the importance of forming a perfectly passivated surface, as evidenced by contact angle measurements, to reduce particle deposition. HF-last cleaning is more beneficial in terms of metallic contamination, as compared to RCA cleans. The improved processing has led to superior oxide integrity (140).

Grundner et al. (143) investigated surface composition and morphology of silicon wafers after HF solution treatments by means of instrumental surface analytical and angle-resolved light-scattering. Hirose et al. (144) studied the chemical stability and oxidation kinetics of H-terminated silicon surfaces after treatments in HF and BHF solutions, and Chabal (145) explored H-termination, atomic structure, and overall morphology.

Anttila and Tilli (146) showed that replacing SC-2 (H_2O-H_2O_2-HCl) with very dilute (1:10^3 to 1:10^6) acids, e.g., 1:10^4 HCl-H_2O, can remove several metallic contaminants (and metal hydroxides) without introducing as many particles as SC-2 does. The benefits of SC-1, which leaves the surface free of particles and organics, are combined with the benefits of immersion in dilute acids at room temperature, which remove metals efficiently without adding particles.

Kniffin et al. (147) showed that the type of chemical bonding of metallic impurities to the silicon surface plays an important role in determining the cleaning efficiency of a wet-chemical processing sequence. Poliak et al. (148) compared the effects of various wet-chemical cleaning sequences for removing metallic contaminants.

Shimono (149) demonstrated that an aqueous solution of 1% H_2O_2 and 0.5% HF has a higher efficiency for removing metallic impurities than conventional cleans and also etches native oxide films on silicon. The improved effectiveness of this mixture is apparently based on the silicon surface etching action. Takizawa and Ohsawa (150) have used a similar approach by employing the classical HNO_3-HF silicon etchant in very dilute form (0.025 - 0.1% HF in HNO_3), so that a silicon etch rate of 3 to 60 nm/min results without a substantial etching of thermal SiO_2 patterns.

Hariri and Hockett (67) showed that replacing the NH_4OH in SC-1 with choline (trimethyl-2-hydroxyethyl ammonium hydroxide) plus a surfactant reduces oxidation-induced stacking faults in silicon better than RCA cleans do, and may also improve the removal of heavy metals, as noted in Sec. 3.3. Lowell (151) used the choline clean after a deglazing etch of doped polysilicon layers with dilute HF. The choline treatment generates a thin oxide film that prevents the formation of a contaminated oxide during rinsing. Menon et al. (152) showed, however, that if megasonic cleaning techniques are used, SC-1 is more effective for removing particles than choline solution. Syverson et al. (153) conducted temperature optimization tests for megasonic particle removal in SC-1/SC-2 which revealed that 55°C is the most effective temperature. Major reductions in wafer particle densities have been achieved by this optimized treatment in an advanced manufacturing environment. Many other aspects of particle contamination and its removal were covered extensively in the volumes of Refs. 10 and 11.

Tong et al. (44) reported that ozonized DI water used with conventional aqueous chemicals has good cleaning efficiency; concentrations of residual metals and particles were found to be equal or lower than after conventional RCA cleans. Matthews (45) carried the preparation of high-purity aqueous chemicals from gaseous precursors a step further by using NH_3 for preparing NH_4OH and HCl gas for HCl solutions, in addition to O_3 for H_2O_2. This method of reagent synthesis will undoubtedly become an important future technology.

Finally, another important area related to wafer cleaning should be mentioned: reagent recycling and repurifying. Davison et al. (41) described the reprocessing, properties and application results of high-purity aqueous HF. Doshi et al. (154) state that impurity levels in this ion-exchange purified HF were routinely below 1 ppb for thirty-four elements and that it performed significantly better than the highest available purity of HF. (An MOS memory fab producing 1-Mbit DRAMs showed a 5% improvement in the wafer probe yield.) Davison (155)(156) and Hsu (42) have reviewed the technology of reprocessing or ultra-purifying both H_2SO_4-H_2O_2 (piranha etch) and aqueous HF.

Vapor-Phase Cleaning Methods. The basic approaches to dry or vapor-phase cleaning have been described in Sec. 3.6. Many of these approaches have been developed in this time period to the point of showing feasibility, as noted in Table 17. Only some selected outstanding highlight results will be briefly noted in this section. Many papers on this subject were included in the Proceedings volumes of Refs. 1 and 2.

Table 17. Gas-Phase Cleaning Approaches

1. **Removal of Organic Contaminants by**
   - ▲ UV/O_3 reaction
   - ▲ NO/HCl/N_2 thermal reaction
   - ▲ Remote (downstream) plasma in O_2
   - ▲ Plasma glow-discharge in H_2
2. **Removal of Native Oxide Films**
   - ▲ HF vapor etching
   - ▲ Reduction annealing in H_2
   - ▲ Low-energy ECR plasma etching in Ar
   - ▲ Remote plasma exposure
3. **Removal of Metallic Contaminants by**
   - ▲ UV/Cl_2 microwave formation of Cl radicals
   - ▲ NO/HCl/N_2 thermal reaction to form volatile nitrosyl compounds
   - ▲ Remote plasma discharge in HCl/Ar/O_2
   - ▲ Formation of volatile metal halides
   - ▲ Formation of volatile metalorganics
   - ▲ Reaction with vapor-phase analogs of SC-1, SC-2

The removal of native, grown, or deposited oxide films on silicon was accomplished before this review period by use of HF gas or vapor etching processes (126)-(128). The chemical mechanisms underlying these vapor-phase etching processes has been elucidated by Helms and Deal (157). Techniques and systems for implementing these processes and optimization of the vapor-phase etching reactions for oxide removal have been developed, investigated and reported during this period by many authors, e.g., Deal et al. (158), Ohmi et al. (159), Onishi et al. (160), Iscoff (161), Wong et al. (162), Nobinger et al. (163), and Deal and Helms (164). The addition of methanol instead of water vapor to anhydrous HF minimizes the formation of the solid reaction products encountered with the HF-H_2O vapor etching systems, as observed by Izumi et al. (165).

Significant progress has been made by physical-chemical methods, such as glow discharge plasma reactions, for removing thin oxide films and certain contaminants on semiconductor wafers. Comfort (166) examined the thermodynamic parameters governing high-temperature thermal desorption and low-temperature oxide removal for pre-epitaxial surface cleaning of silicon. Reif (167) discussed in situ low-temperature cleaning for pre-epitaxial silicon growth; Liehr (168) examined the impact of silicon surface treatments prior to epitaxy and gate oxide growth, and Kalem et al. (169) reported on surface cleaning prior to the formation of SiO_2/Si interfaces.

Tasch et al. (170) reviewed recent results on low-temperature in situ pre-epi cleaning of silicon by remote plasma-excited hydrogen in ultrahigh vacuum. Hattangady et al. (171) applied hydrogen, dissociated with remote noble-gas discharge, for the low-temperature cleaning of Ge and GaAs surfaces. Frystak and Ruzyllo (172) used remote plasma cleaning as a pre-oxidation treatment for silicon. Gas mixtures of O_2, HCl/Ar, and NF_3/H_2/Ar were used to remove organics, metallic impurities, and thin oxide films, respectively. Finally, Chang (173) described an in situ plasma cleaning process for GaAs surfaces and device passivation.

Removal of metallic contaminants is best accomplished by thermal, chemical or photochemical vapor-phase reactions. Gluck (174) reported that nitric oxide can volatilize copper from silicon surfaces at 500°C, and gold reacts with a mixture of nitric oxide and HCl at 900 - 1000°C to form a volatilizable compound. The formation of volatile metal nitrosyl complex compounds with various metals and the NO-HCl-N_2 thermal reaction may be a promising approach to vapor-phase metal removal in general. Formation of volatile organometallic complexes with other reactants is quite possible. Ivankovits et al. (175) reported that reacting iron and copper on silicon surface with 1,1,1,5,5,5-hexafluoro-2,4-pentanedione followed by volatilization at 300°C can reduce their concentrations. These early results look promising, but a great deal more work is needed to develop predictable and efficient processes.

Wong (176) reported results on a pre-oxidation treatment of silicon wafers with HCl/HF vapor mixtures, which were effective in reducing the detrimental effects caused by traces of heavy-metal contaminants; the oxide lifetime improved by 25%.

Low-temperature photochemical reactions have a great potential for transforming metallic contaminants into volatilizable compounds. Ultraviolet radiation is usually employed as the radiation source, for practical and energy considerations, with halogens as the reactants to generate highly reactive halogen radicals. Ito et al. (177)(178) utilized highly purified chlorine radicals to reduce the surface concentrations of Fe, Mg, Al, and Cu to levels lower than was possible with conventional wet cleans. Native oxide layers on silicon can be etched off with fluorine radicals generated by the same photo-activation process (178).

The well-known ultraviolet/ozone reaction for removing organic impurities from surfaces, reviewed by Vig (179), was applied to the cleaning of GaAs in epitaxial deposition processes by Pearton et al. (180) and by Kopf et al. (181) for the reduction of defects. Bedge and Lamb (182) studied the kinetics of the process and reported on experiments and modeling.

Finally, the status of particle removal by especially promising dry cleaning techniques is noted. McDermott et al. (183) described an argon-aerosol jet technique where frozen particles of argon are created and impinged at a high velocity on the wafer surface. Micron-size and submicron-size contaminant particles are dislodged by the collision energy and are entrained in the gas flow and removed from the system.

Bok et al. (184) reported on the theory and practice of a sophisticated method that is based on supercritical fluids, such as supercritical CO_2. Their unique properties enables them to penetrate into deep IC structures and effect complete removal of particles and other contaminants. The supercritical liquid is first forced into trenches and crevices during compression in the pulsating pressure cycle. Subsequent expansion between supercritical and subcritical pressures dislodges particles and causes their ejection with a tremendous force.

Removal of particles in a vacuum system compatible with a dry cleaning sequence is technologically a great deal more difficult. Particle detachment by electrostatic techniques is ineffective. A promising technique is based on the use of laser radiation. Allen (185)(186) has demonstrated that pulsed laser radiation is capable of effectively removing particles from surfaces. A moisture film is condensed between the particles and the wafer surface and is then explosively evaporated by a laser beam of an appropriately tuned wavelength. The dislodged particles can then be swept out of the system with a jet of inert gas.

5.0 SUMMARY AND CONCLUSIONS

In the first part of this chapter we presented an overview of semiconductor wafer contamination aspects, discussing several important areas of technology that are directly or indirectly associated with wafer cleaning. We have shown that wafer cleaning technology in the broadest sense is interdisciplinary in that it involves not only surface chemical reactions in liquid and gas phases, but also chemical engineering to implement cleaning processes, contamination analysis by advanced instrumental methods for detecting and measuring trace impurities on surfaces and in chemicals, electrical measurements to determine the effects of contaminants on semiconductor devices, and the science of ultra-fine particles and their metrology. An understanding of the principles underlying these various disciplines is, therefore, important for successfully solving wafer cleaning problems in the laboratory and the fab.

In the second part we presented an overview of wafer cleaning technology in which we discussed in broad terms principles of liquid-phase cleaning processes and classifications of dry and vapor-phase cleaning methods.

The third part of the chapter was devoted to a chronological review in which we described the evolution of wafer cleaning science and technology from the early beginnings in the 1950s to the present time, mid-1992. The developments in the past three years have been truly "explosive", as judged by the number of scientific conferences and publications devoted to this topic.

We can conclude that the most widely used wafer cleaning methods in VLSI and ULSI silicon circuit fabrication are still, after twenty-five years, the hydrogen peroxide-based wet-chemical processes. High-purity reagents are now available, such as aluminum-free H_2O_2, that have led to improved performance results. However, the concentration of ammonium hydroxide in the original RCA SC-1 solution (5:1:1 H_2O-H_2O_2-NH_4OH) has been reduced by at least four-fold to avoid micro-roughening of the silicon surface by nonuniform micro-etching, resulting in improved gate oxide integrity and increased yields of MOS capacitors. It is also advisable not to exceed 70°C for 10 min in the RCA SC-1/SC-2 wafer cleaning treatment. Removal of the native or chemical oxide film before and after SC-1 and SC-2 by optimized etching with dilute (1:50 - 1:100) ultrapure HF solution can be beneficial.

Remarkable results have also been achieved by wet-chemical cleaning of silicon wafers with aqueous solutions of choline-H_2O_2-surfactant, H_2O-HF-HCl, H_2O-H_2O_2-HF, very dilute acids, as well as with ozonized water. New techniques of wafer drying have been devised of which isopropyl alcohol vapor drying after cold DI water megasonic rinsing is one of the best.

While the use of advanced wet-chemical cleaning techniques for producing ultrapure silicon wafers will persist for at least several more years, the trend is toward a shift from liquid to gaseous reactants for several reasons. Removal of oxide layers by HF vapor-phase etching is now well established, and the elimination of organic contaminants by UV/ozone has been amply demonstrated. Processes for removing trace metals by vapor-phase analogs of SC-1 and SC-2 are being pursued vigorously. Possible solutions comprise photochemical complex formation with chlorine radicals generated from Cl_2 by UV radiation, thermal reaction and volatilization with NO/HCl/N_2 to produce volatile nitrosyl compounds, exposure to remote or downstream plasma discharges in HCl/Ar, and the formation of vacuum volatilizable metalorganic chelates or complexes. Electron cyclotron resonance plasma techniques have also been successful for removing contaminants. Particle elimination could be achieved by cryogenic techniques or by a new pulsed laser method that effectively dislodges particles so that they can be removed by a gas stream. Eventually and ideally, the entire cleaning sequence will be conducted in situ by a sequence of gas-phase reactions at low or reduced pressure and elevated temperature in a cluster tool that can be integrated with other cluster modules for film deposition, annealing, dry etching, and other processing steps.

ACKNOWLEDGMENT

I wish to thank Dr. George L. Schnable for critically reviewing the draft of this chapter and for offering his many helpful comments.

REFERENCES

1. *Proc. First International Symp. on Cleaning Technology in Semiconductor Device Manufacturing,* (J. Ruzyllo and R. E. Novak, eds.) Vol. 90-9, Electrochemical Society, Pennington, NJ (1990)
2. *Proc. Second International Symp. on Cleaning Technology in Semiconductor Device Manufacturing,* (J. Ruzyllo and R. E. Novak, eds.) Vol 92-12, Electrochemical Society, Pennington, NJ (1992)
3. *Semicon/Korea '91 Tech. Proc.,* Seoul, Korea. SEMI, Mountain View, CA (Sept. 26-27, 1991)
4. *Semicon/Europe '92 Tech. Proc.*, Zurich, Switzerland. SEMI, Mountain View, CA (March 10-11, 1992)
5. *Proc. on Chemical Surface Preparation, Passivation, and Cleaning, Growth and Processing, Symp. B,* Spring Mtg. of the Materials Research Society (MRS), San Francisco, CA (April 12-16, 1992)
6. *Proc. of the Ann. Tech. Mtgs. of the Institute of Environmental Sciences,* Mount Prospect, IL; and *Ann. Microcontamination Conf. Proc.,* sponsored by Microcontamination Magazine (1990, 1991, 1992)
7. *Proc. of the Semiconductor Pure Water and Chemicals Conf.,* Santa Clara, CA. Balazs Analytical Laboratory, Sunnyvale, CA (Feb 11-13, 1992)
8. *Electrochemical Society (ECS) Ext. Abstr.* (a) Vol. 90-1 (1990); (b) Vol. 90-2 (1990); (c) Vol. 91-1 (1991); (d) Vol. 91-2 (1991); (e) Vol. 92-1 (1992). Electrochemical Society, Pennington, NJ
9. *Handbook of Contamination Control in Microelectronics,* (D. L. Tolliver, ed.), Noyes Publications, Park Ridge, NJ (1988)
10. *Particles on Surfaces,* Vols. 1 and 2, (K. L. Mittal, ed.), Plenum Publishing Corp., New York (1988 and 1989); *Particles in Gases and Liquids,* Vols. 1 and 2, (K. L. Mittal, ed.), Plenum Publishing Corp., New York (1989 and 1990)
11. *Particle Control for Semiconductor Manufacturing.* (R. P. Donovan, ed.), M. Dekker Inc., New York (1990)
12. Slusser, G. J. and MacDowell, L., *J. Vac. Sci. Technol.* A-5 (4):1649-1651 (1987)
13. Monkowski, J. R., *Treatise on Clean Surfaces Technology,* (K. L. Mittal, ed.), Vol. 1, Ch. 6, pp. 123-148, Plenum Press, New York (1987)
14. Khilnani, A., *Particles on Surfaces 1: Detection, Adhesion, and Removal,* (K. L. Mittal, ed.) pp. 17-35, Plenum Press, New York (1988)

15. Burkman, D. C., Peterson, C. A., Zazzera, L. A., and Kopp, R. J., *Microcontamination* 6(11):57-62, 107-111 (1988)

16. Atsumi, J., Ohtsuka, S., Munehira, S., and Kajiyama, K., pp. 59-66 in Ref. 1

17. Hattori, T., *Solid State Technol.* 33(7):S1-S8 (1990)

18. Jastrzebski, L., *ECS Ext. Abstr.,* 90-1:587-588, Electrochemical Society, Pennington, NJ (1990)

19. Riley, D., and Carbonell, R., *Proc. of The Institute of Environmental Sciences Ann. Tech. Mtg.,* pp. 224-228, New Orleans, LA (1990)

20. Ohsawa, A., Honda, K., Takizawa, R., Nakanishi, T., Aoki, M. and Toyokura, N., *Proc. Sixth International Symp. on Silicon Materials Science and Technology,* (Huff, Barraclough, and Chikawa, eds.) 90-7:601-613, Electrochemical Society, Pennington, NJ (1990)

21. Matsushita, Y., and Tsuchya, N., *ECS Ext. Abstr.,* 90-2:601, Electrochemical Society, Pennington, NJ (1990)

22. Kern, F., Jr., Mitsushi, I., Kawanabe, I., Miyashita, M., Rosenberg, R. W., Ohmi, T., *Proc. of the 37th Ann. Tech. Mtg.,* Institute of Environmental Sciences, San Diego, CA (1991)

23. Riley, D. J., and Carbonell, R. G., *Proc. of the 37th Ann. Tech. Mtg.,* pp. 886-891, Institute of Environmental Sciences, San Diego, CA (1991)

24. Meuris, M., Heyns, M., Küper, W., Verhaverbeke, S., and Philipossian, A., *ECS Ext. Abstr.*, 91-1:488, Electrochemical Society, Pennington, NJ (1991)

25. Bergholz, W., Zoth, G., Gelsdorf, F., and Kolbesen, B., *ECS Ext. Abstr.* 91-1:227-228, Electrochemical Society, Pennington, NJ. (1991)

26. Meuris, M., Heyns, M., Mertens, P., Verhaverbeke, S., and Philipossian, A., pp. 144-161 in Ref. 2

27. Verhaverbeke, S., Meuris, M., Mertens, P. W., Kelleher, A., Heyns, M. M., De Keersmaecker, R. F., Murrell, M., and Sofield, C. J., pp. 187-196 in Ref. 2

28. Tonti, A., pp. 409-417 in Ref. 2

29. Gupta, P., Van Horn, M., and Frost, M., in *Proc. of the Semiconductor Pure Water and Chemicals Conf.,* Santa Clara, CA., pp. 191-198, Balazs Analytical Laboratory, Sunnyvale, CA (Feb 11-13, 1992)

30. Anttila, O. J., Tilli, M. V., Schaekers, M., and Claeys, C. L., *J. Electrochem. Soc.* 139:1180-1185 (1992)

31. Verhaverbeke, S., Mertens, P. W., Meuris, M., Heyns, M. M., Schnegg, A., and Philipossian, A., in *Semicon/Europe '92 Tech. Proc.*, Zurich, Switzerland. SEMI, Mountain View, CA (March 10-11, 1992)

32. Kern, W., in *Semicon/Korea '91 Tech. Proc.,* Seoul, Korea, Session III, pp. 39-88, SEMI, Mountain View, CA (Sept. 26-27, 1991)

33. Osburn, C. M., *Microcontamination* 9(7):19-27, 76 (1991)

34. Dillenbeck, K., *Particle Control for Semiconductor Manufacturing.* (R. P. Donovan, ed.), Ch. 15, pp. 255-261, M. Dekker Inc., New York (1990)

35. Ohmi, T., Inaba, H., and Takenami, T., *Microcontamination* 7(10):29-32, 86-97 (1989)

36. Fisher, W. G., *Particle Control for Semiconductor Manufacturing.* (R. P. Donovan, ed.), pp. 415-428, M. Dekker Inc., New York (1990)

37. Kern, W., Keynote lecture presented at the Semiconductor Pure Water and Chemicals Conf., Santa Clara, CA. Balazs Analytical Laboratory, Sunnyvale, CA, (Feb 11-13, 1992)

38. Hashimoto, S., Kaya, M., and Ohmi, T., *Microcontamination,* 7(6): 25-28, 98-106 (1989)

39. Harder, N., *Solid State Technol.*, 33(10):S1-S4 (1990)

40. Naggan, M., *Handbook of Contamination Control in Microelectronics,* (D. L. Tolliver, ed.), Ch. 11, Noyes Publications, Park Ridge, NJ (1988)

41. Davison, J., Hsu, C., Trautman, E., and Lee, H., pp. 83-91 in Ref. 1

42. Hsu, C., *Chemical Proc. of the Semiconductor Pure Water and Chemicals Conf.,* pp. 44-62, Santa Clara, CA. Balazs Analytical Laboratory, Sunnyvale, CA (Feb 11-13, 1992)

43. Krusell, W. C., and Golland, D. I., pp. 23-32 in Ref. 1: see also original O_3-work by W. C. Krusell et al., *ECS Ext. Abstr.,* 86-1:133 (1986)

44. Tong, J. K., Grant, D. C., and Peterson, C. A., pp. 18-25 in Ref.2

45. Matthews, R. R., *Chemical Proc. of the Semiconductor Pure Water and Chemicals Conf.,* pp. 3-15, Santa Clara, CA. Balazs Analytical Laboratory, Sunnyvale, CA (Feb 11-13, 1992)

46. Faylor, T. L., and Gorski, J. J., *Handbook of Contamination Control in Microelectronics,* (D. L. Tolliver, ed.), Ch. 6, Noyes Publications, Park Ridge, NJ (1988)

47. Sinha, D., *Solid State Technol,* 35(3):S9-S12 (1992)

48. Monkowski, J. R., Freeman, D. W., *Solid State Technol,* 33(7):S13-S17 (1990)

49. Hockett, R. S., pp. 227-242 in Ref. 1. See also: *Semicon/Korea '91 Tech. Proc.,* Seoul, Korea. Ch. III, pp. 89-98, SEMI, Mountain View, CA (Sept. 26-27, 1991)
50. Kamieniecki, E., pp 273-279 in Ref. 1
51. Zoth, G., and Bergholz, W., *ECS Ext. Abstr.,* 91-2:643-644, Electrochemical Society, Pennington, NJ. (1991)
52. Hahn, S., Eichinger, P., Park, J-G., Kwack, Y-S., Cho, K-C., and Choi, S-P., *Semicon/Korea '91 Tech. Proc.,* Seoul, Korea. Session III, pp. 60-78, SEMI, Mountain View, CA (Sept. 26-27, 1991)
53. Shimura, F., *Semicon/Korea '91 Tech. Proc.,* Seoul, Korea. Session III, pp. 23-34, SEMI, Mountain View, CA (Sept. 26-27, 1991)
54. Gupta, P., and Frost, M., *Proc. on Chemical Surface Preparation, Passivation, and Cleaning, Growth and Processing, Symp. B,* Paper B5.10, Spring Mtg. of MRS, San Francisco, CA (April 12-16, 1992)
55. Jastrzebski, L., Milic, O., Dexter, M., Lagowski, J., DeBusk, D., Nauka, K., Mitowski, R., Gordan, R., and Persson, E., pp 294-313 in Ref. 2
56. Rathmann, D., pp 338-343 in Ref. 2
57. Grundner, M., Hahn, P. O., Lampert, I., Schnegg, A., and Jacob, H., pp. 215-226 in Ref. 1
58. Kern, W., and Deckert, C. A., *Thin Film Processes,* (J. L. Vossen and W. Kern, eds.) Ch. V-1, pp. 401-496, Academic Press, New York (1978)
59. Kern, W., and Schnable, G. L., *The Chemistry of the Semiconductor Industry,* (S. J. Moss and A. Ledwith, eds.) Ch. 11, pp. 223-276, Chapman and Hall, New York (1987)
60. Walker, P., and Tarn, W. H., *CRC Handbook of Etchants for Metals and Metallic Compounds,* CRC Press, Inc., Boca Raton, FL (1990)
61. *Quick Reference Manual for Silicon Integrated Circuit Technology,* (W. E. Beadle, J. C. C. Tsai and R. D. Plummer, eds.) John Wiley and Sons, New York (1985)
62. Kern, W., *Semicond. International* 7(4):94-99 (1984)
63. Kern, W., *J. Electrochem Soc.* 137:1887-1892 (1990); also pp. 3-19 in Ref. 1
64. Kern, W., Puotinen, D., *RCA Review* 31:187-206 (1970)
65. Muraoka, H., Kurosawa, K. J., Hiratsuka, H., and Usami, T., *ECS Ext. Abstr.,* 81-2:570-573 (1981)
66. Park, J. G., and Raghavan, S., pp. 26-33 in Ref. 2

67. Hariri, A., and Hockett, R. S., *Semicond. International* 12(9):74-78 (1989)

68. Skidmore, K., *Semicond. International,* 10(9):80-85 (1987)

69. Skidmore, K., *Semicond. International,* 12(8):80-86 (1989)

70. Burggraaf, P., *Semicond. International*, 13(11):52-58 (1990)

71. Busnaina, A. A., and Kern, F. W., Jr., *Solid State Technol,* 30(11):111-114 (1987)

72. Menon, V. B., Clayton, A. C., and Donovan, R. P., *Microcontamination* 7(6):31-34, 107-108 (1989)

73. Tonti, A., pp. 41-47 in Ref. 2

74. Burkman, D., *Semicond. International,* 4(7):103-116 (1981)

75. McConnell, C. F., *Microcontamination,* 9(2):35-40 (1991)

76. Mayer, A., Shwartzman, S., *J. Electronic Materials,* 8:885-864 (1979)

77. Shwartzman, S., Mayer, A., and Kern, W., *RCA Review,* 46:81-105 (1985)

78. Ohmi, T., Mishima, H., Mizuniwa, T., and Abe, M., *Microcontamination,* 7(5):25-32,108 (1989)

79. Mishima, H., Ohmi, T., Mizuniwa, T., and Abe, M., *IEEE Trans. on Semiconductor Manufacturing,* 2(4):121 (1989)

80. Mishima, H., Yasui, T., Mizuniwa, T., Abe, M., and Ohmi, T., *IEEE Trans. on Semiconductor Manufacturing,* 2(3):69 (1989)

81. McConnell, C. F., *Microcontamination,* 9(2):35-40 (1991)

82. Olesen, M. B., *Proc. Institute of Environmental Sciences, Ann. Tech. Mtg.,* pp. 229-241, Mount Prospect, IL (1990)

83. Anabuki, K., Yamashita, Y., and Nawata, T., *ECS Ext. Abstr.*, 90-2:443, Electrochemical Society, Pennington, NJ. (1990)

84. Marra, J., *Ext. Abstr., Third Symp. on Particles in Gases and Liquids: Detection, Characterization and Control,* p. 52, San Jose, CA (1991)

85. Oki, I., Biwa, T., Kudo, J., and Ashida, T., pp. 215-222 in Ref. 2

86. Koontz, D. E., Thomas, C. O., Craft, W. H., and Amron, I., *Symp. on Cleaning of Electronic Device Components and Materials,* ASTM STP No. 246, pp. 136-145 (1959)

87. Feder, D. O., and Koontz, D. E., *Symp. on Cleaning of Electronic Device Components and Materials,* ASTM STP No 246, pp. 40-65 (1959)

88. Wolsky, S. P., Rodriguez, P. M., and Waring, W., *J. Electrochem Soc.* 103:606 (1956)
89. Sotnikov, V. S., and Belanovskii, A. S., *Russian J. of Phys. Chem.* 34:1001-1003 (1960)
90. Larrabee, G. B., *J. Electrochem Soc.* 108:1130-1134 (1961)
91. Kern, W., *Semiconductor Products* (early name for *Solid State Technology*), Vol. 6, Part I, 22-26 (Oct. 1963); Part II, 23-27 (Nov. 1963)
92. Kern, W., *RCA Engineer* 9(3): 62-66 (1963)
93. Kern, W., *RCA Review,* Part I: 31:207-233, (1970); Part II: 31:234-264 (1970); Part III: 32:64-87 (1971)
94. Kern, W., *Solid State Technol.* 15, Part I: (1)34-38 (1972); Part II: (2)39-45 (1972)
95. Kern, W., *J. Electrochem Soc.* 109:700-705 (1962)
96. Henderson, R. C., *J. Electrochem. Soc.* 119:772-775 (1972)
97. Meek, R. L., Buck, T. M., and Gibbon, C. F., *J. Electrochem. Soc.,* 120:1241-1246 (1973)
98. Amick, J. A., *Solid State Technol.* 47(11):47-52 (1976)
99. Murarka, S. P., Levinstein, H. J., Marcus, R. B., and Wagner, R. S., *J. Appl. Phys.* 48:4001-4003 (1979)
100. Gluck, R. M., *ECS Ext. Abstr.* 78-2:640 (1978)
101. Peters, D. A., and Deckert, C. A., *J. Electrochem. Soc.,* 126:883-886 (1979)
102. Phillips, B. F., Burkman, D. C., Schmidt, W. R., and Peterson, C. A., *J. Vac. Sci. Technol.,* A-1(2):646-649 (1983)
103. Goodman, A. M., Goodman, L. A., and Gossenberger, H. F., *RCA Review,* 44(2):326-341 (1983)
104. Kern, W., *RCA Engineer,* 28(4):99-105 (1983)
105. Watanabe, M., Harazono, M., Hiratsuka, Y., and Edamura, T., *ECS Ext. Abstr.* , 83-1:221-222 (1983)
106. Bansal, I. K., *Microcontamination,* 2(4):35-40 (1984)
107. Bansal, I. K., *Solid State Technol.,* 29(7):75-80 (1986)
108. Ishizaka, A., and Shiraki, Y., *J. Electrochem. Soc.,* 133(4):666-671 (1986)
109. Wong, C. Y., and Klepner, S. P., *Appl. Phys. Lett.,* 48(18):1229-1230 (1986)
110. Grundner, M., and Jacob, H., *Appl. Phys.,* A-39:73-82 (1986)

111. Becker, D. S., Schmidt, W. R., Peterson, C. A., and Burkman, D., *Microelectronics Processing: Inorganic Materials Characterization,* (L. A. Casper, ed.), Ch. 23, pp. 368-376, ACS Symp. Series No. 295, American Chemical Society (1986)

112. Kawado, S., Tanigaki, T., and Maruyama, T., *Semiconductor Silicon 1986, Proc. Fifth International. Symp. on Silicon Mater. Sci. Technol.,* (H. R. Huff, T. Abe, and B. Kolbesen, eds.), pp. 989-998, Electrochemical Society, Pennington, NJ (1986)

113. McGillivray, I. G., Robertson, J. M., and Walton, A. J., *Semiconductor Silicon 1986, Proc. Fifth International Symp. on Silicon Mater. Sci. Technol.,* (H. R. Huff, T. Abe, and B. Kolbesen, eds.), pp. 999-1010, Electrochemical Society, Pennington, NJ (1986)

114. Lampert, I., *ECS Ext. Abstr.,* 87-1:381-382 (1987)

115. Gould, G., and Irene, E. A., *J. Electrochem. Soc.,* 174(4):1031-1033 (1987)

116. Ruzyllo, J., *J. Electrochem. Soc.,* 174(4):1869-1870 (1987)

117. Probst, V., Bohm, H. J., Schaber, H., Oppolzer, H., and Weitzel, I., *J. Electrochem. Soc.,* 135(3):671-676 (1988)

118. Peterson, C. A., *Particles on Surfaces 1: Detection, Adhesion, and Removal,* (K. L. Mittal, ed.), pp. 37-42, Plenum Press, New York (1988)

119. Morita, E., Yoshimi, T., and Shimanuki, Y., *ECS Ext. Abstr.,* 89-1:352-353 (1989)

120. Yoshimi, T., Morita, E., and Shimanuki, Y., *ECS Ext. Abstr.,* 89-1:354-355 (1989)

121. Gould, G., and Irene, E. A., *J. Electrochem. Soc.,* 136(4):1108-1112 (1989)

122. Bowling, R. A., *J. Electrochem. Soc.,* 132(9):2208-2214 (1985)

123. Menon, V. B., Michaels, L. D., Donovan, R. P., Hollar, L. A., and Ensor, D. S., *Proc. Institute of Environmental Sciences, Ann. Mtg.,* King of Prussia, PA. (May 2-6, 1988)

124. Vig, J. R., *Treatise on Clean Surface Technology,* (K. L. Mittal, ed.), Vol. 1, pp. 1-26, Plenum Press, New York (1987)

125. Kaneko, T., Suemitsu, M., and Miyamoto, N., *Jpn. J. of Appl. Phys.* 28(12):2425-2429 (1989).

126. Claevelin, C. R., and Duranko, G. T., *Semicond. International,* 10(12):94-99 (1987); Novak, R. E. *Solid State Technol.,* 31(3):39-41 (1988)

127. Clements, L. D., Busse, J. E., and Mehta, J., *Semicond. Fabrication Technology and Metrology, ASTM STP 990,* (D. C. Gupta, ed.), ASTM, Philadelphia, PA (1988)

128. Duranko, G., Syverson, D., Zazzera, L., Ruzyllo, J., and Frystak, D., *Physics and Chemistry of* SiO_2 and *Si-*SiO_2 *Interface* (B. E. Deal and C. R. Helms, eds.), pp. 429-436, Plenum Publishing Corp., New York (1988)

129. Fountain, G. G., Hattangady, S. V., Rudder, R. A., Posthill, J. B., and Markunas, R. J., *MRS Symp. Proc.* 146:139 (1989)

130. Burrows, V. A., Chabal, Y. J., Higashi, G. S., Raghavachari, K. and Christman, S. B., *Appl. Phys. Lett.,* 53(11):998-1000 (1988)

131. Hahn, P. O., Grundner, M., Schnegg, A., and Jacob, H., *Appl. Surf. Sci.,* 39:436 (1989)

132. Zazzera, L. A., and Moulder, J. F., *J. Electrochem Soc.,* 136(2):484-491 (1989)

133. Chabal, Y. J., Higashi, G. S., Raghavachari, K., and Burrows, V. A., *J. Vac. Sci. Technol.* A7(3):2104-2109 (1989)

134. van den Meerakker, J. E. A. M. and van der Straaten, M. H. M., *J. Electrochem. Soc.* 37:1239-1243 (1990)

135. Tanaka, K., Sakurai, M., Kamizuma, S., and Shimanuki, Y., *ECS Ext. Abstr.* 90-1:689-690, Electrochemical Society, Pennington, NJ. (1990)

136. Miyashita, M., Itano, M., Imaoka, T., Kawanabe, I., and Ohmi, T., *ECS Ext. Abstr.* 91-1:709-710, Electrochemical Society, Pennington, NJ (1991)

137. Ohmi, T., Tsuga, T., and Takano, J., *ECS Ext. Abstr.* 92-1:388-389, Electrochemical Society, Pennington, NJ (1992)

138. Ohmi, T., *ECS Ext. Abstr.* 91-1:276-277, Electrochemical Society, Pennington, NJ. (1991)

139. Heyns, M., Hasenack, C., De Keersmaecker, R., and Falster, R., *Microelectronic Engineering* 10:235-257 (1991); also: Heyns, M. M., *Microcontamination,* 9(4): 29-34, 87-89 (1991)

140. Verhaverbeke, S., Alay, J., Mertens, P., Meuris, M., Heyns, M., Vandervorst, W., Murrell M., and Sofield, C., *Proc. on Chemical Surface Preparation, Passivation, and Cleaning, Growth and Processing, Symp. B,* Spring Mtg. of MRS, San Francisco, CA (April 12-16, 1992)

141. Sakurai, M., Ryuta, J., Morita, E., Tanaka, K., Yoshimi, T., and Shimanuki, Y., *ECS Ext. Abstr.,* 90-1:710-711, Electrochemical Society, Pennington, NJ (1990)

142. Rubloff, G. W., *SEMICON/Korea '90 Tech. Proc.,* pp. 2-3 to 11, SEMI, Mountain View, CA (Dec, 6-7, 1990)

143. Grundner, M., Gräf, D., Hahn, P. O., and Schnegg, A., *Solid State Technol,* 34(2):69-75 (1991)

144. Hirose, M., Yasaka, T., Kanda, K., Takakura, M., and Miyazaki, S., pp. 1-9 in Ref. 2

145. Chabal, Y. J., *Proc. on Chemical Surface Preparation, Passivation, and Cleaning, Growth and Processing, Symp. B,* Paper 6.1, Spring Mtg. of MRS, San Francisco, CA (April 12-16, 1992)

146. Anttila, O. J., and Tilli, M. V., pp. 179-189 in Ref. 2; *J. Electrochem. Soc.* 139:1751-1756 (1992)

147. Kniffin, M. L., Beerling, T. E., and Helms, C. R., *J. Electochem. Soc.,* 139:1195-1199 (1992)

148. Poliak, R., Matthews, R., Gupta, P. K., Frost, M., and Triplett, B., *Microelectronics,* 10(6):45-49, 93-94 (1992)

149. Shimono, T., *ECS Ext. Abstr.,* 91-1:278-279, Electrochemical Society, Pennington, NJ. (1991)

150. Takizawa, R., and Ohsawa, A., pp. 75-82 in Ref. 1

151. Lowell, L., *Solid State Technol,* 34(4):149-152 (1991)

152. Menon, V. B., and Donovan, R. P., pp. 167-181 in Ref. 1. Also: *Microcontamination*, 8(11):29-34, 66 (1990)

153. Syverson, W. A., Fleming, M. J., and Schubring, P. J., pp. 10-17 in Ref. 1

154. Doshi, V., Hall, L., and Davison, J., *Defect Reduction in DRAM Manufacture using Ion Exchange-Purified HF,* Paper to be presented at the Fall Mtg. of the Electrochemical Society, Toronto, Ontario, Canada (Oct. 12-16, 1992); *ECS Ext. Abstr.,* 92-2, Abstract No. 410 (1992)

155. Davison, J., *Solid State Technol.*, 35(3):S1-S5 (1992)

156. Davison, J., *Solid State Technol.,* 35(7):S10-S14 (1992)

157. Helms, C. R., and Deal, B. E., pp. 267-276 in Ref. 2

158. Deal, B. E., McNeilly, M., Kao, D. B., and deLarios, J. M., pp. 121-128 in Ref. 1: also: *Solid State Technol.,* 33(7):73-77 (1990)

159. Ohmi, T., Miki, N., Kikuyama, H., Kawanabe, I., and Miyashita, M., pp. 95-104 in Ref. 1

160. Onishi, S., Matsuda, K., and Sakiyama, K., *ECS Ext. Abstr.*, 91-1:519-520, Electrochemical Society, Pennington, NJ. (1991)

161. Iscoff, R., *Semicond. International,* 14(12):50-54 (1991)

162. Wong, M., Moslehi, M. M., and Reed, D. W., *J. Electrochem. Soc.,* 138:1799-1802 (1991)

163. Nobinger, G. L., Moskowitz, D. J., and Krusell, W. C., *Microcontamination,* 10(4):21-26, 68-69 (1992)

164. Deal, B. E., and Helms, C. R., *Proc. on Chemical Surface Preparation, Passivation, and Cleaning, Growth and Processing, Symp. B,* Paper 6.2, Spring Mtg. of MRS, San Francisco, CA (April 12-16, 1992)

165. Izumi, A, Matsuka, T., Takeuchi, T., and Yamano, A., pp. 260-266 in Ref. 2

166. Comfort, J. H., pp. 428-436 in Ref. 2

167. Reif, R., *Proc. on Chemical Surface Preparation, Passivation, and Cleaning, Growth and Processing, Symp. B,* Paper 7.1, Spring Mtg. of MRS, San Francisco, CA (April 12-16, 1992)

168. Liehr, M., *Proc. on Chemical Surface Preparation, Passivation, and Cleaning, Growth and Processing, Symp. B,* Paper 1.1, Spring Mtg. of MRS, San Francisco, CA (April 12-16, 1992)

169. Kalem, S., Lamb, H. H., Yasuda, T., Ma, Y., and Lucovsky, G., *Proc. on Chemical Surface Preparation, Passivation, and Cleaning, Growth and Processing, Symp. B,* Paper 2.2, Spring Mtg. of MRS, San Francisco, CA (April 12-16, 1992)

170. Tasch, A., Banerjee, S., Hsu, T., Qian, R., Kinosky, D., Irby, J., Mahajan, A., and Thomas, S., *Proc. on Chemical Surface Preparation, Passivation, and Cleaning, Growth and Processing, Symp. B,* Paper 1.4, Spring Mtg. of MRS, San Francisco, CA (April 12-16, 1992); see also: pp. 418-427 in Ref. 2

171. Hattangady, S. V., Rudder, R. A., Mantini, M. J., Fountain, G. G., Posthill, J. B., and Markunas, R. J., *MRS Symp. Proc.,* 165:221-226 (1990)

172. Frystak, D., and Ruzyllo, J., pp. 58-71 in Ref. 2

173. Chang, E. Y., *Proc. on Chemical Surface Preparation, Passivation, and Cleaning, Growth and Processing, Symp. B,* Paper 5-18, Spring Mtg. of MRS, San Francisco, CA (April 12-16, 1992)

174. Gluck, R. M., pp. 48-57 in Ref. 2

175. Ivankovits, J. C., Bohling, D. A., Lane, A., and Roberts, D. A., pp. 105-111 in Ref. 2

176. Wong., Liu, D., Moslehi, M., and Reed, D., *Electron Dev. Lett.,* 12:425 (1991)

177. Ito, T., Sugino, R., Watanabe, S., Nara, Y., and Sato, Y., pp. 114-120 in Ref. 1

178. Ito, T., Sugino, R., Sato, Y., Okuno, M., Osawa, A., Aoyama, T., Yamazaki, T., and Arimoto, Y., *Proc. on Chemical Surface Preparation, Passivation, and Cleaning, Growth and Processing, Symp. B,* Paper 3.1, Spring Mtg. of MRS, San Francisco, CA (April 12-16, 1992). Also: *Semicon/Korea '91 Tech. Proc.,* Seoul, Korea. Session III, pp. 44-52, SEMI, Mountain View, CA (Sept. 26-27, 1991)

179. Vig, J. R., pp. 105-113 in Ref. 1

180. Pearton, S. J., Ren, F., Abernathy, C. R., Hobson, W. S., and Luftman, H. S., *Appl. Phys. Lett.,* 58(13):1416-1418 (April 1, 1990)

181. Kopf, R. F., Kinsella, A. P., and Ebert, C. W., *J. Vac. Sci. Technol.,* B9(1):132-135 (1991)

182. Bedge, S., and Lamb, H. H., *Proc. on Chemical Surface Preparation, Passivation, and Cleaning, Growth and Processing, Symp. B,* Paper 3.2, Spring Mtg of MRS, San Francisco, CA (April 12-16, 1992)

183. McDermott, W. T., Ockovic, R. C., Wu, J. J., and Miller, R. J., *Microcontamination,* 9(10):33-36, 94-95 (1991)

184. Bok, E., Kelch, D., and Schumacher, K. S., *Solid State Technol.,* 35(6):117-120 (1992)

185. Allen, S. D., *Scientific American,* 26(6):86-87 (1990)

186. Allen, S. D., "Laser Assisted Particle Removal," Tech. Report, University of Iowa (1990); see also: Lee, S. J., Imen, K., and Allen, S. D., Paper EM-WeM10, presented at the American Vacuum Society National Symp., Seattle, WA (Nov. 11-15, 1991)

2

Trace Chemical Contamination on Silicon Surfaces

Don Tolliver

1.0 INTRODUCTION

Trace chemical contamination on surfaces of silicon integrated circuit (IC) devices is a continuous limitation to a chip manufacturer's ability to produce high yields, reasonable profits, and long term reliability. This type of contamination is a silent, often invisible killer to even the most sophisticated optical or holographic inspection methods now in use on silicon IC production lines. Since trace chemical and metallic contamination on silicon devices is not always readily visible to the manufacturing engineer due to the lack of online analysis equipment, there is a tendency to defer engineering attention on the chemical contaminant problem. Trace metallic contamination from wafer processing and residuals after wafer cleaning steps is an insidious manufacturing threat that has continued to reduce yield and reliability of advanced circuits. The general impact of contamination on device yield has been recognized as a major threat to profitability (1).

Exact correlation of surface contamination to device yield is not been easily assessed and modeled, whereas particle contamination on surfaces has been systematically attacked and measured for the last ten years. Trace chemical contamination is now targeted and will receive significant attention during the next few years as device complexity increases and very low levels of surface contamination are shown to alter device functionality and reliability. Figure 1 tries to capture the inevitable drop in *killer* defect size as a function of the minimum feature size on an integrated circuit.

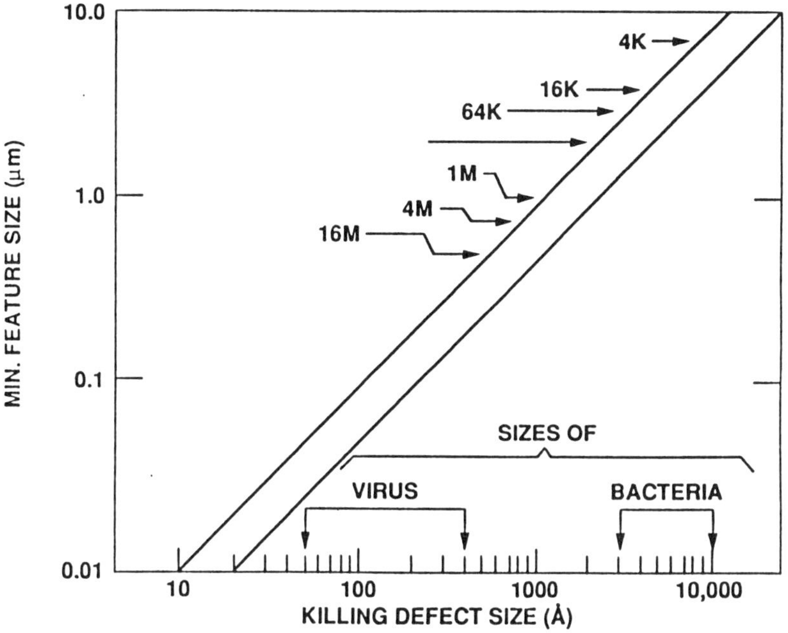

Figure 1. Shrinking design rules and the impact of defects.

During the 1990s the particle size of some contaminants will approach the molecular level. Figure 2 displays the size range of aerosol particles, gas molecules, viruses, and typical bacteria. Killer defects can clearly reach 0.01µm during the second half of the 1990s.

Table 1 outlines the legacy of killer particle defects by mask level from the 1-megabit device level of a DRAM product to the 1-gigabit device level. The last line of this rather aggressive and perhaps perilous picture shows that the defect density, or contamination improvement learning curve per critical mask level, must improve approximately one thousand times over the span of time from the 1-megabit device to the 1-gigabit device. This line follows a square power law.

Some ambiguity exists with respect to particles on the surface of the wafer and chemical contamination. Particles are normally referenced and measured independently of their chemical composition. Nonetheless particles are a specific form of chemical contamination, and as the device geometries shrink, the visibility of particles as discrete objects disappears. Particles represent some significant portion of the total chemical load on the device surface. The semiconductor producer will continue to independently specify and control the particle and the trace chemical contaminants as a function of the device requirements.

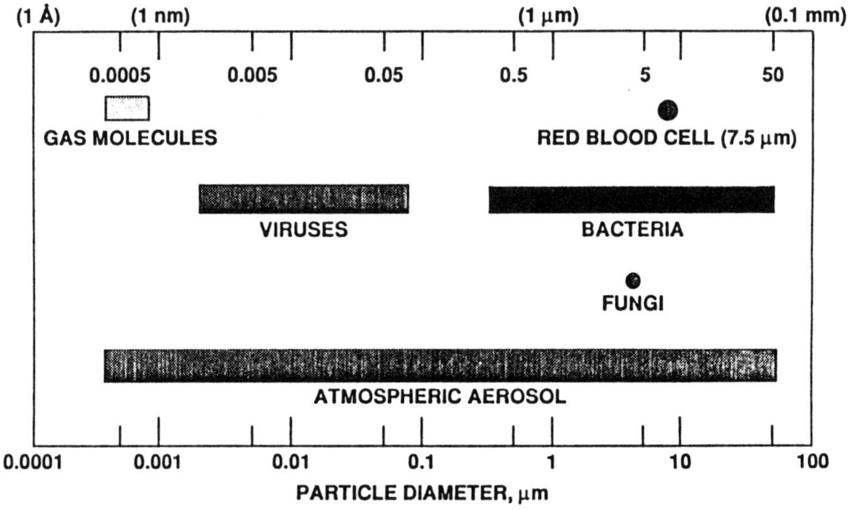

Figure 2. Size of some common particles.

Table 1. The Yield Killers: Killer Defect Density per Critical Mass Level for 70% IC (Chip) Yield (*Univ. of Ariz., SEMATECH Center of Excellence)*

	Killer Defect Density per critical Mask Level for 70 percent IC (Chip) Yield					
Dram Bits	**1 Mbit**	**4 Mbit**	**16 Mbit**	**64 Mbit**	**256 Mbit**	**1 Gbit**
Mask Levels	12	14	16	18	20	22
IC Area (cm^2)	0.60	0.90	1.50	2.45	3.85	6.10
CD (µm)	1.25	0.80	0.60	0.40	0.25	0.15
D/cm^2/level (size > 0.1xCD)	0.05	0.028	0.015	0.008	0.0045	0.0025
D/cm^2/level (particle size > 0.125 µm)	0.05	0.012	0.0035	0.0008	0.00018	0.000036

Metal contaminants, such as iron and copper, introduce electron energy states close to the middle of the silicon energy bandgap on both p-type and n-type silicon. These active sites create centers for electron generation and recombination. Such sites are often called electron traps that in turn cause increases in reverse bias junction leakage current of p-n junctions of a silicon device. A typical example is iron at concentrations on the order of 10^{13} atoms/cm^3, which cause leakage currents in DRAM (dynamic random access memory) devices and restrict the refresh performance of the device due to the excess leakage. Iron introduces these bulk-related traps at 0.4 eV and 0.55 eV above the Si valence band; this also increases the surface recombination velocity of the device.

The electrical breakdown strength of thin oxide films on the surface of silicon devices is subject to metallic contamination from metallic inclusion in the oxide during silicon oxidation or deposition processes (3)(4).

Process and device literature is now available that cites numerous accounts of the impact on device performance that may occur for a specific device that was produced in a process where metallic contaminants had deposited on the silicon wafer.

Each liquid bath presents a unique condition that may or may not cause impurity deposition on the surface. Hydrofluoric acid and buffered (NH_4F) hydrofluoric acid provide entirely different conditions for the silicon surface than an oxidizing medium, such as hydrogen peroxide.

1.1 Horizontal and Vertical Dimensions

Normally, device design criteria are referenced to the minimum line width or the critical dimension of an MOS device feature. These horizontal design criteria infer the degree of device integration and complexity of the chip. In a 4-megabit DRAM chip, this feature size is typically 0.6 to 0.8 µm in the horizontal dimension. (See Fig. 1.) The vertical dimensions of the device are not normally referenced. The corresponding vertical device dimensions that scale with the horizontal dimension are considered to be the most sensitive to surface contamination. Gate oxide thickness on a 4-megabit DRAM will be at or below 100 Å.

Vertical dimensions (in critical gate oxide layers) below 100 Å are prominent in advanced device structures. This thickness requirement creates a significant impact from both non-particle and particle contamination. At gate oxide thicknesses of 200 Å, some surface problems are not observed due to the nature of the oxide-surface interface stability. For this reason, attention to chemical contamination on the surface of an MOS device structure was not a significant yield and process limitation in the 1980s when the lowest gate oxide thicknesses were often specified between 150 and 200 Å on leading edge devices.

The minimum allowable surface metal contamination was not always referenced as a requirement in advanced device manufacturing during the 1980s. Depending on the device and the manufacturer, it is still not always specified for today's leading edge devices. This is due in part to the lack of exact knowledge as to the tolerable level of a specific surface contaminant in terms of device performance and yield. If iron and sodium are under control, can aluminum and chromium be ignored? Not necessarily.

1.2 A Manufacturing Issue, Not a Device Design Specification

Some semiconductor companies may let their production engineering personnel determine how to reduce surface contamination when they implement the device process. Guidelines for manufacturing contamination control are not always clear. In general, surface contamination levels are not readily measured or correlated to specific process steps and process variations.

In the past design limits and processing priorities were often skipped in this particular area, because the correlation of effects and turn-around time on device analysis were slow to reach the design and manufacturing personnel. An example of this type of process design is the need for additional ion implantation steps with the concurrent photoresist processing. Whereas the understanding of the device design need is clear, the impact of the surface contamination left on the device is not. Currently, process additions of this type may not be instituted due to lack of confidence that surface contamination can be eliminated by a subsequent resist strip and cleaning process. The incorporation of these steps must be factored into the process with the awareness that surface contamination is a hidden and decisive factor in real manufacturablilty of the device (5).

1.3 Cleaning Technology in the United States

In the United States, but not in Japan, wafer cleaning technology has not been viewed as a first level process technology in most US based semiconductor manufacturing lines. As a result, focused and dedicated wafer cleaning engineering studies and manufacturing process control are not available. Most often, wafer cleaning technology is viewed as a subset of the diffusion, oxidation, etch, photoresist strip, or deposition processes.

Cleaning processes are not normally given the support necessary to be fully compatible with all process steps.

Khilnani noted the need of cleaner chemicals for semiconductor manufacturing in the mid 1980s (6). At that time Japanese manufacturers were well on the way to enforcing stringent purity controls on liquid cleaning processes. In the United States, lack of recognition of the issue has caused major delays in implementing the production of high-quality liquid chemicals.

Another reason for this delay is that wafer cleaning has not been considered "glamorous" if compared to submicron lithography that is organizationally more visible to higher levels of management.

1.4 Visibility of the Problem

Difficulty of Measurements. Usually, high-sensitivity surface analysis of trace metal contamination has not been performable in the manufacturing line itself. In comparison, oxide thicknesses and sheet resistance measurements are nearly always performed in the immediate vicinity of processing. Using outside analysis, the feedback naturally slows turn-around times, and the rate of defect learning is significantly reduced. Typically, the device is analyzed only after the entire process sequence has been completed. This provides an accumulative picture of surface contamination, but does not isolate those steps which have contributed significantly to the contamination.

Expense of Measurements. Measurements and analysis of silicon surfaces is not only slow but clearly expensive for the device manufacturer. Unless focused effort and experimentation are seen as part of a process evaluation sequence, routine analysis may not be requested due to the cost. Sample preparation and transportation add to the difficulty of rapid analysis and short-term feedback loops. Only recently have certain semiconductor manufacturing lines begun to include surface analysis capability in the manufacturing line itself. Where internal analysis is not available, external analytical laboratories have to be utilized where both the correct equipment and the analytical knowledge are available.

Expertise of Measurements. Expertise is mandatory in highly specialized measurement technology, such as SIMS, TXRF and methods such as vapor-phase decomposition (VPD) and surface photovoltage measurements (SPV), for reliable manufacturing control in the 1990s (2)(6). Experienced judgment is required to carry out low-level trace metal detection (7)(8)(9). For these reasons, the device manufacturer cannot easily provide in-house advanced analytical expertise and response capability to adequately support the analytical procedures that are now required for advanced IC manufacturing.

1.5 Management of Surface Contamination in Manufacturing

Lack of management sensitivity and engineering focus are clearly part of the microcontamination problem in IC manufacturing. Generally, there is a lack of credibility of the importance of the problem. Engineering and research efforts have been limited. Correlation has been poor.

Processing materials, such as wet process chemicals and specialty gases, are commodity items for most fabrication lines producing ICs. Since the cost of these materials is relatively minor (typically 2 to 4% of the entire operating budget), little interest has been shown by management. Very often, no correlation that is based on parametrics and functional data at the end of the process or at unit probe is available to the engineering team.

In some cases, further loss of differentiation of contamination issues has occurred due to the passivation processing. Metal ion contamination was not seen to be serious if adequate final passivation was incorporated. This situation may further obscure the real problems that have occurred earlier in the process sequence and may create the illusion of lack of need for focused engineering efforts.

1.6 Gallium Arsenide and other III-V Compound Semiconductors

Trace chemical contamination on semiconductor surfaces has been driven by silicon-based technology. At this writing GaAs device manufacturing in the United States is becoming a necessary manufacturing technology for emerging high-density and high-speed devices.

GaAs and III-V compounds do not adapt directly to silicon based cleaning technology and contaminants have different effects than for silicon. For example, in the case of GaAs devices, silicon can be a killer defect and has, therefore, serious yield limitations. Aqueous based chemistry, including ultrapure water, can etch GaAs surfaces.

Sensitivity of the manufacturing process to trace chemical contamination on GaAs surfaces obviously changes with time both in acceptable levels of contamination and in regard to which species are of greatest concern to specific devices. Specific processes, chemicals and process tools are needed to provide effective cleaning steps for GaAs surfaces, but the general trends and controls developed for silicon devices are applicable here as well. Specific details appropriate only to compound semiconductor surfaces are beyond the scope of this chapter.

2.0 SOURCES OF CHEMICAL CONTAMINATION

Unless rigorous effort is applied, trace contaminants can be deposited or accumulated on silicon surfaces from a broad variety of sources in the manufacturing process and process systems. Table 2 is a brief listing of a variety of these sources. Focused research and engineering is now being applied on a regular basis in known areas. Other sources are subtle and not readily understood nor easily detected. The purpose of this section is to itemize, delineate, and describe many of the possible sources of chemical contaminants that are likely to arrive on the surfaces of the wafer. The detailed transfer mechanisms are not always known. The exact removal processes to the required level of control are often not known. The engineering expertise, which we have come to call "contamination control technology" in the microelectronics field, has developed largely to understand how to measure and control these expansive lists of chemical and physical species that cause contamination on the silicon wafers and semiconductor devices.

Table 2. Sources of Particle Contamination (*Source: Carlton M. Osburn, Ref. 1)*

Inherent in Process	Extrinsic to Process
Flaking of deposited films from sidewalls	Wafer transport
Gas-phase reaction of CVD films	Valve Actuation
Polymer formation in RIE processes	Vacuum roughing
Back sputtering	Mechanical vibration
Spurious nucleation	Bacterial growth in DI piping
Redeposition of metal particles in lift off	Screw caps on chemicals
Redeposition of resist during Piranha cleans	Corrosion/leaching of piping
Heating that crtosslinks resists	Temperature gradients/ramping
	Outgassing of plasticizers from boxes/ carriers

2.1 Cleanroom Air

Traditionally, the semiconductor cleanroom is specified and monitored exclusively on the basis of airborne particle density and particle size. No attempt has been made to qualify the type of particle, organic vapor concentration, or ionic concentration in the air.

In future manufacturing systems and facilities, wafer environments will become sensitive to specific levels of organic and ionic contamination. With this as a given control requirement, the cleanroom or wafer environment will continue to be seen as a source of chemical contamination for the wafer surface and must not be excluded on the basis of low priority.

Sodium ions in the cleanroom air are known to exist at specific levels, but monitoring of sodium in cleanrooms is not universally practiced. Organic vapors and material outgassing can also be serious, but are not normally monitored.

2.2 Cleanroom Personnel

Unless complete isolation technology in the context of mini- or micro-environments is practiced in the manufacturing line, cleanroom personnel will continue to be a major source of chemical and particle contamination for semiconductor processing. Airborne transport, proximity transport and direct contact with gloved hands or cleanroom garments are sources of chemical contamination that are to this day poorly documented and often discounted. Correlation studies are difficult to conduct and hard to document. A dilemma in leading edge IC manufacturing is that personnel contamination will continue to be a serious source in conventional cleanrooms as the device structures increase in complexity and sensitivity. Many manufacturing lines de-prioritize their operating personnel as a source of contamination.

2.3 Liquid Sources and Photolithographic Resists

Semiconductor device manufacturing consists of a repetitive series of chemical and physical process steps. Since most processes are chemical in nature, many possibilities exist to transfer and deposit chemical contaminants on the front and on the back of silicon wafers. The number of independent wet processing steps for a typical 1-Mbit DRAM process is approximately sixty-three. A typical 4-Mbit process increases this number to approximately 65 - 70 cleaning steps. Accumulation of chemical impurities is a major reason for increased concern in manufacturing due to the continued increase in the number of wet-chemical cleaning steps and lithographic process steps required in advanced device manufacturing (10). Mechanism studies of chemical and particle deposition from liquid medium to the surfaces of semiconductor wafers have become a recent subject of various researchers, especially in laboratories in Japan and in the United States (11)(12)(13).

These metal deposition mechanisms are highly dependent on the nature of the chemistry of solution in use. For example, the hydrogen ion concentration (pH) is a major variable in the deposition of particles and chemical species. The charge state of a particle is related to the pH of the solution in which it resides as well as its composition. The exact type and state of the wafer surface, (silicon, silicon dioxide, silicon nitride, hydrophilic, hydrophobic) are critical to the deposition mechanism. Often times all of these surfaces may be present at one time on the surface an IC during the process. Exact transfer or deposition mechanisms may allow deposition to occur that depend on the circumstances and the nature of the process. Mechanisms of particle deposition will be discussed further by other contributors to this volume.

Water as a Source of Chemical Contamination. Ultrapure water is the most common chemical used in semiconductor manufacturing. Delivery of ultrapure water to the wafer surface is expensive and ever challenging to the total manufacturing system. Chemical contamination can arrive at the silicon surface from ultrapure water in form of ions, bacteria, organic particles, and low level inorganic residues such as silica. Table 3 shows an analysis of ultrapure water samples from a semiconductor manufacturing line during the late 1980s. The point of distribution to the wafer fabrication line may not always represent the quality of the water which reaches the wafer in the process tool. Figure 3 depicts an analysis of ultrapure water upstream of a Teflon™ cartridge, just after the cartridge is installed, and after 200 liters of ultrapure water has flowed through the filter cartridge.

Contamination may or may not be metallic in nature. However, any type of contaminant can impede and impact the quality of a device. Due to the aggressive nature of ultrapure water to dissolve materials it contacts, control and measurement of low level impurities in water is a constant challenge to device manufacturing. Table 4 is an example of an ultrapure water specification which was given at The Semiconductor Pure Water Conference held in 1991 (15). Such specifications and analytical methods of measurement require rigorous attention and increased cost to the IC manufacturer.

Table 3. Ion Analysis in Ultrapure Water (Concentrations in ppb.) (*Source: Balazs Analytical Laboratory)*

MOS 6 LOCATION	F	Cl	NO2	Br	NO3	HPO4	SO4	Na	NH4	K	B	Ni	Fe
PT. OF DISTRIBUTION	*	1.5	0.21	*	*	*	0.41	1.7	2.0	1.5	4.4	.06	<.1
SUPPORT HOOD	*	0.22	0.15	*	*	*	0.30	<.05	0.16	*	2.9	.07	<.1
ETCH CLEAN HOOD	*	0.45	0.11	*	*	*	0.35	<.05	0.35	*	3.7	.11	<.1

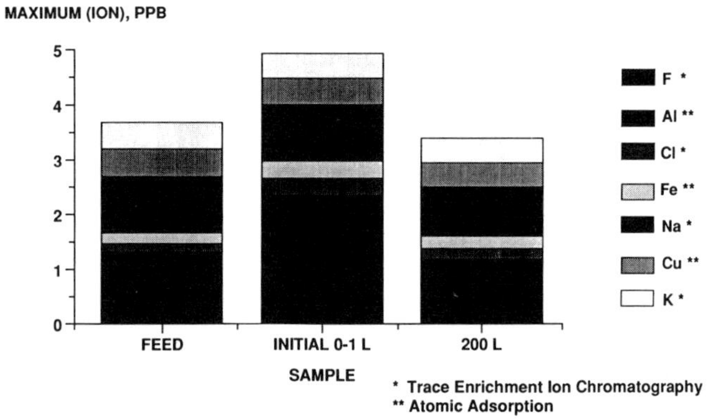

Figure 3. Teflon cartridge ionic extractables/pure water dynamic extraction. *(Source: Millipore Corporation)*

Table 4. Example of Water Quality Requirements for Ultrapure Water (15)

Degree of Integration		256K	1 M	4 M (Estimate)	16 M (Estimate)	Measurement Method
Design Rule (μm)		1.5 ~ 2.0	1.0 ~ 1.2	0.8	0.6	
Resistivity (MΩ cm at 25°C)		over 18	over 18	over 18	over 18.1	specific resistance meter
Fine Particles	Diameter (μm)	0.1	0.1	0.1	0.05	SEM method
	(number/ml)	less than 50	less than 10	less than 10	less than 10	
Live Bacillus (number/ml)		less than 0.05	less than 0.01	less than 0.01	less than 0.005	M-TGE culture method
TOC (μg/l)		less than 50	less than 30	less than 10	less than 5	Wet oxidation TOC meter
Dissolved oxygen meter(μg/l)		less than 100	less than 100	less than 50	less than 10	Dissolved oxygen meter
SiO_2 (μg/l)		less than 10	less than 10	less than 5	less than 1	Ion Chromatograph
Cr (μg/l)				less than 0.007	less than 0.007	
Fe (μg/l)		less than 1	less than 0.1	less than 0.003	less than 0.003	
Mn (μg/l)				less than 0.05	less than 0.05	
Zn (μg/l)				less than 0.02	less than 0.02	
Cu (μg/l)		less than 1	less than 0.1	less than 0.002	less than 0.002	
Na (μg/l)		less than 1	less than 0.1	less than 0.1	less than 0.1	
K (μg/l)				less than 0.1	less than 0.1	
Cl (μg/l)		less than 1	less than 0.1	less than 0.1	less than 0.1	

The production and control of ultrapure water is an independent technology serving the microelectronic manufacturing industry. Extensive research and development in all aspects of ultrapure water has been conducted around the world. Specific conferences and forums are routinely sponsored and attended to support the growth and understanding in the challenging requirement for ultraclean semiconductor manufacturing.

Liquid Process Chemicals. Liquid process chemicals are one of the most visible and most common sources of particle and chemical contamination in the manufacturing process (12). Chemicals such as sulfuric acid, hydrogen peroxide, aqueous ammonium hydroxide, aqueous hydrochloric acid, and isopropyl alcohol are common semiconductor process chemicals. The actual list is much longer, and may include chemical compounds and ions that can be catastrophic to silicon ICs. Each individual chemical can contribute a unique contaminant that may not be present in another chemical. Intense efforts were made to deliver high-quality liquid chemicals during the 1980s. In 1987 and 1988 industry workers at SEMATECH gathered to forecast the needs of ultrapure chemicals for the 1990s (Table 5); actual performance at these levels is not easily obtained. Figure 4, from SEMATECH in 1990, shows the strong trend to improve the quality of process chemicals. The complexity and economic demands for ultrapure chemicals at the point of use are major manufacturing and facility issues. Table 6 defines another attempt to capture the industry needs and clearly contrasts the high differential in ultrapure water and high-purity chemicals in the 1991 period.

Table 5. Maximum Impurity Guidelines for High-purity Chemicals (*Source: SEMATECH Working Group Forecast, Nov. 1987)*

CATEGORY	1988	1990	1993
METALS INDIVIDUAL	< 25 PPB	< 1-3 PPB	< 1PPB
METALS TOTAL	< 200	< 50	< 25
ANIONS INDIVIDUAL	< 10	< 5	< 2
ANIONS TOTAL	< 50	< 25	< 12
PARTICLES	< 1000/L ≥ 0.2 μm	< 5600/L ≥ 0.05 μm	<2700/L ≥ 0.03 μm

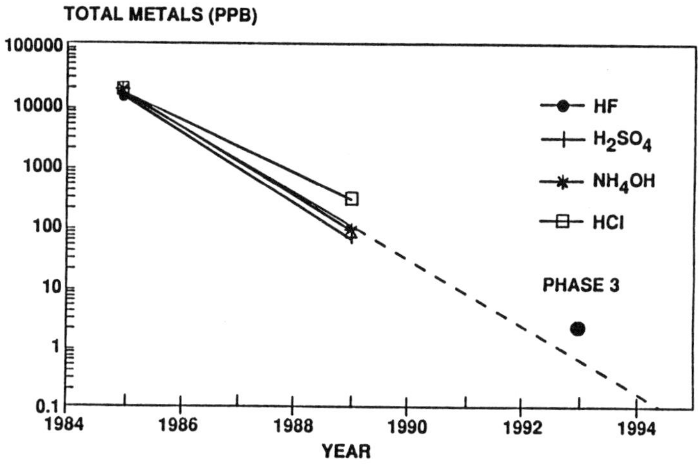

Figure 4. Trends of metallic impurity concentrations in liquid chemicals. *(Courtesy V. Menon, Microcontamination Conference, Oct, '90)*

Table 6. Particles and Metallic Ions in Some Liquid Chemicals. *(Source: Millipore Corp.)*

	49% HF 1991	H_2SO_4 1991	U.P. Water 1991
Particle Counts	10-50/ml ≥ 0.2μm	100 - 500/ml ≥ 0.2μm	6/ml≥ 0.05μm
Particle Concentration	50-300 ppq	0.5-3 ppt	0.5 ppq
Specific Metallic Ions	1 ppb	1-5 ppb	< 100 ppt
Total Metallic Ions	≤ 40 ppb	≤ 70 ppb	<1 ppb

Figure 5 is a representation of sulfuric acid system testing over a period of purification or reprocessing. The starting grade, noted as Semigrade sulfuric acid, is the acid which was typical in the industry as late as 1985 as a leading edge product. Acids of this type are still in use in certain manufacturing lines as of this printing. Sulfuric acid reprocessing is a technology that has been available to IC manufacturing since 1986.

Sulfuric acid is the most common acid used in reprocessors. Hydrofluoric acid reprocessing is also available and has been used for semiconductor manufacturing since 1990.

Figure 6, showing organic contamination in semiconductor process chemicals, is the latest concern in ultrapure chemicals to be addressed. Normally, low level concentrations of organic contamination in process chemicals such as sulfuric acid and hydrogen peroxide, have not been specified. During the 1990s continued attention on this type of contamination will be seen.

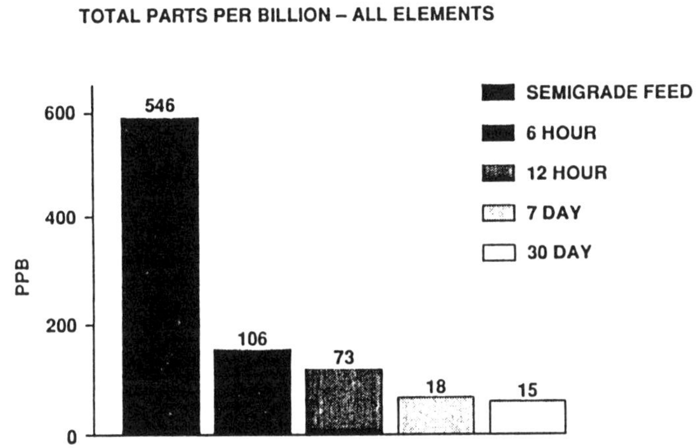

Figure 5. Results of sulfuric acid testing. *(Courtesy of Alameda Instruments Inc.)*

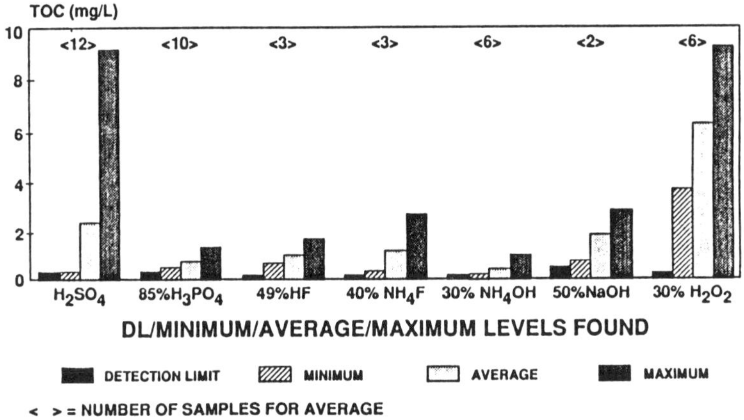

Figure 6. Summary of TOC in chemicals (20).

Photolithographic Process Fluids, Solvents, and Strippers. The semiconductor IC processes are highly interdependent. First, chemical contamination in the form of trace metals from a typical photolithographic process sequence can be a potent source of undesirable contamination (13). Polymeric photoresist materials contain significant levels of metallic contaminants that can be subsequently transferred to the surface of the semiconductor wafer. As an example, Table 7 shows the concentrations of metallic contaminants in positive photoresist. Each column indicates the targeted level in the resist products for three different periods.

Table 7. Metal Impurities in Photoresist. Concentration in ppb. (*Source: Motorola Forecast 1991.)*

Typical Metals	1990 Levels	1992 Target	1995 Target
Na	200	10	1
K	200	10	1
Mg	200	10	1
Ca	200	10	1
Fe	200	10	1
Cu	200	10	1
Mn	200	10	1
Particles	100/ml ≥ 0.2μm	10/ml ≥ 0.2μm	1/ml ≥ 0.1μm

Second, the inorganic- or organic-solvent based lithographic developers can also contribute cationic and anionic species which may be trapped or deposited on the surface and into the complex surface structure of the device patterns. Metal ion free developers for positive acting photoresist are now a mandate for advanced processes. Low ppb trace metals and anions are as carefully specified for the lithographic process as they are for cleaning steps.

Third, as part of the wet-chemical cleaning and resist removal process, efforts have been increased to optimize wet-chemical stripping solutions and processes to effectively and efficiently remove photolithographic polymers from the silicon wafer after etch, deposition, or ion implant steps following the patterning process. The nature of resist stripping has not led to a sophisticated and well understood process by use of liquid stripping solutions. Assumptions that the wafer will be more adequately cleaned at a later step in the process have led to de-prioritization of resist removal in many manufacturing lines.

Wafer Rinsing and Drying. All wet processing steps are subject to some method of wafer drying. In spite of careful control in the wet sequence, wafer drying, if not properly executed, can kill a sensitive device. Centrifugal or on-axis spin drying is the most common drying step in semiconductor manufacturing. Normally, this is a dry step, but a spray rinse cycle prior to the spin dry is also frequently used. Highly controlled wafer handling is necessary at this step. Other methods, such as alcohol vapor drying and hot nitrogen drying are also used, but to a lesser extent. Perfect wafer drying processes are difficult to sustain and control. Particle deposition results frequently from the wafer drying step. Further discussion is presented in a later section on wafer process equipment.

2.4 Materials, Components, and Systems for Process Liquids

In spite of efforts to control chemical contamination in the production and use of liquid chemicals and water, piping materials and their components can be a major source of contamination (14).

Piping materials such as polyethylene, polypropylene, PVDF, (polyvinyladiene fluoride) and PFA (perfluoroalkoxy) are common piping materials in semiconductor manufacturing lines. All joints (i.e., glued joints, socket welds, butt welds, sanitary welds, elastomers and "O" ring seals in the piping system often contribute to the very contamination that the material supplier and the IC producer have attempted to remove prior to distributing the fluid to the wafer process system, bath, chamber, or vessel. Lack of careful specification of piping materials and all components of the liquid delivery system can spell disaster for the user (16).

Filters and filter systems for liquids are common in semiconductor manufacturing. What is intended to be a control and contamination reduction step can be a contributor to the process fluids if not properly engineered, managed and understood (17). Figure 3 shows the leaching effects of ionic contaminants from some filter components. Most filters and filter media for liquid applications are polymeric in nature and specifically designed to be compatible with various fluid compounds such as water, hydrofluoric acid, or sulfuric acid. Misused or abused filtration devices may contribute more contamination than they remove. Examples of this are:

1. Improper wetting processes used on filter cartridges prior to installation.
2. Poor rinse down of filter systems after installation.
3. Use of wrong elastomer or O-rings in the filter system.
4. Lack of, or inadequate integrity testing of each filter housing or filter system prior to returning to service. Avoidance or lack of post installation monitoring methods to test the performance of the filter or filter system that is intended to improve contamination levels in the fluid.
5. Extended use of filters and filter systems beyond their adequate life in the system leading to low flow rates, rupture of membranes, particle leakage, and seal degradations.

2.5 Storage and Transport of Liquids and Chemicals

Use of "high purity materials" to store and transport liquid chemicals and ultrapure water is necessary, mandatory and accepted practice. Chemical contamination from storage and distribution of these liquids are common horror stories in semiconductor manufacturing.

Leaching of Contaminants. In spite of the inherent inert nature of many polymeric materials, such as PVDF and PFA, it cannot be assumed that they do not contribute chemical contamination, especially if the liquid specifications are in the ppb range (18). The porous nature of these materials can lead to chemical background contamination at subsequent use points.

Ultrapure water by its very nature is an aggressive solvent. Leaching of metallic impurities from PVDF and PFA materials is known. One particularly common contaminant from piping materials is fluorine. Relatively large quantities of fluorine can be readily leached from polymer materials, such as PVDF and PFA. Hot water as an ultra pure rinse liquid accelerates the leaching of undesirable chemical contaminants from the very system that provides the hot "ultrapure water".

Certain chemicals, such as sulfuric acid, by the nature of their corrosive capability, can attack side walls and components of a sulfuric acid polyethylene container, an all-PVDF filter, an all-PFA lined drum, a PFA distribution line, and a centrifugal or diaphragm pump designed to transport the acid (19)(20).

Storage Vessels and Tanks. Storage of ultrapure water and ultrapure chemicals for any extended length of time can produce contamination of undetermined magnitude. Construction and fabrication of storage vessels for water and chemicals cannot be considered to be a trivial task. One seam or one minor loss of integrity in the entire vessel surface will spell disaster for the chemical and its ultimate use at the wafer level. Case studies of just this nature are not uncommon. Gas phase permeation from chemicals such as dilute HF and aqueous HCl, through polymers (e.g., polyethylene and PFA), can interact with metal outer packages such as stainless steel drums. Corrosion products from the gaseous chemical can then re-penetrate the polymeric lining of the container and contribute to the trace metal contamination. "Just-in-time" delivery and distribution are strong attributes of a high purity delivery system for any semiconductor facility.

Point-of-Use Generation of Chemicals from Distributed Pure Gases. Due to problems of production, storage, transportation, and delivery of ultrapure liquids to the point of use in a semiconductor fabrication line, the use of high-purity gases for generation of ultrapure chemicals can provide an opportunity to control chemical contamination in liquid sources. Two excellent examples of this type are aqueous HCl and NH_4OH. Distribution to the point of use of high-purity anhydrous HCl gas and NH_3 in ultrapure gas systems is perceived by the author and his co-workers to be easier to engineer and control than their liquid counterparts. If this assumption is true, then delivery of controlled amounts of high-quality just-in-time chemicals is a possibility in semiconductor manufacturing systems designed for wafer cleaning and wafer stripping applications.

2.6 High-Purity Process Gases in the Semiconductor Fab as a Source of Surface Contamination

Manufacturing Processes. The majority of all process steps in semiconductor device manufacturing requires some type of capability for gas delivery. In the simplest case, it may be the activation of a pneumatic cylinder. Gases may deliver key reactants for the deposition of critical thin film materials or provide etching species for patterning of dielectrics and metals. In all cases of gas delivery, a high degree of gas system integrity is required to minimize contamination that can be transported to the wafer or reaction chamber. A serious and pervading contamination is water vapor originating from the system itself or through low level leaks, allowing contamination (air) to enter from the operating environment. Corrosion of internal surfaces of stainless steel tubing and components, as well as seal degradation can cause both particle and metallic ion contamination to the wafer, the process chamber or the deposition system.

Table 8 lists a number of specific process applications where trace contaminants may cause harmful process interactions if the contaminant level is sufficiently high or if variations over time continue to occur.

Table 8. Types and Effects of Gas and Chemical Contaminants in Processes (1)

TYPE OF CONTAMINATION	EFFECT OF CONTAMINATION
Wafer Cleaning residues	Oxidation Rate, lifetime, breakdown
Water vapor in oxygen	Oxidation Rate
Oxygen during silicide annealing	High silicide resistivity, nonuniformity
Oxygen, water vapor during evaporation	High metal resistivity stress, hillocking
Oxygen/hydrocarcarons during epitaxy	Poor Epi Quality, higher deposition temp.
Surface contaminants during selective tungsten depositon	Spurious nucleii formation
HCl depletion during oxidation	Nonuniform oxides
Nitrogen during post oxidation anneal	Degraded oxide breakdown
Metal impurities in WF6	High tungsten resistivity
Nitrogens vs. Argon for silicide anneals	Improved thermal stability of silicide

Gas Storage Cylinders and Containers. Most specialty gas cylinders used for the storage and delivery of semiconductor reactive and toxic process gases are constructed from carbon steel. Significant effort is being made to control moisture and contamination on the internal side walls of these gas storage cylinders. Stainless steel and aluminum cylinders are used for specific applications. Cost and safety issues have not allowed the use of these types of cylinders to grow at a significant rate. Normally, if the cylinder filling is well managed, high pressure carbon steel cylinders do not fundamentally limit the ability to transport gases which are free of metallic contaminants.

Gas Integrity Management to the Point of Connection. Poor management of the integrity of the gas delivery system down stream from the gas storage cylinder or tank is a key factor in transporting chemical contaminants to the wafer. Lack of due diligence and professional attention to detail can produce a condition or source of variability in the delivery system that can cause severe device contamination. Case studies of this sort are known but not well documented. Normally, catastrophic events, such as major leaks or loss of pressure to the process tool, are easily recorded. Subtle contamination in the form of submicron particles and vaporized ionic species is not readily caught or monitored in the process characterization. Tables 9 and 10 show analyses conducted by a major gas component supplier. Table 9 gives elemental surface analysis of type 316 stainless steel surfaces after four independent surface cleaning steps. Subtleties of component cleaning are one additional source of chemical contamination that are difficult to account for in normal process evaluations. Moisture is an accepted major contaminant in ultrapure gas systems. Table 10 shows water vapor outgassing rates as a function of surface treatment of type 316 L stainless steel. Each component and pretreatment sequence can generate moisture and incipient chemical contaminants that may cause major variations in sensitive processes.

Table 9. Metal Analysis of 316L Stainless Steel Surfaces after Various Cleaning Methods; Atomic Concentration (%) Using XPS/ESCA Analysis (*Source: MKS Instruments*)

ELEMENT	ETHANOL	TRICHLOROETHYLENE	DETERGENT	ELECTROPOLISH
O	17.55	13.93	35.06	42.15
N	0.95	1.13	3.81	3.56
C	74.36	76.09	49.10	40.24
Ca	0.02	0.17	0.81	0.66
S	0.31	0.56	0.17	0.49
P	0.00	0.00	0.62	7.61
Na	0.33	0.14	0.25	0.15
Cl	0.17	0.16	0.00	0.00
Fe	0.24	0.09	1.94	2.83
Cr	0.39	0.20	3.63	2.31
Si	1.88	4.88	4.60	0.00
F	0.05	0.75	0.00	0.00
Mg	3.75	1.91	0.00	0.00

Table 10. Water Vapor Outgassing Rates of 316L Stainless Steel Surfaces in torr-liter/sec-cm^2 *(Source: MKS Instruments)*

TYPE	RATE
316 L	10^{-9}
316 L baked for 24 hrs. @ 150° C	10^{-12}
316 L VIM or VAR	10^{-11}
316 L VIM/VAR	10^{-13}
316 L VIM Passivated (O_2)	10^{-14}

Types of Homogeneous Impurities in the Gas Distribution Lines. Generation of unacceptable corrosion products and chemical interaction with supplied gases in the delivery system and wafer chamber make water the most critical contaminant to control. Tables 11, 12, and 13 are shown as examples of the purity requirements of three separate ultrapure-grade gases in the late 1980s and early 1990s. Significant emphasis was placed on inert or bulk gases during the 1080s because of the capabilities for analyzing very low concentrations of trace homogeneous impurities. These gases are typical of some of those used in semiconductor manufacturing. Specifications such as these are expected to continue to be tightened to the parts per trillion (ppt) level during the course of the 1990s.

Table 11. Ultrapure Nitrogen (all concentrations are in ppb) *(Source: Air Products and Chemicals Inc.)*

CONTAMINANT	PRE 1988 TRADITIONAL	1988 1MEG	1990 4MEG
Oxygen	<1000	<100	<10
Total Hydrocarbon	<1000	<100	< 5
Carbon Monoxide	<1000	<100	<10
Carbon Dioxide	<1000	<100	<10
Water	<1000	<100	<10
Hydrogen	<3000	<100	<10
Helium	< 500	<100	N.S.
Neon	<20,000	<100	N.S.
Particles	<10 ≥0.2μm	<1 ≥0.2 μm	<20 ≥.02 μm

Table 12. Ultrapure Oxygen (all concentrations are in ppb) *(Source: Air Products and Chemicals Inc.)*

CONTAMINANT	PRE 1988 TRADITIONAL	1988 1MEG	1990 4MEG
Water (H20)	<1000	<100	<10
Total Hydrocarbon	<30,000	<100	< 30
Carbon Monoxide	<1000	<100	<20
Carbon Dioxide	<2000	<100	<20
Nitrogen	100,000	<100	<40
Argon	500,000	< 500,000	<55
Hydrogen	N.S.	<100	<20
Krypton	11,000	<11,000	<20
Helium	< 500	<100	<20
Neon	<20,000	<100	<20
Particles	<5 ≥0.2 μm	<1 ≥0.2 μm	<20 ≥ .02 μm

Table 13. Ultrapure Ammonia (all concentrations are in ppb) *(Source: Air Products and Chemicals Inc.)*

CONTAMINANT	PRE 1988 FOR NITRIDE	1988 1MEG	1990 4MEG - Japan
Oxygen	<2,000	<1,000	<500
Nitrogen	<5,000	<5,000	< 500
Total Hydrocarbons	<1,000	< 500	<500
Carbon Monoxide	N.S.	<1,000	<100
Carbon Dioxide	N.S.	<1,000	<100
Hydrogen	N.S.	< 500	<500
Water (H20)	<5,000	<2,000	<500
Metals:			
Na			5
Mg			10
Zn			2
Fe			5
Co			5
Ni			5
Cd			2
Cr			5
P			2

Oxygen, CO_2, CO, CH_4, C_2H_6 and other hydrocarbons are additional critical impurities in gases. Inert impurities, such as argon in nitrogen, are often specified as contaminants. However, there is no evidence that low levels of inert gases act as contaminants. At this writing, analytical capability is not available in the semiconductor process laboratory or in the manufacturing line to detect low level contaminants in reactive and corrosive gases in the 1 to 50 ppb range. Instruments to measure moisture and oxygen in the 1 to 10 ppb range in inert gases are available and can be employed in large distribution systems. Cost of individual instruments limit widespread use of these monitoring instruments.

Control of gas distribution systems and their chemical contaminants is primarily achieved through gas integrity management. Actual contamination levels in quantitative terms at the point of connection to the tool are not normally known for many process gases. Certainly, low level moisture in inert gases can be measured to the ppt level using state of the art APIMS (Atmospheric Pressure Ionization Mass Spectroscopy) technology (21).

Particles in the Gas Distribution Systems. Particle control and removal is better understood and more easily measured in inert gas systems. As long as the particle exists as a discrete particle in an aerosol gas stream, filtration control and quantitative counting down to 0.02 µm sized particles and smaller can be employed. In corrosive gas systems compatible particle counting is less common below the 0.1 µm particle size. Particle measurements below 0.1 µm is technically feasible by use of state-of-the-art light scattering systems.

Point-of-Use Purification of Gases. Purification of both inert and reactive gases at the point of use has become a common practice at the process tool connection point or inside the gas delivery system. This technology is being used even though the analytical measurements are not normally available at the location of the purifier near the process tool. If proper management and control of the upstream side of the system is not employed on a rigorous basis, the point-of-use purification system can itself become a source of catastrophic chemical contamination. Complete reliance on point-of-use purification as the primary method of impurity control is not the solution to chemical contamination. Both economic and uptime considerations dictate that trace homogeneous gas impurities cannot be controlled over long timeframes by a single control point at the end of the delivery line. Multiple, systematic control points will continue to be necessary to sustain control and low variability in the gas distribution system.

Dry Cleaning Technology Based on Reactive Gas Chemistries. Control of chemical contamination from process gases continues to be a fundamental need in advanced wafer processing. Wafer cleaning technologies as described in Part III of this volume may use plasmas or non-aqueous reactive gases and is a developing technology for this industry. Unless ultrapure gas systems and purification methods are available at the cleaning chamber itself, chemical contaminants on the surface of the wafer will not be entirely removed. What is intended to be removed could be delivered and deposited by the very process intended as purification.

2.7 Chemical Contamination from Thermal Process Tools and Systems

The intent of the following discussion is to remind the reader of the wide potential of silicon surface contamination possible in today's processing environment. A variety of systems and process environments are discussed that have the potential of generating a high level of contamination that may not be removed during the course of conventional wafer cleaning (22).

High-Temperature Process Tools. Horizontal oxidation and diffusion furnaces, vertical furnaces, annealing tubes, epitaxial reactors, rapid thermal annealing tools and various chemical vapor deposition systems are typical high temperature process tools used in semiconductor wafer operations. Perhaps too often, they are not seen or treated as sources of contamination in the course of normal wafer processing.

Reactor Tubes and Vessels. Quality of fused quartz tubes used in high temperature oxidations, annealing and deposition can be the source of serious contamination of the wafer during processing. Sodium as a contaminant in quartz is a known yield killer. Traces (ppb) of alkali cations, heavy metals, and transition metals can be vaporized and transported to the silicon surface and diffuse in the silicon bulk during high temperature processing. Table 14 depicts an analysis of quartz used in semiconductor manufacturing. Assumptions that metallic contamination from this source are not significant in device manufacturing cannot be accepted. Incoming quality control and tight specification limits for these critical materials should be mandatory for leading edge semiconductor products.

Chambers constructed of graphite, and silicon carbide crucibles also cannot be ignored as potential sources of metallic contamination. If the temperature or chemistry of the process is sufficient to transport metallic or organic species to the wafer, in-depth analysis and process characterization are needed to qualify the use and the supplier of these materials.

2.8 Plasma Etch, Strip, and RIE Processing Tools

Dry Resist Stripping Tools. Plasma process systems are typically categorized as low-temperature process tools and can be misused in the context of chemical transfer of contamination due to the lack of understanding of the impurity transfer mechanisms.

Table 14. Impurities (ppm) in Fused Quartz *(Typical Analysis by Supplier)*

Contaminant	VENDOR A	VENDOR B	VENDOR C
Al	8	15.2	20.0
B	0.03	2	2
Cu	0.01	ns	ns
Fe	0.4	0.3	1
K	0.2	1.5	1.5
Na	0.2	1.3	1.2
OH	200	10	ns

Removal of processed resist from the wafer surface may be the least appreciated (therefore under-engineered) contamination source in the entire process sequence. A variety of aqueous based and organic based stripping solutions are used in silicon processing. Glow discharge plasma and activated (microwave) oxygen plasma systems are commonly used for initially stripping organic polymers. UV/ozone systems are also used for removal of organics in cleaning. Major chemical contamination problems occur due to:

a. Inadequate and non-uniform removal of the resist polymer caused by hot spots on the wafers that may leave non-uniform chemical contaminants on the surface.

b. Severe resist damage previously caused by high-energy bombardment during ion implantation and high-energy reactive ion etching steps.

c. Subsurface damage to the silicon lattice and the formation of thin oxides by localized high-energy plasmas and ion bombardment in the stripping process itself.

d. Diffusion of metallic contaminants from the resist or previously contaminated surfaces into the oxide structure or to the Si-SiO_2 interface due to in situ heating and thermal generation during resist removal.

e. Vaporization of previous process condensates that can re-deposit on the wafer surface.

f. Inadequate control of the reactive gases used in the plasma strip process. Oxygen and etchant gases, such as CCl_4 and CF_4 are not carefully controlled with respect to transport of metallic contaminants into the reaction chamber. (See gas contamination sources above).

Reactive Ion Etching Process and Equipment. Reactive ion etching is a prerequisite to advanced wafer process technology. Along with the advancements that this etching technology provides, surface contamination from the process and equipment are a major concern.

a. High-energy ion bombardment of the resist layers can deform and polymerize (acylate) the resist layer, making its removal extremely difficult in localized etch areas. The removal depends on the type of resist, the geometric pattern, the chemical species in the etch process, and the temperature of the process.

b. The etch process surface temperatures are often high enough to diffuse metallic contaminants from the resist to the silicon surface.

c. Residues from etch processes are created on the sidewalls of device structures and inside the etching chamber. Removal and control of these residues have become another fundamental requirement in submicron wafer cleaning and resist removal. Table 15 is an example of chlorine contamination transferred to the surface of a silicon wafer as a result of reactive ion etch of aluminum-copper metallization (23). Table 15 shows that removal of chlorine after the etch process is difficult and that certain cleaning steps can add and redistribute chlorine on the wafer surface.

d. Internal sputter deposition of metal surfaces in the etch chamber, which include the anode, cathode, the platen materials and covers, the side walls of the etch chambers, and the wafer-holding contact pins or tabs, are additional sources of concern.

Table 15. TXRF Results of Chlorine Etching Processes (Contamination: 10^{12} atoms/cm^2) *(Source: R. S. Hockett) (24)*

Wafer Number	Chlorine Concentration	Process
1	43	Aluminum - Copper
5	I*	Aluminum - Copper + Photoresist
6	1	Aluminum - Copper + Photoresist + Plasma Ash
23	120	Aluminum - Copper + Clean
15	4100	Etch A
20	1900	Etch A + Ultra Pure Water Rinse
19	3000	Etch A + Ultra Pure Water Rinse + Water Rinse
21	5	Etch A + Ultra Pure Water Rinse + Plasma Ash
18	60	Etch A + Ultra Pre Water + Plasma Ash + Acid 1
17	7000	Etch A + Spin Rinse and Dry
22	7500	Etch B
16	6800	Etch B + Ultra Pure Water Rinse
13	6500	Etch B + Ultra Pure Water Rinse + Water Rinse
14	10	Etch B + Ultra Pure Water Rinse + Plasma Ash
12	96	Etch B + Ultra Pure Water Rinse + Ash + Acid 2
24	120	Etch B + Ultra Pure Water Rinse + Ash + Acid 2 + Acid 3

* I indicates interference from high sulfur

2.9 Wet Etching, Wet Cleaning, and Drying Equipment

Wet etching and cleaning equipment, which includes hardware components of these systems, are part of the solution of wafer cleaning and part of its problem. Depending on the critical nature of the wet etching or cleaning process, the equipment itself can cause contamination of the wafer. In situ analysis of the performance of wet process tools is not conventional. The economics and reliability of diagnostic monitors makes instrument specification and justification difficult.

Liquid chemicals are known to transport and deposit metallic and ionic contaminants on the wafer by various processes. For example, Figs. 7 and 8 display a common deposition relationship of copper and iron in cleaning solutions such as NH_4OH and H_2O_2. Native oxides, formed during an oxidizing cleaning process using NH_4OH and H_2O_2 show higher concentrations of copper and iron than thin thermal oxides.

Table 16 outlines six experimental cleaning solutions and the resulting surface contamination on the silicon after the clean and after gate oxidation. These results were obtained even though ppb-level chemicals were employed. High-purity choline solutions in lieu of NH_4OH were used in these experimental tests. The reference wafer has seen no cleaning in this series. The initial clean used a dilute H_2O/HF dip and rinse. The samples were analyzed by the vapor phase decomposition method, which has been shown to correlate with analysis methods such as SIMS.

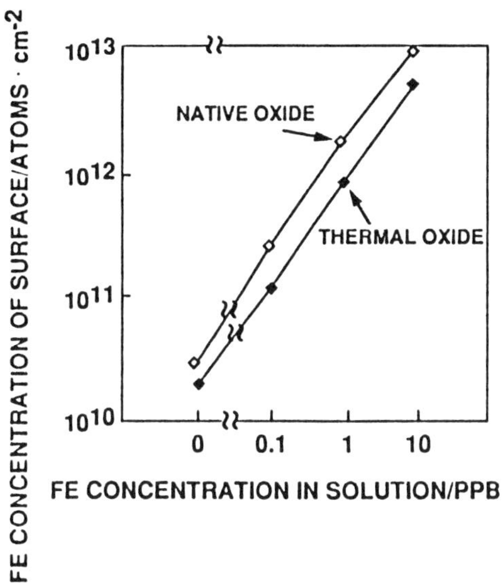

Figure 7. Iron deposition on silicon from cleaning solutions (7).

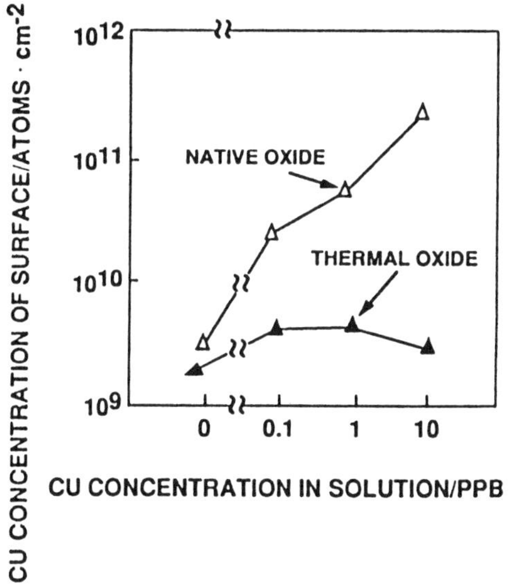

Figure 8. Copper deposition on silicon from cleaning solutions (7).

Table 16. Motorola Experimental Cleaning Results: VPD Surface Analysis of Metallic Species in 10^{10} atoms/cm^2 *(Source: Internal Motorola Experiments)*

WAFER SAMPLE	METALLIC SPECIE (X E10 ATOMS/CM2)				
	IRON	ALUMINUM	COPPER	CHROMIUM	NICKEL
REFERENCE	1.00	<2.8	<0.1	<0.1	<0.6
INITIAL CLEAN W/(HF)	0.60	<2.8	<0.1	<0.1	<0.6
H20 + CHOLINE + H202 (70°C)	11.00	25.60	0.30	<0.1	<0.6
With Gate Oxidation	<0.5	10.10	<0.1	<0.1	<0.6
H20 + CHOLINE + H202 (RT)	15.10	110.00	<0.1	<0.1	<0.6
With Gate Oxidation	<0.5	<2.8	<0.1	<0.1	<0.6
CHOLINE ONLY (70°C,2 min.)	0.70	<2.8	<0.1	<0.1	<0.6
With Gate Oxidation	<0.5	<2.8	<0.1	<0.1	<0.6
H2O2 ONLY (70°C)	0.80	13.30	<0.1	<0.1	<0.6
With Gate Oxidation	<0.5	14.70	<0.1	<0.1	<0.6
ULTRA PURE WATER ONLY (70°C)	10.10	<2.8	<0.1	2.20	<0.6
With Gate Oxidation	1.80	12.00	<0.1	<0.1	<0.6

Table 17 shows five additional wet cleaning solutions and their resulting metallic surface contaminants. This table again supports the observation that deposition of metallic impurities occurs at a higher rate when the pH of the solution is high. This analysis method was conducted with polysilicon-coated wafers by SIMS analysis.

Table 17. PC/SIMS Analysis of Metal Contamination in atoms/cm^2 on Silicon after Polysilicon Coating Resulting from Several Cleaning Processes (4)

	Aluminum	Calcium	Magnesium
1% HF	1.1×10^{9}	2.0×10^{9}	1.1×10^{9}
H2SO4 + H2O2, and 1% HF	1.3×10^{9}	2.1×10^{9}	$<1.0 \times 10^{9}$
HCl + H2O2 (SC2)	1.3×10^{9}	1.7×10^{10}	1.7×10^{9}
NH4OH + H2O2 (SC1) and HCl + H2O2 (SC2)	1.5×10^{10}	7.5×10^{9}	1.0×10^{9}
NH4OH + H2O2 (SC1)	3.5×10^{11}	1.1×10^{11}	1.8×10^{11}

Environmental Influences on Wet Baths. The quality of the wet clean and wet etch systems is subject to the influences of the cleanroom or exposed facility environment. Their impact is seldom measured and the influence is not normally noticed. If wet cleaning baths are open to the cleanroom environment for extended periods, then not only particles can be transported to the wet process but additional organic contaminants, bacteria, and ions from the working environment can be transferred. This occurs primarily by inadequate balancing of the strong exhaust requirements of exposed wet clean and wet etch baths. Avoidance is primarily accomplished by isolation of the wet process bench or implementation of a full robotic interface to remove people from the immediate environment of the process baths, the fluids, and the wafers.

Chemical Baths and Vessels. Independent of the liquid chemicals themselves, materials of construction used for the chemical baths, tanks, rotors, vessels, lids, seals, robotic arms, and dispense heads, are a source of contamination. Residence time, chemical compatibility, temperature, electrostatics, and vibration may all potentially contribute to the contamination in the bath or liquid stream. Intense quality engineering review of these systems and their components in relation to the objectives and chemistries of the process in question is required. Many times, cross use of chemical cleaning, rinsing and etching systems in the same chemical bath or container is deemed necessary for reasons of economies. This decision, if not correctly controlled, can cause significant cross-contamination of wafers (24).

Components, such as pumps, filters, valves, and seals, retain chemical residues and background levels of the previous chemistries. Different chemical processes used subsequently may leach these deposits from the surface of the equipment materials.

Wafer Cassettes in Wet Processes. Killer defects and significant cross-contamination can be caused by the use of universal wet cleaning cassettes so common in semiconductor processing. In some cases the wafer clean cassette will travel as the transportation carrier with the material. This avoids an important wafer transfer step, but may re-deposit contaminants on the wafer during a subsequent process. Long storage times in cassettes used in liquid chemicals is considered poor process practice. PFA cassettes are highly standardized but the porous nature of these materials is usually ignored. Chemical outgassing from the PFA or PTFE material can spell disaster if the materials are heated in the presence of the wafers (25). For these reasons, exacting control and process history of each wafer cassette is mandatory for contamination control. Many times phantom contamination episodes will come and go. Process cassettes are an undetermined source of random and uncontrolled contamination in wafer fabrication lines.

Wafer Drying Equipment and Processes. Wafer drying as a source of contamination from the process was mentioned briefly above. It is the critical final process step after ultrapure water rinsing. Equipment and material interactions are critical. Significant progress has been made in engineering and design of wafer drying equipment during the past ten years.

Spin Drying. This method of drying is the most common in the semiconductor industry. Normally this is a batch process using single cassette or multiple cassettes. Spin-rinse dryers (SRDs) normally include ultrapure water rinse cycles, heated N_2 gas, antistatic generators in the N_2 gas stream, and system heating. Currently, horizontal wafer spin drying technology is being introduced as the wafer size reaches 200 mm. Unless properly designed and controlled, conventional SRDs can contribute 10 to 1000 particles per wafer per treatment. Particle size specifications vary depending on the analysis system in use. Microbiological contamination and carbon particles from the water delivered to an SRD are often observed. Metallic contamination at this point may occur if internal metal parts are corroded or are not designed for compatibility with the acid or water environment of the spin dryer.

Vapor drying. Vapor drying of any type requires a water miscible high vapor pressure organic chemical, such as isopropanol or mixed organic solvents. Vapor drying technology for silicon wafer cleaning is known primarily through the use of isopropyl alcohol. System design is a key factor in success of vapor drying. Some systems utilize wafer immersion in the liquid phase. Others rely on high temperature vapor immersion only or displacement of water rinses with hot alcohol vapor (azeotropes). The fundamental mechanisms of particle deposition or removal from the wafer is not yet fully defined. Surface tension of the alcohol vapor at the alcohol/water interface in contrast to that of ultrapure water is a major attribute of the alcohol system. Residues from the alcohol are of primary concern, but have not been detected or quantified.

Hot Ultrapure Water Capillary Drying. Capillary drying is accomplished by use of hot ultrapure water delivered to the wafer rinse bath. Removal of the chemically cleaned hydrophilic wafers or the cassette of wafers is performed with high-precision robotics at a controlled and specified rate. The rate of removal determines the controlled drying of the wafers. Separation of the wafers from the grooves in the cassette is necessary to avoid localized spotting. Tight control of water purity, water temperature, and the wafer environment during the removal period is mandatory for best results. Transfer of contaminants comes primarily from the water itself and any water spotting due to remaining water drops on the edge of the wafer. Aerosol contamination to the wafer is likely if air flow control is not well managed at this step.

Nitrogen Drying. Hot nitrogen wafer drying (100 - 200°C) is one of the earliest techniques of wafer drying, and is normally a batch (cassette) process. Nitrogen drying can be made to be a very clean and a high performance process, but the efficiency, and therefore productivity, of the process may not compete well with spin drying. Turbulence in the drying system can induce particle deposition from the drying chamber. A preferred method for 200 mm, 250 mm or 300 mm single wafers is horizontal spin drying with extremely pure nitrogen.

2.10 Vacuum Processing Equipment

Loadlocks. Organic contaminants and particles are often redistributed during the wafer transport in and through a loadlock that normally takes place at room temperature. Wafers that are oriented face down or vertically in the vacuum transport chambers are less likely to receive front side deposition of particles. Venting and pumping cycles are known to be subject to increased contamination on process wafers. Control of the turbulence during pump down and venting has proven helpful in contamination control of vacuum processing. Wafer temperature during the transfer steps plays an important role in controlling the deposition velocity and particle flux at the wafer surface. The process of thermophoresis created by elevating the wafer temperature will reduce particle deposition. Electrostatic enhanced deposition effects are less important here since the individual wafer is more likely to be grounded. However, high static charges are well established for polyethylene and PFA storage cassettes.

High-Vacuum Chambers. Water vapor is a prime contaminant in any vacuum system. Moisture entrapment from the room air or the back fill gases contributes to background moisture in the vacuum chamber. Adiabatic expansion during pump-down can freeze out moisture and cause crystals to deposit on surfaces.This moisture can redistribute surface metallic contaminants due to residues, oxidation and corrosion of the interior piping surfaces or chamber walls and the various shielding parts. These interactions occur at very low moisture levels. Transfer of mobile species, such as sodium, to the wafer surfaces is increased at elevated temperature in the presence of moisture. Cleanliness of system hardware, purity of materials, ultrapure gases, and good base vacuum pumping practices to less than 10^{-8} torr are needed for ultrapure film depositions in vacuum. Where this practice is common to good vacuum processing, it may not be seen as a source of surface contamination on silicon wafers.

Sputtering Gases. A high purity sputtering gas (usually argon) is a prerequisite for clean processing. Excess water vapor, oxygen, and hydrocarbons in the sputtering gas will directly contribute to film impurity and loss of film integrity. Current VLSI processes must have sputtering gases with moisture and oxygen concentrations less than 10 ppb. By 1995, control down to 1 - 10 ppb will be necessary for advanced devices. (See previous discussion on contamination from ultrapure gases and systems.)

2.11 Wafer Handling and Storage Systems

Wafer handling and storage systems are usually ignored and weakly engineered for contamination control. Manual handling and manual wafer transport in and out of cassettes is not an acceptable method for advanced manufacturing processes. Nonetheless, the need for manual intervention often occurs, resulting in the possibility of a transfer of contaminants.

Wafer-Transfer Hardware. Wafer boxes, plastic storage bins, plastic and metal wafer cassettes, wafer slide transfer systems, robotic transfer arms, SMIF storage pods, tweezers, vacuum wands, robot gripper arms, elastomeric transfer bands and vacuum or electrostatic chucks on process systems are able to add and transfer both front and backside contaminants to wafers during single-wafer and batch processing. Precision cleaning, storage and control of the surfaces of these detailed hardware pieces are now demanded in well controlled processes. Wafer cleaning and scrubbing subsequent to wafer handling steps must be designed to eliminate contaminants from these steps.

Cleaning of Cassettes and Storage Boxes. Automated systems for precision cleaning of plastic and metal cassettes and wafer storage boxes for silicon wafer processing are commercially available. Specific cleaning treatments for these support items are not rigorously specified and few standards exist to guide the process engineers on what levels of cleanliness are needed. Cleaning systems are typically aqueous based (but can be based on isopropyl alcohol) and may be based on spray washing with ultrapure water and perhaps some type of organic surfactant. Particle specifications on cassettes and the inside of storage boxes are not typical, but particle control is practiced. Contaminants comprised of organic and metallic species can be expected to get transferred to the wafers while the wafers reside inside these enclosures. Again, it is believed that washing, wiping and cleaning will improve the cleanliness of the system and reduce the likelihood of significant contamination transfer.

New wafer pods for mini-environments and SMIF-related systems are now in use. Materials of construction for these newer systems have received focused attention to minimize transfer of contaminants during the automated transfer of wafers in what is intended to be an ultraclean environment. These systems are designed to completely isolate the wafers from human and cleanroom interactions. Nitrogen, argon and vacuum are being used to eliminate or reduce the sources of oxygen and water vapor from the interior surfaces of transfer and storage systems.

3.0 KILLER CONTAMINANTS

Any foreign material, organic or inorganic in nature, that comes in contact with a semiconductor surface during the manufacturing process has the potential of being a killer contaminant for that device or devices on that wafer. The exact type, size, and quantity of a contaminant that reduces yield and impacts device reliability is a predominant question for research in microcontamination control. A "reasonable" understanding of the tolerance levels for surface contamination provides certain specifications and controls for the manufacturing processes. The general specifications that have been developed specify limits for both particle and metallic contaminants on the wafer surface.

3.1 Particle Concentrations

Roadmaps for contamination control technology feature surface particle concentrations in density and size as a function of device complexity and time frame. Table 18 lists particle densities as a function of the critical device dimension in semiconductor manufacturing. These numbers should not be confused with aerosol numbers for cleanroom classifications, which are given in particles per cubic foot. Projections, such as those in Table 18, represent the technologists' and engineers' best estimates of the upper tolerance levels for a particular leading edge device. Extensive research and experimentation is rarely done to confirm these roadmaps beyond one to three years. Roadmaps are routinely referenced to DRAM devices due to the accepted level of sensitivity and complexity of these devices, as was shown in Table 1. Each semiconductor device has its own particular sensitivity to particles, contaminants, and other defects.

Table 18. Surface Contamination Projections on Silicon from Particles and Metals *(Source: Motorola)*

Device Critical Dimension	0.65	0.5	0.35	0.25	0.15-0.10
Parameters					
PARTICLE SIZE, μm	0.3	0.16	0.11	0.08	0.05
PARTICLE DENSITY CM-2, Target / Limit	.03/.09	.06/.10	.05/.10	.04/.08	.02/.05
CONTAMINATION LEVELS					
IONIC SPECIES	<E11	<E10	<E9	<E8	<E7
TRANSITION METALS	<E12	<E11	<E10	<E9	<E8

Surface particle concentrations for leading edge devices in the late 1980s were normally referenced at 0.5 μm in size. Manufacturing specifications would be referenced from 10 to 100 particles per wafer for 100 mm, 125 mm, and 150 mm wafers. These were referenced as specifications, not targets.

By the end of the 1980s, 0.3 μm particle specifications were common in manufacturing lines as surface metrology tools were introduced that allowed 0.3 μm measurements with improved signal to noise ratios. In this time period, particle concentration specifications had slightly higher densities for the some device complexity, but some technologists insisted on lower particle densities with smaller particle sizes. Particle specifications of 10 particles per 150 mm wafer could be found for a 0.3 μm particle size. Typically, 100 particles for 150 mm wafer were specified as the upper limit while levels below 50 per wafer are typical today.

At this writing, 0.2 μm specifications are being generated for advanced semiconductor manufacturing lines and 0.1 to 0.2 μm size particles are being measured and reported by various laboratories and equipment suppliers. Particle size sensitivity is key in that the lower the detection limit of the measurement tool, the more discriminating power the device producer and tool supplier have to minimize process defects. Leading edge processes and tools would be well served to include performance levels at or below 10 particles per process step on 200 mm wafer at 0.1 μm. This level of particle control is expected in 1995 on the wafers for leading edge devices.

3.2 Trace Metal Concentrations

Specifications for trace metal contamination on the surfaces of IC wafers is a recent addition to manufacturing requirements. In Table 18 projections for "allowable" metallic contaminants on the silicon surface are shown. Specifications and process limitations are not possible without qualified metrology systems. Only recently have semiconductor manufacturing lines begun to include trace metal analysis, such as TXRF, in the manufacturing line's required list of wafer analysis instruments. Use of such complex surface analysis systems, in house, is difficult to support by the manufacturing line for rapid analytical feedback.

On certain devices, some metallic species can be tolerated at relatively high concentrations, for instance, at 10^{13} atoms/cm^2. On other devices the same contaminant cannot be tolerated above 10^9 atoms/cm^2, as predicted in Table 18.

3.3 Killers Other than Particles and Metallics

We cannot control what we do not measure. Generally, semiconductor process and microcontamination engineers are measuring organic contaminants, native oxide thicknesses, and surface microroughness. Although not confirmed in manufacturing, some of these represent potential killer contaminants on leading edge devices. As surface metrology improves, better understanding of surface cleanliness and morphology as a function of wafer cleaning will continue to evolve.

4.0 FUTURE DIRECTIONS AND NEEDS

Chemical contaminants on IC surfaces are recognized as a major limitation to high yield and high quality. IC manufacturing is essentially a sequence of chemical processes. How to execute cost effectively and within the limits of tolerance of a specific device performance has become the question. What is not clear is the appropriate trade-off in cost of manufacturing and wafer yield.

4.1 Equipment Design

The requirements to achieve low levels of contamination can be addressed through multiple paths operating in parallel. Ultimately, maximum process capability should be attained in each process chamber or system where the individual processes occur. Control of contamination must start at the design of the system to achieve contamination-free processing.

Wafer size and number of wafers processed per batch become critical design factors that affect contamination control, process control, throughput, and product quality. These factors then drive the cost effectiveness of the processing system.

4.2 Material Specifications

Rigorous material specifications and systems performance characteristics are now understood to be required for most processes. Process interaction, reactivity, adsorption, desorption, sputtering, and leaching are all related to material compatibility and, ultimately, chemical contamination. Significant research has been conducted in recent years on high-purity stainless steel, alloys of stainless steel, Hasteloy™ metals, gas and water system components, anodized aluminum, improved seals and elastomers, and polymeric piping and connection materials.

4.3 Fabrication and Installation

With improved design concepts and highly specified material performance, the knowledge and skill levels needed for fabrication and installation cannot be overstated. Less than superclean methods for joining, assembly, and installation of precision components can compromise a contamination-free process. One filter inadvertently installed backwards in a gas or liquid fluid stream can have devastating effects.

4.4 Characterization and Evaluation

Due to their degree of complexity and facility demands, advanced processing systems are difficult to fully characterize and evaluate at the equipment supplier's factory. This puts more demand on the supplier to fully understand the semiconductor manufacturing processes. What may work well in the supplier's laboratory may not do so in a manufacturing environment.

4.5 Safety and Environmental Requirements

As if the requirements on wafer manufacturing are not stringent enough, the environmental and safety demands on chemical processes are being continually tightened and regulated. The chemical nature of semiconductor processes places them in a highly vulnerable position to be concurrently cost effective, non-polluting, and fully compatible with current environmental and safety requirements. What is compatible and desirable with low contamination levels is frequently in direct conflict with safety or environmental demands. A recent example of this conflict is the desire to support safety regulations and fire insurance codes inside the IC facility by placing conventional (galvanized or chrome-plated steel) fire sprinkler heads above open tanks of oxidizing and etching acid baths located inside chemical cleaning processors. Strategic planning of mutual objectives of these two disciplines is severely lacking in IC manufacturing.

4.6 Directions in Research and Development

Key attributes of semiconductor manufacturing requirements with respect to specific contamination issues are being sought. Research and modeling are needed to explore the relationships between contamination, wafer cleaning and device performance. In many cases contamination control is being misled by limited observations, measurements at the low end of detection methods, best guess estimates, and closely held device defect densities monitored inside manufacturing lines.

Correlation studies are now being designed and executed to better link the level of chemical contamination on the device surface and the ultimate functional performance or functionality of a specific device. These studies will continue to be diversified between particle and very low level molecular, ionic, and elemental contaminants on the device wafer surface.

Research and development activities in the 1990s are necessary to focus resources on cost effective, clean, and contamination-free processes. Far reaching methods and processes to measure and eliminate chemical contamination in processes and equipment may not necessarily be cost effective for the product. The challenge will be to effectively combine the needs of the devices with appropriate manufacturing costs to remain competitive in the ever increasing demands of the semiconductor market.

REFERENCES

1. Osburn, C., *Microcontamination,* pp. 19-24 (July 1991)
2. Penka, V. and Hub, W., *Fresenius Z. Anal. Chem.,* 333:568-589 (1989)
3. Takizawa, R., Nakanishi, T., and Ohwawa, A., *J. Appl. Phys.,* 62 (12):4933-4935 (Dec. 1987)
4. Uchida, H., Aikawa, I., Horashita, N., and Ajioka, T., *Proc. 1990 IEDM,* pp. 405-408 (Dec. 1990)
5. Hiratsuka, H., Tanaka, M., Tada, T., Yoshimura, R., Matsushita, Y., *Proc. of the 11th Workshop on ULSI Ultra Clean Tech.: Advanced Wet Chemical Proc. 11,* pp. 5-20, Ultra Clean Soc. of Japan (June 6, 1991)
6. Khilnani, A., *Microcontamination,* pp. 24-29 (Nov. 1986)
7. Shiraiwa, T., Fujino, N., Sumita, S., Tanizoe, Y., *Semiconductor Fabrication: Tech. and Metro.,* ASTM STP 990 (D. C. Gupta, ed.) American Soc. for Testing and Materials (1989)
8. Fujino, N., Sumita, S., Murakami, K., Kitagawa, K., Sano, M., and Johnston, R., *Proc. on Automated Integrated Circuits Manuf.* (V. E. Akins and H. Harada, eds.), 91-5:301-315, Electrochem. Soc., Pennington, NJ. (1991)
9. Hourai, M., Naridomi, T., Oka, Y., Murakami, K., Sumita, S., Fujino, N., and Shiraiwa, T., *Jpn. J. Appl. Phys.,* 27(12):L2361-L2363 (1988)
10. Morita, E., Yoshimi, T., and Shimanuki, Y., *Electrochem. Soc. Ext. Abstr. #237,* Spring mtg., 89-1:352-353 (1989)
11. Kern, F. W., Jr., Itano, M., Kawanabe, I., Miyashita, M., Ohmi, T., *Proc. of the 11th Workshop on ULSI Ultra Clean Tech. Adv. Wet Chem. Processing II,* pp.21-37, Ultra Clean Soc. of Jpn. (June 6,1991)
12. Tonti, A., *Electrochem. Soc. Ext. Abst. #552,* Fall Mtg., 91-2:824 (1991)
13. Fujimura, S., and Yano, H., *J. Electrochem. Soc.,* 135(5):1195-1201 (1988)
14. Pengelly, S., *Ultrapure Water,* pp.24-32 (May/June 1987)
15. Miki, N., Yonezawa, T., Watanabe, A., Maeno, M., Kawanabe, T., Ohmi, T., *Conf. Proc., 10th Annual Semiconductor Pure Water Conf.,* pp.335-353 (Feb.1991)
16. Matthews, R., *Proc. Millipore 9th Annual Microelectronics Tech. Symp.,* (May 1991)

17. Farrell, L., Penniman, T., and Schmidt, M., *Proc. Millipore 9th Annual Microelectronics Tech. Symp.,* (May 1991)

18. Camenzind, M., Tan, S., Balazs, M., *Microcontamination '91 Proc.,* pp 401-411 (Oct 1991)

19. Hackett, T., and Dillenbeck, K., *Microcontamination '91 Proc.,* pp 427-440 (Oct 1991)

20. Gruver, R., Silverman, R., and Kehley, J., *1990 Proc., Inst. of Environmental Sciences,* pp.312-315 (May 1990)

21. Ohmi, T., Murota, J., Kanno, Y., Mitsui, Y., Sugiyama, K., and Kawano, H., *Proc. 1st International Symp. on Ultra Large Scale Integration Science and Tech.,* Electrochem. Soc. (May 1987)

22. Kabayashi, M., Yamazaki, K., Okui, Y., and Ogawa, T., *Electrochem. Soc. Ext. Abstr. #543,* Fall Mtg., 91-2:813-814 (1991)

23. Hockett, R., *Proc., Semicon/Korea '91 Tech. Symp.,* pp.III-89-98, Seoul, Korea (Sept. 1991)

24. Rathman, D., Fabry, L., *Electrochemical Soc. Ext. Abstr. #546,* Fall Mtg., 91-2:817 (1991)

25. Goodman, J., *Proc., 14th Symp. on ULSI Ultra Clean Tech., Adv. Semiconductor Manuf. System,* pp.53-70 (Oct 1991)

Part II.

Wet-Chemical Processes

3

Aqueous Cleaning Processes

Don C. Burkman, Donald Deal, Donald C. Grant, and Charlie A. Peterson

1.0 INTRODUCTION TO AQUEOUS CLEANING

The purpose of this chapter is to draw attention to the important considerations in aqueous cleaning of semiconductor surfaces.

The basic requirement for cleaning processes is the removal of contamination. Such removal is of paramount importance to semiconductor chip manufacturers since it is generally accepted that over 50% of yield losses in integrated circuit fabrication are due to microcontamination. Furthermore, any metals left on the surface may spread and diffuse into the semiconductor interior and cause yield loss and/or loss of chip function reliability. Whether the contaminants are specific or general, or whether the source of the contaminants is known or unknown, the successful removal of contamination is the essence of cleaning.

Aqueous chemistries involve a variety of solutions which can be made by dissolving a gas, liquid, or solid in water. Aqueous cleaning solutions are currently the most widely used due to their many advantages over alternative processes. Alternatives include cleaning with organic solvents, and the application of vapor phase chemistries (both organic and inorganic), as well as the use of various physical and thermal methods of contaminant removal. Some of the advantages and disadvantages of aqueous cleaning are listed below.

1.1 Advantages of Aqueous Cleaning:

- Rinsing is easily accomplished in water
- Residues left after drying can be avoided by deionized water rinsing and suitable drying
- Flammability hazard is low or non-existent
- Disposal of a large variety of aqueous chemicals has low environmental impact
- A wide range of chemicals is available
- Many aqueous chemicals are low cost
- Aqueous chemistries are capable of removing organics and inorganics to very low levels
- Aqueous solutions generally have lower vapor pressure than organic solvents
- In most cases aqueous chemicals have very high reaction selectivity between the contaminants to be removed and surfaces being cleaned

1.2 Disadvantages of Aqueous Cleaning:

- Drying is not as fast or as easy as with organic solvents due to the low vapor pressure of water. Incorrect drying can result in recontamination
- Some organic solvents are more efficient for removing certain organic contaminants
- Aqueous chemicals can be dangerous to handle, breathe, etc.
- Disposal of some aqueous chemicals is difficult and can be expensive
- Aqueous systems are difficult to couple to vacuum systems

Since there are many applications requiring aqueous cleaning in semiconductor manufacturing, with widely differing demands for surface cleanliness, not all of these considerations are important to each application. Nevertheless, in an attempt to address the most pertinent issues, the information in this chapter has been organized under the following headings:

1. Introduction to Aqueous Cleaning
2. Considerations of Contaminants and Substrates

3. Factors Affecting Aqueous Cleaning
4. Cleaning Chemistries
5. An Example of an Aqueous Chemical Cleaning Process
6. The Effect of Process Variables on Aqueous Chemical Cleaning
7. Semiconductor Wafer Drying
8. Equipment Used For Aqueous Cleaning
9. Conclusion

2.0 CONSIDERATIONS OF CONTAMINANTS AND SUBSTRATES

In order to obtain the desired clean surface, it is necessary to remove all contamination and to prevent it from re-establishing residence on the wafer surface prior to use or the next operation. Contaminants may occur as films or as particles on the surface. They may also be incorporated in the top layers of the wafer surface. The contaminants may be either organic or inorganic in composition. The cleaning procedure must be designed to address both the chemical type (organic or inorganic) and the physical form (film, particulate, or surface incorporated) of the contamination.

Organic contaminants may originate from a variety of sources, such as lubricants, coolants, cutting oils, fatty materials from human handling, airborne particles, detergents, corrosion inhibitors, and organic residues. These organic residues are most commonly found after evaporation of organic solvents on the wafer surface. This surface deposition can also occur during evaporation of aqueous solutions and during the dilution associated with water rinsing. Aqueous solutions are likely to leave behind inorganic species as well.

Evaporative deposition of organics can occur during solvent evaporation. A carboxyl group, if present in the residue, is generally physically adsorbed to the surface whereas the hydrocarbon portion orients itself away from the surface. Following the physical adsorption of the molecule, it is possible to have a chemical reaction between the functional group (carboxyl group in this example) and the wafer surface. When this happens, the molecule is said to be chemisorbed (1).

Physical methods of scrubbing, spraying, or ultrasonics alone are generally ineffective in removing chemisorbed materials. These materials must be removed chemically or by removal of some of the surface.

Inorganic contamination can be "brought along" with the organic residues or can be deposited independently. Inorganic residues can be either charged (ionic) or uncharged. Ionic contaminants can be physically or chemically adsorbed. It can also be introduced into wafer subsurface layers through adsorption or by exposure to thermal energy, with resultant diffusion into the wafer.

Uncharged metallic contamination can result from replacement plating in acid etchant solutions. These contaminants are difficult to remove because they are less soluble than ions and generally require oxidation to render them soluble. The actual forces that hold contaminants on surfaces are covered in more detail in the next subsection.

A good example of a contaminated surface is shown in Fig. 1 (2). This illustration represents a typical situation on a metal surface. Many metals and semiconductors, including silicon, form thin oxide layers on the surface even under room atmospheric conditions. During this oxidation, it is possible to incorporate airborne contamination into the film. The thin "native SiO_2" on a freshly sawed silicon wafer can be contaminated with aluminum, which comes from the aluminum oxide lapping compound introduced into the room air during the lapping process (3).

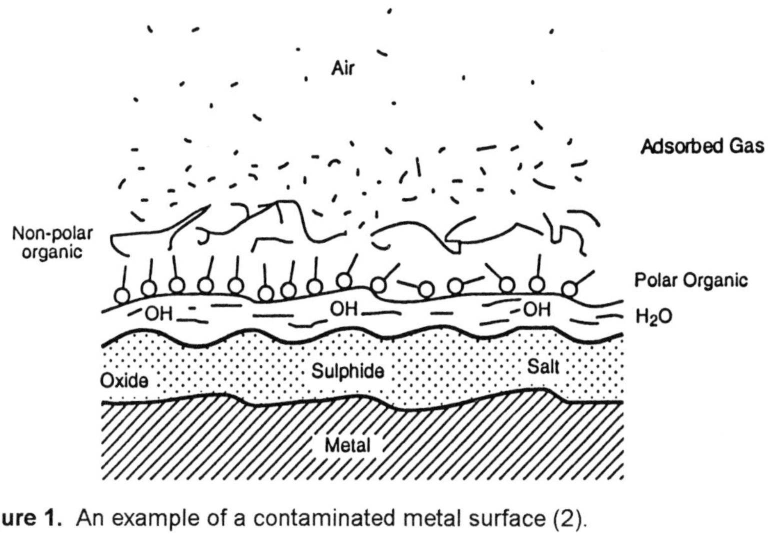

Figure 1. An example of a contaminated metal surface (2).

2.1 Surface Effects—Forces Holding the Contaminants

The forces which attract and/or hold contaminants to the surface to be cleaned are multifarious in both type and extent. Although a complete discussion of these forces is beyond the scope of this chapter, a basic understanding will help in determining how the contaminants can best be removed. The forces holding an adsorbed contaminant on the surface can be divided into two classes, depending on whether the forces are chemical or physical in character.

2.2 Chemical Adsorption

Chemical adsorption, or chemisorption, involves the formation of a chemical bond between the surface (adsorbent) and the contaminant (adsorbate). Chemical bonds can further be classified as *ionic, covalent,* or *metallic*. An ionic bond is formed when an electron is completely transferred from one atom to another, creating a positive ion and a negative ion. These ions are then held together by a coulombic force. An example of a compound of this type is NaCl. In a covalent bond, there is no permanent transfer of electrons. Instead, electrons are shared between atoms. Examples of this type include O_2, N_2, H_2, and Cl_2. The metallic bond also involves a sharing of electrons. However, in contrast to the short range sharing associated with the covalent bond, the metallic bond involves a long range sharing in which the valence electrons of each atom are shared by all atoms in the metal crystal.

It is important to remember that chemisorption is irreversible in a thermodynamic sense. Since it is necessary to break a chemical bond, it is not possible to merely rinse off (dissolve) the contaminant with a solvent.

2.3 Physical Adsorption

The forces involved in physical adsorption are the van der Waals or intermolecular type. These include the following:

Dipole-Dipole. Asymmetrical charge distribution within polar molecules results in attractive-repulsive forces on the molecules. This is termed the dipole-dipole force. Hydrogen bonding is a special form of this type of force in which a hydrogen atom forms a "bond" with a highly electronegative atom such as oxygen.

Dipole-induced Dipole. A polar molecule in the vicinity of a symmetrical molecule can induce a momentary dipole in the symmetrical molecule, giving rise to dipole-induced dipole forces.

London Dispersion Forces. London dispersion forces are explained by Hirschfelder et al. (4) in the following way: "At any instant the electrons in molecule A have a definite configuration, so that molecule A has an instantaneous dipole moment (even if it possesses no permanent electric moment). This instantaneous dipole in molecule A induces a dipole in molecule B. The interaction between these two dipoles results in a force of attraction between the two molecules. The dispersion force is then this instantaneous force of attraction averaged over all instantaneous configurations of the electrons in molecule A."

Physical adsorption is reversible in the thermodynamic sense. Also important for aqueous cleaning is the fact that water forms strong hydrogen bonds, which can outweigh other physical adsorption forces holding the contaminants on the surface. The approximate magnitudes of the forces discussed so far are listed in Table 1 (5).

Table 1. Relative Strengths of Binding Forces (5)

Type of Force	Energy (kcal/mol)
Chemical Bonds:	
Ionic	140-250
Covalent	15-170
Metallic	27- 83
Intermolecular (van der Waals) forces:	
Hydrogen bonds	< 12
Dipole-dipole	< 5
Dipole-induced dipole	< 0.5
Dispersion	< 10

Other physical forces. In addition to those already mentioned, other physical forces attract contaminants to the surface and keep them there. Gravity can deposit and hold particles on the top surface of a substrate. Electrostatic forces and the environment surrounding the surface can also be important in attracting particles.

Whitfield has shown that relative humidity can be a critical factor in surface particle retention (6). At elevated humidity levels, condensed water can fill the space between a particle and the surface. The water exerts a binding force on the particle. This phenomenon, sometimes referred to as a capillary force, increases the effect of the physical adsorption forces discussed above. To decrease the number of retained particles, it is recommended by Whitfield that surfaces not be exposed to environments with relative humidity above 50%, even for short periods of time.

The forces which are of major importance in a particular situation are dependent on many factors. One of the most important is the chemical nature of the substrate. Not only are we concerned with its bulk composition, e.g., a silicon wafer, but also the chemical nature of the surface. In the case of a silicon wafer, the surface may be bare silicon, or oxidized silicon with a simple oxide, siloxane, silanol, or dehydrated silanol structure (See Fig. 2) (7). The exact chemical nature of the surface will determine its chemical reactivity or tendency to form chemical bonds with contaminants.

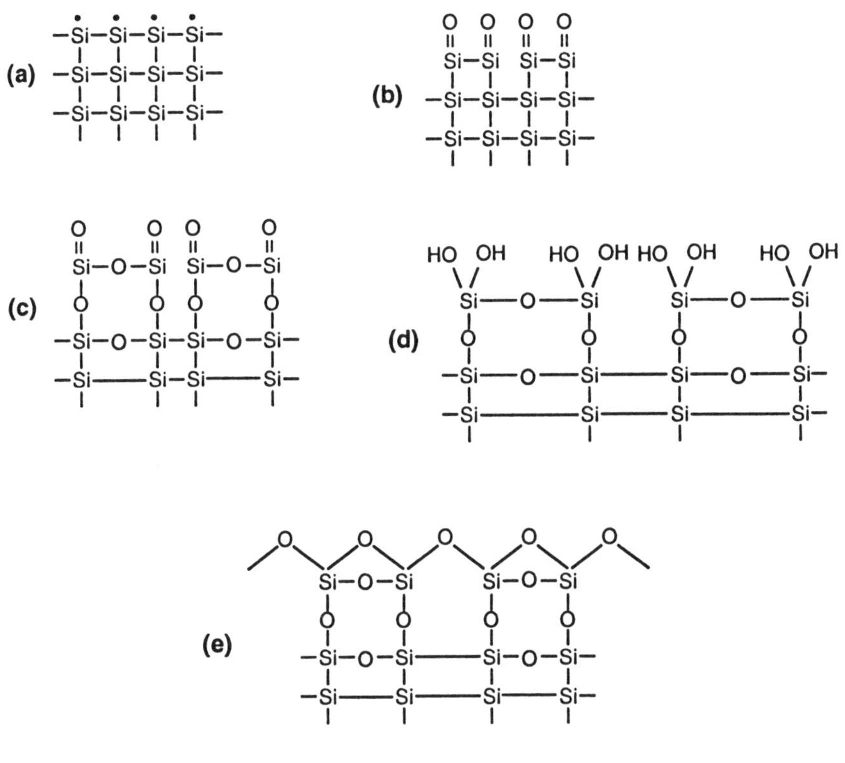

Figure 2. Silicon surface chemical states: (a) Bare silicon, (b) Oxidized silicon with a simple oxide, (c) Silane, (d) Silanol, (e) Dehydrated silanol (7).

Another factor is the substrate's surface texture, which may be porous, rough, or smooth on a microscale. The forces important to adsorption are short range forces. The surface texture determines how much of a particle will come into close contact with the substrate where these forces become appreciable.

As an example of the effect of surface condition, Kern has reported that thermally steam grown SiO_2 on silicon wafers retained less than 3 x 10^{13} Na^+/cm^2 after being immersed in 0.7N NaOH at 100°C and rinsed with water for 60 seconds. However, pyrolytically deposited SiO_2, which is more porous, retained 1.4 x 10^{16} Na^+/cm^2 after the same treatment (8).

The nature of the contaminant also determines which forces are most important. Is the contaminant a particle, an atom, a molecule or an ion? Is it organic or inorganic? What is its exact chemical makeup? Identification of the contaminant improves the chances of developing a cleaning process to successfully remove it.

3.0 FACTORS AFFECTING AQUEOUS CLEANING

3.1 Predicting and Enhancing Contaminant Solubility

Aqueous solutions are generally polar solvents and thus are most effective in dissolving charged contaminants. The "like dissolves like" rule of thumb is used to predict the ability of a solution to dissolve a contaminant. However, relying on this principle can lead to error. For instance, because water and sugar are both rich in hydroxyl groups, water should be a good solvent for sugar, and it is. Chloroform and beryllium chloride contain about the same amount of chlorine. Chloroform could be predicted to be a solvent for this salt. In fact, beryllium chloride is insoluble in chloroform.

A better prediction of solubility is based on the solubility parameter of the two materials, solute (contaminant) and solvent (proposed cleaning solution). This solubility parameter is a measure of the "internal pressure" of a substance and can be defined as the square root of the energy of vaporization per cubic centimeter:

$$\sigma = (H_v/V)^{1/2}$$

where: σ = solubility parameter
H_v = molar heat of vaporization
V = molar volume

The solubility parameter can also be calculated from other properties, such as surface tension, van der Waals constant for a gas, critical pressure, and air viscosity. Regardless of the calculation procedure used, when the values of the solute and solvent solubility parameters are close, the likelihood of high solubility increases (9)(10).

The solubility of a contaminant can be increased by modifying one or more of several parameters, such as concentration, pH, temperature, or by the addition of another solution component. Process parameters such as these are discussed in more detail in Sec. 6. Another method of improving the solubility of contaminants is the formation of soluble complexes. Many metal ions can be made more soluble this way. Thus, the anion used in an acid solution can be important, since it may contribute to complex formation. The metal ion acts as an electron acceptor (Lewis acid) and the solution anion acts as the electron donor (Lewis base), or "ligand", in the formation of an inorganic complex or coordination compound. For example, when copper is to be removed, nitric acid is a better choice than hydrochloric acid, possibly because nitrate (NO_3^-) is a better ligand for copper than chloride (Cl^-). Change in acidity alone cannot explain this effect, since hydrochloric acid and nitric acid are similar in strength.

The ability of negative ions to form complexes or coordination compounds can be ranked. As seen in the example with copper, Cl^- ranks below NO_3^-. Although both NO_3^- and NO_2^- are present in nitric acid solutions, complexes will form predominantly with NO_2^-. The negative ion ligand preference series tends to rate relative base strength and corresponding preference as a ligand (11):

$$NO_2^- > F^- > NO_3^- > Cl^- > Br^- > I^- > ClO_4^-$$

Contaminants may be made more soluble, and thus removable, by allowing a solution component to react with the contaminant to form a more soluble species. This is a very prominent mechanism in organic removal, where the insoluble material is oxidized to a more soluble species such as a carboxylic acid. In order to take advantage of this phenomenon, an oxidizing agent, such as hydrogen peroxide, ammonium persulfate, or ozone can be added to the solution. Also, a strong oxidizing acid, such as nitric acid, could be used.

3.2 Etching as a Means of Contaminant Removal

A contaminant which is partly absorbed into the subsurface of a wafer can be very difficult to remove with cleaning agents. For example, sodium ions can be "trapped" when the base silicon surface forms a native oxide in an environment containing sodium. Furthermore, surface sodium ions can migrate through the oxide during a high temperature operation.

Removal of a subsurface contaminant can be facilitated by a light surface etch which carries off some of the contaminant in the process. In the case of removing a surface oxide, any contaminants left after the etching process may now reside at the surface and thus are much easier for the cleaning solutions to remove.

Another instance in which etching is useful is for the removal of silicon particles. When initially deposited, the particles can be removed by rinsing with water, particularly if a wetting agent is added. However, if allowed to dry, both the silicon surface and the particles begin to oxidize. It is likely that this oxidation leads to an intermingling, and the two oxides become interlocked (12). Besides physical means, etching appears to be the only effective way of removing these particles.

4.0 CLEANING CHEMISTRIES

Many chemical solutions have been used for cleaning wafers. The most prevalent have been the RCA solutions, aqueous mixtures of unstabilized hydrogen peroxide with ammonia and hydrochloric acid. These cleans were developed before 1970 and have been the primary cleaning method used by the industry ever since (8)(13)(14).

The original RCA clean consisted of two cleaning solutions:

$$NH_4OH\ (29w/w\%) + H_2O_2\ (30\%) + DI\ H_2O \text{ at } 70 - 80°C$$

$$HCl\ (37w/w\%) + H_2O_2\ (30\%) + DI\ H_2O \text{ at } 75 - 80°C$$

The purpose of the first step, known as Standard Clean 1 or SC-1, was to oxidize surface organic films and remove some metal ions. The second step, known as Standard Clean 2 or SC-2, was to remove alkali cations and other cations like Al^{+3}, Fe^{+3} and Mg^{+2}. The solutions were mixed typically in the ratio 1:1:5.

The RCA cleaning chemistries, as well as other cleaning solutions, are listed in Table 2. The ratios of the chemicals have been omitted since they can vary greatly between fabs. The solutions listed in Part 2 of Table 2 have been reported only recently.

Table 2. Silicon Wafer Cleaning Solutions

Solution	Chemical Symbols	Common Name	Purpose or Removal of:
Part 1.			
Ammonium hydroxide/ hydrogen peroxide/ water	$NH_4OH/H_20_2/H_2O$	RCA-1, SC-1 (Standard Clean-1), APM (ammonia/peroxide mix), Huang A	Light organics, particles, and metals; protective oxide regrowth (14)
Hydrochloric acid/ hydrogen peroxide/ water	$HCl/H_2O_2/H_2O$	RCA-2, SC-2 (Standard Clean-2), HPM (hydrochloric/peroxide mix), Huang B	Heavy metals, alkalis, and metal hydroxides (14)
Sulfuric acid/ hydrogen peroxide	H_2SO_4/H_2O_2	Piranha, SPM (sulfuric/ peroxide mix), "Caros acid"	Heavy organics (15)
Hydrofluoric acid/water	HF/H_2O	HF, DHF (dilute HF)	Silicon oxide
Hydrofluoric acid/ ammonium fluoride/ water	$HF/NH_4F/H_2O$	BOE (buffered oxide etch), BHF (buffered hydrofluoric acid)	Silicon oxide
Nitric acid	HNO_3		Organics and heavy metals
Choline	$(CH_3)_3N^+$-$CH_2CH_2OH \cdot OH$	trimethyl(2-hydroxyethyl) ammonium hydroxide	Metals and organics (16)
Choline/ hydrogen peroxide/water	$(CH_3)_3N^+$-$CH_2CH_2OH \cdot OH^-/H_2O_2/H_2O$	Choline/ peroxide	Heavy metals, organics and particles (16)
Ammonium persulfate/ sulfuric acid	$(NH_4)_2SO_4/H_2SO_4$	SA-80	Organics (17)
Part 2. Silicon Wafer Cleaning Solutions Developed Since 1987			
Peroxydisulfuric acid/sulfuric acid	$H_2S_2O_8/H_2SO_4$	PDSA, "Caros acid," Piranha	Organics (18)
Ozone dissolved in deionized water	O_3/H_2O	Ozonized water	Protective oxide regrowth; organics (19)
Sulfuric acid/ozonized water	$H_2SO_4/O_3/H_2O$	SOM (sulfuric/ozone mix)	Organics (19)
Hydrofluoric acid/ nitric acid	HF/HNO_3		Slight Si etch; metals (20)
Hydrofluoric acid/ hydrogen peroxide	HF/H_2O_2		Slight Si etch; metal (21)

5.0 AN EXAMPLE OF AN AQUEOUS CHEMICAL CLEANING PROCESS

The sequence of chemical solutions used to clean a wafer depends upon the contaminants present and the requirements of the clean. A general discussion of the effects of chemical sequence is beyond the scope of this chapter. Possible sequences of the chemistries used in a common modified RCA clean will be discussed as an example. This clean, which can be used to remove most contaminants, consists of the following steps:

- Organic removal
- Native oxide removal
- Particle removal with simultaneous oxide regrowth
- Metal removal

5.1 Organic Removal

Organic removal is often the first step in cleaning because the presence of organic films on wafer surfaces can render the surface hydrophobic and prevent other cleaning solutions from reaching the surface. Two solutions are commonly used for removing organic films. If heavy organic contamination like photoresist is present, mixtures of H_2SO_4 and H_2O_2 at temperatures >100°C are often used (15). Light organics can be removed at 80°C using the NH_4OH/H_2O_2 chemistries described above (13). If very heavy contamination is present the H_2SO_4/H_2O_2 solution followed by the NH_4OH/H_2O_2 solution may be effective.

5.2 Native Oxide Removal

The native oxide removal step is included because a thin layer of oxide is always present on a silicon surface and inorganic contaminants are often trapped in this layer. When the oxide is removed the contaminants are also removed, resulting in a surface with very low metallic contamination (12).

Oxide can be removed using either dilute solutions of HF or buffered oxide etch (BOE). BOEs are used instead of dilute HF because they provide a more stable etch rate and prevent photoresist liftoff which can occur in dilute HF. However, the formulation of BOEs is not straightforward. Surfactants are often added to improve wetting of the wafer surface since silicon becomes hydrophobic when the oxide is removed. The surfactants tend to make the solution foam, so defoamers are added. Optimization of this multicomponent solution is a very complex task (22).

HF solutions need to have extremely low metal levels to be effective. Metals like copper and gold, which have a lower electronegativity than silicon, can plate onto the wafer surface (23)(24). The solutions should also be free of organics, as should all subsequent cleaning or rinsing solutions used before a clean oxide is grown, because the hydrophobic silicon surface is very prone to hydrocarbon adsorption (25). In addition, the hydrophobic surface following this clean is very susceptible to particulate contamination (26)(27) that may result when it is exposed to gas-liquid interfaces (25). Hence, the methods used to rinse and dry the wafer following this step are critical in controlling particle contamination (28).

5.3 Particle Removal With Simultaneous Oxide Regrowth

Because the oxide removal process tends to add particles to the wafer surface, it is often followed by an NH_4OH/H_2O_2 step. This step is effective in both removing particles and growing a thin oxide film. The oxide "passivates" the surface by making it hydrophilic and less susceptible to organic and particulate contamination (29)(30). Particles are presumably removed by slowly etching the surface of the silicon from under the particles. The etch rate is a function of the type of oxide present and is in the range of 0.09 - 0.4 nm/min at 80°C when the ratio of $NH_4OH:H_2O_2:H_2O$ is 1:1:5 (31). This step is extremely effective in removing particles when it is combined with sonic cleaning (32). The use of sonic cleaning for particle removal is discussed in more detail in Sec. 8.4.

One disadvantage of using the NH_4OH/H_2O_2 solution to grow an oxide is that some metals are insoluble in this highly basic solution and, if present, have a high tendency to precipitate onto the wafer surface. Aluminum is an example of a metal of this type which, when present in sub-ppm concentrations can cause a substantial shift in the flat band voltage of a dual dielectric (33). Also, because aluminum is one of the few metals which does not cause H_2O_2 to decompose, the equipment used to make and store H_2O_2 was historically made of aluminum. The resulting H_2O_2 contained significant levels of aluminum (34). Recent advances in the technology of making H_2O_2 have virtually eliminated this source of contamination (35).

5.4 Metal Removal

Metal removal is usually accomplished using the HCl/H_2O_2 solution described above. This solution effectively removes metals and prevents them from plating back onto the surface by complex formation. It has been shown to be effective for removing cobalt, copper, iron, lead, magnesium, nickel and sodium as well as aluminum precipitated from the NH_4OH/H_2O_2 solutions (14) and other metals.

6.0 EFFECTS OF PROCESS VARIABLES ON AQUEOUS CHEMICAL CLEANING

The material presented so far in this chapter has noted that many aqueous chemistries are available for cleaning wafers. Their effectiveness in providing a contamination-free surface depends upon a number of variables, including the sequence of chemistries used, the ratio of the chemicals, the processing temperature, the age of the solutions, etc. Hence, aqueous cleaning is a complex process which often must be tailored to specific needs. For example, the structure and growth rate of oxides is highly dependent upon the cleaning procedures used to prepare the surface prior to oxide growth (36)-(38). This section describes the effects of some of these variables on wafer surface properties.

6.1 The Effect of Changing the Sequence of the Chemical Cleaning Steps

The four-step cleaning sequence described in Sec. 5 results in a silicon surface with low metal, organic and particle contaminant levels and a thin layer of chemical oxide. If the oxide removal step is eliminated, the metal levels following the complete clean are higher, as shown in Table 3 (39). The purpose of the HF is to remove small amounts of surface oxide, which improves removal of inorganic species. This suggests that the HF acts by a surface etching mechanism. When the order of the HF and NH_4OH steps are switched, metal levels are reduced. However, particle levels are increased by approximately a factor of four.

Because the wafer surface following native oxide removal has very low metal levels, it has been considered for a final step in cleaning. However, its use did not become practical as a final step until recently because of the contamination issues associated with particles, organics and metals. Improvements in the purity of HF solutions and in the quality of the water used to rinse the surface following the cleaning step has made its use practical (40)(41). This clean is now capable of producing surfaces with very low metal levels and low particle counts, as shown in Table 4.

Table 3. The Effect of Cleaning of Silicon Wafers With and Without HF Solution on Metallic Contamination (39)

Cleaning Sequence	SIMS Result (metal/Si x 10^{-9})		
	Na	K	Cu
Complete*	26	156	233
Without HF**	90	246	658

* Cleaning sequence was:
96% H_2SO_4:30% H_2O_2 (4:1)
0.5% HF
29% NH_4OH:30% H_2O_2:DI water (1:1:5)
37% HCl:30% H_2O_2:DI water (1:1:5)

** Cleaning sequence was:
96% H_2SO_4:30% H_2O_2 (4:1)
29% NH_4OH:30% H_2O_2:DI water (1:1:5)
37% HCl:30% H_2O_2:DI water (1:1:5)

Table 4. Metal Contamination on Silicon Wafer Surfaces After Oxide Removal in an HF-Last Clean (40)

Metal	Concentration (atoms/cm^2)
Cr	$< 2 \times 10^{10}$
Fe	$< 2\text{-}3 \times 10^{10}$
Ni	$< 2\text{-}4 \times 10^{10}$
Cu	$10\text{-}45 \times 10^{10}$
Zn	$< 2 \times 10^{10}$

If HCl is added to the HF solution in this "HF-last" clean, still lower metal levels can be achieved. The chloride ion forms complexes with many metals as described in Sec. 3.1, thereby reducing plate back, as shown in Table 5 (42).

Ozonized water can be used instead of the NH_4OH/H_2O_2 to control growth of a protective oxide layer following the oxide strip (19)(42). Ozone has an oxidation potential similar to that of H_2O_2 and can be used to grow a very clean oxide. Unlike the NH_4OH/H_2O_2 step, ozonization does not remove particles. However, if the wafer is handled properly after the oxide is stripped, particle removal is not necessary. In addition, because this treatment results in a surface with lower metal content than the modified RCA clean described above, the HCl-based metal removal step is not required, as shown in Table 6 (19).

Table 5. The Effect of HCl Addition into an HF Oxide Stripping Solution on Metallic Contamination on Silicon Wafer Surfaces (42)

Stripping Solution	SIMS Relative Element Concentration					
	Na	K	Al	Ca	Mg	Fe
10% HF	6	10	7	60	8	7
10% HF + HCl*	1	10	1	10	4	2

* Concentration not defined in reference.

Table 6. The Effect of Ozone Oxide Growth on Metallic Contamination (19)

Cleaning Sequence	SIMS Relative Element Concentration				
	Na	K	Cu	Ca	Mg
Modified RCA*	22	6	<2	11	2
Ozone Oxide Growth**	10	2	<2	10	1

* Cleaning sequence was:
    96% H_2SO_4:30% H_2O_2 (4:1)
    0.5% HF
    29% NH_4OH:30% H_2O_2:DI water (1:1:5)
    37% HCl:30% H_2O_2:DI water (1:1:5)

** Cleaning sequence was:
    96% H_2SO_4:Ozonized water
    0.5% HF
    Ozonized water

Another method of removing and regrowing the thin native oxide is through the controlled etch method. In this technique chemicals which etch and oxidize silicon are combined. These two competing reactions slowly etch the surface and form a new oxide layer without exposing a hydrophobic surface. Two solutions have been used to achieve controlled etch: mixtures of HF/HNO_3 and mixtures of HF/H_2O_2 (43). Both are very effective in producing surfaces with low metal contamination.

6.2 The Effect of Concentration

The concentrations of reactants in the chemical solutions used to clean wafers can significantly alter the effectiveness of the solution in creating a contaminant-free silicon surface. For example, it was originally thought that the ratios in the NH_4OH/H_2O_2/water (SC-1) clean were not very important. Subsequent studies have revealed that if the ratio of NH_4OH to H_2O_2 is too high the wafer will be etched (44). Another study indicated that the amount of NH_4OH should be substantially reduced to increase the removal efficiency of small particles and to prevent formation of surface microroughness (45). In this study a ratio of 0.05:1:5 was recommended. However, this study did not address the removal of microcontaminant films which might require higher concentrations (14).

The etchant to oxidizer ratio used in controlled etch cleans is very important. If the amount of etchant is too high the surface is roughened and device electrical properties are affected. If the etchant concentration is too low the metal impurities in the original native oxide are not removed.

The concentration of HF used for oxide removal also plays a role in the effectiveness of cleans performed. Decreasing the HF concentration decreases the ratio of F-terminated to H-terminated silicon atoms on the wafer surface. The decreased F concentration reduces the rate at which oxide grows on the silicon wafer surface when it is exposed to either water or air (46).

6.3 The Effect of Temperature

Changing the temperature of a cleaning solution can have several important effects:

1. Increasing the temperature increases the rate of reactions. A rough rule of thumb is that an increase of 10°C doubles the reaction rate. This rule applies to both desired reactions and undesired side reactions.

2. Increasing the temperature usually increases the solubility of contaminants and reaction products. Table 7 lists the relative solubility of various metal salts as a function of solution temperature (11). Some contaminants (Group A) become considerably more soluble as the water temperature is increased (e.g., $FeSO_4$, NH_4HCO_3). However, Group B chemicals (e.g., sodium chloride) show very little increase in solubility with increasing water temperature and Group C compounds actually decrease in solubility with increased temperature. Therefore, knowledge of the solute and the solvent are necessary to select the best conditions to promote solubility. When the solubility increases, the cleaning rate increases.
3. Increasing the temperature can increase the rate and probability of contaminants plating onto the wafer surface.
4. Increasing the temperature increases the decomposition rate of unstable reactants. For example, H_2O_2 decomposes to form water and oxygen. The rate of decomposition approximately doubles with every 10°C increase.

Although higher temperatures are preferred, it is often necessary to make tradeoffs. For example, the 80°C temperatures normally used for the SC-1 and SC-2 cleans was chosen to maximize the cleaning effectiveness without decomposing the H_2O_2 too rapidly (14).

The manufacturers of commercially available chemical blends often recommend specific temperature ranges that give optimal performance. If the recommended temperature ranges are not used, cleaning may be inadequate or may require significantly longer than expected.

6.4 Wetting

Chemicals must wet a surface before they can clean it. The wettability of a solid surface depends upon the surface tension of both the solid and the solution. When the surface tension of the solid is greater than or equal to the surface tension of the solution it wets. If it is lower it does not wet. Many common organic solvents have surface tensions which are lower than water. Many inorganic solutions have surface tensions which are greater than water.

The surface tension of a wafer surface is a function of its chemical nature. Surfaces with a native oxide are hydrophilic (wet with water), whereas those with no oxide are hydrophobic (do not wet with water). In addition, the presence of contaminants on the wafer surface can change its surface tension. Organic contaminants in particular can drastically decrease the surface tension of a hydrophilic wafer surface rendering it hydrophobic.

Table 7. Solubilities of Inorganic Compounds in Water at Various Temperatures (11)

Substance	Formula	0°C	20°C	40°C	60°C	80°C	100°C
GROUP A							
Ammonium oxalate	$(NH_4)_2C_2O_4$	2.2	4.4	8.0	---	---	---
Ferrous sulfate	$FeSO_4$	15.65	26.5	40.2	---	---	---
Ammonium bicarbonate	NH_4HCO_3	11.9	75.5	91.1	107.8	126	145.6
Potassium chloride	KCl	27.6	34.0	40.0	45.5	51.1	56.7
Calcium chloride	$CaCl_2$	59.5	---	---	136.8	147.0	159
Cupric chloride	$CuCl_2$	70.7	77.0	83.8	91.2	99.2	107.9
Sodium nitrate	$NaNO_3$	73	88	104	124	148	180
Potassium iodide	KI	127.5	144	160	176	192	208
Ammonium manganese phosphate	NH_4MnPO_4	---	192	297.0	421.0	580.0	871.0
Calcium iodate	$Ca(IO_3)_2$	---	208.8	242.5	284.6	354.5	426.3
Zinc iodide	ZnI_2	429.4	---	445.2	467.2	490	510.5
GROUP B							
Sodium chloride	NaCl	35.7	36.0	36.6	37.3	38.4	39.8
GROUP C							
Sodium carbonate	Na_2CO_3	---	---	48.5	46.4	45.8	45.5
Calcium sulfate	$CaSO_4$	0.1759	0.1688	0.0973	0.0576	---	0

Group A - Solubility increases as temperature increases.
Group B - Solubility shows no change as temperature increases.
Group C - Solubility decreases as temperature increases.

Wetting agents, or surfactants, are sometimes added to cleaning solutions to promote wetting. These surface active agents reduce the interfacial tension between the solution and the wafer surface, allowing the solution to spread evenly across the wafer. The effectiveness of a surfactant to promote wetting is a function of its structure (47).

6.5 The Effect of Solution Degradation

There are several effects besides reactant decomposition which can cause solution cleaning effectiveness to change with time.

Reactant Depletion. Most of the reactions that remove contaminants from wafer surfaces reduce the concentration in the solution as the reaction takes place. A decrease in concentration usually results in a decrease in reaction rate. The rate of change in concentration is usually very slow unless large amounts of contaminants need to be removed. However, if the concentration of reactant is low, the contaminant concentration is high and the solution is used to process many wafers, the change in reactant concentration can result in a decrease in the reaction rate. A dilute HF bath is an example of a solution in which this might occur.

Solution Contamination. The function of cleaning solutions is to remove contaminants. When contaminants are removed from wafers they are sometimes converted into harmless byproducts. However, in many cases contaminants are simply removed from the wafer surface and remain in an active form in the cleaning solution. In this case, the concentration of contaminants in the solution increases as additional wafers are processed. Eventually, the concentration can increase to the point where the bath can contaminate subsequent wafers. Hence, the bath must be replenished or the contaminants must be removed before their concentrations reach unacceptable levels.

Component Removal. Many of the chemical cleaning processes described above are performed in recirculated baths in which filtration is employed to remove particles generated in the cleaning process. Unfortunately, filters can sometimes remove components from the cleaning solution. For example, some surfactants added to oxide etchants to enhance wetting are removed by adsorption (48). Filtration cannot be used unless the filter is compatible with all components in the solution.

CO_2 Absorption. Some of the solutions used to clean semiconductor wafers are basic. These solution can absorb CO_2 from the atmosphere. The CO_2 forms carbonic acid which reacts with the bases present in the solution, thereby decreasing the solution strength. Although the change in concentration is fairly slow, it can change the solution effectiveness if the bath is not changed frequently. Absorption can be prevented by blanketing the bath with an inert gas like nitrogen or argon.

Evaporation. Chemical solutions exposed to the atmosphere can release volatile components. This release results in changes in the chemical properties of the bath, thereby changing cleaning effectiveness.

Since evaporation increases exponentially with temperature, this effect is more pronounced at higher temperatures.

6.6 Carrier Effects

Wafers are often cleaned in batches of twenty-five or fifty or more. When cleaning is performed in this manner a carrier is required to hold the wafers. The carrier must be made of material that is compatible with the cleaning solutions to which it will be exposed. Carriers are often made of Teflon* because of its excellent chemical resistance. However, Teflon is somewhat porous and can absorb tangible amounts of chemicals (49) that can be carried to subsequent processes, resulting in contamination. In addition, some grades of Teflon slowly outgas trace amounts of HF that can etch oxide-coated wafer surfaces (50). Hence, care must be taken to ensure that the carriers used do not contribute to contamination; they should be frequently cleaned to remove absorbed impurities.

7.0 SEMICONDUCTOR WAFER DRYING

Much attention has been given to the wafer cleaning process, but the drying of the clean wafers is equally critical. In fact, wafer drying may be the most important step for ensuring that a cleaning process is successful in eliminating contamination. The drying process must remove water from the surface before it can evaporate, leaving residue behind (14)(51). There are three basic drying mechanisms: physical separation as in centrifugal drying, solvent displacement of DI water followed by solvent removal as in vapor drying, and evaporation as in hot water drying techniques.

7.1 Centrifugal Drying

Centrifugal or spin dryers are very common in the semiconductor industry. The centrifugal force resulting from spinning wafers at high speed eliminates the major portion of water from the wafer surface. The thin layer left behind evaporates. Because the evaporating water layer is very thin, deposition of residuals is minimal (52).

Two types of drying equipment are commonly used: horizontal and downflow. In horizontal systems wafer cassettes are inserted with the

---

* Teflon is a registered trademark of E. I. Dupont de Nemours, Wilmington, Delaware.

wafers oriented vertically. In downflow dryers the wafers are positioned horizontally. A diagram of a downflow dryer is shown in Fig. 3 (52). In downflow dryers water is removed both by centrifugal force and by air which is drawn across the wafer surface by the rotating action.

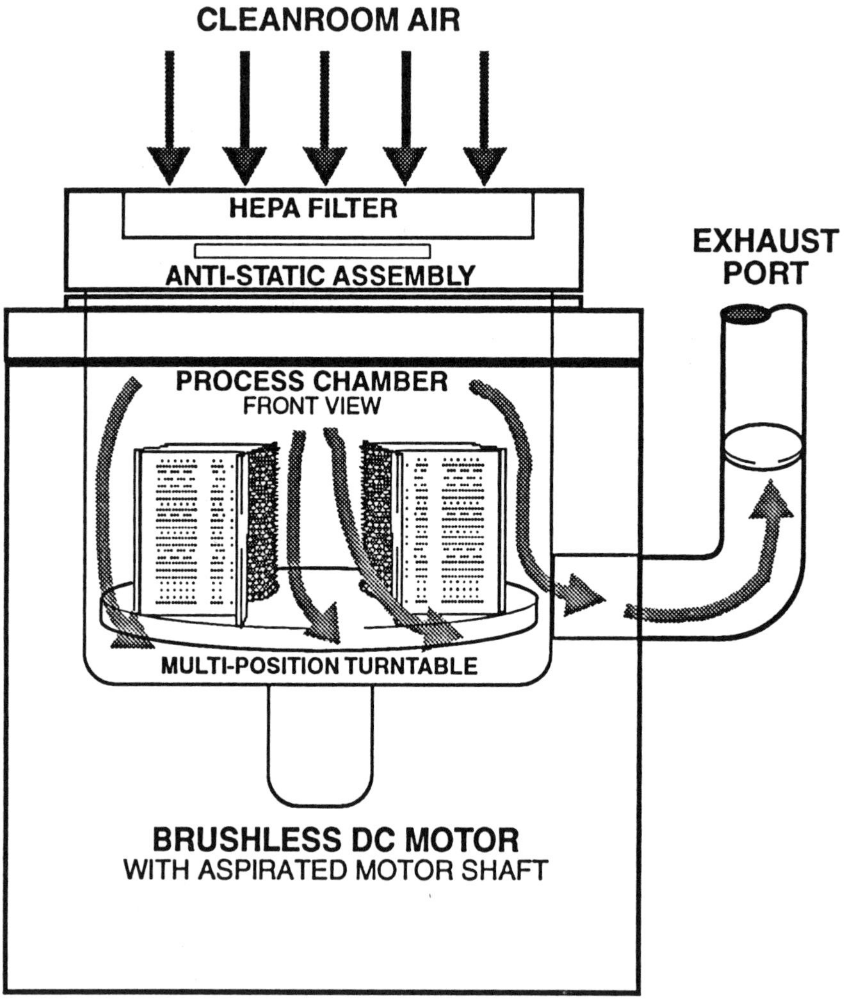

Figure 3. Cross section of a downflow centrifugal dryer (52).

7.2 Vapor Drying

In vapor drying, wafers wet with deionized water are suspended in a solvent vapor emanating from a heated bath. The vapor displaces the water on the wafer, leaving a vapor "coated" surface. When displacement is complete, wafers are raised above the vapor cloud and dry very quickly due to the high volatility of the solvent (53)(54). Vapor dryers are attractive because they eliminate some of the problems associated with spinning wafers, however, acceptance has been slow. In part this is due to concern over potential organic residues and safety problems (3). A schematic diagram of a vapor dryer is shown in Fig. 4 (55).

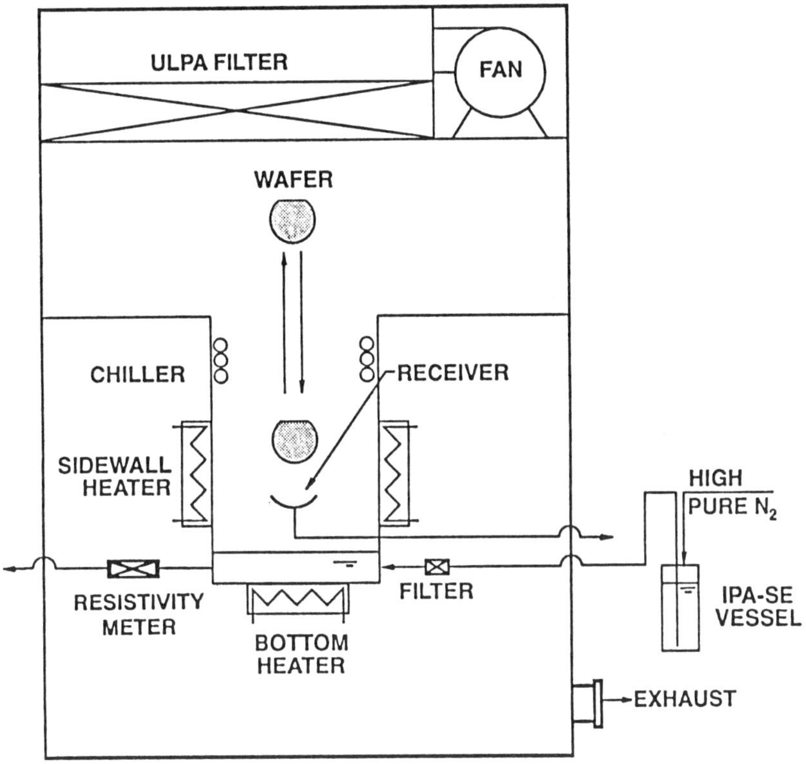

Figure 4. Cross section of an isopropyl alcohol vapor dryer (55). *(Copyright 1989, IEEE)*

The solvent used in vapor dryers has been primarily isopropyl alcohol (IPA). Mixtures of other solvents are under investigation but are not yet commercially available. The solvents being investigated usually have high flash points, thus offering greater safety than IPA.

The design of the vapor dryer has been hampered by the thermal mass of the carrier and platform, especially for larger wafers. When introduced into the hot vapor zone from the room ambient temperature, the large thermal mass can cause the vapor cloud to collapse. The result is ineffective drying, water evaporation from the wafers, and greater residuals (particles, etc.) on the wafer surface.

7.3 Hot Water Drying Techniques

A drying system based on the surface tension and capillary action of water is illustrated in Fig. 5 (56). In this technique, sometimes termed slow pull drying, wafers in a cassette are immersed into a tank of hot DI water where rinsing takes place. When rinsing is complete, the wafers are lifted slowly and at a controlled velocity from the bath. Because the wafers are pulled slowly and constantly, the water is not broken into droplets. Capillary action and surface tension pull the deionized water from the surface of wafers back into the bath. The wafers are dry once the cassette is removed. Although the technique's advantages of minimal stress on the wafer and simplicity of the tool's mechanisms are recognized, the technique has not yet been widely accepted (51).

8.0 EQUIPMENT USED FOR AQUEOUS CLEANING

8.1 General Design Considerations

The equipment used for cleaning semiconductor wafers must be capable of supplying a number of chemicals and chemical mixtures to the wafer surface in a defined sequence. The equipment must provide for rinsing between chemicals to prevent cross contamination. Also, drying capabilities must be included.

High-purity chemicals are required for cleaning because the wafers are very susceptible to damage by contamination. Hence, the materials of construction used in the equipment must be compatible with the chemicals and selected to minimize the equipment as a source of contamination.

Parts that contact liquids must be free of dead spots where contamination or bacteria may collect. Materials must be carefully selected to avoid chemical leaching and flaking.

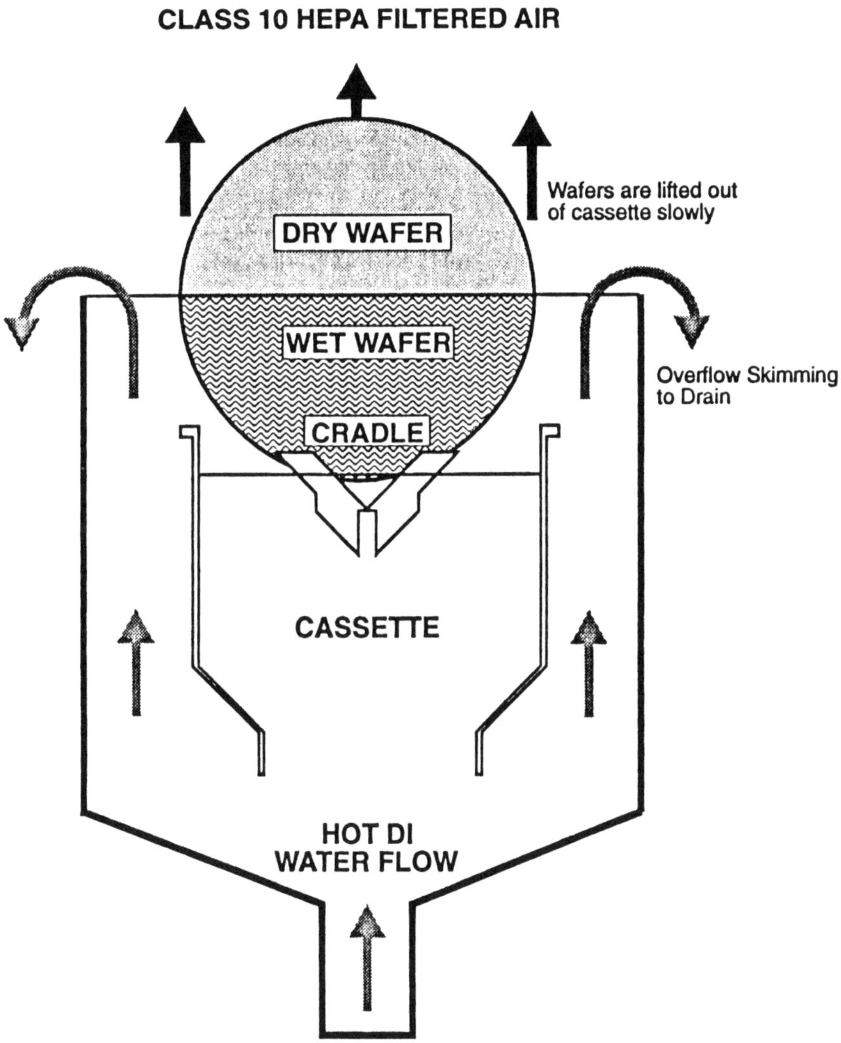

Figure 5. Slow pull dryer (56).

Contamination resulting from a poor choice of materials may be relatively obvious or fairly subtle. A more subtle example would be low levels of boron contamination which can originate from Pyrex and deposit on semiconductor wafer surfaces. Boron can cause electrical anomalies in the final chip if the contamination is high enough or occurs in a sensitive area of the device.

Metal leaching from stainless steel can also occur. The alkaline and acidic chemicals used in cleaning are highly corrosive. Although stainless steels are relatively corrosion resistant, they are not corrosion proof. They are formulated to slow the rate of attack, but they do not completely stop it. Corrosive solutions coming into contact with these surfaces will pick up Fe, Ni, and Cr, as well as other metals used in their manufacture. Even ultrapure water will dissolve stainless steel to the extent that 18 megohm-cm water will degrade to unacceptable levels after a brief exposure.

A very common set of materials used in construction of wet cleaning equipment is the Teflon family of fluorinated polymers and other polyfluorocarbons. These materials are used because of their extensive chemical resistance. Teflon polytetrafluoroethylene (PTFE) is a white, opaque material which must be shaped initially by techniques similar to those of powder metallurgy. Teflon fluorinated ethylenepropylene (FEP) and Teflon perfluoroalkoxy (PFA) are melt-processable resins with significantly improved ease of fabrication compared with that of PTFE.

The high chemical resistance of Teflon polymers to semiconductor cleaning chemicals helps make Teflon the chosen material for many processes. Chemical resistance can be estimated by measuring the absorption of chemicals by the material. Although Teflon absorptivities are unusually low compared to other plastics or elastomers, fluoride ions and related compounds are absorbed into the resins. Absorption is a function of temperature and pressure. Prolonged retention of absorbed chemicals can result in their decomposition within the Teflon matrix. Since the exposure time is long in some applications, it is advisable to perform in-process testing to ensure that the Teflon properties have not been altered (57)(58).

8.2 Immersion Processors

Immersion processors, or wet benches, are based on the principle that wafers are immersed sequentially into the proper chemical solutions in separate baths. In a typical wet bench clean operation, a cassette of wafers is immersed in an appropriate solution for a specified period of time, after which it is rinsed in DI water. The rinse terminates the reaction. The cassette of wafers is similarly immersed in subsequent cleaning and rinsing solutions. After a final rinse with filtered DI water, the wafers are dried. The original RCA cleaning process was developed in a simple immersion system of this type (13)(52). An example of a bench system using immersion processing is shown in Fig. 6 (59).

Many cleaning processes are performed at elevated temperatures. Immersion baths can be directly or indirectly heated to achieve the desired temperature. Physical agitation is used often to flush away contaminants and provide fresh solutions to the surface being cleaned. The agitation is accomplished in various ways including ultrasonic agitation (described in Sec. 8.4), nitrogen bubbling through the solution, or mechanical movement of the part in the solution.

Several recent developments in wet bench technology have substantially improved its performance. Automated wet benches with robotics handling have been introduced over the last five years. Although this type of system is considerably more expensive than manual units, it has gained acceptance in many fabs. Filtered recirculation of chemicals has been employed to improve chemical cleanliness and reduce chemical consumption. In the past, the use of circulation systems had been limited to lower temperatures (<80°C) by technology and material requirements. Recirculation tanks made from recently developed polymeric materials are suitable for use at higher temperatures (53).

Cassetteless transport systems have been developed to reduce the mass of material entering the bath. In these systems wafers are most typically transported between baths by use of a silicon or carbide carrier. These systems have the potential advantage that the amount of contact with the wafer and subsequent transfer of contamination is reduced. The vibrational security of the wafers in the cassetteless system is an area that needs to be addressed.

Immersion processors produce very good results and allow a wide selection of cleaning chemistries, however, chemical usage and floor space requirements are high. In addition, the chemicals in the bath must be monitored and replaced or replenished to ensure consistent results and effectiveness. For further information several reference are available (52)-(54)(60)(61).

8.3 Spray Processors

Spray processor systems provide for automatic centrifugal spray cleaning with corrosive or caustic chemicals. They were first introduced by

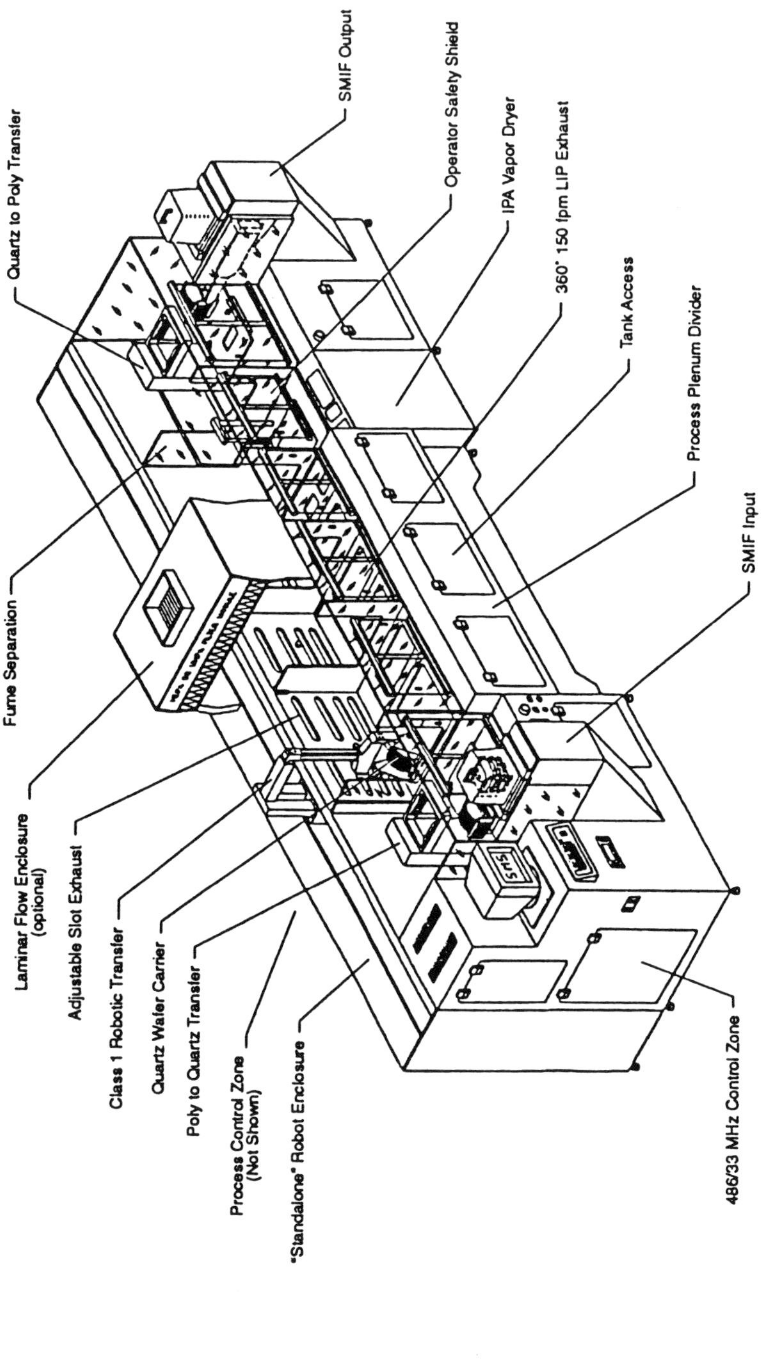

Figure 6. Wet bench for immersion processing (59).

FSI in 1975 (15). In a spray processor, a batch of wafers contained in cassettes is loaded into a turntable that rotates the wafers past a stationary spray post. Filtered acids and reagent solutions are introduced into the chamber through a liquid feed system which includes pressurized spray, a mixing manifold, and nitrogen atomization. The spray is directed uniformly through the spray post at the wafers; the spent chemicals are drained continuously through the bottom of the bowl so that fresh chemicals always contact the wafer. This eliminates solution contamination and degradation problems possible with immersion techniques (52). The system chambers are totally enclosed, with no exposure of the chemicals to the operator or to the environment. Spray processors perform an entire cleaning sequence, including all rinses and a final drying step, without removing the wafers from the equipment. This is achieved by using a microprocessor-based controller that opens and closes pneumatically controlled valves in a programmed sequence. Examples of spray processors are shown in Figs. 7 and 8.

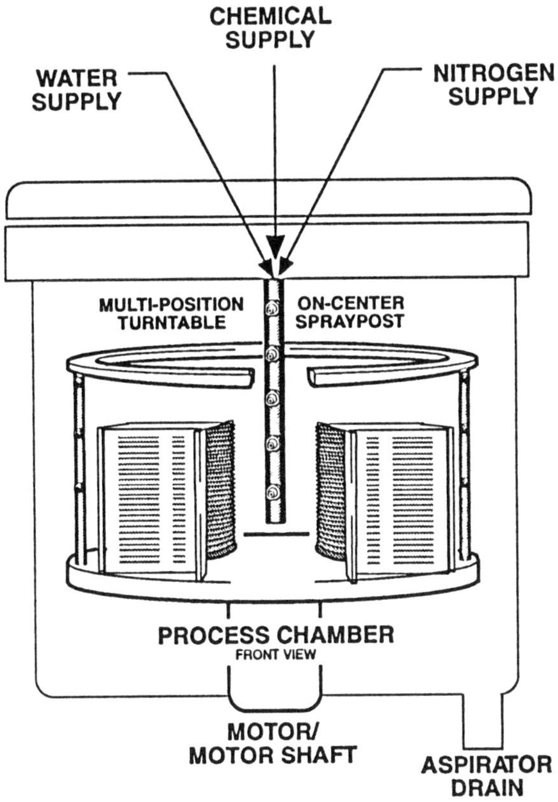

Figure 7. Multiposition spray processor.

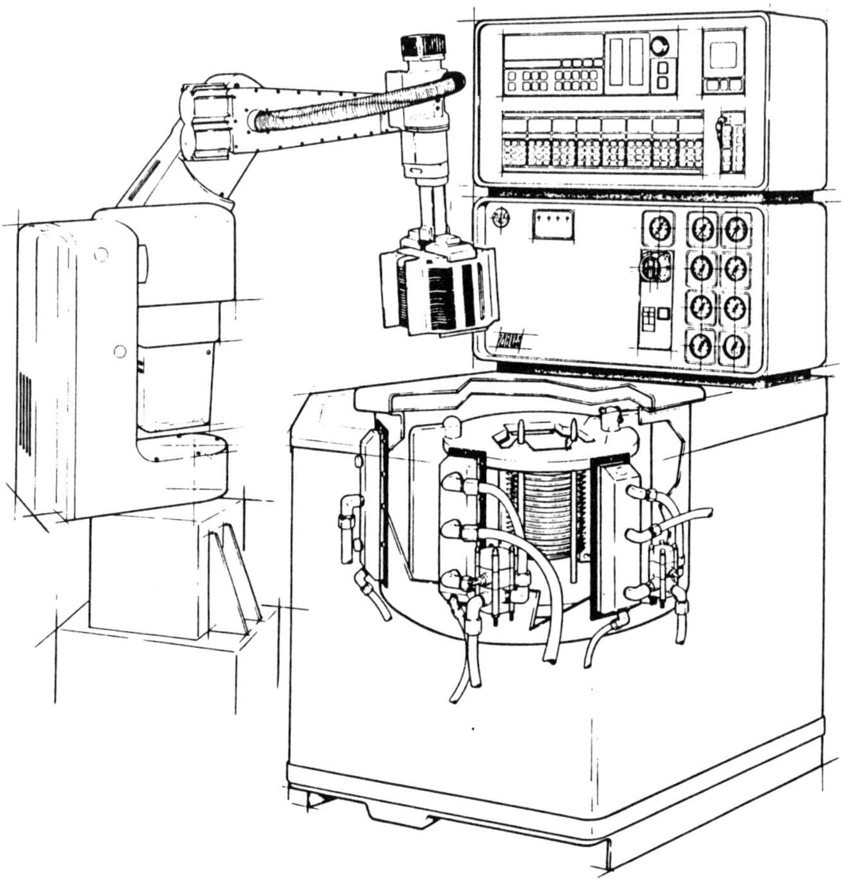

Figure 8. Spray processor with automatic loading robot.

Spray systems take various forms. High pressure spray helps to physically remove contaminants, but can reduce the time available for surface chemical reactions that are sometimes required for removal of contaminants. Atomization of sprays helps distribute the solution to wafers more efficiently. Although dispensing of fresh chemicals is desirable for minimizing cross-contamination, recirculation of chemicals is acceptable for many processes; such systems are available commercially.

A number of improvements have allowed the original spray system to meet changes in process requirements and more stringent specifications for particle addition and metal contamination, while still providing process flexibility. As in immersion processors, robotics systems have been incorporated. Spray processors have gained acceptance in many places where operator safety, ease of use, floor space, chemical usage, and performance for certain applications are important. Maintenance requirements can be high and provisions must be made for moving parts. For further information several references are available (15)(52)(53).

8.4 Ultrasonics and Megasonics

Ultrasonic systems have long been used for particle removal and are still used for some applications. Ultrasonic techniques use sonic energy of 10 to 100 kHz passed through a cleaning fluid to dislodge particles on a wafer surface. High intensity sound waves generate pressure fluctuations that result in the formation of cavitation bubbles. Upon collapsing, the bubbles release enough energy to dislodge and disperse particles but can also lead to surface damage (54). It is theorized that cavitation is the cleaning mechanism in ultrasonic cleaning (62).

The use of ultrasonic energy on wafers has always been suspect because of concerns about surface damage due to high temperatures and cavitation bubble explosions. Recent research has been directed toward determining the parameters that can reduce cavitation but still produce effective cleaning (63)(64).

Recent investigations have examined various cleaning fluids for their ability to remove particles from wafers in ultrasonic and spray jet systems. DI water was most effective for removing polymeric particles, while ethanol-acetone (1:1) was most effective, even better than fluorocarbons, for removing inorganic particles. Cleaning efficiency decreases with decreasing particle size (65).

In 1979, researchers at RCA reported on the use of megasonic energy (frequency 700 - 1000 kHz) for removing particles from wafers (32). Because megasonic frequencies are much higher than ultrasonic frequencies, large cavitation bubbles do not have time to form and surface damage should be reduced. In megasonic systems, the energy is produced by an array of piezoelectric crystals or transducers which usually are mounted at the bottom of the tank. The tank contains wafers carried on cassettes immersed in a cleaning liquid, such as SC-1 solution. The impact of megasonic pressure waves on the wafers enhances the cleaning ability of the cleaning solutions. As stated by Menon et al., "The force required to remove a particle from the wafer surface must be equal to, or exceed, the force of adhesion, and is a function of particle size, particle and wafer surface composition, and the nature of the liquid medium. For a silica particle with a 1 micron diameter (mass = 5×10^{-13} g) that is adhering to a bare silicon surface, the force of adhesion in water is approximately 4×10^{-4} dynes (which is known as the van der Waals force of adhesion). The applied megasonic force acting on this particle can be represented as F_{meg} = mass x acceleration. Since the acceleration produced by a megasonic transducer vibrating at a total power of 300 W is approximately 2.5×10^8 cm/sec, $F_{meg} = 1.25 \times 10^{-4}$ dynes. This megasonic force is approximately of the same magnitude as the force of adhesion; hence, 1 micron particles should be removed from silicon wafers in a megasonic tank containing DI water" (53). Although the higher frequencies inherent to megasonic energy should theoretically be effective for particles smaller than 0.03 microns in diameter, this has not yet been demonstrated.

The enhanced cleaning ability of the SC-1 solution in megasonic systems is due to the combination of its removal capability for organic thin films, its slight Si and SiO_2 etching action, and the force supplied by the megasonic energy for removal of particles. It has been suggested that other chemicals could be used in conjunction with the megasonic energy (28)(29)(52). For example, hydrochloric acid could be used in conjunction with megasonic energy to detach metallic particles, while HF and megasonic energy might work well for removal of silicates.

It should be remembered that a megasonic wafer cleaner can generate particles as well as remove them from wafer surfaces. Deteriorating seals or gaskets and defective transducer-bonding materials can shed particles when the transducers are vibrating. Mechanical movement of the cassette can also be a problem. Improved megasonic systems built under license from RCA have become available in the past few years (52). Equipment makers Verteq, FSI International, and SubMicron Systems have incorporated such innovations as focused spread energy, heated recirculation, liquid coupled transducers, and a variety of tank materials for a wide variety of applications (65)(66).

8.5 Liquid Displacement Processors

In 1986, CFM Technologies designed an immersion system which allowed the wafers to remain stationary and enclosed during the entire cleaning, rinsing and drying process (52). The system was designed to reduce or eliminate potential recontamination resulting from phase boundary crossing, for example, from solution to air. The vessel containing the wafers is hydraulically controlled so that liquids are displaced out of the top as the next process chemical is introduced from the bottom (41). For drying, IPA vapor is introduced through the top to displace the water as it drains out the bottom. Nitrogen drying of the IPA completes the process (67).

8.6 Point of Use Chemicals

Another innovation that has applicability to all types of wet cleaning equipment is the development of point-of-use (POU) chemical generation systems. The aqueous cleaning processes presently used to produce microcircuit devices require chemicals with very low metal ion and particle concentrations. Future devices are expected to require even cleaner chemicals. In the POU process, high-purity filtered gases are bubbled through high-purity water to form solutions such as NH_4OH, HF, or HCl. The chemicals formed have very low metal concentrations because both the water and the gases are essentially metal free and metallic extractables from storage containers are minimized. Particle contamination is also low because of the greater filtering efficiency of gases and DI water. Because the technique produces highly pure chemicals it lowers the probability of introducing contamination. In addition, because it can greatly reduce chemical use and cost, the point of use chemical generation technique has high potential for the future.

8.7 Single-Wafer Cleaners

A number of systems have been introduced in the last five years that permit single-wafer cleaning. Some systems are designed as single-wafer spinning systems where the wafer is mounted on a rotating chuck and chemicals are applied through a nozzle over the surface, as in photoresist coating. One system by Dai Nippon Screen has incorporated a megasonic unit in the nozzle. Semitool has also marketed a single-wafer clean system. The driving force for single-wafer systems has been their potential for individual wafer control and improved uniformity especially on large wafers (200 mm or larger). To date, these single-wafer, all-wet systems have not found wide acceptance (3).

8.8 Alternative Cleaning Techniques

Many other wet techniques for cleaning silicon wafers have been tried over the years with varying degrees of success. Some techniques are limited to specific applications by their undesirable side effects. An example is wet-chemical etching of silicon to remove entire surface layers by etch dissolution. The following techniques have been found viable and, in some cases, can be a desirable addition or alternative to the conventional processes based on hydrogen peroxide solutions.

Brush Scrubbing. Removal of large particles, such as those left after sawing and lapping operations, has been accomplished since the early days with wafer scrubbing machines. Brushes made of a hydrophilic material, such as nylon, dislodge particles hydrodynamically while DI water or isopropyl alcohol is applied to the surface. A thin layer of fluid must be retained between the brush and the wafers by careful mechanical adjustment to prevent surface scratching. While many contradictory claims have been made, if properly maintained, brush scrubbing can be very effective for removing particles larger than 1 micron from planar and preferably hydrophilic wafer surfaces (52)(68).

Fluid Jet. High pressure fluid jet cleaning consists of a high velocity jet of liquid sweeping over the surface at pressures of up to 4000 psi. The liquid can be DI water or organic solvents. The shear forces effectively dislodge submerged particles and penetrate into topography, but damage to the wafer can result with improperly adjusted pressure (52)(68).

8.9 Combined Wet/Dry Systems

As effective "dry cleaning" process steps are developed, systems which combine wet and dry processing may find niches in the market. An example of this type of system is the Excalibur In Situ Rinse*. In this equipment anhydrous HF and water vapor are used to etch the oxide on a wafer surface. The metals and other contaminants that are embedded in the oxide and at the oxide-silicon interface are removed in a subsequent water rinse. Wafers cleaned in this manner have extremely low residual metal contamination levels (69). Both the etching step and the rinse are performed in a single chamber resulting in a compact system.

---

*Excalibur In Situ Rinse is a registered trademark of FSI International, Chaska, Minnesota.

8.10 Rinsing and Drying

Because rinse tanks and dryers can be major sources of particulate contamination, systems and the process procedures used must be investigated thoroughly. The proper design of the rinse and dry steps is especially critical in HF-last type cleaners where hydrophobic surfaces are produced. The quick-dump rinse tanks with top spray rapidly remove chemicals from the wafer surfaces and periodically drain the chemical solution (54). However, problems with bacterial growth in parts of the system, especially the nozzle, and the inherent turbulence which can move particles through the solution make this method of rinsing less desirable from a particle contamination standpoint.

Another common rinse system, the cascade overflow system, is less susceptible to particle contamination. Hot water is often used for cascade rinses when viscous chemicals such as sulfuric and phosphoric acids need to be removed. The hot water adds cost to the system, but potentially reduces the number of particles on the finished dry wafer.

The final rinse technique is most often combined with a spin dryer to form the ubiquitous spin rinse/dryer, shown in Fig. 9. Water is sprayed on the wafers as they rotate in a chamber, as in spray processors. After the rinse, the wafers are spun dry, often with nitrogen flowing through the chamber. The chamber needs to be designed such that water on the walls after the rinse is not deposited on the wafer. Many users, to avoid this possibility, use the rinse dryer to dry wafers that have been rinsed in another system, such as a cascade overflow tank. Recently, dryers have been introduced which eliminate the rinse function. The wafers are spun dry in a chamber that allows massive quantities of HEPA filtered air to flow in the top and out through the exhaust. These have been termed *downflow* devices and have met with some success (55).

Several other techniques for drying have been proposed and investigated, such as hot nitrogen drying, vacuum drying, and slow pull drying (Sec. 7.3), but the technique that has garnered the most interest is IPA vapor drying (Sec. 7.2). Vapor drying systems should potentially leave fewer particles than spin-dry systems, but superiority has yet to be demonstrated. In addition, there is some concern over organic residues being left behind. The technique has been slowly gaining acceptance, especially due to its lack of mechanical motion and small number of moving parts in the system.

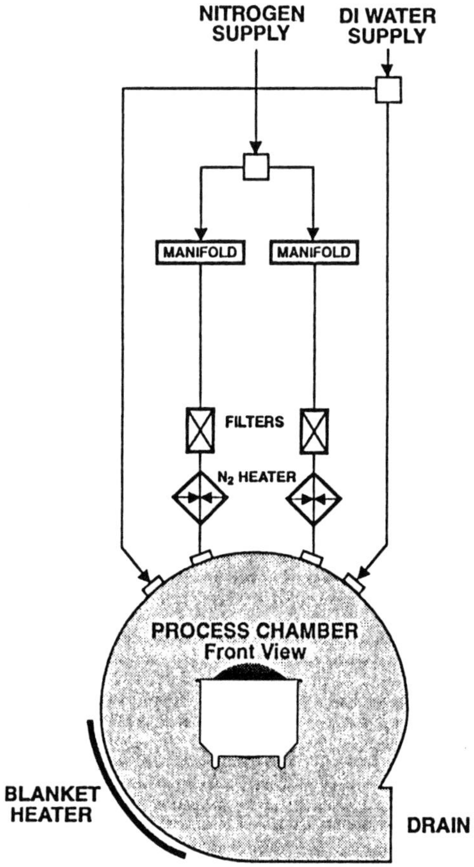

Figure 9. Schematic diagram of a spin rinser/dryer.

9.0 CONCLUSION

Although it may initially appear that aqueous cleaning of semiconductor wafers is a technology that is in the decline phase of its life cycle, it should now be clear that it, like many other technologies, is extending its time horizon of utility for a number of reasons that have been discussed in this chapter. The chief driving forces for the preservation of aqueous cleaning technology include the superior ability to remove metallic impurities, the high level of selectivity between the contaminants and the semiconductor surface, and the improvements in efficiency in use of aqueous cleaning agents to lower the cost and lessen the environmental effects.

It is clear that there is continuing R & D effort being expended for refining aqueous cleaning chemistries and equipment to achieve the needed improvements. Some of these efforts are directed toward greater contaminant removal to keep up with the ever increasing requirements of the semiconductor device manufacturers for cleaner wafers. Likewise, there is a real need to tailor the cleaning process (chemistry, sequence, and apparatus) to the needs of specific applications. This is sometimes referred to as *application specific cleaning*, which is the logical step to take since there is no universal best cleaning process. There are, and will continue to be, trade-offs in choosing the best procedure for a given situation. This logic can be extended to suggest that a combination of dry cleaning and wet (aqueous) cleaning of semiconductor wafers will be used for the foreseeable future.

REFERENCES

1. Feder, D. O. and Koontz, D. E., *Symposium on Cleaning of Electronic Device Components*, ASTM STP No. 246, pp. 40-63 (1959)
2. Mittal, K. L., in: *Surface Contamination: Genesis, Detection, and Control*, (K. L. Mittal, ed.), 1:3-45, Plenum Press, New York (1979)
3. Confidential communication with members of a semiconductor firm.
4. Hirschfelder, J. O., Curtiss, C. F. and Bird, R. B., *Molecular Theory of Gases and Liquids*, John Wiley and Sons, New York (1954)
5. Good, R. J., in: *Treatise on Adhesion and Adhesives*, Vol. 1, Chap. 2, (R. L. Patrick, ed.), Marcel Dekker, New York (1967)
6. Whitfield, W. J., in: *Surface Contamination: Genesis, Detection and Control*, (K.L. Mittal, ed.), 1:73-81, Plenum Press, New York (1979)
7. Pudvin, J. F., in: *Environmental Control in Electronic Manufacturing*, (P. W. Morrison, ed.), pp. 146-178, Van Nostrand Reinhold Company, New York (1973)
8. Kern, W., *RCA Review*, 31:207-233, (1970)
9. Hildebrand, J. H., in: *The Encyclopedia of Chemistry* (C. A. Hampel and G. G. Hawley, eds.), 3:1018-1021, Van Nostrand Reinhold Company, New York (1973)
10. Mittal, K. L., in: *Surface Contamination: Genesis, Detection, and Control*, (K. L. Mittal, ed.), 1:9-10, Plenum Press, New York (1979), and references therein.

11. Lange, N. A., in: *Lange's Handbook of Chemistry*, Revised 10th Edition, McGraw-Hill, Inc. (1967)
12. Amick, J. A., *Solid State Technology*, 19(11):47-52 (1976)
13. Kern, W. and Puotinen, D., *RCA Review*, 31:87-206 (1970)
14. Kern, W., *Proc. First International Symposium on Cleaning Technology in Semiconductor Device Manufacturing/1989*, 90-9:3-19, (J. Ruzyllo and R. E. Novak, eds.), The Electro-chemical Society, Pennington, NJ (1990)
15. Burkman, D. C., *Semiconductor International,* 4(7):103-114 (1981)
16. Muroakor, H., Kurosawa, K., Hiratsuka, H. and Usami, T., *Electrochem. Soc. Ext. Abstracts*, No. 238, 81-2:570 (1981)
17. van Zant, P., *Semiconductor International*, 7(4):109-111 (1984)
18. Davidson, J. and Hoffman J., *Proc. First International Symposium on Ultra Large Scale Integration Science and Technology*, pp. 798-804, (S. Broydo and C. M. Osburn, eds.), The Electrochemical Society, Pennington, NJ (1987)
19. Tong, J. T., Grant, D. C., and Peterson, C. A., *Proc. Second International Symposium on Cleaning Technology in Semiconductor Device Manufacturing*, 99-12:18-25, (J. Ruzyllo and R. E. Novak, eds.), The Electrochemical Society, Pennington, NJ (1992)
20. Takizawa, R. and Ohsawa, A., *Electrochem. Soc. Ext. Abstracts*, No. 387, 89-2:564 (1989)
21. Imaoka, T., Kezuka, T., Takano, J., Sugiyama, I. and Ohmi, T., *Proc. 38th Annual Technical Meeting Institute of Environmental Sciences*, pp. 466-474, The Institute of Environmental Sciences, Mount Prospect, IL (1992)
22. Kikuyama, H. and Miki, N., *Proc. 9th International Symposium on Contamination Control*, pp. 387-383, The Institute of Environmental Sciences, Mount Prospect, IL (1988)
23. Henderson, R. C., *J. Electrochem. Soc.*, 119(6):771-775 (1972)
24. Kern, F., Jr., Mitsushi, I., Kawanabe, I., Miyashita, M., Rosenberg, R. W. and Ohmi, T., "Metallic Contamination of Semiconductor Devices From Processing Chemicals, The Unrecognized Potential," Presented at 37th Annual Technical Meeting in San Diego, CA, Institute of Environmental Sciences (1991)
25. Riley, D. J. and Carbonnel, R. G., *Proc. 37th Annual Technical Meeting*, pp. 886-891, The Institute of Environmental Sciences, Mount Prospect, IL (1991)

26. Milner, T. A. and Brown, T. M., *Proc. Microcontamination Conf. and Exposition*, pp. 146-156 (1986)

27. Menon, V. B. and Donovan, R. P., *Proc. First International Symposium on Cleaning Technology in Semiconductor Device Manufacturing/ 1989*, 90-9:167-181, (J. Ruzyllo and R. E. Novak, eds.), The Electrochemical Society, Pennington, NJ (1990)

28. Atsumi, A., Ohtsuka, S., Munehira, S. and Kajiyama, K., *Proc. First International Symposium on Cleaning Technology in Semiconductor Device Manufacturing/1989*, 90-9:59-66, (J. Ruzyllo and R. E. Novak, eds.), The Electrochemical Society, Pennington, NJ (1990)

29. Menon, V. B., Clayton, A. C. and Donovan, R. P., *Microcontamination*, 7(31):31-34, 107-109 (1989)

30. Peterson, C. A., Schmidt, W. R., Burkman, D. C. and Phillips, B. F., *Proc. Technical Programme Semiconductor 1983 International*, presented in Birmingham, England (27-29 September, 1983). Also available as FSI Technical Report 217 from the authors.

31. Watanabe, M., Harazono, M., Hiratsuka, Y. and Edamura, T., *Electrochem. Soc. Ext. Abstracts*, 81-3:221-222 (1983)

32. Shwartzman, S., Mayer, A. and Kern, W., *RCA Review*, 46:81-105 (1985)

33. Slusser, G. J. and MacDowell, L., *J. Vac. Sci. Technology*, A-5(4):1649-1651 (1987)

34. Kawado, S., Tanigaki, T. and Maruyama, T., *Semiconductor Silicon 1986, Proc. Fifth International Symposium on Silicon Material Scientific Technology*, pp. 989-998, (H. R. Huff, T. Abe, and B. Kolbesen, eds.), The Electrochemical Society, Pennington, NJ (1986)

35. Seitaro, S. I. and Tanaka, F., *Proc. 9th International Symposium on Contamination Control*, pp. 374-377, The Institute of Environmental Sciences, Mount Prospect, IL (1988)

36. Lampert, I., *Electrochem. Soc. Ext. Abstracts*, 87-1:381-382 (1987)

37. Gould, G. and Irene, E. A., *J. Electrochem. Soc.*, 174(4):1031-1033 (1987)

38. Ruzyllo, J., *Technical Proc. Semicon/Europa, 1986*, pp. 1869-1870, Zurich (March 3-6, 1986)

39. Becker, D. S., Schmidt, W. R., Peterson, C. A. and Burkman, D., in: *Microelectronics Processing, Inorganic Materials Characterization*, Chap. 23, ACS Symp. Series No. 295, pp. 368-376, (L. A. Casper, ed.), American Chemical Society, Washington, DC (1986)

40. Christenson, K., *Proc. Second International Symposium on Cleaning Technology in Semiconductor Device Manufacturing*, 92-12:286-293, (J. Ruzyllo and R. E. Novak, eds.), The Electrochemical Society, Pennington, NJ (1992)

41. Walter, A. E. and McConnell, C. F., *Microcontamination*, 8(1):35- 61 (1990)

42. Krusell, W. C. and Golland, D. I., *Proc. First International Symposium on Cleaning Technology in Semiconductor Device Manufacturing/ 1989*, 90-9:23-32, (J. Ruzyllo and R. E. Novak, eds.), The Electrochemical Society, Pennington, NJ (1990)

43. Takizawa, R. and Ohsawa, A., *Proc. First International Symposium on Cleaning Technology in Semiconductor Device Manufacturing/1989*, 90-9:75-82, (J. Ruzyllo and R. E. Novak, eds.), The Electrochemical Society, Pennington, NJ (1990)

44. Kern, W., *Semiconductor International*, 7(4):94-99 (1984)

45. Ohmi, T., Mishima, H., Mizuniwa, T. and Abe, M., *Microcontamination*, 7(5):25-32, 108 (1988)

46. Grundner, M., Hahn, P. I., Lampert, I., Schnegg, A. and Jacob, H., *Proc. First International Symposium on Cleaning Technology in Semiconductor Device Manufacturing/1989*, 90-9:215-226, (J. Ruzyllo and R. E. Novak, eds.), The Electrochemical Society, Pennington, NJ (1990)

47. Shaw, D. J., *Introduction to Colloid and Surface Chemistry*, 2nd Edition, p. 118, Butterworth, London (1979)

48. Blum, R., Personal communication with D. Grant (1991)

49. Goodman, J. and Mudrak, L., *Solid State Technology*, 31(10):37-39 (1988)

50. Goodman, J. and Andrews, S., *Solid State Technology*, 33(7):65-68 (1990)

51. Skidmore, K., *Semiconductor International*, 12(8):80-86 (1989)

52. Kern, W., *J. Electrochem. Soc.*, 137(6):1887-1892 (1990) (This paper was originally presented at the 1989 Fall Meeting of The Electrochemical Society, Inc. held in Hollywood, FL.)

53. Menon, V. B. and Donovan, R. P., *Microcontamination*, 8(11):29-34, 66 (1990)

54. Skidmore, K. *Semiconductor International*, 9(9):80-85 (1987)

55. Mishima, H., Yasui, T., Mizuniwa, T., Abe, M. and Ohmi, T., *IEEE Transactions on Semiconductor Manufacturing*, 2(3):69-75 (1989)

56. Figure provided by Robert Orr, Trebor Incorporated, West Jordan, Utah.
57. du Pont Bulletin, No. E-08572, E. I. du Pont de Nemours & Company, Wilmington, Delaware
58. du Pont Bulletin, No. E-21623-1, E. I. du Pont de Nemours & Company, Wilmington, Delaware
59. Figure provided by R. Novak, SubMicron Systems, Inc., Wayzata, MN.
60. "Wafer Cleaning Equipment, 1986 Master Buying Guide," *Semiconductor International*, 8(13), pp. 76-77 (1986)
61. Singer, P. H., *Semiconductor International*, 11(8):42-48 (1988)
62. Suslick, K., *Sci. American*, 260(2):80-86 (1989)
63. Kashkoush, I., Busnaina, A., Kern, F. and Kunesh, R., *Proc. 36th Annual Technical Meeting*, pp. 407- 413, Institute of Environmental Sciences, Mount Prospect, IL (1990)
64. O'Donoghue, M., *Microcontamination*, 2(5):63-67 (1984)
65. Menon, V. B., Michaels, L. D., Clayton, A. C. and Donovan, R. P., *Proc. 35th Annual Technical Meeting*, pp. 320-324, Institute of Environmental Sciences, Mount Prospect, IL (1989)
66. Mayer, A. and Shwartzman, S., *J. Electronic Materials*, 8(6):855-863 (1979)
67. McConnell, C. F., *Proc. 36th Annual Technical Meeting*, pp. 269-272, Institute of Environmental Sciences, Mount Prospect, IL (1990)
68. Burggraaf, P. S., 4(8):71-102 (1981)
69. Syverson, D., *Proc. 37th Annual Technical Meeting*, pp. 829-833, Institute of Environmental Sciences, Mount Prospect, IL (1991)

4

Particle Deposition and Adhesion

Robert P. Donovan and Venu B. Menon

1.0 INTRODUCTION

Avoiding or minimizing contamination in the first place should be a major part of any wafer cleaning program—"an ounce of prevention is worth a pound of cure." This precept of Benjamin Franklin applies to particulate control for wafers since the time, energy, and resources needed to remove surface particles frequently exceed those that are required to avoid the particulate contamination to begin with. Minimizing particulate contamination by understanding particle deposition mechanisms is the subject of this chapter.

Minimizing wafer particulate contamination consists of two steps: *(i)* minimizing the concentration of particles at their source of generation or introduction to the wafer environment; and *(ii)* minimizing the transport of particles from their source of generation to the wafer surface and their deposition on the wafer. The second step will be discussed here, not that the first step is less important, but primarily that the second step lends itself to general description whereas the first step is often source specific.

This chapter considers two wafer environments: *(i)* gaseous atmospheres such as ambient air, and *(ii)* liquid baths. The former is reasonably well understood, although not always well controlled; the latter is more complex and is neither adequately understood nor controlled, although considerable progress has been made over the last several years. The two main divisions of this chapter are entitled "Aerosol Particle Deposition" and "Particle Deposition from a Liquid Bath." A third chapter division, entitled

"Particle Adhesion," discusses particle adhesive forces applicable to both environments and how they can be measured.

2.0 AEROSOL PARTICLE DEPOSITION (1)

An aerosol is at least a two-phase dispersed system consisting of particles (liquid or solid) suspended in a gas, such as air. An aerosol particle is a particle quasi-stably suspended in the gas. If the definition of quasi-stable is arbitrarily taken to mean a settling velocity of less than 0.01 cm/s, the upper size limit of a unit density, spherical aerosol particle in still air, is about 2 µm.

Settling velocity is the velocity at which the Stokes drag force on an aerosol particle is equal to the gravitational force on the particle:

Eq. (1) $$\underset{\text{(Drag)}}{3\pi\eta V_T d_p} = \underset{\text{(Gravitational)}}{\rho_p \pi d_p^3 g/6}$$

or

Eq. (2) $$V_T = \rho_p d_p^2 g/18\eta = \tau g$$

where*:

- V_T = particle terminal velocity (gravitational settling velocity) $[lt^{-1}]$
- ρ_p = particle mass density $[ml^{-3}]$
- d_p = particle diameter $[l]$
- η = fluid viscosity $[ml^{-1}t^{-1}]$
- g = gravitational acceleration $[lt^{-2}]$
- τ = particle relaxation time $[t]$

In still air, the settling velocity equation can be used to predict the rate at which particles will accumulate on a wafer surface because of gravitational settling:

Eq. (3) $$J_G = V_T C_0$$

where:

- J_G = particle flux attributable to gravitational settling $[l^{-2}t^{-1}]$
- C_0 = aerosol particle concentration $[l^{-3}]$
- V_T = gravitational settling velocity $[lt^{-1}]$

---

*Symbols in brackets give the dimensions of the parameter: l = length; t = time; m = mass; q = electrical charge; T = absolute temperature; – = dimensionless.

The values of V_T and hence J_G depend directly on d_p^2, and thus for small particles become small. For particles smaller than about 1 μm, a slip correction factor, called the Cunningham correction factor, C_S, must be included in the numerator of Eq. (2):

Eq. (4) $$V_T = C_S \rho_p d_p^2 g / 18\eta$$

This empirical correction factor compensates for departures from the assumption of zero gas velocity at the particle surface (Stokes law assumption) that occur as the particle diameter approaches the mean free path of air molecules. These small particles "slip" between the air molecules and settle faster than predicted by Eq. (2). The correction factor appearing in Eq. (4) extends the usefulness of the Stokes-derived settling velocity equation to regions where it otherwise would not be valid. The correction factor is always greater than one and can be approximated by:

Eq. (5) $$C_S = 1 + [2.514 + 0.800 \exp(-0.55 d_p / \lambda)] \lambda / d_p$$

where:

λ = the mean free path of the air or gas molecules [l] (λ = 0.066 μm for air at atmospheric pressure and 20°C)

d_p = particle diameter [l]

Values of V_T, as calculated by Eq. (4), are listed in Table 1. While V_T decreases significantly with particle diameter, it does so at a rate somewhat less than d_p^2 because of C_S.

Table 1. Settling Velocity, V_T, for Unit Density Spheres at 20°C and 1 Atm in Still Air

d_p(μm)	V_T(cm/s)
0.01	6.74×10^{-6}
0.1	8.68×10^{-5}
1.0	3.50×10^{-3}
10.0	3.05×10^{-1}

Values of J_G that are calculated from V_T in Table 1 for 0.1 μm particles exposed to Class 1 air (<1.24×10^{-3} particles/cm^3 > 0.1 μm) predict a relatively modest buildup of such particles on a wafer surface because of gravitational settling:

$$J_G = V_T C_0$$
$$= (8.68 \times 10^{-5}\ \text{cm/s})(1.24 \times 10^{-3}\ \text{particles/cm}^3)$$
$$= 1.1 \times 10^{-7}\ \text{particles/cm}^2 \cdot \text{s}$$

or about 2 particles/day on a 150 mm wafer.

Unfortunately, gravitational settling is not the only mechanism whereby particles deposit on a wafer surface and, for 0.1 µm particles, it is not the most important mechanism. The dominant aerosol particle deposition mechanism for small particle sizes is diffusion. The particle diffusion coefficient, D, depends on particle diameter as follows:

Eq. (6) $$D = C_s kT / 3\pi\eta d_p$$

where:

- D = particle diffusion coefficient $[l^2/s]$
- k = the Boltzmann constant $[ml^2/s^2T]$
- T = absolute temperature $[T]$

and all other symbols are as before.

Equation (6) shows that the diffusion coefficient increases as the particle diameter decreases. Typical values are listed in Table 2. Particle flux to a surface attributable to diffusion depends on the width of the concentration boundary layer adjacent to the surface, across which a concentration gradient exists. At the wafer surface, the aerosol particle concentration is zero; particle contact with a surface is assumed to result in particle capture (100% sticking coefficient). At some distance, d, from the surface, the aerosol concentration is the steady-state value, C_o, characteristic of the ambient wafer environment. The distance, d, is a measure of the concentration boundary layer width which affects particle flux attributable to diffusion as follows:

Eq. (7) $$J_{DIF} = DC_0/d = V_{DIF}C_0$$

where:

- J_{DIF} = particle flux attributable to diffusion $[l^{-2}t^{-1}]$
- C_0 = aerosol particle concentration at the edge of the concentration boundary layer $[l^{-3}]$

d = the width of the concentration boundary layer [l]

V_{DIF} = D/d, a term with dimensions of velocity which is descriptive of particle deposition attributable to diffusion.

Table 2. Diffusion Coefficient, D, for Aerosol Particles at 20°C

d_p(μm)	D(cm²/s)
0.01	5.31×10^{-4}
0.1	6.84×10^{-6}
1.0	2.76×10^{-7}
10.0	2.40×10^{-8}

The particle flux contributions of gravitation and diffusion are assumed to be additive:

Eq. (8) $$J_{ToT} = J_G + J_{DIF} = (V_T + V_{DIF})\, C_0 = V_D C_0$$

or

$$V_D = J_{ToT}/C_0$$

V_D is defined by Eq. (8) and is called the deposition velocity. It simply represents particle flux to a surface divided by the ambient aerosol particle concentration. It does not always have the clear physical interpretation that is evident when gravitational settling dominates the deposition process. When gravitational settling dominates, the deposition velocity is equal to the settling velocity (Eq. 4). When other mechanisms become important, the deposition velocity must be thought of simply as the ratio defined in Eq. (8).

Using Eq. (8) to calculate values of V_D as a function of particle diameter requires knowledge of the width of the aerosol particle concentration boundary layer, d. This quantity can vary with air flow field surrounding the wafer surface. For a wafer resting horizontally on a table located in a clean room protected by vertical laminar flow, predicted mean values of V_D are plotted in Fig. 1. (These values are called mean values because they have a small radial dependence due to variation in d across the wafer.) These predicted curves quite clearly show branches corresponding to the two deposition mechanisms. The right-hand side of the plot—the large particle diameter branch ($d_p > 1$ µm)—is that portion of the curve dominated by gravitational settling. The deposition velocity has the values of V_T predicted by Eq. (4). The left-hand side of the Fig. 1 plot—the small particle diameter branch ($d_p < 0.1$ µm)—reflects values of V_{DIF} (Eq. 7) calculated for boundary layer widths that are appropriate for the different velocities of vertical laminar flow (gravitational settling is independent of air flow).

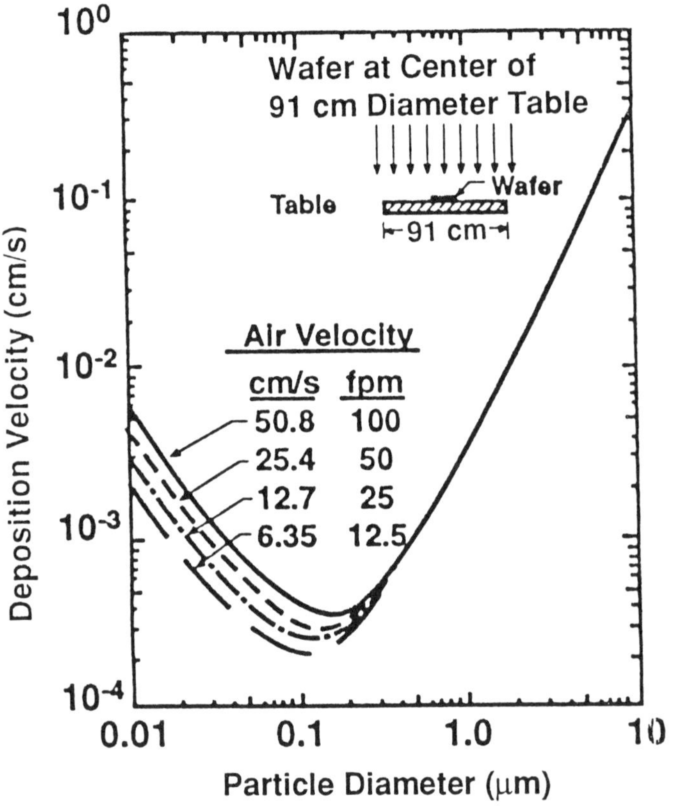

Figure 1. Mean deposition velocity for a 125 mm diameter wafer in the center of a solid 75 mm diameter table (2). *(With permission B. Y. H. Liu and K. H. Ahn. Copyright 1987, University of Minnesota)*

The transition region from gravity-dominated deposition to diffusion-dominated deposition (0.1 to 1.0 µm in Fig. 1) represents a superposition of the two velocity terms calculated independently (Eq. 8). No rigorous justification for this simple additive combination assumption exists other than the predictions agree reasonably well with observations.

Equation (8) and Fig. 1 consider only two mechanisms: gravity and diffusion. Electrical forces are also very important, especially at particle diameters in the transition region and below. The contribution of electrical forces is calculated by adding a third term to Eq. (8) based on electrical forces, as follows (3):

Eq. (9) $$J_{ELEC} = C_0 q E_0 D/kT = C_0 V_{ELEC}$$

where:

J_{ELEC} = particle flux attributable to electrical forces

q = particle charge

E_0 = applied electric field at wafer surface

V_{ELEC} = qE_0D/kT, the electrostatic contribution to deposition velocity.

Equation (9) is based on a Coulomb interaction which typically is the dominant interaction between a charged particle and a charged wafer.

Figure 2 is a plot of calculated values of V_D including electrical contributions in addition to the gravitational and diffusional contributions previously considered. The electrical force modeled is a Coulomb force, depending on both the electric field at the wafer surface and the electrical charge on the particle (Eq. 9). The surface field strength considered in Fig. 2 is the relatively modest value of 100 V/cm. Values many times greater than this field strength have been measured in both laboratory and manufacturing operations.

Deposition velocities for three different levels of particle charge are plotted in Fig. 2. The zero charge case is the same as the gravitation-diffusion case previously presented (Fig. 1) except that the air flow field is that for a horizontal wafer freely suspended in space rather than resting on a table. This altered flow around the wafer reduces d slightly, resulting in somewhat higher values of V_D in the diffusion branch of the curve.

In practice, a zero charge case never exists or exists only under the most unusual or special circumstances, namely, the deliberate removal of all charged particles from a region by the application of an electric field. Otherwise, the minimum charge state of an aerosol particle ensemble is approximated by a Boltzmann distribution:

Eq. (10) $$N_n/N_o = \exp(-n^2e^2/d_pkT)$$

where:

N_n = number of particles of diameter d_p carrying plus or minus n units of electrical charge

N_o = number of electrically neutral particles

e = elementary unit of charge; the charge of 1 electron (4.8×10^{-10} stat Coulombs).

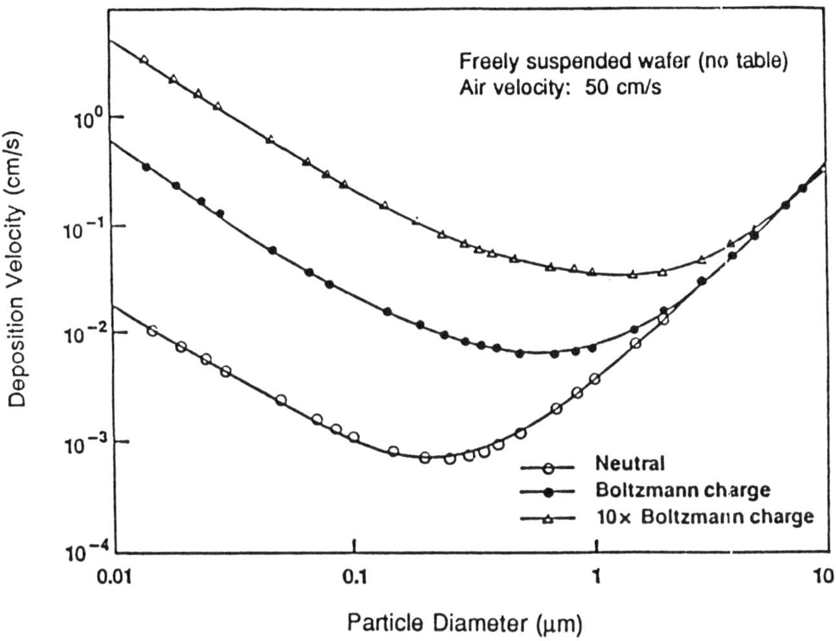

Figure 2. Calculated deposition velocity including effects of diffusion, settling and electric field (E = 100 V/cm) (4). *(With permission B. Y. H. Liu, B. Fardi, and K. H. Ahn. Copyright 1987, Institute of Environmental Sciences)*

Table 3 shows the predicted size distributions of particles according to the Boltzmann equation (Eq. 10). The Boltzmann equation predicts a charge distribution that is symmetric about zero charge; equal numbers of positive and negative particles are predicted for each particle size. No fundamental reason supports the Boltzmann distribution of particle charge. It is another one of those fortuitously simple relationships in aerosol technology that has been found useful. In particular, the Boltzmann distribution is the charge distribution that results from passing an aerosol stream by a radioactive source, such as a Kr^{85} neutralizer, or the charge distribution that results from bipolar air ionization (clean room air ionizers). The Boltzmann charge distribution represents the minimum practical particle charge condition achievable by these charge neutralizing actions. The predictions of Eq. (10) and Table 3 appear accurate down to about 0.1 μm. Below 0.1 μm diameter significant errors arise and corrections to the predicted relationships are needed (5).

Table 3. Boltzmann Charge Equilibrium Equation
$Nn_p = N_o \exp(n_p^2 e^2/d_p kT)$

	Percent of particles carrying n_p elementary charge units								
d_p, μm$_p$	n_p=-4	-3	-2	-1	0	+1	+2	+3	+4
0.01				0.34	99.32	0.34			
0.02				5.23	89.53	5.23			
0.04			0.23	16.22	67.10	16.22	0.23		
0.06		0.01	1.25	21.30	54.88	21.30	1.25	0.01	
0.08		0.08	2.78	23.37	47.53	23.37	2.78	0.08	
0.10		0.26	4.39	24.09	42.52	24.09	4.39	0.26	
0.20	0.32	2.33	9.66	22.63	30.06	22.63	9.66	2.33	0.32
0.40	2.19	5.92	12.05	18.44	21.26	18.44	12.05	5.92	2.19
0.60	3.82	7.41	11.89	15.79	17.36	15.79	11.89	7.41	3.82
0.80	4.83	7.94	11.32	14.00	15.03	14.00	11.32	7.94	4.83
1.00	5.42	8.06	10.71	12.70	13.45	12.70	10.71	8.06	5.42

Source: Liu, B. Y. H. and D. Y. H. Pui, *J. Colloid Interface Sci* 49(2), pp. 305-312, November 1974.

The curves of Fig. 2 labeled Boltzmann charge are calculated for a particle having the rms value of the charge distribution. The 10x Boltzmann curve is simply 10x the Boltzmann rms value of charge. The predicted curves of V_D in Fig. 2 show that electrical forces are very important for small particles and can easily dominate aerosol particle deposition in the submicrometer size range, increasing the deposition velocity by one to two orders of magnitude under the calculation conditions. Experimental verification of these predictions requires careful control of the independent variables, C_0, d_p, particle electrical charge and electric field at the wafer surface. Such control is typically only possible in laboratory settings with customized apparatus built to facilitate such experiments. In addition, the dependent variable of the increase in surface particle concentration per unit time, J_{ToT}, must be measured adequately. In the present state of the art of instrumentation for surface particle counting, only particles larger than about 0.1 μm diameter can be counted, thus limiting verification experiments primarily to the gravitational branch of the deposition velocity curve. Given these restrictions, however, experimental verification of the calculated deposition velocity curves is generally good. The observed V_D shows the type of behavior predicted and is of the predicted magnitude.

For example, in experiments in which the surface electric field was varied by an external potential applied to the wafer (Fig. 3), the measured deposition velocities were as given in Fig. 4. In these plots the average surface electric field in volts/cm was approximately equal to the voltage applied to the wafers; i.e., a wafer voltage of 1000 V in Fig. 4 implies that the average surface electric field strength was about 1000 V/cm (6). While particle charge was designed to represent a Boltzmann distribution (the aerosol stream was passed through a Kr^{85} neutralizer), the lack of symmetry observed with respect to the direction of the electric field shows that the concentration of negatively charged particles exceeded that of the positively charged aerosol particles introduced into the deposition process. Presumably, either triboelectric charging in the aerosol lines downstream of the particle charge neutralizer caused more positively charged particles than negatively charged particles to be removed from the subsequent aerosol stream, or the higher mobility of the negative ions neutralized more positively charged particles. In any event, the data of Fig. 4 show that at about 100 V/cm the deposition velocities are about 10^{-2} cm/s, which agrees with the calculated values of Fig. 2, even though the wafers used to collect the depositing particles were inside a well-mixed chamber rather than in the vertical laminar flow of the clean room. This lack of dependence on flow field is predicted to be true for the gravitational branch of deposition velocity where the Fig. 4 data fall.

Figure 4 also shows that applying the wafer bias with an AC field effected no enhanced deposition. This observation implies that a Coulomb force rather than a polarization effect is causing the electrically enhanced deposition under DC bias (the dielectrophoretic polarization force depends on the direction of the electric field gradient rather than the direction of the field). In an AC field, the direction of the field gradient does not change on the alternate half cycles even though the direction of the field itself does. On the other hand, the average field strength in an AC field is zero. Since changing the applied bias from DC to AC eliminated the electrical enhancement, the conclusion must be that the field-direction-dependent Coulomb force is the electrical force causing the electrical enhancement.

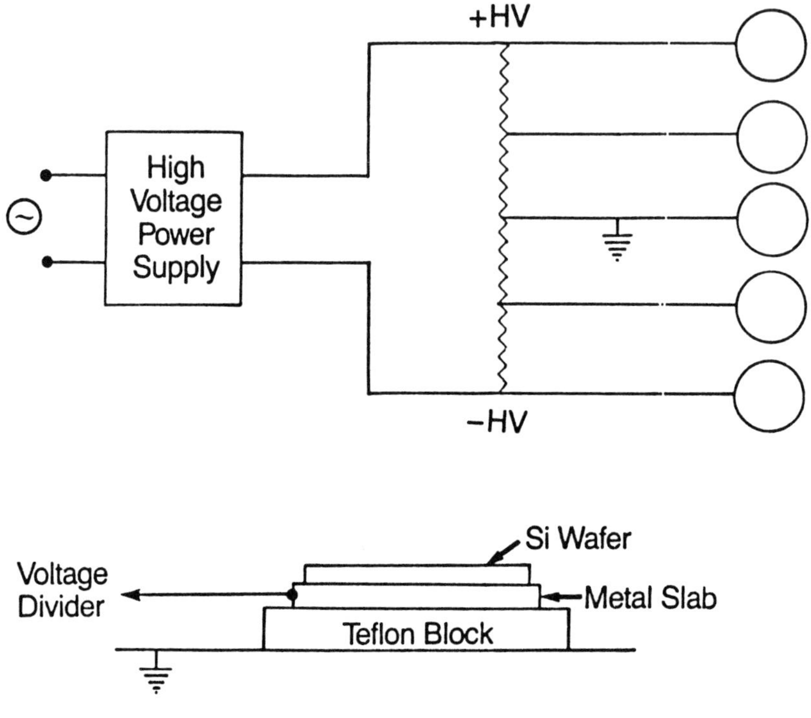

Figure 3. Bias network for controlling wafer potential.

When the aerosol stream bypassed the Kr^{85} neutralizer, the average electrical charge on the challenge aerosol particle was increased and the deposition velocities observed also increased, as shown in Fig. 5. Although the magnitude of the increase in aerosol particle charge was not measured, the factor of ten increase in deposition velocity at all field strengths agrees with the increase predicted in Fig. 4 for a 10x increase in average particle charge.

The overriding conclusion emerging from these analyses and experiments is that achieving minimum particle deposition on wafers requires minimizing the electrical charge on both the aerosol particles and the wafers. Allowing either one to rise above minimum levels increases particle deposition, and allowing both to do so is doubly degrading.

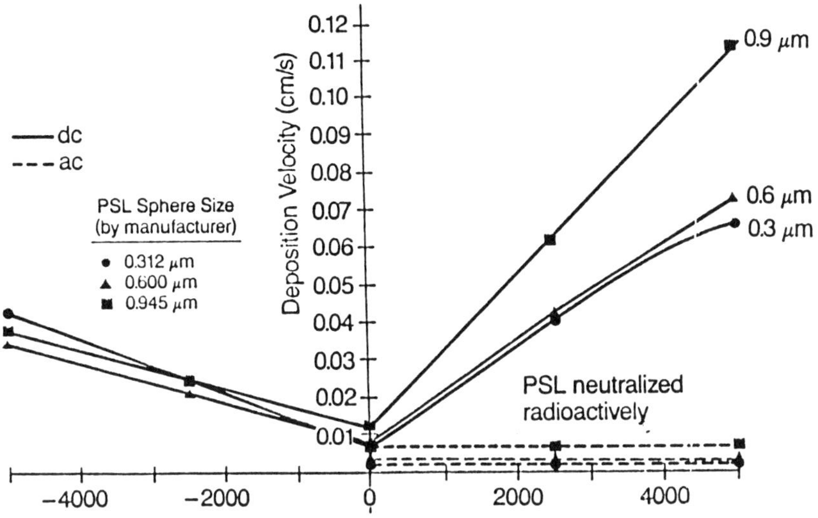

Figure 4. Mean values of deposition velocity as a function of wafer surface potential.

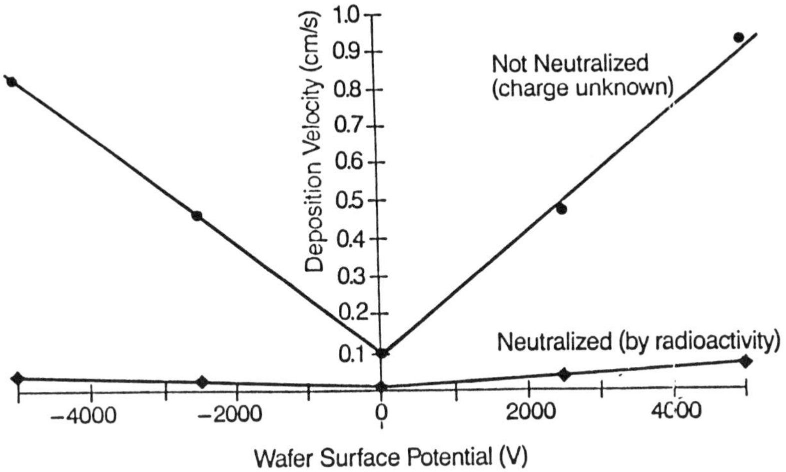

Figure 5. Increase in deposition velocity brought about by increased electrical charge on 0.6 μm polystyrene latex spheres.

2.1 Thermal Effects

Electrical charges in an aerosol deposition system increase aerosol particle deposition velocity. Thermal effects, on the other hand, can reduce particle deposition velocity, as indicated in Fig. 6. Two independent effects can explain this interaction:

1. Thermophoresis: a force exerted on a particle in the direction of decreasing temperature.
2. Thermal convection: a drag force created on a particle by thermally induced buoyancy.

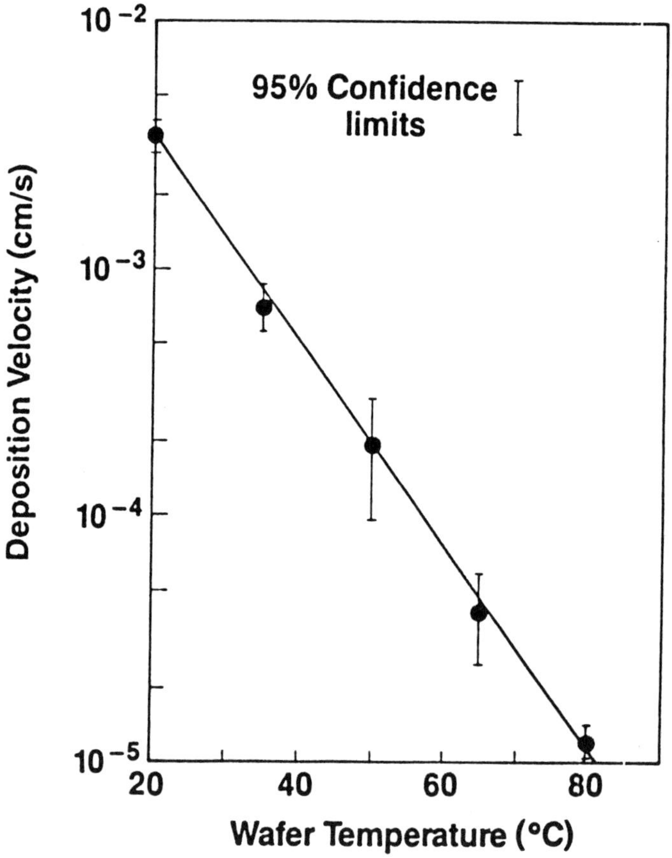

Figure 6. Effect of wafer temperature on the particle deposition velocity of neutralized 0.6 μm polystyrene latex spheres.

As can be seen in Fig. 6, only modest increases in wafer temperature (30°C) can produce reductions on the order of 10x in V_D. This shielding effect has been recognized for over a century, the particle-free region surrounding a hot surface held in a smoke flow having been observed in the 19th century. In this century, wafers at 50°C in semiconductor operations have been shown to collect fewer particles than wafers at room temperature (Table 4).

Table 4. Decrease in Particle Deposition Brought About by Heating Wafers

	Particles Added to Horizontal Wafer (cm^{-2})	
Wafer Exposure	Room Temperature (20°C)	Heated (50°C)
Cabinet top, photolithography bay, 24 h	1.44 ± 0.32	0.46 ± 0.24
Electron beam deposition chamber during vacuum pumpdown and vent	5.66 ± 2.73	0.61 ± 0.42

Unfortunately, the size range over which thermal effects are effective is narrow, as shown by the calculated curves in Fig. 7. In addition, their protective action can be easily reduced or eliminated by the electrical forces previously discussed. Figure 7 plots a normalized particle flux for five cases:

1. A standard diffusion-gravitation curve
2. A gravitation-diffusion deposition curve with the addition of thermal forces (wafer 30°C above ambient)
3. A gravitation-diffusion curve with the addition of electrical forces
4. The same as #3 with the addition of the same thermal forces as in #2
5. The same as #4 except that the magnitude of the electrical forces has been increased

Curve 4 shows that modest electrical forces are predicted to reduce the thermal effect, and Curve 5 shows that strong electrical forces are predicted to eliminate the effect of thermal forces altogether. Data in Fig. 8 confirm these predictions. At zero wafer voltage, the decrease in V_D of a wafer heated 30°C above ambient was an order of magnitude smaller than that for an unheated wafer. As wafer voltage increased, the reduction in V_D became smaller until at a wafer voltage of about 2000 V the thermal effect had disappeared altogether.

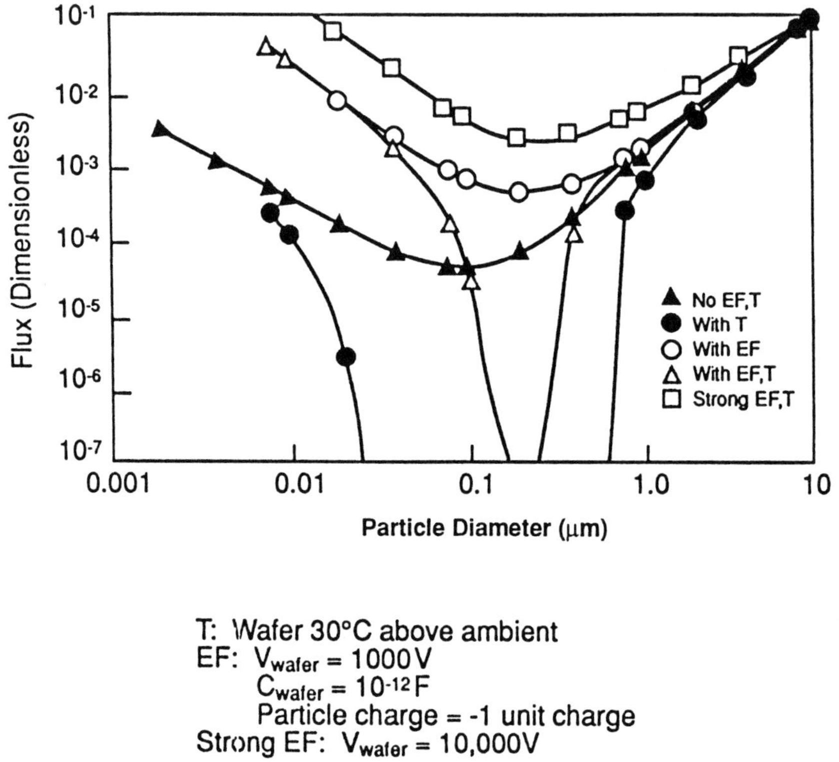

Figure 7. Dimensionless particle flux for diffusion, convection, settling, electrical forces (EF) and thermophoresis (T) (7). *(With permission T. W. Peterson, F. Stratmann, and H. Fissan. Copyright 1989, Pergamon Press)*

Thus, while thermal effects can reduce V_D, their presence does not remove the need to minimize electrical effects. That requirement remains a cardinal rule for minimizing particle deposition velocity.

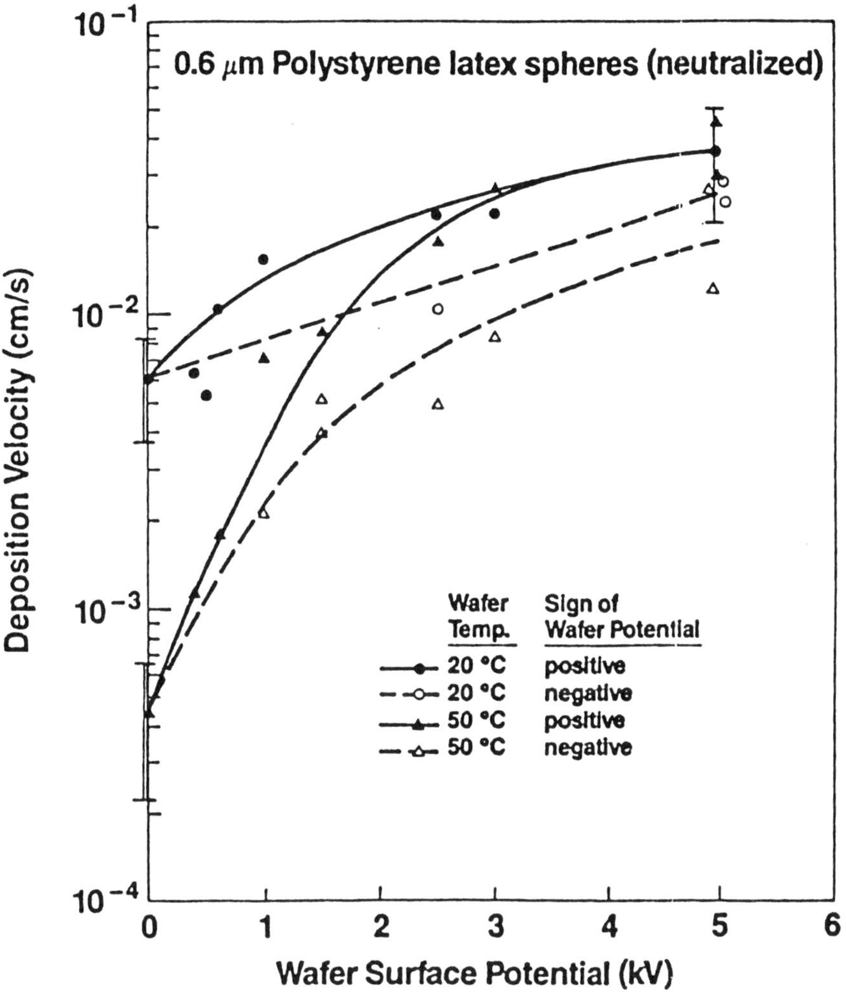

Figure 8. Electrically enhanced deposition velocity of heated wafers.

3.0 PARTICLE DEPOSITION FROM A LIQUID BATH

Wafer cleaning remains primarily a wet chemical process in spite of the growing interest in dry cleaning methods reported elsewhere in this book. Thus, liquid baths, especially aqueous baths, remain an important environment for wafers during processing. In spite of the cleaning role often assigned to liquid baths, such baths can also constitute a source of particulate contamination. Cleaning solutions, designed for metallic ion removal or oxide removal, have sometimes been reported to increase particulate contamination. Even those liquid baths or rinses designed to remove particles reach limiting values below which the number of particles remaining on the surface does not fall. Either these last few remaining particles adhere unusually well or the number of particles removed is counterbalanced by the number depositing. That the pattern of remaining particles continually changes during successive wet cleans (but does not decrease) favors the latter description, although particle motion along the surface cannot be ruled out.

3.1 Comparison of Hydrosol and Aerosol Particle Deposition Mechanisms

As described earlier, the primary mechanisms, whereby aerosol particles deposit on wafers are gravitational settling, transport by diffusion and electrical attraction. Particle deposition was measured in terms of an aerosol particle deposition velocity which implies a deposition process constant in time.

In liquid baths, the first two mechanisms prove much less important than for aerosol deposition. Gravitational settling is resisted by the buoyancy of the liquid so that the density factor ρ_p in Eq. (2) should be replaced by $\rho_p - \rho_f$ where ρ_f is the fluid density. Liquid densities are much greater than that of air and are generally not negligible with respect to particle density. Fluid density sometimes exceeds particle density so that the particles float on the liquid surface rather than settle onto the surface of a submersed wafer. Even when ρ_p exceeds ρ_f, surface tension forces can hold particles at the liquid surface.

Particles immersed in a liquid bath do diffuse but with significantly lower values of D than in air. Thus, particle diffusion across a boundary layer in a liquid bath is expected to have lower magnitude than in air—doubly so because the higher densities of liquids probably produce wider concentration boundary layers across which a concentration gradient must drive diffusion deposition. Nonetheless, deposition behavior similar to that observed in aerosol systems can also be observed sometimes in hydrosol systems such as illustrated in Fig. 9. Plots of surface particles as a function of time do not always exhibit the linear behavior illustrated in Fig. 9, but when they do, a particle deposition velocity can be calculated. The deposition data in Fig. 9 correspond to a V_D value of 1 - 2 x 10^{-6} cm/s, a full three orders of magnitude lower than the values reported in air (exclusive of thermal effects).

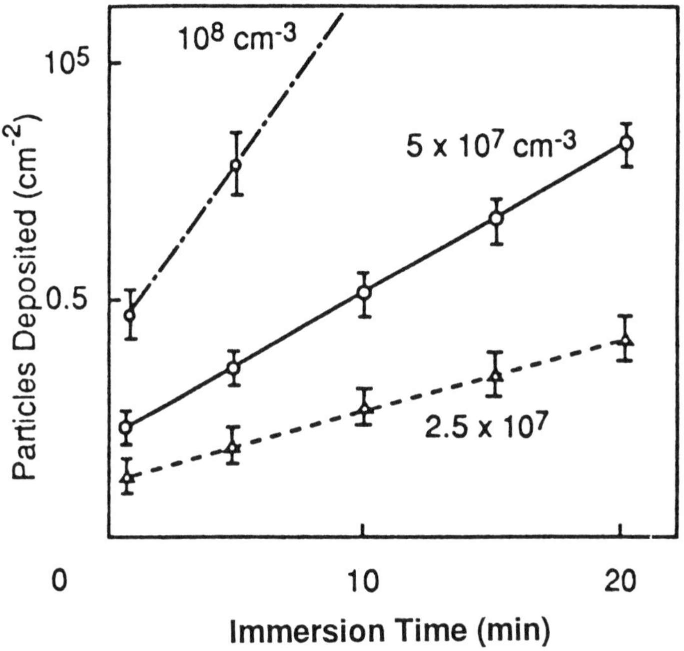

Figure 9. Deposition of 0.9 µm PSL spheres on Si wafers immersed in DI H_2O (8). *(With permission A. Saito, K. Ohta, M. Watanabe, and H. Oka. Copyright 1991, Canon Communications, Inc.)*

Electrical effects are important deposition mechanisms for both aerosol and hydrosol systems, but the particle charging mechanisms are different, and a different vocabulary is used to describe the electrical interactions in the two systems. High electric field strengths measured in thousands of volts/cm and extending over large ranges often exist in aerosol systems. In liquids, the electric fields act over much shorter ranges but can have even more significant deposition effects.

Dependencies unique to the hydrosol problem include those associated with the liquid-air interface that does not exist in aerosol systems. In addition, there is the role of the wafer surface that itself acquires a charge when immersed in a liquid.

3.2 Concepts of Colloid Chemistry

Finely divided particles dispersed or suspended in liquid systems are called colloids. Colloids typically acquire an electrical charge when immersed in a liquid medium (9). Most often these charges result from selective ion adsorption from the liquid, but colloid charging could also result from ionization of surface groups or from electronic charge exchange across the liquid-colloid interface. Electronic charge exchange across an interface, an important bulk semiconductor effect, is relatively unimportant in colloid charging. Ionic effects almost always dominate.

Charge on a colloid is often measured in terms of a zeta potential, the potential at the shear plane between the colloid and the liquid. A typical charge distribution surrounding a colloid is sketched in Fig. 10. This sketch shows a colloid that acquires a net negative charge as a result of adsorption on its surface of, say, OH^-. The now negatively charged colloid attracts an excess of positive ions to the liquid layers adjacent to the colloid to counterbalance the negative colloid charge. Electrical potential at the colloid surface thus is a maximum and decreases through the surrounding layers of excess positive ions until reaching the value of the bulk solution where equilibrium concentrations of positive and negative ions exist.

Relative motion of the colloid with respect to the solution—either the colloid moves through the liquid or liquid flows past the colloid—divides the positive ion sheath into two regions: *(i)* the ion region near the charged colloid that moves with the colloid; and *(ii)* the diffuse region remote from the colloid which moves with the bulk liquid. The boundary between these two regions is the shear plane, and the potential at this shear plane is called the zeta potential.

Zeta potential is important because it is the simplest potential of the colloid system to measure. By applying an electric field across a region of the liquid and measuring the resulting drift velocity of the colloid, the colloid mobility and potential—the zeta potential—can be determined (11). The zeta potentials of some common colloids in deionized water are listed in Table 5.

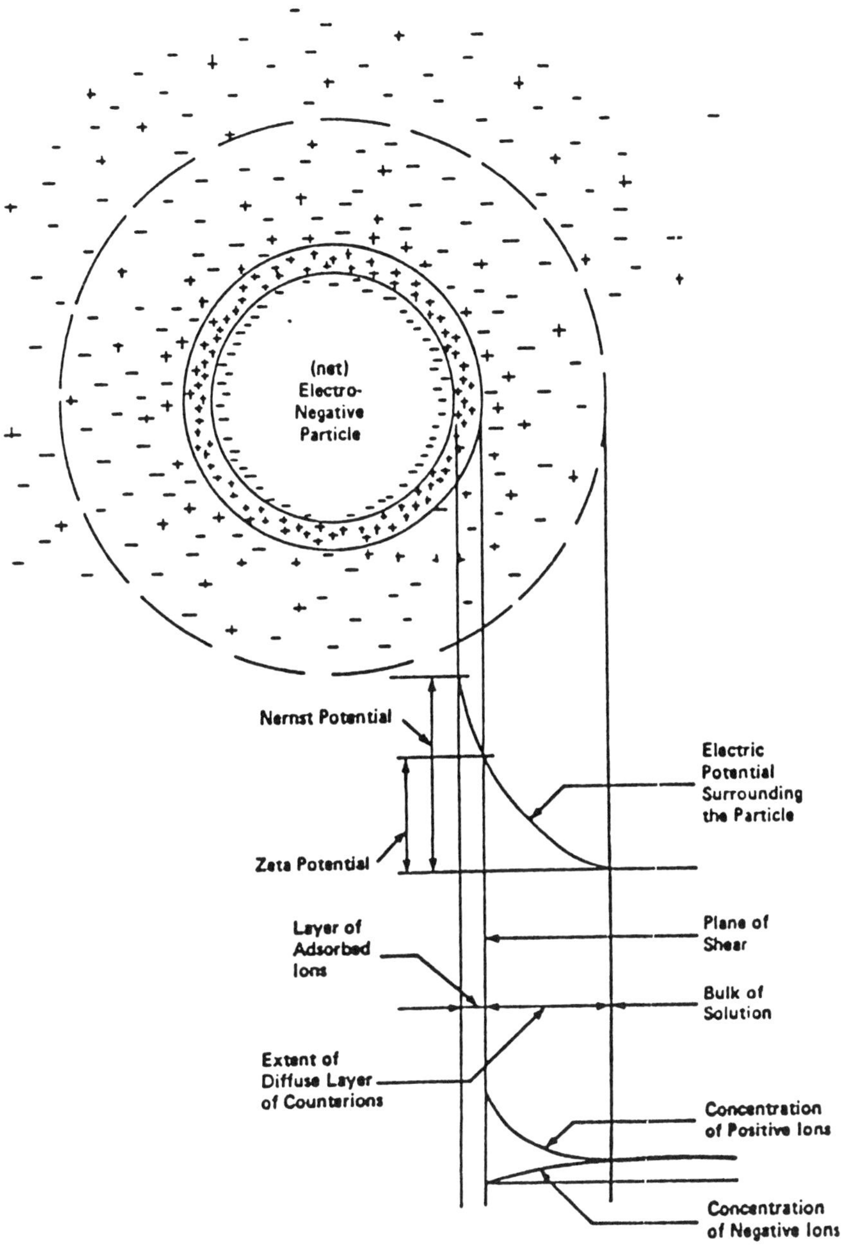

Figure 10. Electrostatic double layer around a particle (10). *(With permission M. B. Ranade. Copyright 1987, Particle Technology, Inc.)*

Table 5. Zeta Potential of Test Colloids

Particle Type	ζ(mv)*
2 µm glass beads	38.8 ± 1.7
1 µm PSL spheres	29.7 ± 5.1
Silicon dust	
coarse supply (type unknown)	24.6 ± 5.6
finely ground n-type	22.7 ± 4.3
finely ground p-type	22.1 ± 5.6

*Measured at pH ~6 - 7

For measuring the zeta potential of wafer surfaces, the streaming potential technique is more practical. In this configuration, the potential created by forcing liquid flow through a narrow channel created by, say, two wafers held parallel and face-to-face, as in Raghavan's experiments, is measured (12). The liquid flow sweeps the excess positive charge of the diffuse region through the channel along with the flow. This charge motion creates a current and a potential drop across the channel that can be measured and related to the zeta potential of the stationary surfaces (11).

Zeta potential generally depends on the ionic concentration of the liquid in which the surface is immersed, although the exact interaction is surface and ion specific. In aqueous systems, the zeta potential typically varies with pH as shown in Fig. 11. Increasing the concentration of OH^- makes the zeta potential more negative, and vice versa. The pH at which the zeta potential crosses zero is called the isoelectric point (i.e.p.). Table 6 lists the i.e.p.'s of some common colloidal particles.

Colloid surface charge, as opposed to the zeta potential, can be measured directly by determining the number of ions removed from a solution. The technique is to plot the pH of a test solution (KNO_3 in Fig. 12) as a function of number of ionic charges added (NaOH in Fig. 12). In a second titration, the surface or colloid whose point of zero charge (p.z.c.) is being measured is placed in the KNO_3 solution before adding the OH^-.

The additional number of ionic charges required to be added to achieve a given pH is a measure of the number of charges adsorbed by the surface. This number of surface charges is plotted against pH for at least three different ionic strengths of test solutions (Fig. 13). The intersection of the three curves defines the p.z.c., and the relative scale of surface charge in Fig. 13 can be adjusted to read zero at the intersection. Failure of the curves to intersect at the same point implies that changing the ionic strength of the test solution changes the zeta potential of the surface being analyzed; i.e., the ions of the test solution are themselves potential determining ions (p.d.i.'s). In that event another test solution must be selected for the measurement.

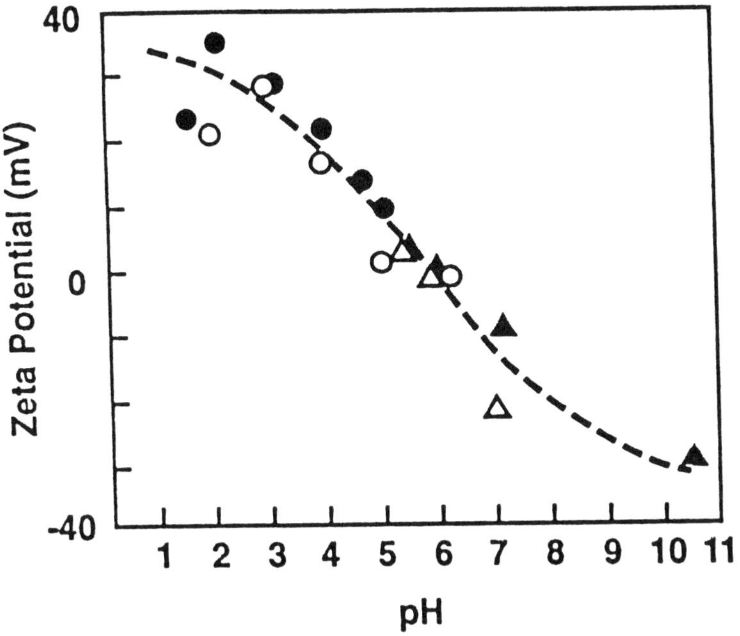

Figure 11. Zeta potential of Fe_2O_3 particles in solution of various pH (13). *(With permission M. Itano, et al. Copyright 1991, Plenum Press)*

Table 6. Isoelectric Points of Some Common Metal Oxides

Metal Oxide	Isoelectric Point (i.e.p.)
SiO_2	pH 1.5 - 3.7
Fe_2O_3	5.7 - 6.9
TiO_2	6.0 - 6.7
Al_2O_3	7.4 - 9.5
ZnO	9.3 - 10.3

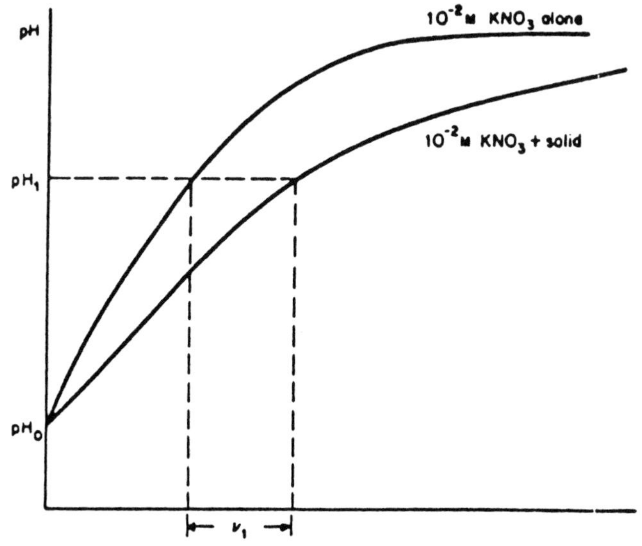

Figure 12. Measurement of surface charge by titration (11). *(With permission R.J. Hunter. Copyright 1981, Academic Press)*

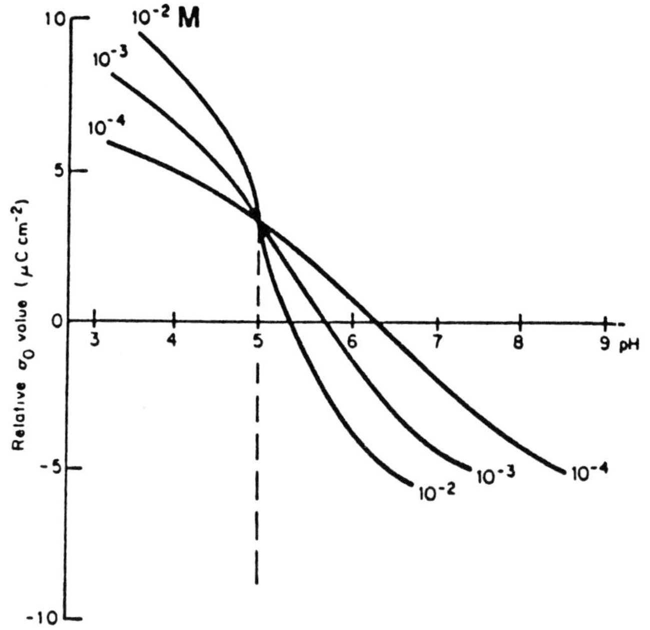

Figure 13. Measurement of point of zero charge (p.z.c.) (11). *(With permission R.J. Hunter. Copyright 1981, Academic Press)*

3.3 Zeta Potential and Particle Deposition

That colloids and surfaces in liquids become electrically charged implies that electrical forces are present that can affect the deposition process. Figure 14 supports this suggestion by showing that colloid deposition on a hydrophilic wafer surface exhibits a dependence on pH similar to zeta potential.

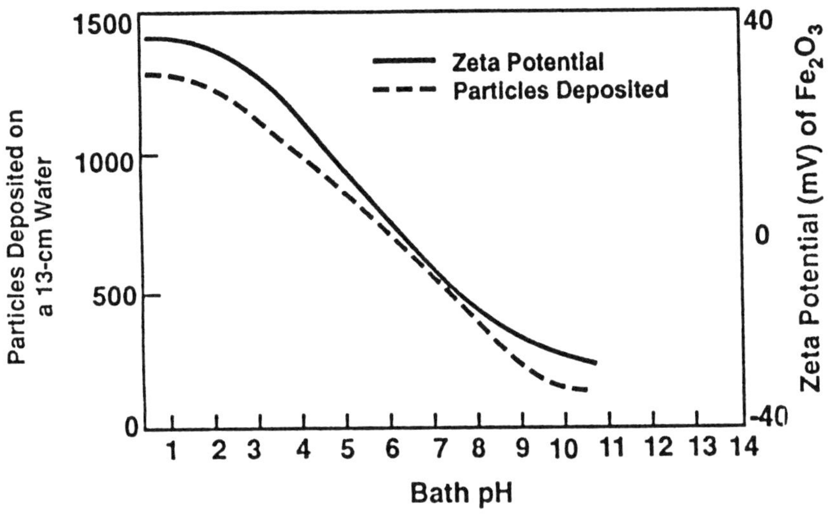

Figure 14. The influence of bath pH on particle deposition and zeta potential (13). *(With permission M. Itano, et al. Copyright 1991, Plenum Press)*

The model invoked is sketched in Fig. 15. A negatively charged colloid is repelled from a negatively charged wafer surface by the charges adsorbed on the surfaces of each. If these repulsive forces dominate the interaction between the colloid and the surface, no contact between the two takes place, and no colloid deposition can occur. This interaction is called electrostatic double layer repulsion (EDR).

When the zeta potentials of the colloid and the wafer are of opposite sign of the wafer, as sketched in Fig. 16, no repulsive barrier to colloid deposition exists so that significant wafer contamination should occur in this system during wafer immersion in the bath.

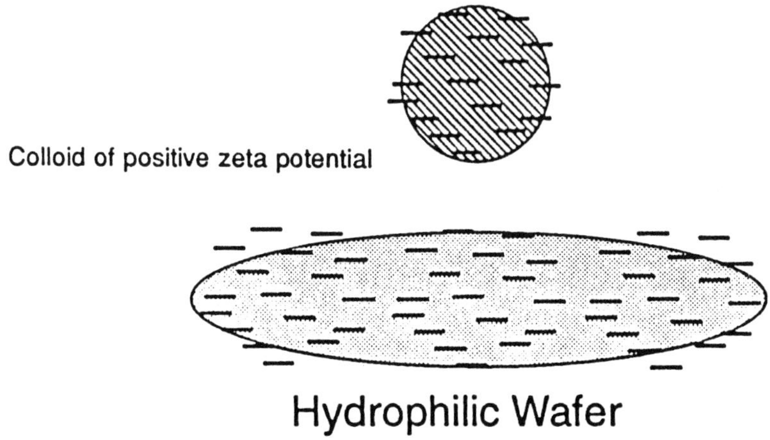

Figure 15. Electrostatic double layer repulsion (EDR).

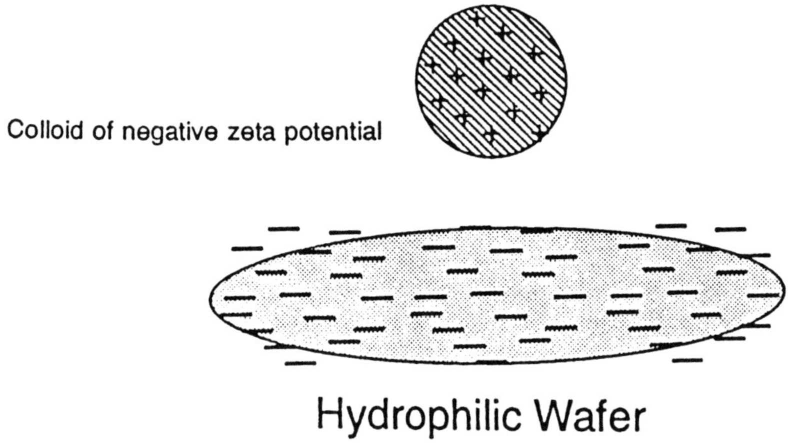

Figure 16. Barrier-less deposition condition.

These predictions apply to the interactions between a colloid and a wafer surface while the wafer is immersed in the liquid bath. This exposure is just one phase of the total exposure between a wafer and a liquid bath.

Colloid deposition can occur during bath entry and exit as well as during immersion. Experiments to isolate the deposition contributions during each of these three exposure phases show that the withdrawal step is the most important deposition phase for systems consisting of colloids with negative zeta potential and hydrophilic silicon wafers that also exhibit negative zeta potentials in aqueous baths of pH >2.5 (12). In experiments comparing colloid deposition during a total immersion time of 3t achieved by either: *(i)* a one-time exposure of 3t, or *(ii)* three successive exposures of t time each in which the wafer was immersed for time t, withdrawn, and dried three times, deposition has been shown to correlate with the number of withdrawals not the total time of immersion (14); that is, colloid deposition in the second case was three times that of the first. In addition, varying the time of the single exposure made no difference on the deposition (Fig. 17).

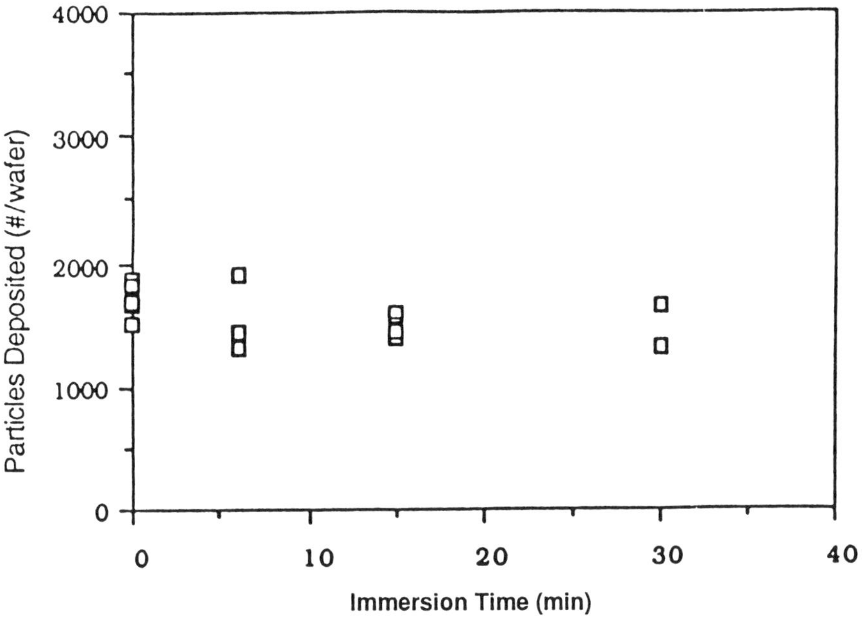

Figure 17. Time dependence of negatively charged colloid deposition (14). *(With permission D. J. Riley and R. G. Carbonell. Copyright 1990, Canon Communications, Inc.)*

In contrast, when the zeta potential of the colloid was positive, deposition on a hydrophilic wafer in deionized water showed the time dependence characteristic of bulk convective diffusion (Fig. 18). The data of Fig. 18 were collected under conditions identical to those of Fig. 17, including colloid composition except that the zeta potential of the colloid was made positive by the addition of appropriate surface groups to the colloid surface. The time dependence shown in Fig. 18 is similar to that previously presented in Fig. 9 and described as deposition by convective diffusion. This mechanism refers to the continuous transport of colloids through the bulk solution and across the boundary layer at the wafer surface.

The relationship between deposition on a hydrophilic wafer and the zeta potential of a colloid is summarized in Fig. 19. This plot suggests that electrostatic double-layer repulsion was an effective mechanism for shielding these hydrophilic wafers from colloid deposition as long as the zeta potential of the colloid was below about -10 mV. When the zeta potential of the colloid became more positive than -10 mV, the repulsive forces became too small to retard deposition.

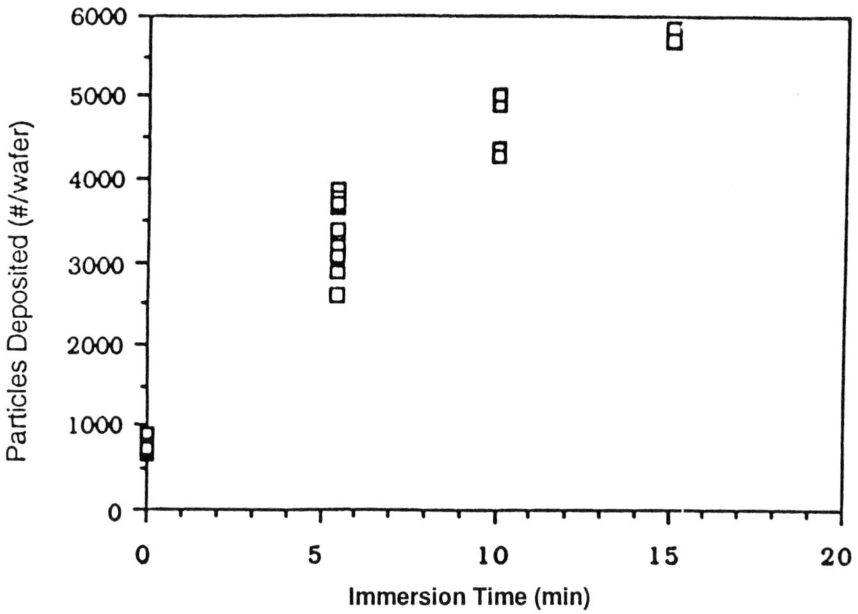

Figure 18. Time dependence of positively charged particle deposition (14). *(With permission D. J. Riley and R. G. Carbonell. Copyright 1990, Canon Communications, Inc.)*

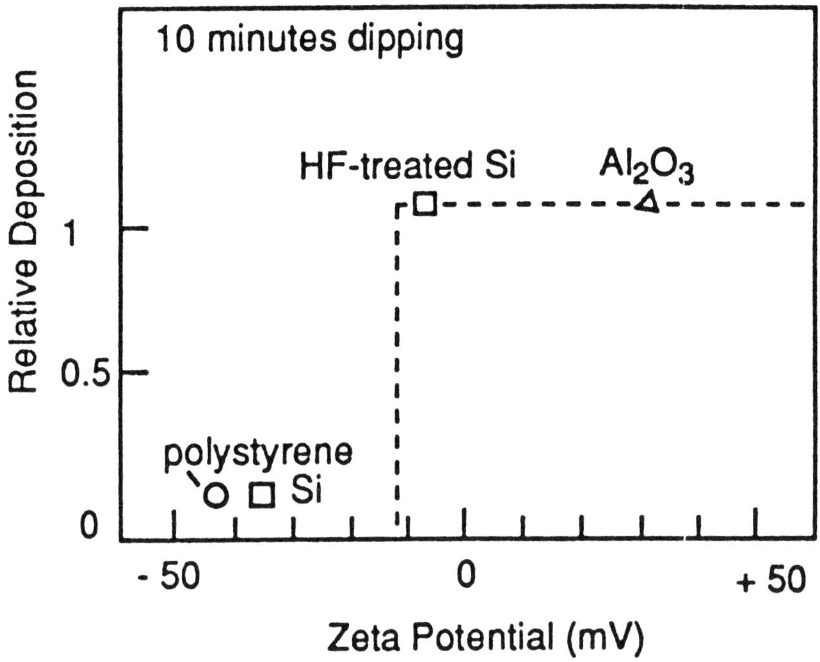

Figure 19. Correlation of zeta potential with colloid deposition on wafers in aqueous solutions of low ionic strength (8). *(With permission A. Saito, K. Ohta, M. Watanabe, and H. Oka. Copyright 1991, Canon Communications, Inc.)*

Note that deposition still occurred even when effective EDR was present; wafers still collected some colloids as a result of immersion and withdrawal from a liquid bath. As noted previously, the source of these colloids is associated with wafer withdrawal from the bath. With hydrophilic wafers a thin film of liquid adheres to the wafer as it is withdrawn (15). This film contains colloids from the bath that adhere to the surface during the drying cycle. Once the colloids contact the surface, they adhere strongly and become difficult to remove. If, however, the wafer can be rinsed in water of low colloid concentration prior to drying and without allowing any significant drying to occur, the concentration of surface particles remaining is that characteristic of the rinse water not that of the liquid bath. Once drying occurs, however, the surface concentration is unlikely to be decreased by subsequent clean water rinses (15).

3.4 Effect of Ionic Strength on Electric Double Layer Repulsion

The width of the diffuse layer depicted in Fig. 10 depends upon the ionic strength of the liquid surrounding the colloid. At high ionic concentration, the range of the region of excess positive charge is short because the colloid charge can be balanced by only a small volume of liquid.

The simplest mathematical model of the potential in the diffuse layer is of the exponential form:

Eq. (11) $$\psi = \psi_o e^{\kappa x}$$

where:

ψ = potential [V]

ψ_o = colloid surface potential [V]

κ = the Debye-Hückel parameter [l^1]

x = distance measured perpendicular to the colloid surface [l]

The Debye-Hückel parameter has the dimensions of reciprocal length, and $1/\kappa$ represents the distance from the colloid surface for the potential to decrease to 1/e its surface value. The width of the diffuse layer is often approximated as $1/\kappa$.

Ionic strength determines as follows:

Eq. (12) $$\kappa = (e^2 I)^{1/2}/\varepsilon kT$$

where:

I = ionic strength [l^{-3}]

ε = dielectric constant of the liquid [t^2q^2/ml^3]

The implication of Eq. (12) is that the range of the interacting diffuse layers of a colloid and a wafer surface depends on the ionic strength of the liquid medium. In deionized water at 25°C with minimal ionic concentration, EDR forces are long range, being on the order of tenths of micrometers. At high ionic strengths, they decrease in range and can become ineffective with respect to van der Waals adhesion forces, discussed in the next section.

The form of the classical EDR equation applicable to a spherical colloid adjacent to a planar wafer is as follows (8):

Eq. (13) $$W_{EDR} = \frac{\varepsilon d_p}{8}\left[(\psi_p^2 + \psi_s^2)\ln\frac{\exp(2\kappa Z_o)-1}{\exp(2\kappa Z_o)} + 2\psi_p\psi_s \ln\frac{\exp(\kappa Z_o)+1}{\exp(\kappa Z_o)-1}\right]$$

where:

ψ_p, ψ_s = the surface potential of the particle, colloid (the Nernst potential, Fig. 10)

Z_o = the separation distance between the wafer and the colloid.

3.5 Van der Waals Attraction

Van der Waals attraction between molecules arises from an interaction between the electron clouds surrounding each molecule. The displacement in electron distributions induced by the presence of the other molecule introduces a dipolar electric force which is always attractive. For the case of a spherical particle adjacent to a planar wafer, the force can be described in terms of the Hamaker constant (10)(16):

Eq. (14)
$$F_{Ad} = \frac{A_{132} d_p}{12 Z_o^2}$$

where:

A_{132} = the Hamaker constant of the system composed of a sphere (material 1) adjacent to a plane (material 2) in a medium (material 3) $[ml^2t^{-2}]$

Z_o = the separation between the sphere and the surface

The magnitude of the system constant depends upon the Hamaker constants of the materials making up the system according to the following rule:

Eq. (15)
$$A_{132} = A_{12} + A_{33} - A_{13} - A_{23}$$

where:

$$A_{ij} = (A_{ii}A_{jj})^{1/2}$$

and A_{ii} is the Hamaker constant of the material i.

Table 7 lists Hamaker constants of some common materials; Table 8 tabulates values of A_{132} calculated from Eq. 15 and the Table 7 values. These values show that van der Waals attraction varies with system composition and that some liquids support larger van der Waals forces than others. Particles in liquid media invariably adhere with less van der Waals force than in air. Thus, immersing a wafer in a liquid––most any liquid––is a good first step toward removing the particle and an important reason for the success of liquid baths in particle removal.

Table 7. Hamaker Constants of Selected Materials (17)(18)

	A_{11} (picoergs)
Graphite	4.70
Gold	4.55
Silicon	2.48
Glass beads	0.34
Polystyrene latex	0.65
Deionized water	0.44
Ethanol-acetone (50% v/v)	0.33
Freon-TMS (TF + 6% methanol)	0.18
Freon-TF	0.17

Table 8. A_{132} for Selected Particle/Plane Systems (picoergs)

	Deionized water	Ethanol-acetone	Freon-TMS	Freon-TF
Glass beads on silicon wafer with native oxide	-0.07	0.009	0.18	0.20
1 μm PSL spheres on silicon wafer with native oxide	0.13	0.23	0.44	0.46

4.0 DLVO THEORY

As outlined in the previous section, the two most important long-range forces between a colloid and a wafer surface are electrostatic double-layer repulsion (EDR) and van der Waals attraction, as stated by Eqs. (13) and (14), respectively. As the separation distance, Z_o, between the colloid and the wafer decreases, van der Waals attraction, exhibiting a Z_o^{-2} dependence, eventually dominates the EDR, which has a Z_o^{-1} dependence or less. This dominance of van der Waals force at low Z_o is indicated by the deep trough on the left-hand side of Fig. 20, which is a plot of the net interaction energy (van der Waals plus EDR) between a colloid and a wafer. The abscissa in this plot is separation distance, Z_o. The various curves represent different values of W_{EDR} (Eq. 13). Curve "a" illustrates an interaction dominated by EDR until low values of Z_o at which the van der Waals force always exceeds the EDR force. Curve "e" represents a pure van der Waals attraction without any EDR. Curves "b" to "d" represent net interactions intermediate between "a" and "e", characterized by reduced EDR because of lower colloid or wafer zeta potentials and/or higher bath ionic strengths. Lower zeta potentials correspond to reduced values of in Eq. (13), reducing W_{EDR}. Higher ionic strengths reduce the range over which the EDR forces operate. Van der Waals attraction, on the other hand, is independent of zeta potential and ionic strength.

The primary capture trap is that at low values of Z_o in Fig. 20. However, a secondary trap is evident in curves "b" to "d" which provides only weak trapping of colloids in the presence of EDR repulsion at separation distances that would normally correspond to stronger trapping. Increasing ionic strength and decreasing surface potential shift the curves toward the pure van der Waals case. The interactions depicted in Fig. 20 were first described independently by Derjaguin and Landau and by Verwey and Overbeek and the model described is called the DLVO model (19).

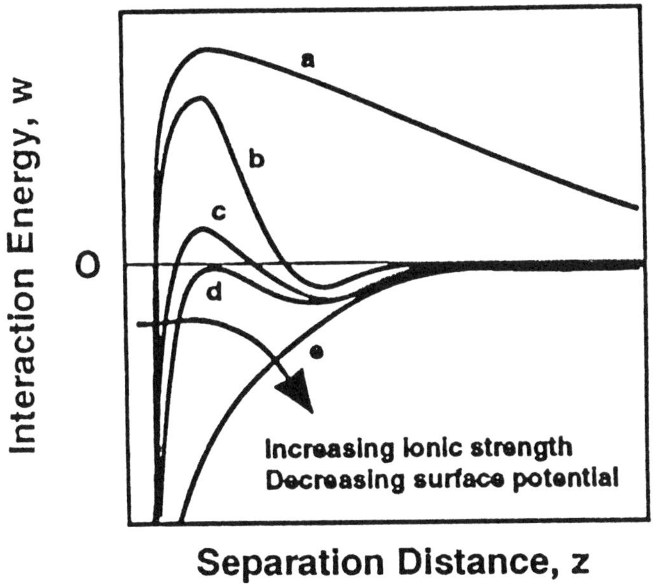

Figure 20. Types of net interaction energies between a colloid and a wafer (19). *(With permission J. N. Israelachvili. Copyright 1985, Academic Press)*

4.1 The Effect of Solution pH on Colloid Deposition

The correlation of colloid deposition and solution pH has been noted previously (Fig. 14). Changes in solution pH generally imply changes in solution ionic strength but not necessarily vice versa, as indicated in Fig. 21 (20). Here solution pH is plotted as a function of ionic strength for three different aqueous solutions. The KCl solution maintains constant pH over a wide range of ionic strength while the HCl and KOH solutions show the expected relationship between ionic strength and pH. Colloid deposition in these three solutions as a function of ionic strength is shown in Figs. 22 to 24. Each plot consists of two deposition observations: *(i)* a 0 min, corresponding to a dip in and out of the solution, and *(ii)* a 5 min immersion. When an effective EDR barrier is present, the observed colloid deposition is the same for these two exposures. Whenever the 5 min exposure produces more colloid deposition than the dip alone, the conclusion is that no EDR barrier, or a reduced barrier, exists to oppose deposition during immersion, i.e., the EDR is too narrow or too small to prevent deposition by van der Waals attraction.

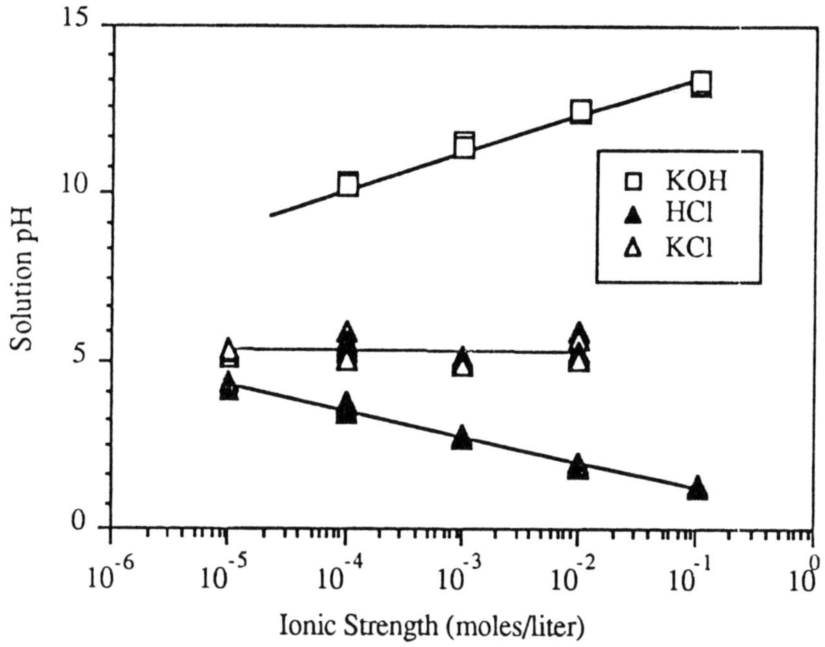

Figure 21. Control of bath properties (20). *(With permission D. J. Riley and R. G. Carbonell. Copyright 1991, Plenum Press)*

In Fig. 22, an effective EDR barrier exists as long as the ionic strength remains below 10^{-2} mol/L where the 0- and 5-min deposition data coincide. At 10^{-2} mol/L and higher ionic strengths, the width of the diffuse layer decreases and van der Waals forces dominate so that deposition can occur during immersion—the 5 min depositions exceed the 0 min depositions.

In the HCl solutions, plotted in Fig. 23, the lower pH values reduce the W_{EDR} values so that higher concentrations of colloids occur in the water film adhering to the wafer surface upon withdrawal. At low ionic strength ($<10^{-2}$ mol/L) this barrier is effective; but at ionic strength of 10^{-2} mol/L and above, the range of the EDR has decreased as in the KCl data, so that colloid deposition occurs during immersion. The primary difference between the HCl and the KCl data is the reduced barrier of the former (attributable to lower pH and zeta potentials) with respect to the latter.

Finally the KOH data of Fig. 24 shows that an effective EDR barrier exists over the entire range of ionic strengths investigated; the height of the EDR barrier is adequate to prevent deposition even at high ionic strength and the colloid deposition observed remains that characteristic of an effective barrier for all conditions investigated.

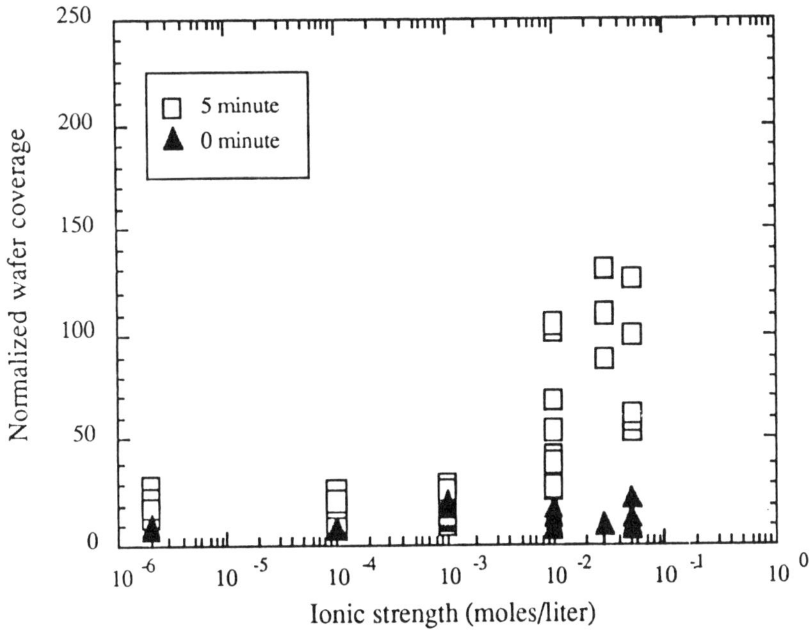

Figure 22. Deposition as a function of ionic strength controlled by KCl addition (21). *(With permission D. J. Riley and R. G. Carbonell. Copyright 1991, Plenum Press)*

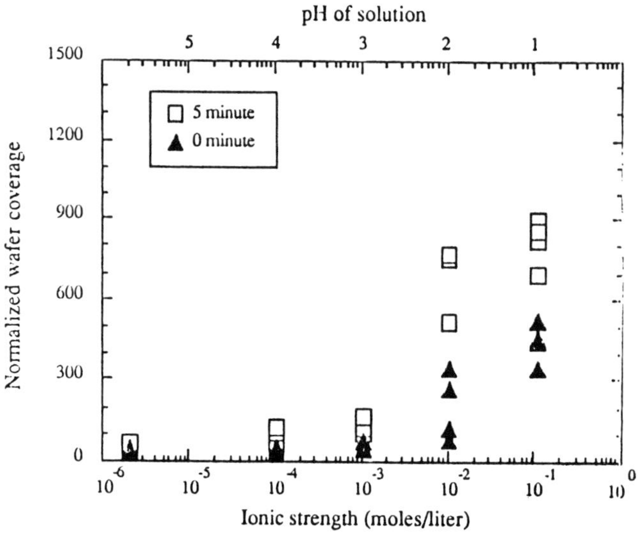

Figure 23. Deposition as a function of ionic strength controlled by HCl addition (21). *(With permission D. J. Riley and R. G. Carbonell. Copyright 1991, Plenum Press)*

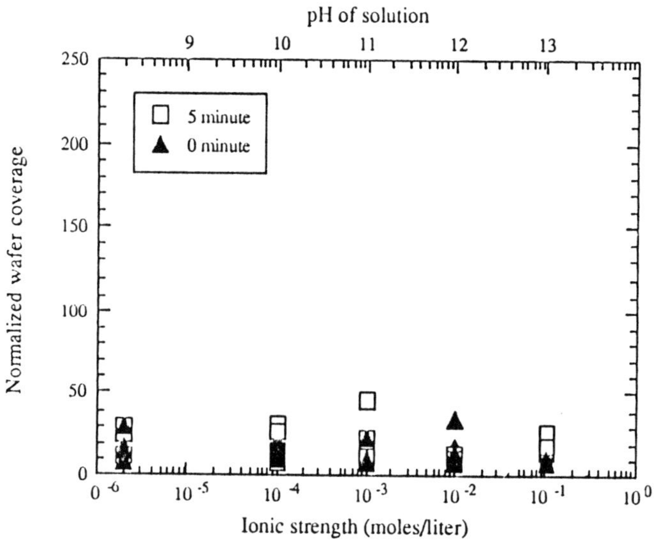

Figure 24. Particle deposition as a function of ionic strength controlled by KOH addition (21). *(With permission D. J. Riley and R. G. Carbonell. Copyright 1991, Plenum Press)*

4.2 Hydrophobic Surfaces

The discussion and data presented in previous sections have been exclusively on bare hydrophilic silicon surfaces with native oxides or chemically grown oxides. These surfaces are high energy surfaces in the sense that water droplets placed on their surface spread out, forming a low contact angle (<10°) with the surface. The water droplet adheres to the high energy surface, an action invoked to explain the particle deposition observed when a hydrophilic wafer is removed from an aqueous bath.

Hydrophobic silicon surfaces have very different properties and behave quite differently. A silicon surface freshly etched in hydrofluoric acid is a low energy surface in the sense that a water droplet placed on its surface beads up rather than spreads, exhibiting a high contact angle (>60°) with the wafer surface. Water does not adhere to a hydrophobic surface and the adhering water film mechanism does not apply to hydrophobic surfaces. The absence of this mechanism means that wafer withdrawal should not dominate particle deposition on a hydrophobic wafer and experiments that isolate deposition times confirm this conclusion. Nonetheless, common experience in laboratory wet operations has shown hydrophobic surfaces to be more heavily contaminated with particles than hydrophilic surfaces, a widely recognized penalty of HF-last cleaning recipes. When and how do these colloids deposit on the surface?

One candidate explanation is that the zeta potential of a freshly HF-etched wafer is positive (or at least less negative than a hydrophilic surface) so that colloid deposition takes place during immersion (22). This explanation has proven unsatisfactory in that regions of HF-etched wafers have been shown to behave as though negatively charged, just like the hydrophilic discussed previously, and, because of the absence of particle deposition from an adhering water film, have emerged from contaminated baths with fewer particles than hydrophilic wafers exposed to the same bath (23). However, other regions of the same hydrophobic wafer surfaces, characterized by Auger electron spectroscopy as being covered with hydrocarbon films (~100Å thick), appeared saturated with scattering centers to the laser scanner. Such hydrocarbon films were far fewer and thinner (~30Å) on the hydrophilic wafers but could never be eliminated from hydrophobic surfaces after water rinsing. This sensitivity to hydrocarbon contamination seems to be related to the bad reputation HF-last cleans have acquired rather than an electrostatic difference in zeta potential. The hydrocarbon presence seems to depend on interfacial interactions and surface exposure to liquid/gas interfaces. It requires further study.

4.3 EDR Effects in the Filtration of Colloids

The same DLVO model that explained the interaction between colloids and immersed wafers also applies to the interactions between particles or colloids and the fibers or collection sites of liquid-based filters. The interaction problem is the same except that the wafer is replaced by a fiber and, of course, with a filter the goal is to maximize particle capture rather than minimize particle capture as with a wafer.

Classical mechanisms of particle capture by a fiber are depicted in Fig. 25. In aerosol systems, diffusion and interception produce grade collection efficiencies that have a minimum in the vicinity of about 0.1 μm. The exact size at which the collection efficiency is the lowest is called the most penetrating particle size (MPPS). At particle sizes greater and smaller than the MPPS, collection efficiency is higher. The performance of both fibrous and membrane filters can be adequately modeled by these mechanisms.

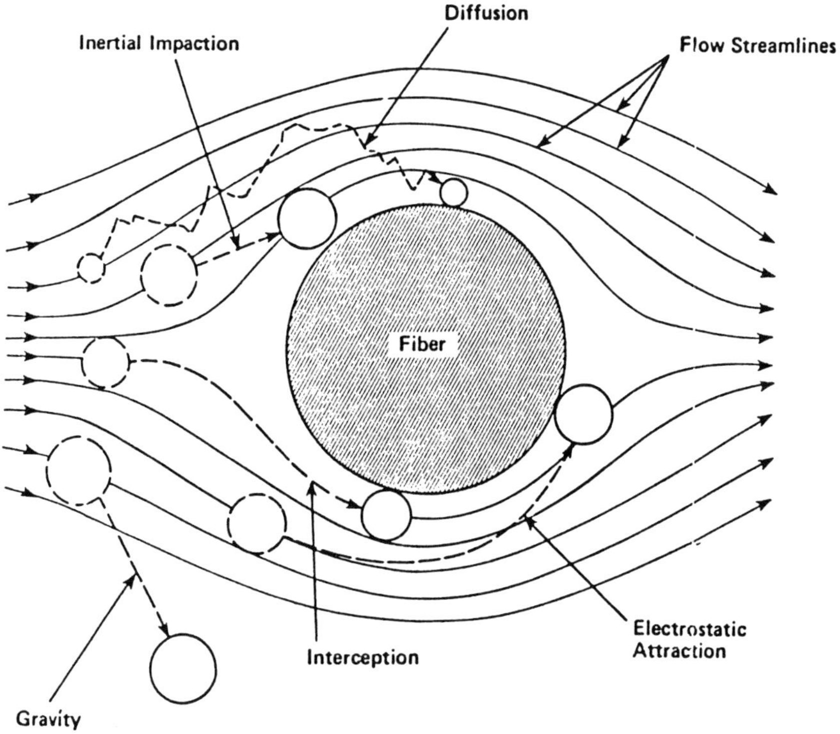

Figure 25. Classical mechanisms of particle capture.

In hydrosol systems, however, the presence of EDR can thwart these capture mechanisms even as it does in wafer baths. In deionized water systems, typically of very low ionic strength, the range of the EDR forces is large and their consequences most pronounced. Filters that are predicted to exhibit high collection efficiency over the entire range of typical colloid sizes often perform instead as sieves, capturing only those colloids too large to fit through the filter pores. Negatively charged colloids smaller than the pore size penetrate the filter in stark contrast to the predictions of the classical filtration models and observed performance in aerosol systems.

Figure 26 contrasts predicted performance and observed performance for a filter in deionized water. The performance predicted from the filter structure and the capture mechanisms of interception and diffusion is indicated by the solid curves on the left of the graph. The observed penetration characteristics are shown as data points in the upper right of the graph. These observed characteristics match up with the performance predicted by the sieving capture mechanism: negatively charged colloids smaller than the pore openings penetrate the filter; the filter captures only those colloids too big to fit through its pores (sieving). The reason that the observed performance falls so short of the predicted high-quality performance is that EDR prevents colloid-fiber contact for all negatively charged colloids except those trapped in pores of smaller physical dimensions. The EDR that proved so beneficial for wafer protection is a distinct liability in filter performance. The same filter that performs as a sieve in deionized water (Fig. 26) performs superbly as an aerosol filter where the diffusion and interception mechanisms operate without interference from EDR.

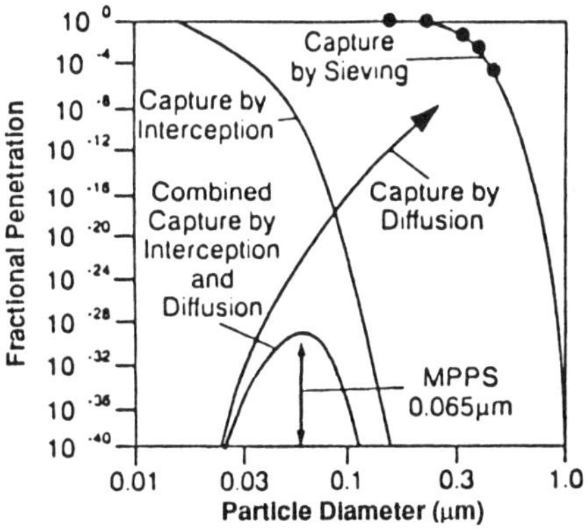

Figure 26. Degradation of predicted filter performance attributable to double layer repulsion (24). *(With permission D. C. Grant, B. Y. H. Liu, W. G. Fisher, and R. A. Bowling. Copyright 1989, Institute of Environmental Sciences)*

A technique suggested by W. Kern (private communication) for minimizing this shortcoming would be to increase the ionic strength of the DI H_2O by, say, saturating it with CO_2. This action should decrease the range of the EDR forces as described previously and allow the diffusion/interception mechanisms to occur. No experimental verification of this approach has yet been carried out.

The alternative approach is to alter the charge on the filter fibers, making them positively charged and hence attractive to negatively charged colloids. This approach has been tried and is commercially available from various filter manufacturers. Additional discussion of their performance appears in Ch. 9 by Menon and Donovan.

5.0 PARTICLE ADHESION

Particle deposition on a wafer has been equated to particle contact with the wafer. For sure, particle contact is a necessary condition for particle capture and indeed particle contact can be defined as reduction of the separation distance between the particle and the surface to the range of the primary van der Waals trap depicted in Fig. 20. At these distances the van der Waals attractive force is binding and the particle is captured. Other forces, such as the EDR force, also contribute to the net force between the particle and the wafer but the van der Waals force is universal and dominating.

The magnitude of the van der Waals force is such that particle elastic or even plastic flattening often alters the contact area between the sphere and the plane wafer surface from the no-deformation configuration assumed in Eq. 14. This deformation can be primarily of the substrate as sketched in Fig. 27, or, when the substrate hardness is greater than that of the particle, primarily of the particle (Fig. 28). In each instance the effect is to modify the area of contact between the particle and the surface, adding a term to the van der Waals adhesive force equation that depends on contact area (16)(24) and making the sphere-planar surface configuration look more like that of two flat plates (26).

Figure 27a explicitly depicts deformation created by a tensile force during particle approach to the surface. The meniscus-like contact regions created by such tensile force (Figs. 27c and 27d) have been observed experimentally (27) and tensile forces are now part of contemporary models of particle-sphere deformation in addition to the forces that compress either the particle or the substrate.

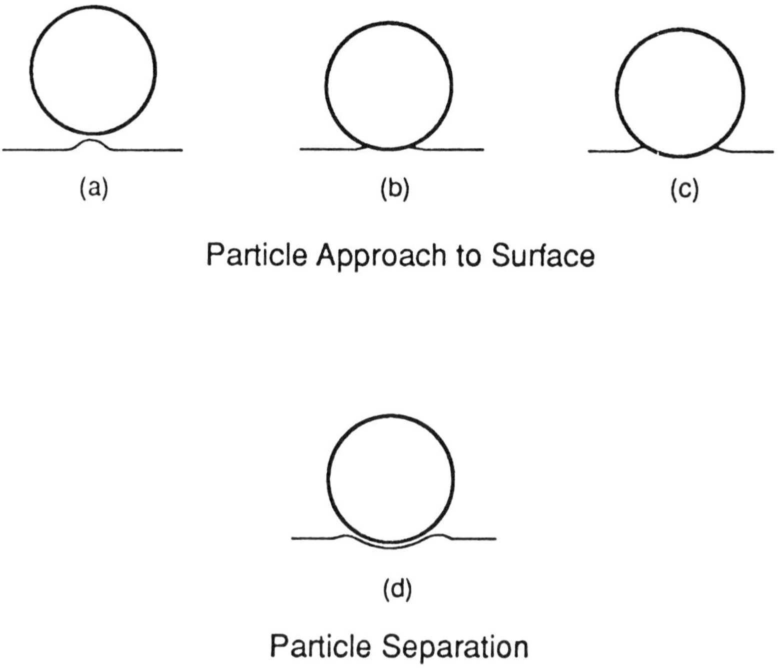

Figure 27. Deformation caused by van der Waals attraction between a sphere and a surface (25). *(With permission H. Krupp. Copyright 1967, Elsevier Science Publishers)*

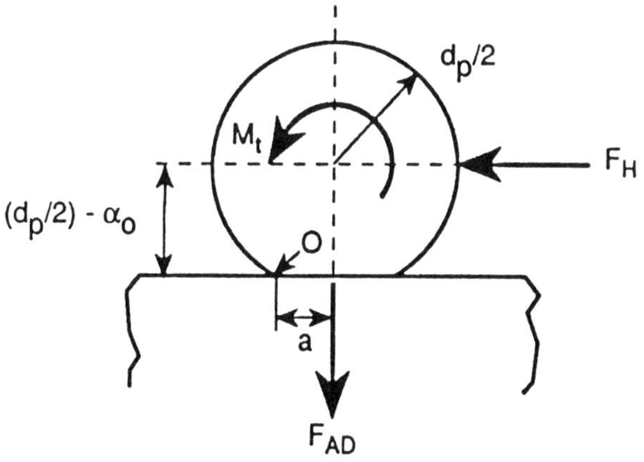

Figure 28. Moment model of particle rotation detachment (26). *(With permission C. J. Tsai, D. Y. H. Piu, and B. Y. H. Liu. Copyright 1991, University of Minnesota)*

Measuring particle-surface adhesion forces experimentally generally means measuring the force required to remove a particle from the surface. A number of techniques have been used to make these measurements, such as centrifuging, vibration, and fluid drag (16). These same techniques are used in wafer rinsing and cleaning and are discussed further in Ch. 9 by Menon and Donovan as spinning (centrifugal force), ultrasonics/megasonics (vibration), and spraying (fluid drag). The force equations used to deduce particle adhesion forces (F_{AD} in Fig. 28) are presented here.

Relating applied force to particle adhesion requires understanding of the mechanisms of particle removal. Three primary mechanisms of particle removal are: *(i)* lift-off, *(ii)* sliding, and *(iii)* rotation (28). Lift-off corresponds to particle motion perpendicular to the wafer surface and occurs when the removal force perpendicular to the surface exceeds F_{AD}. Such forces have been applied using a centrifuge in which the test surface is mounted perpendicular to the plane of centrifugal rotation. The force exerted on the attached particle by the centrifugal motion is given by Visser (16):

Eq. (16) $$F_c = \frac{\pi d_p^3}{6}(\rho_p - \rho_f)(\omega^2 R)$$

where:

F_c = the centrifugal force on the particle [ml/t^2]

d_p = particle diameter [l]

ρ_p = particle density [m/l^3]

ρ_f = fluid density [m/l^3]

ω = angular velocity of rotation [-/t]

R = particle distance from axis of rotation [l]

The speed of rotation, ω, is the primary variable for controlling F_c.

Vibration perpendicular to the surface is also a method by which lift-off can be brought about. The equation describing the force exerted by oscillations perpendicular to the wafer is also given by Visser (16):

Eq. (17) $$F_V = \frac{2\pi^3 d_p^3 f^2 y(\rho_p - \rho_f)}{3}$$

where:

F_V = force exerted on particle by the vibratory motion

f = frequency of vibration

y = amplitude of vibration

By controlling frequency and amplitude, F_V can be varied.

In both these experiments the applied force is applied in a direction opposite F_{AD} so that the value of F_c or F_V at which the particle lifts off is equal to F_{AD}. For forces applied in directions other than opposite F_{AD} a different relationship exists and indeed a motion other than lift-off, i.e., sliding or rotating, takes place.

Consider the force F_H applied as indicated in Fig. 28. This force has no component opposite F_{AD} so lift-off (particle motion perpendicular to the wafer surface) does not occur. What can occur, however, is particle sliding (28) when

Eq. (18) $$F_H > kF_{AD}$$

where k is the static coefficient of friction between the particle and the wafer surface. The static coefficient of friction depends on the composition of the particle and the surface and is not generally known. However, for many particle-wafer combinations, $k < 1$ so that the value of F_H that must be applied for particle removal is less than F_{AD}.

One method of applying a force in this direction is to mount the wafer surface parallel to the plane of centrifugal rotation. F_c (Eq. 18) is then in the direction of F_H shown in Fig. 28. A more common technique, perhaps, for exerting an F_H on an attached particle is to create a hydrodynamic drag force by establishing a relative fluid flow parallel to the wafer surface. Equation (19) describes the force exerted on a particle by relative fluid motion (1):

Eq. (19) $$F_H = C_d \rho_f A_p V^2/2$$

where:

F_H = hydrodynamic drag force on the particle

C_d = drag coefficient [-]

A_p = particle frontal area [l^2]

V = fluid velocity with respect to the particle [l/t]

The primary variable for controlling F_H is typically V.

A second type of particle motion that can be induced by a particle force exerted parallel to the wafer surface is rotation. Rotation occurs when the moment exerted by F_H about the particle contact point (point O in Fig. 28) exceeds that exerted by F_{AD} (26):

Eq. (20) $$F_H\,[(d_p/2) - \alpha_o)] > F_{AD}a$$

Equation 20 requires knowledge of contact area or particle flattening, a value that depends on particle and surface mechanical properties. Only when both are known can it be calculated from material properties. In many examples, however, particle rotation can be caused by values of F_H that are smaller than F_{AD} (28)(29).

5.1 Other Adhesive Forces

The primary force by which particles adhere to a wafer surface have been assumed to be van der Waals forces. Other forces can be important, two of which will be mentioned here. The first is capillary adhesion to a hydrophilic surface caused by a liquid film between the particle and the surface (Fig. 29). The force of capillary adhesion is given by:

Eq. (21) $$F_{CAP} = 2\pi d_p\gamma$$

where γ is the liquid surface tension.

This force is important for hydrophilic wafers removed from aqueous baths or for wafers exposed to high humidity. The magnitude of F_{CAP} can be similar to, or greater than, van der Waals forces (30).

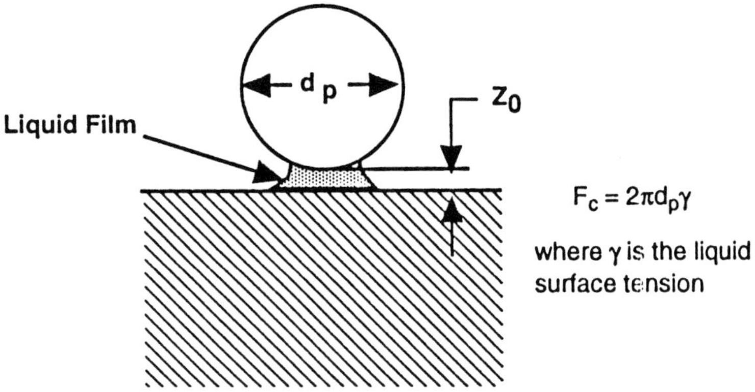

Figure 29. Capillary adhesion (10). *(With permission M. B. Ranade. Copyright 1987, Particle Technology, Inc.)*

Chemical bonds, which are generally much stronger than the physical bonds considered here, are the second type of particle adhering force. They are species-dependent and not generally reversible so that they constitute a different type of particle adhesion but a type that can be of practical importance.

6.0 SUMMARY

This chapter reviewed mechanisms whereby particles deposit on wafer surfaces. Two wafer environments were considered: gaseous environments at atmospheric pressure and liquid baths. Gravity and convective diffusion are important deposition mechanisms in gaseous environments, although electrophoresis (Coulomb forces) often dominates the deposition of submicrometer-sized particles. Thermophoresis can also be important. The parameter used to measure particle deposition rate was particle deposition velocity defined as particle flux to the wafer surface divided by the particle concentration in the gas surrounding the wafer.

Electrical forces, as described by the electrostatic double layer model of a particle in a liquid bath, often create barriers to particle deposition on wafers immersed in a liquid bath. Under these conditions particle deposition occurs primarily through the liquid film that adheres to the hydrophilic wafer surface upon withdrawal from the bath. This particle deposition mechanism does not apply to hydrophobic wafers; the important mechanisms whereby particles deposit on hydrophobic wafers are not understood.

When the electrostatic double layer does not create a barrier to particle deposition, the primary particle deposition mechanism in liquid baths is convective diffusion. However, the particle deposition velocity attributable to convective diffusion in liquid bath is two to three orders of magnitude lower than in atmospheric gases, thus compensation for the two to three orders of magnitude higher particle concentrations that typically characterize liquid baths.

Van der Walls forces are a primary, universal force of adhesion between particles and a wafer surface. They are often of sufficient magnitude to cause deformation of the particle, increasing the contact area between the particle and the wafer, and thus the adhesive force. Electrostatic double layer forces, capillary forces, and chemical bonds can also contribute significant adhesive force under appropriate conditions.

Particle adhesive forces are typically measured by measuring the force required to detach the particle from the wafer surface. Such removal forces can be applied by centrifuging, vibration, or fluid drag. Relating the removal force to the adhesive force requires understanding of the removal mechanism, i.e., lift-off, sliding, or rotation. These same particle removal forces are used in wafer cleaning operations as described in Ch. 9.

REFERENCES

1. Hinds, W. C. *Aerosol Technology,* John Wiley and Sons, New York (1982)
2. Liu, B. Y. H. and Ahn, K. H., *Aerosol Sci. and Tech.* 6(3):215-224 (1987)
3. Cooper, D. W., Peters, M. H. and Miller R. J., *Aerosol Sci. and Tech.* 11(2):133-143 (1989)
4. Liu, B. Y. H, Fardi, B. and Ahn, K. H., in: *1987 Proc.of the 33rd Annual Tech. Meeting of the IES (Institute of Environmental Sciences),* pp. 461-465 (1987) (Available from IES, 940 E. Northwest Highway, Mt. Prospect, IL 60065)
5. Scheibel, H. G., Hussin, A. and Porstendorfer, J., *J. Aerosol Sci.* 15(3):372-375 (1984)
6. Turner, J. R., Fissan, H. J., Clayton, A. C. and Donovan, R. P., in: *1988 Proc. of the 34th Annual Tech. Meeting of the IES*, pp. 433-437 (1988)
7. Peterson, T. W., Stratmann, F. and Fissan, H., *J. Aerosol Sci.,* 20(6):683-693 (1989)
8. Saito, A., Ohta, K., Watanabe, M. and Oka, H., in: *Microcontamination Conference Proceedings*, pp. 562-569 (October 1991)
9. Ali, I., Raghavan, S. and Risbud, S. H., *Semiconductor International* 13(5):92-95 (April 1990)
10. Ranade, M. B., *Aerosol Sci. and Tech.* 7(2):161-176 (1987)
11. Hunter, R. J., *Zeta Potential in Colloid Science: Principles and Applications,* Academic Press, San Diego (1981)
12. Raghavan, S., SRC Video Lecture (Sept. 12, 1991)
13. Itano, M., Kawanabe, I., Kern, F. W., Miyashita, M., Rosenberg, R. W., Ohmi, T., Fukushima, R. and Akiyama, S., Fine Particle Society Meeting, San Jose, CA (July 1991)
14. Riley, D. J. and Carbonell, R. G., *Microcontamination,* 8(12):19-25, 60-61 (December 1990)

15. Riley, D. J. and Carbonell, R. G., in: *Proc. of the 1990 Annual Tech Meeting of the IES,* New Orleans, LA, pp. 224-228 (April 1990)
16. Visser, J., in: *Surface and Colloid Science 8,* (E. Matijevic, ed.) John Wiley and Sons, New York (1976)
17. Menon, V. B., Michaels, L. A., Hollan, L. A., Donovan, R. P. and Ensor, D. S., in: *Proc. of the 1988 Annual Tech Meeting of the IES,* King of Prussia, PA, pp. 382-389 (May 1988)
18. van den Temple, M., *Advances in Colloid and Interface Science* 3:137-159 (1972)
19. Israelachvili, J. N., *Intermolecular and Surface Forces,* Academic Press, San Diego (1985)
20. Riley, D. J. and Carbonell, R. G., in: *Proc of the 38th Annual Tech Meeting of the IES,* p. 455 (1992)
21. Riley, D. J. and Carbonell, R. G., in: *Particles in Gases and Liquids: Detection, Characterization, and Control,* Vol. III (K. L. Mittal, ed.,) (1992)
22. Donovan, R. P., Clayton, A. C., Riley, D. J., Carbonell, R. G. and Menon, V. B., *Microcontamination,* 8(8):25-29 (1990)
23. Riley, D. J. and Carbonell, R. G., in: *1991 Proceedings of the 37th Annual Tech Meeting of the IES,* pp. 886-891 (1991)
24. Grant, D. C., Liu, B. Y. H., Fisher, W. G. and Bowling, R. A., *J. Environ. Sci.* 33(4):43-51 (July/August 1989)
25. Krupp, H., *Advan. Colloid Interface Sci. I,* pp. 111-239 (1967)
26. Tsai, C. J., Pui, D. Y. H. and Liu, B. Y. H., *Aerosol Sci. and Tech.* 15(4):239-255 (1991)
27. Demejo, L. P., Rimai, D. S. and Bowen, R. C., *J. Adhesion Sci. Technol* 5(11):959-972 (1991)
28. Wang, H. C. *Aerosol Sci. and Technol.* 13:386-393 (1990)
29. Tsai, C. J., Pui, D. Y. H. and Liu, B. Y. H., *J. Aerosol Sci.* 22(6)737-746 (1991)
30. Bowling, R. A., *J. Electrochem. Soc.* 132(9):2208-2214 (September 1985)

Part III.

Dry Cleaning Processes

5

Overview of Dry Wafer Cleaning Processes

Jerzy Ruzyllo

1.0 INTRODUCTION

Semiconductor wafer cleaning is the most frequently applied processing step in the IC manufacturing sequence. Therefore, both production yield and device performance depend to a significant extent on the quality of wafer cleaning methods employed. The commonly used technology of liquid-phase, or wet, wafer cleaning, discussed in detail in previous chapters, will certainly remain the workhorse of everyday IC processing, and advances in this area are needed and will be made. On the other hand, wet cleaning technology has a number of inherent shortcomings that may limit its effectiveness in the fabrication of at least some future generation integrated circuits. The need thus exists to replace this type of wafer cleaning in selected applications in high-end silicon IC manufacturing, such as next generation DRAMs (1), with equally effective gas-phase, or dry, wafer cleaning processes. During the last five years, impressive strides have brought dry wafer cleaning technology closer to industrial applications. Still, much needs to be accomplished to make it fully compatible with large scale industrial production.

Part III of this volume is devoted to dry semiconductor wafer cleaning processes, primarily addressing silicon technology. Authors of consecutive chapters detail important issues related to this emerging technology and discuss specific approaches proposed and investigated.

The purpose of this introductory chapter is to give an overview of dry wafer cleaning processes established to date. The reasons why "dry" alternatives to conventional wet wafer cleaning need to be developed are stated, general characteristics of dry wafer cleaning processes are considered, and various possible approaches to dry wafer cleaning are discussed. Finally, an overview of results of representative dry wafer cleaning experiments is presented. This overview is intended to illustrate emerging trends rather than to discuss details of various approaches to dry cleaning which are covered in greater depth in subsequent chapters.

2.0 LIMITATIONS OF WET WAFER CLEANING PROCESSES

Wet wafer cleaning will remain the dominant technique for the removal of trace contaminants from silicon surfaces in IC manufacturing. Several factors, however, indicate the need to develop alternative methods of dry wafer cleaning (2)(3).

First, a clear incompatibility exists between wet wafer cleaning operations and process integration. It is safe to state that full-scale cluster tool processing at reduced pressure cannot be accomplished unless techniques of gas-phase removal of trace contaminants become available (4).

Second, there are problems associated with wet processing of high-aspect-ratio structures. The complications stem not from the difficulty of getting liquid into small openings, as this problem can be dealt with by using surfactants, but from the difficulty of getting it out (5). Once forced into deep trenches, cleaning liquids can only be removed by evaporation which is very likely to leave most of the contaminants behind.

Next, there is the difficulty of controlling particles during wafer treatments in liquid chemicals and rinse/dry cycles. Use of "all dry" cleaning is expected to reduce this problem substantially, since particles are easier to control in gases that in liquids. Furthermore, the very replacement of liquid cleaning agents with their gaseous counterparts is expected to result in a decrease in the overall level of silicon surface contamination, as at least part of the contamination originates from the wet cleaning environment itself (6). Metallic contaminants, for instance, can be plated on silicon surfaces from HF solutions (7) and deposited from NH_4OH and other agents used in wet cleaning, such as H_2O_2 (8)(9). Organic contaminants, on the other hand, originate in part from the containers and utensils used in wet-chemical operations (10).

Finally, the cost of high-purity chemicals and DI water, problems with waste disposal, and safety issues are additional reasons for the rapidly growing interest in dry wafer cleaning technology.

3.0 ANTICIPATED ROLE OF DRY WAFER CLEANING IN IC FABRICATION

The general discussion on the role of dry wafer cleaning operations in IC manufacturing should begin by stating that wafer cleans, although they account for a major part of all operations performed on the wafer, do not serve any constructive purpose. In other words, in contrast to oxidation, CVD, implantation, or etching, we do not need cleaning directly to build device features. We use it only because the wafer surface becomes contaminated in the course of the manufacturing process.

If liquids, resists, equipment, wafer handling utensils, and the process environment in general would not shed contaminants, then there would be no need for wafer cleaning. One may conclude from such reasoning that an ultimate goal in IC manufacturing technology should be a totally "cleaning-less" fabrication process. Following this reasoning, one may see dry cleaning methods playing an important role in bridging the IC technology of today with totally integrated "cleaning-less" technology of the future in which surface treatments will be limited only to processes such as native oxide removal, or surface passivation through oxidation or hydrogenation. It is in this context that both short term and long term roles of gas-phase processes in wafer cleaning technology should be considered.

In general, it is not expected that dry cleaning will replace wet cleaning applications in which the latter is successfully used. Instead, dry cleaning should be considered for those applications in which wet cleaning, due to its inherent limitations, cannot be used. First and foremost, this concerns in situ cleaning operations carried out within the integrated processor systems. An example of the likely early application of dry cleaning in cluster tool manufacturing is shown in Fig. 1. It is concerned with a three-step process for gate structure formation in MOS technology. Other likely applications, some of them already in place for some time, involve cleaning operations applied in two-step integrated processes including pre-contact cleans, pre-epi cleans, and cleans applied prior to polysilicon emitter formation in bipolar technology (4).

Overall, it can be assumed that during the next few years newly developed dry cleaning modules will be used predominantly in conjunction with clusters tool. It remains to be seen whether further refinements of dry cleaning methods will eventually lead to the use of dry cleaning reactors in the stand alone mode.

In further defining the potential role of dry cleaning in IC processing, one should also consider the ability of both dry and wet cleans to remove specific surface contaminants. Clearly the most challenging contaminants are metals, including transition and heavy metals, as well as alkali ions. Dry cleans may not be able to take off large amounts of these contaminants without substantial etching of silicon and/or roughening the surface. Therefore, dry cleans are not expected to be as effective as wet cleans are on surfaces grossly contaminated with metals which, however, should never occur in future microelectronic fabrication.

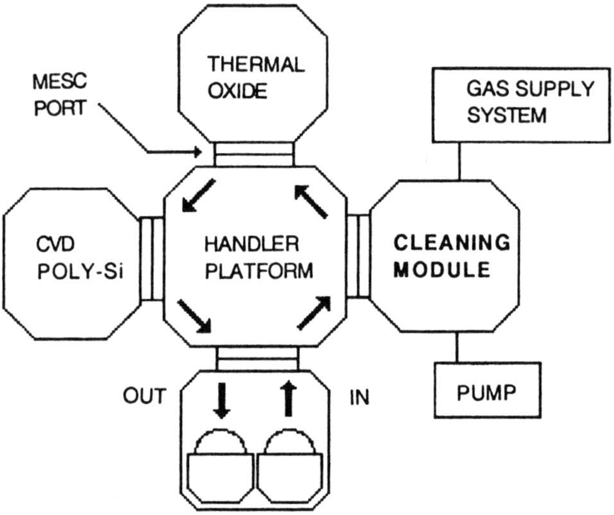

Figure 1. Example of application of dry cleaning in cluster tool processing of MOS gate structure.

The role of dry cleans is seen primarily in terms of operations performed on silicon surfaces in a fully controlled environment immediately prior to a subsequent processing step. In this context, the term dry cleaning is synonymous with in situ cleaning, while ex situ cleaning will be carried out predominantly by wet cleaning operations.

Considering the nature of various cleaning modes, it may be appropriate to separate them into: *(i)* surface treatments that are expected to remove fairly large amounts of surface contaminants, and *(ii)* surface treatments that, besides removing trace contaminants, change in a controlled manner the chemical state of silicon surface, for instance, by removal of native/chemical oxide or surface passivation through oxidation or hydrogenation. In this latter case, the term "surface conditioning" may be more appropriate than "surface cleaning." The nature of such surface conditioning will have to be closely related to the requirements of the subsequent processing step.

Following this reasoning, the role of processes currently referred to as dry cleaning may be limited in the future mostly to the in situ surface conditioning while wet cleans will continue to carry out ex situ surface cleaning operations. If the currently observed trends continue, the emphasis should be gradually shifting toward the former and with improved overall process cleanliness less emphasis will be placed on surface cleaning in the traditional meaning of this term.

4.0 REQUIREMENTS OF DRY WAFER CLEANING TECHNOLOGY

To become technically viable, any dry cleaning method will have to meet certain requirements which can be briefly summarized as follows.

1. Dry cleaning methods have to display experimentally proven capability to remove contaminants comparable to the performance of wet cleans. This requirement applies in particular to the removal of heavy, transition, and alkali metals which, among other contaminants of concern, are the most difficult to volatilize. No general purpose dry clean will be considered effective unless it displays experimentally proven ability to remove trace metals.
2. Dry cleans must not roughen the semiconductor surface nor generate defects in the oxide. This issue is particularly relevant to cleaning processes that are based on aggressive plasma-generated chemistries and/or very short wavelength electromagnetic radiation. Consequently, a careful assessment of gas-phase cleaning environment interactions with surface features must be carried out.
3. Gas-phase cleaning reactions must not generate solid residues that remain on the cleaned surfaces
4. Any dry cleaning method has to feature adequately high throughput, ideally high enough to make the cleaning steps compatible with batch furnace processing. As further discussion in this chapter will reveal, dry cleaning is typically a single-wafer process. Therefore, throughput is an important issue not only for interfacing dry cleans with batch processes, but also for implementing cluster tool processing.
5. Dry cleaning methods have to be effective at fairly low temperatures for two reasons. First, they have to meet an overall requirement of low-thermal budget processing. Second, elevated temperature at the early stage of the dry

cleaning process may have an adverse effect on surface contaminants which, instead of being removed, can diffuse and become permanently embedded in silicon.

6. Dry cleaning modules must be designed to assure compatibility with cluster tool processing and integrated process automation. As mentioned earlier, it is envisioned, at least for the time being, that dry cleaning modules will be mainly used as a part of an integrated systems rather than as stand alone units.

5.0 MECHANISMS OF DRY WAFER CLEANING

An objective of any wafer cleaning operation is to selectively remove from the semiconductor or oxide surfaces elements and compounds other than silicon and some of those that were purposely deposited there, e.g., oxide, nitride, or polysilicon layers. These undesirable impurities are commonly referred to as either physical (particles) or chemical contaminants. This section considers the mechanisms of dry wafer cleaning, focusing on the removal of chemical contaminants.

Chemical contaminants can appear on the silicon surface in either atomic, ionic, or molecular form. Their removal during wet wafer cleaning is accomplished by selective reaction causing either dissolution of the contaminant in the solvent or its conversion into a soluble compound.

Different principles apply to dry wafer cleaning processes involving chemical contaminants. In this case, the contaminant has to be either: *(i)* converted into a volatile compound, *(ii)* lifted off during removal of underlying material, or *(iii)* sputtered off. These principles imply the possibility of either physical or chemical interactions to drive gas-phase cleaning processes, in contrast to wet cleaning operations that depend solely on chemical interactions and dissolution. The wide-range controllability of the pressure of cleaning agents, as well as the much larger flexibility in applying thermal enhancement for the contaminant removal process in the case of gas-phase cleaning, represent other important differences between wet and dry cleaning modes.

Figure 2 illustrates schematically the mechanisms involved in physical and chemical gas-phase cleaning. In the former case (Fig. 2a), removal of contaminants takes place via momentum transfer between ions accelerated toward the surface and contaminant species. Since selective interactions are not possible in this case, and in addition, due to silicon being lighter than most contaminants of concern in silicon device processing, erosion of the silicon substrate is very likely. The same lack of selectivity can create additional problems related to sputter etching of materials that were purposely formed on the surface, and are meant to remain there. Also, redeposition of the sputtered species may create problems, particularly in the case of large diameter wafers and tight geometries.

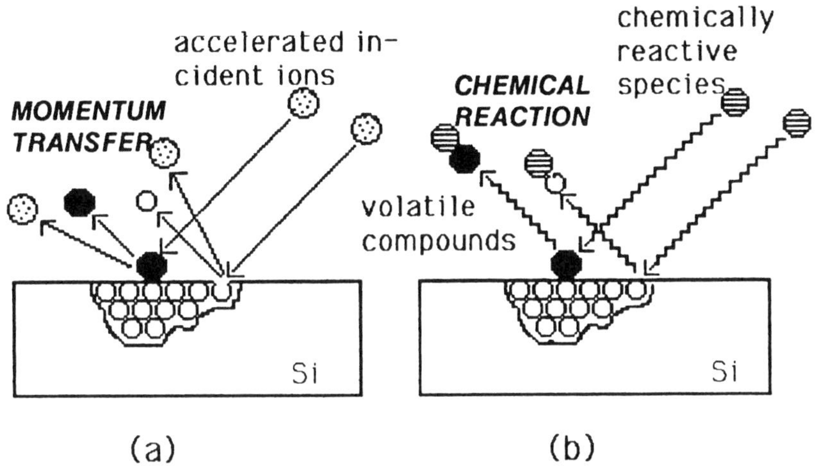

Figure 2. Schematic illustration of (a) physical, and (b) chemical dry wafer cleaning.

In contrast, chemical dry cleaning (Fig. 2b) allows better selectivity and reduces a risk of contaminant redeposition. If adequately performed, it has a potential for incurring less surface damage. In this case, selection of chemically reactive species is based on their ability to selectively form volatile compounds with a specific type of contaminant, for instance organic or metallic. These chemically reactive species are either added to, or generated in the process gases through thermal decomposition, photolysis, or glow discharge. Following the formation of volatile compounds containing contaminants, these volatile compounds are either immediately removed from the surface, or conditions are created to facilitate their volatilization (increased temperature and/or reduced pressure). The need of the latter appears to be difficult to bypass in the case of gas-phase removal of some metallic contaminants, as indicated below.

The process of volatilization of metallic species involves effects specific to this process, not encountered in the case of gas-phase removal of organic species and native/chemical oxides. These effects are schematically illustrated in Fig. 3.

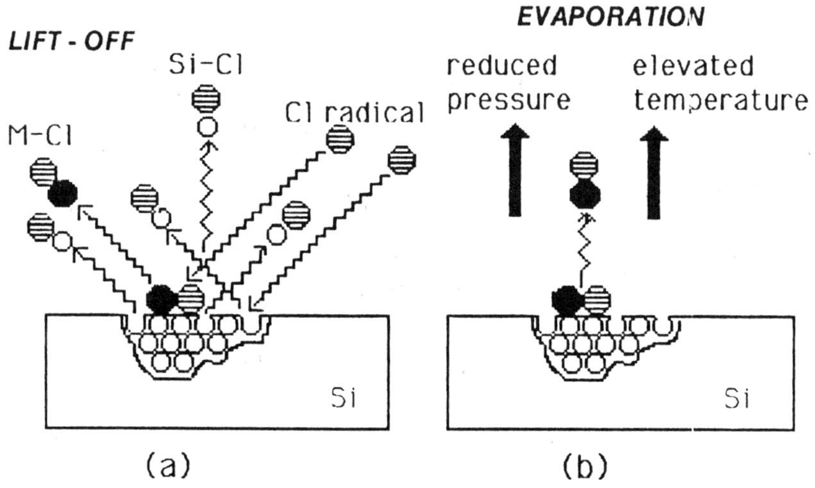

Figure 3. Mechanisms of metal contaminants removal during dry wafer cleaning: (a) formation of metal chlorides and removal through lift off process, (b) evaporation of metal chlorides at reduced pressure and increased temperature.

First, silicon surface is exposed to chemically reactive species that form potentially volatile compounds with metallic contaminants of interest in silicon processing (Fig. 3a). Chlorine is an element of choice in this application because metal chlorides are generally more volatile than other metal halides. This metal complexing operation has to be carried out at low temperature, not to exceed 200°C, to avoid penetration of the metals into silicon.

During this process, chlorine also interacts with silicon forming silicon chlorides, which are more volatile than chlorides of metals contaminating the surface (Table 1). Therefore, it is not possible to prevent volatilization of silicon in an environment in which these metals also become volatile. In general, slight silicon etching does not seem to be a problem as long as it is uniform and does not result in surface roughness. A beneficial aspect of the silicon etching is that along with the stream of volatile silicon chloride molecules leaving the substrate, some metal chlorides are likely to be lifted off the surface (11), as indicated in Fig. 3a.

Table 1. Vapor Pressure of Selected Metal Chlorides (12)

NAME	FORMULA	TEMPERATURE °C		
		760 Torr	100 Torr	1 Torr
Aluminum chloride	$AlCl_3$	180.2s	152.0s	100.0s
Arsenic trichloride	$AsCl_3$	130.4	70.9	-11.4
Boron trichloride	BCl_3	12.7	- 32.4	- 91.5
Boron trifluoride	BF_3	- 110.7	- 123.0	- 154.6
Cuprous chloride	Cu_2Cl_2	1490.0	960.0	546.0
Ferrous chloride	$FeCl_2$	1026.0	842.0	
Lead chloride	$PbCl_2$	954.0	784.0	547.0
Magnesium chloride	$MgCl_2$	1418.0	1142.0	778.0
Nickel chloride	$NiCl_2$	987.0s	866.0s	671.0s
Phosphorus trichloride	PCl_3	74.2	21.0	- 51.6
Silicon tetrachloride	$SiCl_4$	56.8	5.4	63.4
Silicon tetrafluoride	SiF_4	- 144.0s	- 113.3s	- 94.8s
Silver chloride	AgCl	1564.0	1297.0	912.0
Sodium chloride	NaCl	1465.0	1220.0	865.0
Sodium fluoride	NaF	1704.0	1455.0	1077.0
Zinc chloride	$ZnCl_2$	428.0	610.0	428.0
Zinc fluoride	ZnF_2	970.0	1254.0	970.0

Due to the rather random nature of interactions involved, one cannot rely only on the lift-off process to remove all the metal-containing species from the silicon surface. Consequently, additional measures may have to be taken to assure a complete removal of remaining complexed metals trapped at the surface. These measures involve treatment of the wafer simultaneously at the elevated temperature and reduced pressure that are needed to achieve adequate volatility of metal chlorides (Fig. 3b).

The above procedure has to be followed unless new chemistries for gas-phase metal removal are developed and become available. Of special promise in this regard are organic compounds which remove metal via the formation of highly volatile metalorganic compounds.

In contrast to the gas-phase metal removal process which is essentially a two-step operation, volatilization of organic contaminants and residual oxides can be accomplished a single operation. As discussed later in this chapter, chemistries and reaction stimulation methods used in each case will determine to what degree gas-phase removal for all three types of contaminants can be integrated into an overall cleaning procedure.

6.0 GENERAL OUTLINE OF DRY WAFER CLEANING METHODS

As noted already, the goals of any overall semiconductor wafer cleaning process are to remove the following from the surface: *(i)* organic contaminants, *(ii)* metallic contaminants including transition, heavy, and alkali metals, and *(iii)* native or chemical oxides spontaneously formed on the surface during air exposure or during cleaning. Table 2 depicts these contaminants, and links them with gas-phase reactions that can be used to remove them from the surface. Organic contaminants are subdivided in Table 2 into categories depicted as "gross organics" and "fine organics". The former are mechanically coherent layers of organic materials such as photoresist and plastic polymers, while the latter are miscellaneous organic residues and hydrocarbons. This inclusion of "gross organics" removal process in the cleaning sequence is justified in the case of dry wafer cleaning because of the similarity of methods used to volatilize both "gross" and "fine" organics. Therefore, it is likely that operations, such as resist and/or polymer stripping, will be integrated with dry cleaning operations providing adequate ability of dry cleans: *(i)* to remove metallic contaminants that have remained on the surface after the resist strip so that liquid-phase ex situ pre-clean can be avoided, and *(ii)* to remove hardened resists without damaging the surface.

Table 2. Summary of Dry Wafer Cleaning Methods

GAS-PHASE CLEANING MODE	GROSS ORGANICS	FINE ORGANICS	METALS	NAT./CHEM. OXIDE
PHYSICAL				* LOW ENERGY Ar SPUTTER.
THERMAL	* OXIDATION	* NO : HCl : N_2 * OXIDATION	* HCl ANNEAL * FORMATION OF METAL-ORGANICS	* H_2 ANNEAL * HIGH T/UHV * MID T/UHV * $GeH_4 : H_2$
VAPOR		* HCl : HF : H_2O VAPOR		* $HF:H_2O$ VAPOR * HF : CH_3OH
PHOTO.	* UV/OZONE	* UV/OZONE * UV/O_2 . H_2O VAPOR	* UV/Cl_2	* UV/HF:CH_3OH * UV/ $NF_3:H_2:Ar$
PLASMA	* DIRECT PLASMA O_2 * REMOTE PLASMA O_2	* REMOTE PLASMA O_2	* REMOTE PLASMA HCl	* REMOTE PLASMA H_2 * ECR PLASMA NF_3 or H_2 * REMOTE PLASMA $NF_3:H_2$

Not addressed in the Table 2 is the crucial issue of particle removal. The reason for excluding particle removal processes from discussion in this chapter is three-fold. First, the most important aspects of particle control and removal are covered in other chapters of this volume. Second, technically viable methods of particle removal in the gas-phase are yet to be developed. The experiments with some promising techniques, such as the cryogenic argon-aerosol method (13), need to be developed further before their feasibility will be fully assessed. Third, it is assumed that by the time dry cleaning methods are fully ready for industrial implementation, adequate control of particles in gas-phase cluster tool processing will be achieved.

As seen in Table 2, besides purely physical cleaning used primarily to sputter off residual oxides spontaneously grown on the silicon surface, four different approaches to dry cleaning are possible, depending on the method used to generate an adequately reactive chemical environment. It should be noted, that in the case of plasma-driven cleaning, chemical reactions can be supplemented with physical interactions to the degree determined by the plasma technique used. Moreover, the classification of cleaning methods presented in Table 2 was derived solely for the purpose of this discussion. It is very likely that most of the practical applications will use simultaneously a variety of modes, e.g., thermal enhancement in conjunction with photochemical interactions. We now briefly define each dry cleaning method listed in Table 2.

In purely thermally enhanced cleaning, externally generated thermal energy delivered to the wafer and process gases is used to break down gaseous molecules. The atomic species released in this process subsequently react with metallic and other impurities on the wafer surface. Typically, fairly high temperatures are needed to enforce these interactions. If combined with very low pressure, thermal treatments can, in some cases, result in the removal of surface contaminants by sublimation. An example of such a process is the thermal decomposition of silicon dioxide throughout the formation of volatile SiO (14). One important consideration in thermally enhanced cleaning is, as indicated earlier, to trap, or preferably, to volatilize metal contaminants at the surface before they are driven into the silicon or the oxide. Consequently, for as long as metallic contaminants remain on the surface, the silicon wafer should not be subjected to elevated temperature treatments during cleaning operations.

The cleaning mode defined in Table 2 as a "vapor-phase cleaning" reflects an effort to replace chemistries used in liquid-phase cleaning with their gas-phase analogs. Instead of using water solutions of given chemicals, water vapor is added to the gas stream. A typical example of this type of process is a vapor HF etching of native/chemical silicon oxides. In some cases, moisture can form as a product of the cleaning reaction and can control the process, although the original reactant gases may have been entirely anhydrous.

Listed next in Table 2 is photochemically enhanced cleaning. This approach is gaining in popularity due to its effectiveness and conceptual simplicity, resulting in relatively simple equipment. Reactant species are generated through the photolysis of gaseous compounds. Radiation in the wavelength range of UV (ultraviolet) is commonly used, as it corresponds to a wide spectrum of binding energies in compounds of interest in silicon processing. Thus, as indicated in Table 2, a variety of cleaning chemistries can be achieved with UV-enhanced processes.

Finally, Table 2 lists plasma as still another source of energy with a potentially wide range of applications in dry wafer cleaning. Here, cleaning chemistries are generated through electrical discharge in process gases. In wafer cleaning operations, preference is being given to remote plasma processing, over direct plasma processes that typically result in appreciable damage of the semiconductor surface structure.

7.0 REVIEW OF EXPERIMENTAL RESULTS

This section briefly reviews the results of experiments with various dry wafer cleaning processes to exemplify trends and approaches emerging in this area. This review will follow the guidelines established earlier in Table 2, and thus, will be concerned with the methods used to stimulate dry wafer cleaning as outlined in this table. Since most of these methods are discussed in detail in other chapters, the main purpose of the discussion that follows is to establish a framework for those more in-depth treatments.

7.1 Cleaning Through Physical Interactions

An application of purely physical cleaning in IC fabrication typically comes down to argon ion (Ar^+) sputtering of native/chemical oxides prior to contact metallization (e.g., Ref. 15), and silicon epitaxy (e.g., Ref. 16). Such sputter cleaning is carried out at the energy ranging from 20 eV to 1000 eV. Surface damage incurred during etching at higher energies is substantial (17). Lower ion energy, in combination with increased temperature of the wafer during sputtering (17), appears to produce surfaces adequate for subsequent epitaxial deposition of silicon (16).

In the recently published method of low-energy (25 eV) sputter etching of ultra-thin oxide prior to Ti contact formation (15), hydrogen is added to argon to enhance etching through hydrogen reduction of the oxide. This method, however, cannot be classified as purely physical because of the concurrence of chemical interactions in the etching process.

Overall, purely physical dry wafer cleaning is expected to be used in IC manufacturing only in some specialized applications related to the removal of native or chemical oxides. As far as other contaminants are concerned, it is reasonable to assume that during Ar^+ sputter etching some metallic and organic contaminants are also sputtered off the surface. However, most of the metals of concern are heavier than silicon, and hence, their complete removal during low-energy sputtering is unlikely. Sputtering at higher energy, on the other hand, may damage the silicon surface beyond repair. Also, as mentioned earlier, a possible redeposition of sputtered species remains of concern in purely physical cleaning.

7.2 Thermally Enhanced Cleaning

Among methods of thermal wafer cleaning (Table 2), i.e., cleaning methods in which no energy other than heat is used to stimulate cleaning reactions, the above noted high-temperature anneals performed in vacuum for the purpose of oxide decomposition and volatilization (14)(18) and the hydrogen reduction of surface oxide at elevated temperatures are well established. These methods are still being improved with the aim of temperature reduction (e.g., Refs. 19, 20). In fact, low pressure H_2 annealing has been demonstrated to be effective prior to epitaxial silicon deposition at temperatures as low as 850°C (21) and even 700°C (22). The temperature of this process can be lowered substantially if plasma stimulation of cleaning is used (see Sec. 7.5 and Ch. 8).

Another well established method of thermal silicon wafer cleaning consists of HCl addition to the process gases applied prior to, or during, thermal oxidation. A specific goal of this method is to either getter metallic impurities, in particular sodium, or to remove metallic contaminants from the oxidation chamber. Effective removal of organic contaminants during thermal oxidation of silicon has also been experimentally demonstrated (23).

More innovative approaches to thermally stimulated wafer cleaning are represented in Table 2 by the NO:HCl:N_2 process and the reactions for forming volatile metalorganics. The former method achieves simultaneous removal of both organic and metallic contaminants through the formation of highly volatile nitrosyl compounds (24). This method has been applied with promising preliminary results in pre-gate oxidation cleaning (25). As an example, Fig. 4 illustrates the capability of this process to remove Cu from silicon surfaces. A potentially attractive feature of this approach is its compatibility with conventional atmospheric pressure batch furnace processing, such as thermal oxidation.

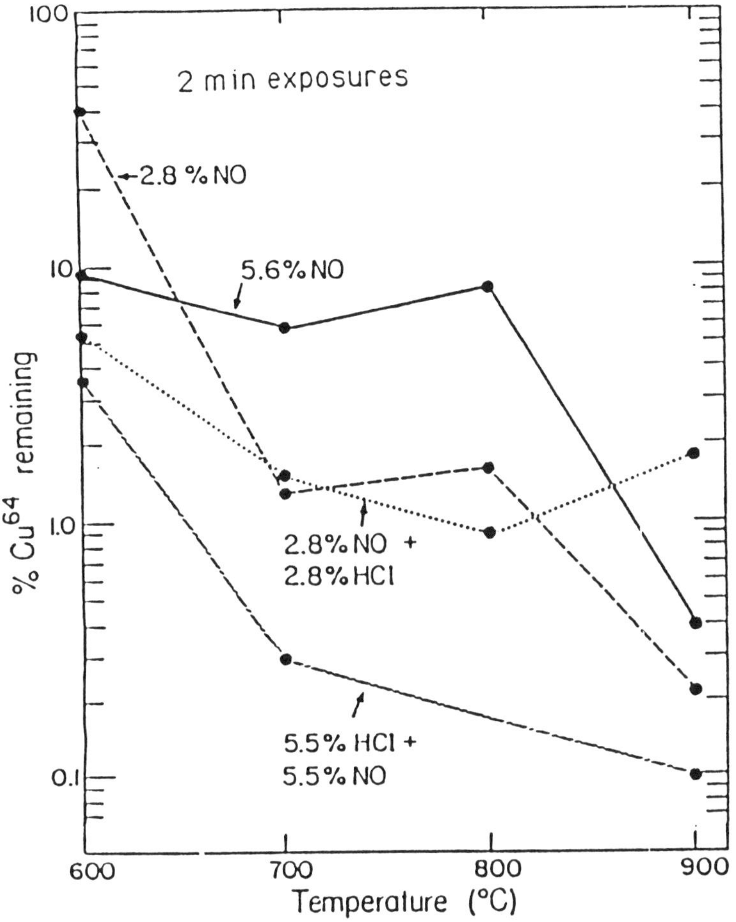

Figure 4. Effect of temperature and composition of reactants on Cu removal using NO:HCl:N_2 chemistry (24).

The same potential compatibility concerns a process using a germane/ hydrogen (GeH_4/H_2) mixture for native and chemical oxide etching (26). It is carried out at the temperature of 600°C - 750°C and is said to offer noticeable advantages in several applications (4).

The conversion of surface metallic contaminants into highly volatile metalorganic compounds as a method of silicon surface cleaning has been considered for some time, but only recently have the results of an attempted practical implementation been published (27). In this work, an effective removal at 300°C or below of trace iron and copper from a silicon wafer surface utilizing 1,1,1,5,5,5-hexafluoro-2,4-pentanedione as a volatilizing reactant has been demonstrated.

In general, experiments with thermally enhanced cleaning appear to be gaining momentum mainly because of the potential benefit of in situ cleaning prior to epitaxial film deposition and gate oxidation. In both these cases, heating capabilities are already in place, and no major changes in reactor design are needed. The main thrust of these efforts focuses on the reduction of temperature and is a driving force behind experimentation with novel cleaning chemistries.

7.3 Vapor Phase Cleaning

According to the definition formulated earlier in this chapter, vapor phase cleans are preferably performed at about room temperature, and proper chemical reactivity of the cleaning ambient is accomplished through the addition of moisture to the process gases. The relevant approaches and methods of implementation are discussed in detail in Ch. 7.

Among vapor phase cleaning chemistries outlined in Table 2, HF vapor treatments carried out for the purpose of native/chemical oxide removal are certainly the best explored. The feasibility of vapor HF etching was first demonstrated in 1966 (28); since then, significant strides have been made in understanding the chemical reactions involved. This is reflected by the number of papers published in this area by various research groups (29)-(34), as well as by the availability of commercial tools developed specifically for the purpose of HF vapor etching.

Various refinements of vapor HF etching techniques have so far not been able to eliminate entirely certain problems associated with this process. The most persistent one involves solid residues, formed as a product of etching, that remain on the surface. An additional water rinse needs to be applied following vapor HF exposure to wash these residues off. The positive side of this last treatment has been experimentally demonstrated to consist of the ability to remove certain of the metallic residues in the form of soluble fluorides (35). Still, the need for water rinsing seems to defeat the purpose of what was intended to be an entirely gas-phase operation. Also, a strong dependance of the vapor HF process on the moisture content (36), the degree of hydration of etched oxides, and the amount of surface hydrocarbons, can all potentially lead to etch nonuniformity and difficulties in controlling etch selectivity, as well as to very high etch rates resulting in undesired or excessive etching of some oxides on silicon, in particular BPSG. As a result, moisture content has to be very precisely controlled to achieve adequate process controllability (Fig. 5). Moreover, the need to add moisture to anhydrous gaseous HF in order to control etching reactions leads to the corrosion of stainless steel gas lines typically used in the state-of-the-art reduced pressure processing.

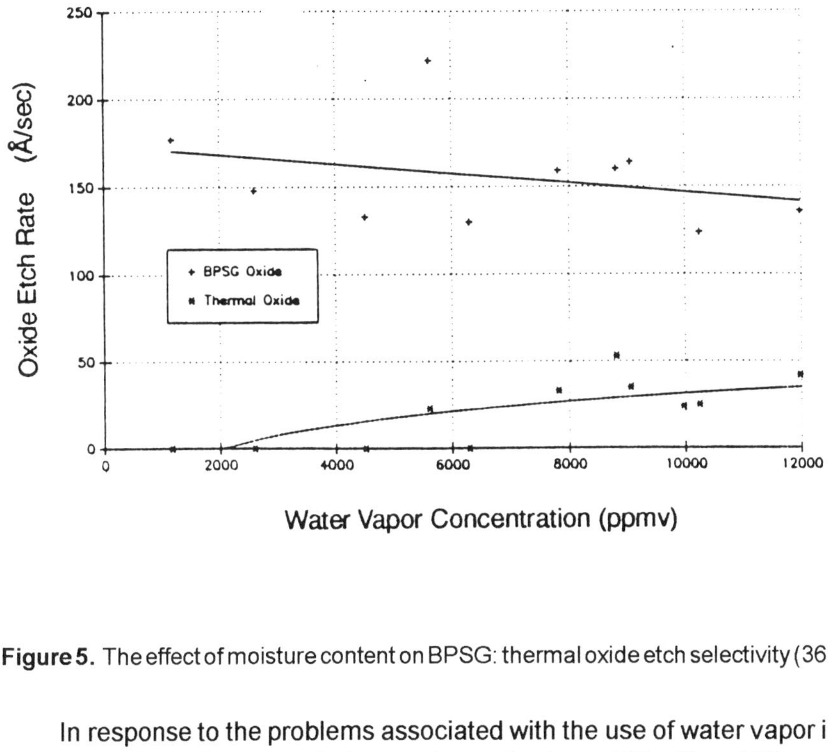

Figure 5. The effect of moisture content on BPSG: thermal oxide etch selectivity (36).

In response to the problems associated with the use of water vapor in gaseous HF etching, a technique using anhydrous HF with addition of methanol, CH_3OH, instead of water vapor has been investigated (38). It was demonstrated that if properly controlled, the HF/CH_3OH gas process suppresses formation of solid residue on the etched surface (Fig. 6), and results in improved etch selectivity as compared to HF/H_2O vapor process. At the same time, lower oxide etch rates were observed for this process as compared to etching with HF/H_2O vapor. The same HF/CH_3OH process was found suitable for controlled etching of thermal oxide (39) using an adequately-designed commercial reactor (40).

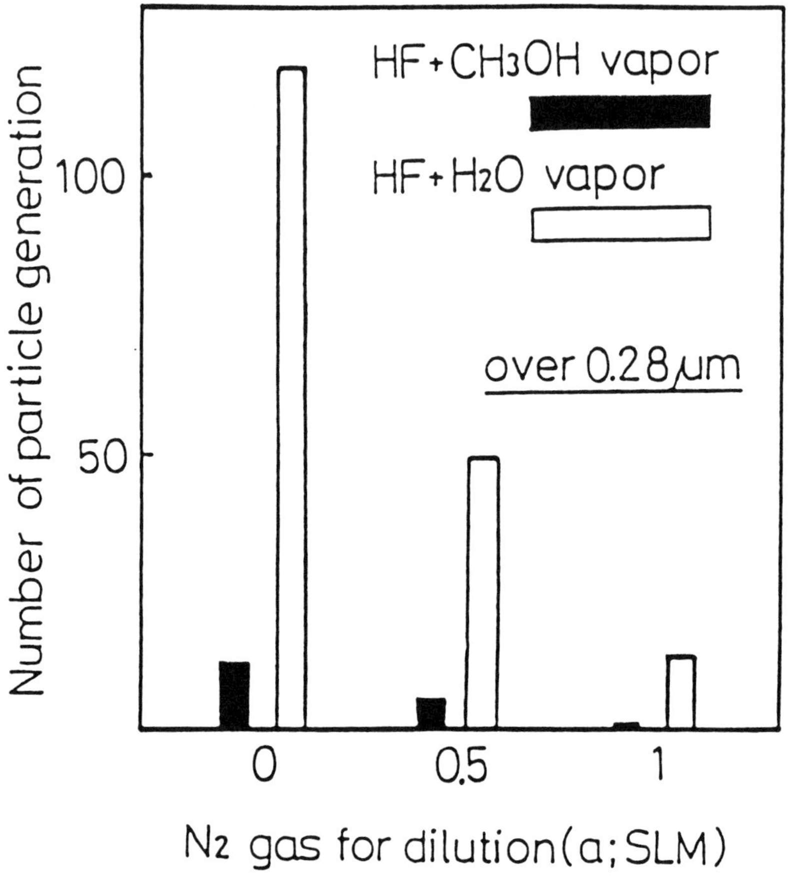

Figure 6. Suppression of solid residue formation during HF/CH_3OH vapor process as compared to HF/H_2O vapor process (38).

Early vapor phase etching processes were concerned mostly with etching of native/chemical oxide on silicon wafer by HF vapor. Only recently was this cleaning technique expanded to include other operations involved with surface cleaning. An interesting illustration of this trend is the use of HCl:HF:H_2O gas mixture as a pre-gate oxidation clean (41). This process was investigated as a one-step replacement for the conventional two- step sequence of aqueous HF dummy oxide strip and wet clean applied prior to gate oxidation. The results obtained, including electrical characteristics of 12 nm thick gate oxides indicate an adequate performance of this process and illustrate the not yet fully explored potential of vapor phase cleaning.

7.4 Photochemically-Enhanced Cleaning

As seen in Table 2, photochemically enhanced dry wafer cleaning reactions find applications in all areas of silicon surface cleaning. In many cases, low pressure mercury vapor UV lamps as the radiation sources are sufficient to generate useful chemistries. The general rule is to match the output of the UV source to the energies required for efficient photolysis. The wide range of reactants potentially used in photochemical cleaning may display a broad range of binding energies. Consequently, in order to allow adequate process versatility, the UV sources for photochemical wafer cleaning should ideally feature uniform high intensities for wavelengths ranging from 150 nm to 600 nm. This means that, for a wide range of photochemical cleaning applications, UV sources other than low pressure mercury vapor lamps are required. In the case of any UV source considered for these applications, it is important to verify its inertness with regard to defect generation in oxides existing on the exposed surfaces.

The best established photochemically enhanced cleaning process is the UV/ozone removal of organic contaminants discussed in detail in Ch. 6 of this volume. It was first studied as a method to oxidize and volatilize organics from solid surfaces (42)(43). It was later introduced to silicon IC processing for a variety of applications, including pre-oxidation surface treatments (44)(45), organic removal prior to metallization (46), and epitaxial deposition (47). Overall, the efficacy of this process for surface carbon and hydrocarbons removal is well documented (48). It is illustrated by Fig. 7 which shows removal of carbon from the silicon surface prior to Ti contact formation.

Besides effective removal of residual organics, the UV/ozone process can also be used to strip photoresists and to remove polymer films remaining on the surface after RIE processes (49). For all these applications, it is essential that 185 nm and 254 nm wavelength radiation be present in the lamp spectrum to generate atomic oxygen and ozone. These strong oxidizing agents are responsible for both decomposing organic compounds to form volatile carbon oxides, as well as causing a slight oxidation of the silicon surface. At room temperature, such "UV oxide" typically does not grow thicker than about 1.5 nm (Fig. 8) which makes it very useful as a passivating/protective oxide (44)(45). If needed however, substantially thicker oxides can be grown on silicon by UV enhancement at temperatures not exceeding 500°C (50).

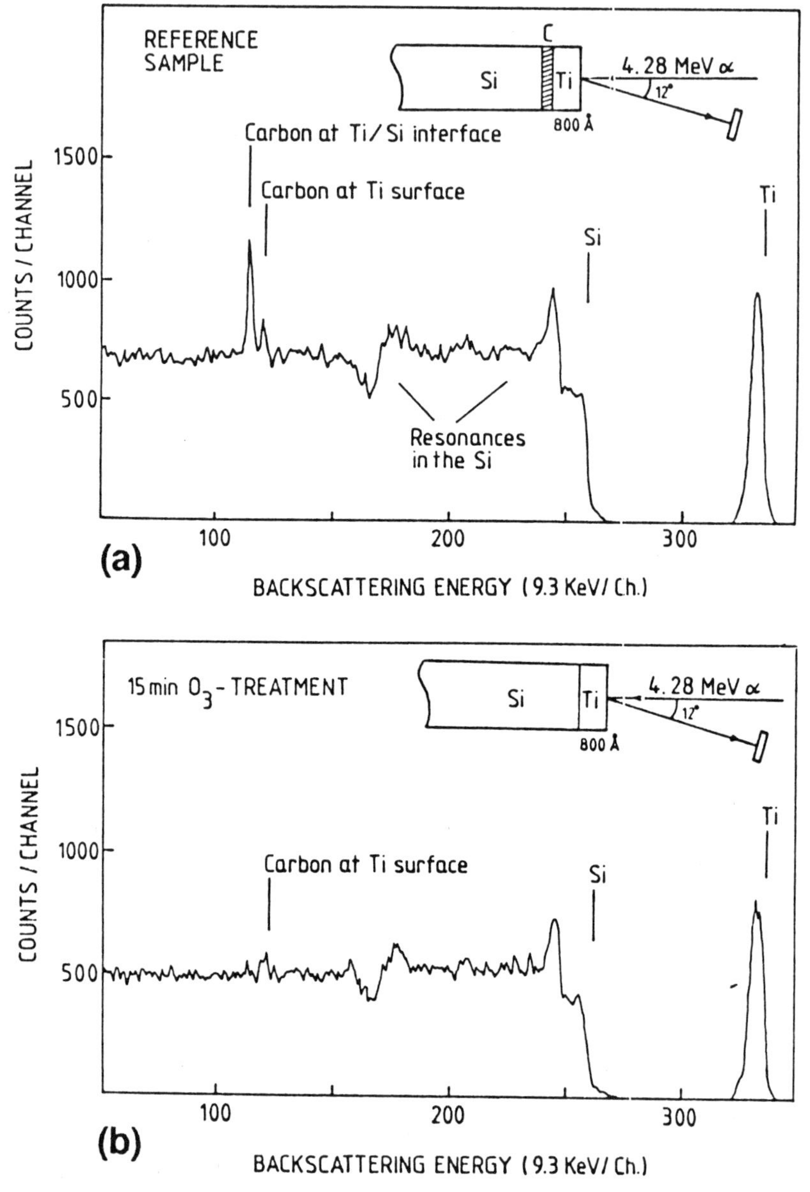

Figure 7. Backscattering data indicating UV/ozone removal of carbon from silicon surface: (a) no UV/ozone exposure prior to Ti contact formation, (b) 15 min. UV/ ozone exposure prior to Ti contact formation (46).

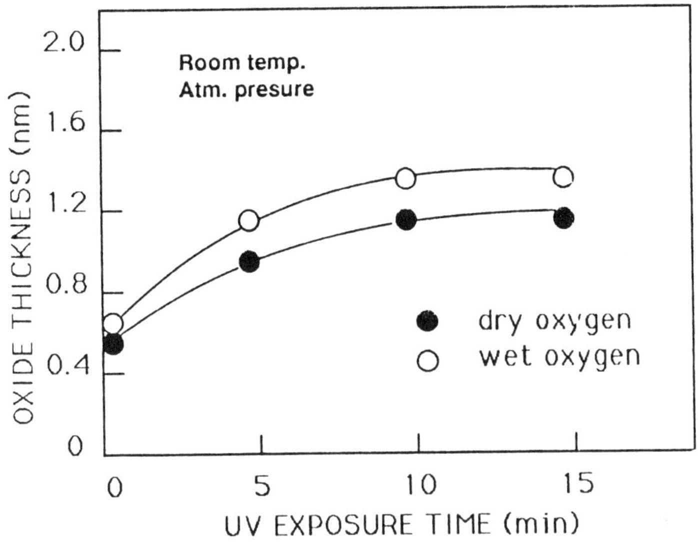

Figure 8. Growth of an oxide on the silicon surface during exposure to UV irradiation in the oxygen ambient.

Besides UV/ozone treatments, there is a growing interest in applying UV stimulation to gas-phase metal removal and native/chemical oxide etching processes. In both these applications, the potential of UV enhanced processes has already been established.

UV exposure of silicon in a chlorine ambient can be used to remove metallic contaminants from the silicon surface first by complexing them and then volatilizing the resulting chlorides (11)(51)(52). The same complexing and volatilization cycle also concerns silicon itself, as indicated in Fig. 3, which may help lift metal contaminants off the surface. The role of the lift-off process was confirmed in an investigation in which selected trace metals were removed without additional elevated temperature-reduced pressure treatment (53). However, lift-off only is clearly not enough to remove alkali metals such as Ca and Na. Figure 9 demonstrates overall effectiveness of UV/Cl_2 treatment in trace metal removal, but at the same time illustrates problems with volatilization of Ca. It is evident that additional treatments following UV/Cl_2 exposure will be needed to accomplish volatilization of these metals. Alternatively, different chemistries, based for instance on the process of formation of metalorganics, will have to be developed. The only problem with this approach is a need to avoid non-uniform, excessive etching of silicon potentially resulting in rough surfaces.

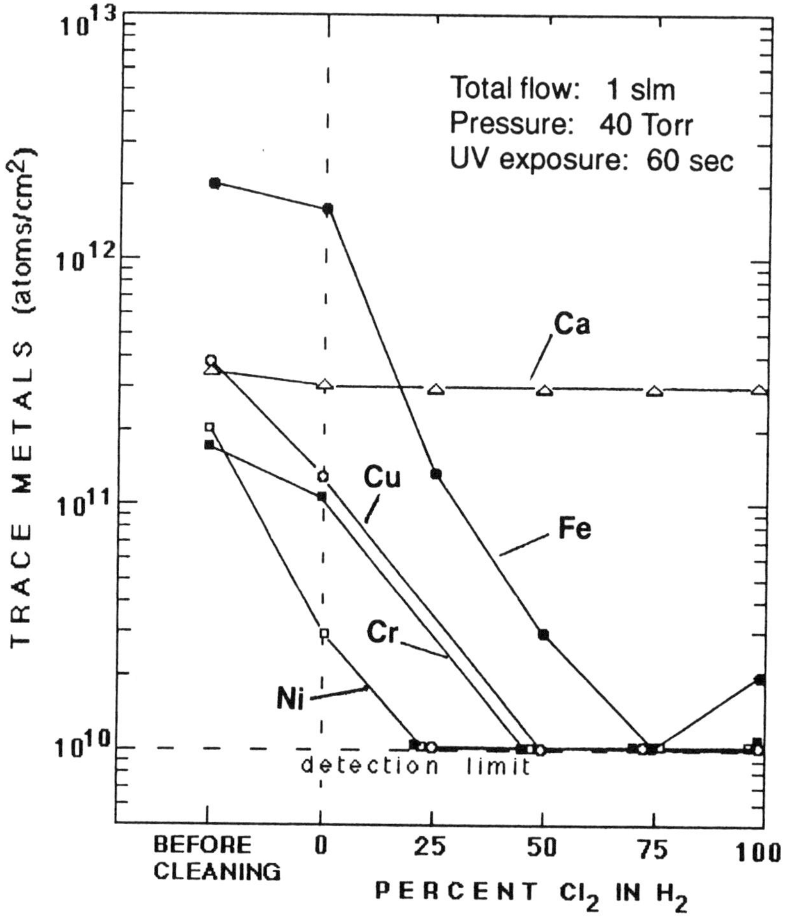

Figure 9. TXRF data showing changes in surface concentration of metallic contaminants on the Si surface as a function of Cl_2 content in H_2 during UV exposure (adapted from Ref. 53).

Recently, UV/NF_3:H_2:Ar etching of silicon oxides was proposed and investigated (54). This method was found to be suitable for native/chemical oxide etching, including pre-metal oxide etch. As the I-V characteristics in Fig. 10 indicate, this process is able to replace the standard HF dip conventionally applied immediately prior to contact metallization without compromising the characteristics of Al-Si ohmic contacts. Similar chemistry was used previously by ArF excimer laser (193 nm) enhanced etching of native oxide (55).

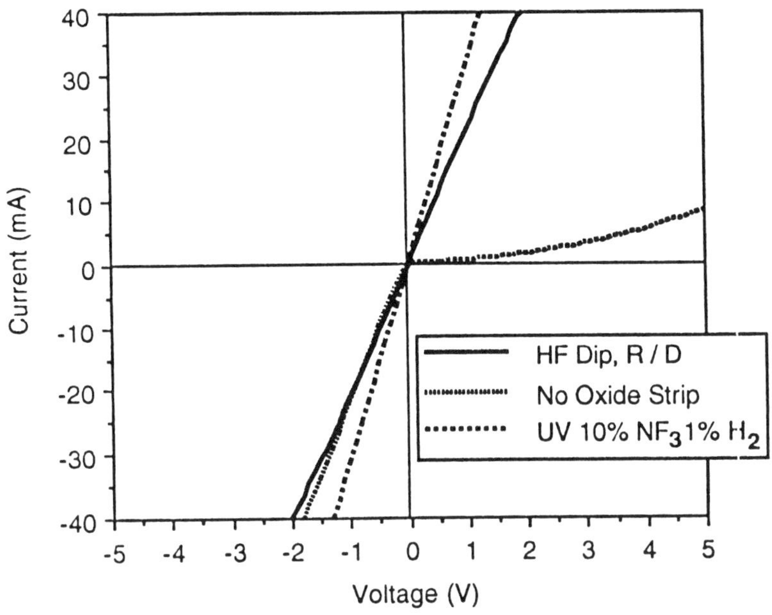

Figure 10. I-V characteristics of Al-Si contacts with chemical oxide not removed prior to Al deposition and removed by HF dip and UV/NF_3:H_2:Ar exposure.

Another recent development concerns the use of UV exposure in conjunction with anhydrous HF/CH_3OH chemistry (56). In this process, the UV irradiation can be used to control oxide etching reactions, leading to increased process versatility. The UV irradiation can also be used to control moisture level on the silicon surface. In addition, UV/ozone treatment can be applied *in situ* prior to native oxide etching to remove hydrocarbons, and hence, to assure adequate etch uniformity. In other respects, this procedure preserves advantages of the straight anhydrous HF/CH_3OH process discussed in the previous section.

Overall, UV-enhanced processing shows great promise in a variety of dry wafer cleaning applications. Consequently, dry cleaning modules with UV capabilities have become prime candidates for commercialization. An example is a recently introduced cluster tool compatible module designed to support various gas-phase cleaning modes with or without UV enhancement (40). The schematic diagram in Fig. 11 identifies features of this module. An attractive characteristic of this reactor is its compatibility with essentially any chemistry potentially of interest in gas-phase silicon cleaning. Moreover, the system is equipped with both reduced pressure and increased temperature capabilities. These features allow adequate control of moisture on the processed Si surfaces during native oxide etching. They are also needed to facilitate volatilization of metallic and organic contaminants.

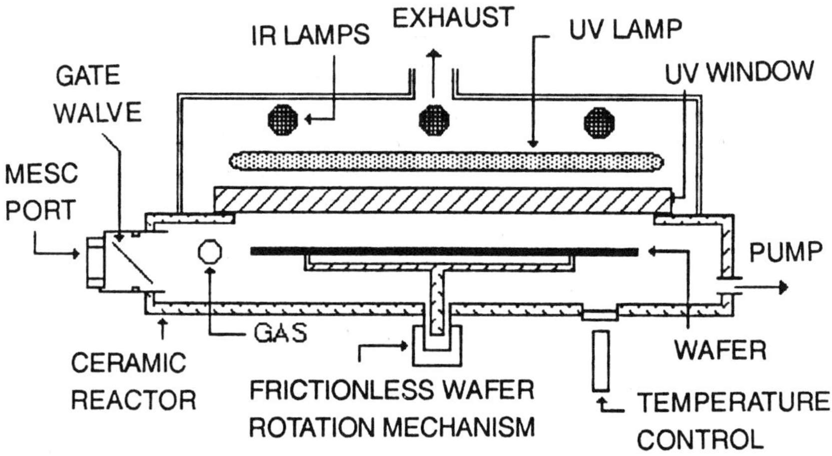

Figure 11. Schematic diagram of the dry cleaning module of Submicron Systems, Inc.

7.5 Plasma-Enhanced Cleaning

By virtue of being capable of generating high densities of chemically reactive species essential in surface cleaning and allowing for a broad range of chemistries, plasma offers potentially significant advantages for wafer cleaning applications (Table 2). The best established is plasma ashing of photoresist in oxygen, commonly applied in IC manufacturing. Recently, more refined applications of plasma enhancement in trace contaminant removal processes have emerged.

An important consideration in plasma-enhanced cleaning is the selection of reactor configuration. Direct plasma systems, commonly used in many film deposition and etching applications, are not suitable for wafer cleaning. This is because of the energetic interactions between the plasma and the wafer positioned in the glow discharge region in the direct plasma configuration. Specifically, ion bombardment can result in the surface contaminants, particularly metals including alkali ions, being "knocked" into the silicon substrate and/or the oxide. These interactions may also lead to substantial damage of the surfaces subjected to cleaning. Moreover, high-energy radiation generated in plasmas, to which the wafer within the plasma would be exposed, is likely to adversely alter features of the processed surface. For instance, radiation induced oxide defects can result.

To alleviate problems associated with direct plasma exposures, a remote, or downstream plasma processing mode has become the method of choice in dry wafer cleaning applications. This trend is exemplified by the variety of approaches to plasma wafer cleaning technology.

Choices in plasma-enhanced cleaning concern not only reactor configuration and process chemistry, but also plasma excitation modes. Both RF and microwave plasmas are used in remote plasma wafer cleaning. Also, magnetically confined microwave, or ECR (Electron Cyclotron Resonance), plasmas are explored in cleaning applications.

While the brief general outline of relevant methods is presented below, more detailed discussion of remote plasma enhanced cleaning is contained in Ch. 8 of this volume.

Several attempts at remote plasma wafer cleaning have been reported. Remote plasma etching of native/chemical silicon oxide in a hydrogen ambient prior to low-temperature silicon vapor-phase growth appears to be well explored (57)-(59). Also, recently reported ECR plasma NF_3 (60) and H_2 (61) processes have shown good potential in removing oxide and polymer films from silicon surfaces.

In the most far-reaching application of remote plasma cleaning, this technique was used to implement a complete, fully integrated, cleaning process applied after photolithography and prior to gate oxidation (62). In this process, without breaking vacuum, an O_2 plasma was used to strip the photoresist and to remove fine organics; HCl:Ar plasma was applied to remove metallic contaminants; and NF_3:H_2:Ar plasma was used to etch native/chemical oxide in the sequence shown in Table 3. After problems with surface pitting during HCl plasma exposure (63) were put under control (64), this cleaning process was applied prior to silicon gate oxidation and resulted in overall electrical characteristics of gate oxides comparable with those achieved by using conventional wet clean applied in the same process sequence. The results of this comparison are presented in Fig. 12 which shows the breakdown statistics of oxides grown on wet and dry cleaned silicon surfaces.

Table 3. Processing Steps in Integrated Remote Plasma Cleaning of Silicon

STEP	Gas comp.	T (°C)	p(Torr)	Etch rate (nm/min)	Process time(sec)
1. Resist strip (1000 nm)	O_2	200	1	650	120
2. Residual oxide (1.4 nm) etch	NF_3 1% H_2 2% Ar 97%	200	1	2.5	40
3. Fine organics removal/oxide regrowth (1.3 nm)	O_2	200	1	—	30
4. Metal complexing	HCl 10% Ar 90%	200	1	Si etch 30 nm	30
5. Residual oxide etch/partial volatilization of metal compounds	NF_3 1% H_2 2% Ar 97%	200	1	2.5	40
6. Final volatilization of metal compounds	Temperature ramp				*)
	reduced pressure anneal	750	0.02	—	120
7. Protective oxide regrowth (2 nm)	O_2	750	1	—	25

*) TIME OF TEMPERATURE RAMP DEPENDS ON HEATING ELEMENTS USED

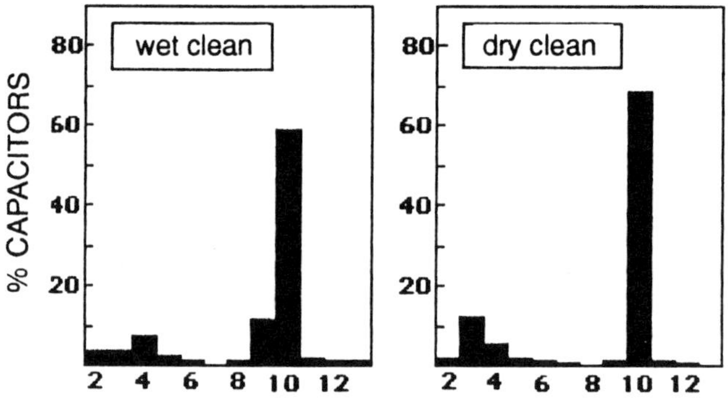

Figure 12. Distributions of breakdown events in oxides thermally grown on wet and dry (remote plasma) cleaned silicon surfaces.

In conclusion, it is expected that further refinements of remote plasma cleaning will eventually secure for this method a useful role in future dry wafer cleaning technology.

8.0 SUMMARY

The goal of this chapter has been to present an overview of issues involved in current and future dry wafer cleaning technology. It is important to recognize that this technology will cover only part of the anticipated future needs in IC manufacturing. In applications that are less challenging from the point of view of process integration, wet wafer cleaning methods are expected to maintain their present role. At the same time, however, the number of cleaning operations performed on the wafer in the course of IC fabrication is expected to decrease as a result of the gradually improving overall cleanliness of IC manufacturing processes.

Taking into account all the features outlined in this overview, dry wafer cleaning should be classified as a single wafer process, and, as such, should be considered for practical applications.

In terms of performance, certainly the greatest challenge remaining is the ability of dry cleans to remove all metallic contaminants of concern in silicon technology. While the ability of various dry cleans to remove several of them has been experimentally confirmed, their effectiveness in volatilization of some others, such as Na and Ca, still need to be unequivocally proven.

Dry wafer cleaning technology is maturing rapidly. As revealed in this volume, various dry cleaning techniques are being investigated, and some of them show very good potential for practical implementation. As for now, however, no dry cleaning method has been proven to perform clearly better than wet cleans, although, some have been shown to perform in selected applications at least equally well. Bear in mind, however, that dry cleans do not necessarily have to perform better than wet cleans to find useful applications in advanced device processing. This is because dry wafer cleaning methods are not meant to replace its wet counterpart in the areas where the latter is performing well, but rather to be used in those applications in which, due to the inherent limitations, the use of wet cleans will be restricted.

REFERENCES

1. Larrabee, G. and Chatterjee, P., *Semiconductor International* 14(6):90 (1991)
2. Ruzyllo, J., *Microcontamination* 6(3):39 (1988)
3. Ruzyllo, J., *Solid State Technology* 33:S1 (1990)
4. Moslehi, M. M., Chapman, R. A., Wong, M., Paranjpe, A., Najm, H. N., Kuehne, J., Yeakley, R. L. and Davis, C. J., *IEEE Trans. Electron Dev.* 39:4 (1992)
5. Tipton, C. M. and Bowling, R. A., in: *Proc. First Intern. Symp. on Cleaning Technol. in Semicond. Dev. Manufacturing,* (J. Ruzyllo and R. Novak, eds.), pp. 33-42, The Electrochem. Soc., Pennington, N. J. (1990)
6. Burkman, D., Peterson, C. A., Zazzera, L. A. and Kopp, R. J., in: *Semiconductor Processing Microcontamination* 6(11):57 (1983)
7. Hsu, E., Parks, H. G., Craigin, R., Tomaoka, S., Ramberg, J. S., and Lowry, R. K., in: *Proc. Second Intern. Symp. on Cleaning Technol. in Semicond. Dev. Manufacturing,* (J. Ruzyllo and R. Novak, eds.), pp. 170-178, The Electrochem. Soc., Pennington, N.J. (1992)
8. Kern, W., *J. Electrochem. Soc.* 137:1887 (1990)
9. Meuris, M., Heyns, M., Mertens, P., Verhaverbeke, S. and Philipossian, A., in: *Proc. Second Intern. Symp. on Cleaning Technol. in Semicond. Dev. Manufacturing,* (J. Ruzyllo and R. Novak, eds.), pp. 144-161, The Electrochem. Soc., Pennington, N.J. (1992); also Heyns, M. M., *Microcontamination*, 9(4):29 (1991)
10. Licciardello, A., Puglisi, C., and Pignataro, S., *Appl. Phys. Lett.* 48:41 (1986)
11. Ito, T., Sugino, R., Watanabe, S., Nara, Y., and Sato, Y., in: *Proc. First Intern. Symp. on Cleaning Technol. in Semicond. Dev. Manufacturing,* (J. Ruzyllo and R. Novak, eds.), pp. 114-120, The Electrochem. Soc., Pennington, N. J. (1990)
12. Stull, D. R., *Industrial and Engineering Chemistry*, 39:517 (1947)
13. McDermott, W. T., Ockovic, R. C., Wu, J. J., and Miller, R. J., *Microcontamination* 10:33 (1991)
14. Ghidini, G. and Smith, F. W., *J. Electrochem. Soc.* 131:2924 (1984)
15. Ohmi, T. and Shibata, T., *The Electrochem. Soc. Extended Abstracts*, No. 528, Vol. 91-2 (1991)

16. Comfort, J. H., Garverick, L. M., and Reif, R., *J. Appl. Phys.*, Part I, 62:3388 (1987); also Comfort, J. H., in: *Proc. Second Intern. Symp. on Cleaning Technol. in Semicond. Dev. Manufacturing,* (J. Ruzyllo and R. Novak, eds.), pp. 428-436, The Electrochem. Soc., Pennington, N.J. (1992)
17. Bean, J. C., Becker, G. E., Petroff, P. M., and Seidel, T. E., *J. Electrochem. Soc.*, 48:907 (1977)
18. Racanelli, M. and Greve, D. W., in: *Proc. Second Intern. Symp. on Cleaning Technol. in Semicond. Dev. Manufacturing,* (J. Ruzyllo and R. Novak, eds.), pp. 461-468, The Electrochem. Soc., Pennington, N.J. (1992)
19. Donahue, T. J. and Reif, R., *J. Appl. Phys.*, 57:2757 (1985)
20. Meyerson, B. S., Ganin, E., Smith, D. A., and Nguyen, T. N., *J. Electrochem. Soc.*, 133:1232 (1986)
21. Silvestri, V. J., Nummy, K., Ransheim, P., Bendernagel, R., Kerr, D., Phan, V. T., Borland, J. O., and Hann, J., *J. Electrochem. Soc.*, 137:2323 (1990)
22. Tsukune, A., Miyata, H., Mieno, F., Shimizu, A., and Furumura, Y., in: *Proc. Second Intern. Symp. on Cleaning Technol. in Semicond. Dev. Manufacturing,* (J. Ruzyllo and R. Novak, eds.), pp. 469-476, The Electrochem. Soc., Pennington, N.J. (1992)
23. Hossain, S., Pantano, C., and Ruzyllo, J., *J. Electrochem. Soc.*, 137:10 (1990)
24. Gluck, R. M., U.S. Patent 4,159,917, July 3, 1979; also Gluck, R. M., in: *Proc. Second Intern. Symp. on Cleaning Technol. in Semicond. Dev. Manufacturing,* (J. Ruzyllo and R. Novak, eds.), pp. 48-57, The Electrochem. Soc., Pennington, N.J. (1992)
25. Ridley, R. and Ruzyllo, J., in: *Proc. Second Intern. Symp. on Cleaning Technol. in Semicond. Dev. Manufacturing,* (J. Ruzyllo and R. Novak, eds.), pp. 98-104, The Electrochem. Soc., Pennington, N.J. (1992)
26. Moslehi, M. and Davis, C., *J. Mat. Res.*, 5:1159 (1990)
27. Ivankovits, J. C., Bohling, D. A., Lane, A., and Roberts, D. A., in: *Proc. Second Intern. Symp. on Cleaning Technol. in Semicond. Dev. Manufacturing,* (J. Ruzyllo and R. Novak, eds.), pp. 105-111, The Electrochem. Soc., Pennington, N.J. (1992)
28. Holmes, P. J. and Snell, J. E., *Microelectronics Reliab.*, 5:337 (1966)
29. Claevelin, C. R. and Duranko, G. T., *Semicond. Intern.*, 10(12):94 (1987)
30. Novak, R. E., *Solid State Technol.*, 31(3):39 (1988)

31. Ohmi, T., Miki, N., Kikuyama, H., Kawanabe, I., and Miyashita, M., in: *Proc. First Intern. Symp. on Cleaning Technol. in Semicond. Dev. Manufacturing,* (J. Ruzyllo and R. Novak, eds.), pp. 95-104, The Electrochem. Soc., Pennington, N.J. (1990); also *IEDM Tech. Digest,* p. 730 (1988)

32. Deal, B. E., McNeilly, M., Kao, D. B., and deLarios, J. M., in: *Proc. First Intern. Symp. on Cleaning Technol. in Semicond. Dev. Manufacturing,* (J. Ruzyllo and R. Novak, eds.), pp. 121-128, The Electrochem. Soc., Pennington, N.J. (1990); also *Sol. St. Technol.*, 33(7):73 (1990); also Helms, C. R. and Deal, B. E., in: *Proc. Second Intern. Symp. on Cleaning Technol. in Semicond. Dev. Manufacturing,* (J. Ruzyllo and R. Novak, eds.), pp. 267-276, The Electrochem. Soc., Pennington, N.J. (1992)

33. Wong, M., Moslehi, M. M., and Reed, D. W., *J. Electrochem. Soc.*, 138:1799 (1991).

34. Duranko, G., Syverson, D., Zazzera, L., Ruzyllo, J., and Frystak, D., in: *Physics and Chemistry of* SiO_2 *and the Si-*SiO_2 *Interface*, (B. E. Deal and C. R. Helms, eds.), pp. 429-436, Plenum Publishing Corp. (1988)

35. Witowski, B., Chacon, J., and Menon, V., in: *Proc. Second Intern. Symp. on Cleaning Technol. in Semicond. Dev. Manufacturing,* (J. Ruzyllo and R. Novak, eds.), pp. 372-397, The Electrochem. Soc., Pennington, N.J. (1992)

36. McIntosh, R., Kuan, T. S., and Defresart, E., *J. Electron. Mat.*, 21:57 (1992)

37. Jenson, M. and Syverson, D., *FSI Technical Report DC/DE*, TR 362 (Jan. 29, 1991)

38. Izumi, A., Matsuka, T., Takeuchi, T., and Yamano, A., in: *Proc. Second Intern. Symp. on Cleaning Technol. in Semicond. Dev. Manufacturing,* (J. Ruzyllo and R. Novak, eds.), pp. 260-266, The Electrochem. Soc., Pennington, N.J. (1992)

39. Ruzyllo, J., Torek, K., Daffron, C., Grant, R., and Novak, R., *J. Electrochem. Soc.* (in press)

40. Product Information, ClusterClean SP 200, SubMicron Systems, Inc., Allentown, Pennsylvania

41. Wong, M., Liu, D. K. Y., Moslehi, M. M., and Reed, D. W., *IEEE El. Dev. Lett.*, 12(8):425 (1991)

42. Sowell, R. R., Cuthrell, R. E., Mattox, D. M., and Bland, R. D., *J. Vac. Sci. Technol.*, 11:474 (1974)
43. Vig, J. R., *J. Vac. Sci. and Technol.*, A3:1027 (1985); also in: *Proc. First Intern. Symp. on Cleaning Technol. in Semicond. Dev. Manufacturing,* (J. Ruzyllo and R. Novak, eds.), pp. 105-113, The Electrochem. Soc., Pennington, N.J. (1990)
44. Ruzyllo, J., Duranko, G., and Hoff, A., *J. Electrochem. Soc.*, 134:2052 (1987)
45. Liehr, M., Offenberg, M., Kasi, S., Rubloff, G., and Holloway, K., *Proc. 22nd Intern. Conf. Sol. St. Dev. and Materials*, Sendai, Japan, pp. 1099-1102 (1990)
46. Norstrom, H., Ostling, M., Buchta, R., and Peterson, C. S., *J. Electrochem. Soc.*, 132:2285 (1985)
47. Tabe, M., *Appl. Phys. Lett.*, 45(10):1073 (1984)
48. Kasi, S. R. and Liehr, M., *Appl. Phys. Lett.*, 57:2095 (1990)
49. Ruzyllo, J., Duranko, G., Kennedy, J., and Pantano, C., in: *ULSI Science and Technology/1987*, (S. Broydo and C. M. Osburn, eds.), pp. 281-289, The Electrochemical Society, Pennington, N.J. (1987)
50. Ishikawa, Y., Takagi, Y., and Nakamichi, I., *Japan. J. Appl. Phys.*, 28:1453 (1989)
51. Sugino, R., Okuno, M., Shigeno, M., Sato, Y., Ohsawa, A., Ito, T., and Okui, Y., in: *Proc. Second Intern. Symp. on Cleaning Technol. in Semicond. Dev. Manufacturing,* (J. Ruzyllo and R. Novak, eds.), pp. 72-79, The Electrochem. Soc., Pennington, N.J. (1992)
52. Sato, Y., Sugino, R., Okuno, M., and Ito, T., *Proc. 22nd Conf. Solid State Dev. and Materials*, pp. 1103-1110, Japan Society of Physics, Sendai, 1990
53. Bat, S., deLarios, J., Doris, B., Gordon, M., Krusell, W., McKean, D., and Smolinsky, G., to be published in *Microcontamination.*
54. Torek, K. and Ruzyllo, J., in: *Proc. Second Intern. Symp. on Cleaning Technol. in Semicond. Dev. Manufacturing,* (J. Ruzyllo and R. Novak, eds.), pp. 80-86, The Electrochem. Soc., Pennington, N.J. (1992)
55. Hirose, M., Yokoyama, S., and Yamakage, Y., *J. Vac. Sci. Technol.*, B3:1445 (1985)
56. Grant, R., Torek, K., Novak, R., and Ruzyllo, J., Patent Application, (1992)

57. Anthony, B., Breaux, L., Hsu, T., Banerjee, S., and Tasch, A., *J. Vac. Sci. and Technol.*, B7:621 (1989); also Tasch A., Anthony, B., Banerjee, S., Hsu, T., Qian, R., Irby, J., and Kinosky, D., in: *Proc. Second Intern. Symp. on Cleaning Technol. in Semicond. Dev. Manufacturing,* (J. Ruzyllo and R. Novak, eds.), pp. 418-427, The Electrochem. Soc., Pennington, N.J. (1992)
58. Schneider, T. P., Cho, J., Aldrich, D. A., Chen, Y. L., Maher, D., and Nemanich, R. J., in: *Proc. Second Intern. Symp. on Cleaning Technol. in Semicond. Dev. Manufacturing,* (J. Ruzyllo and R. Novak, eds.), pp. 122-132, The Electrochem. Soc., Pennington, N.J. (1992)
59. Rudder, R. A., Fountain, G. G., and Markunas, R. J., *J. Appl. Phys.*, 60:3519 (1986)
60. Chung, B.-C., Delfino, M., Tsai, W., Salimian, S., and Hodul, D., in: *Proc. Second Intern. Symp. on Cleaning Technol. in Semicond. Dev. Manufacturing,* (J. Ruzyllo and R. Novak, eds.), pp. 87-97, The Electrochem. Soc., Pennington, N.J. (1992)
61. Ditizio, R., Fonash, S., and Leary, H., to be published.
62. Ruzyllo, J., Frystak, D. C., and Bowling, R. A., *IEDM Tech. Digest*, pp. 409-412 (1990)
63. Ruzyllo, J., Hoff, A., Frystak, D. C., and Hassain, S., *J. Electrochem. Soc.*, 136:1474 (1989)
64. Frystak, D. C. and Ruzyllo, J., *Proc. Seventh Symp. on Plasma Processing*, (G.S. Mathad, ed.), pp. 125- 136, The Electrochem. Soc., Pennington, N.J. (1988); also in: *Proc. First Intern. Symp. on Cleaning Technol. in Semicond. Dev. Manufacturing,* (J. Ruzyllo and R. Novak, eds.), pp. 129-140, The Electrochem. Soc., Pennington, N.J. (1992)

6

Ultraviolet-Ozone Cleaning Of Semiconductor Surfaces

John R. Vig

1.0 INTRODUCTION

The capability of ultraviolet (UV) light to decompose organic molecules has been known for a long time, but it is only since the mid-1970s that UV cleaning of surfaces has been explored (1)-(6). Since 1976, use of the UV/ozone cleaning method has grown steadily. UV/ozone cleaners are now available commercially from several manufacturers.

2.0 HISTORY OF UV/OZONE CLEANING

That ultraviolet light causes chemical changes has been generally known for a long time. Commonly known manifestations are the fading of fabric colors and changes in human skin pigmentation (i.e., sun tanning) upon exposure to sunlight. The chemical changes produced by short wavelength UV light inside the cells of living organisms can damage or destroy the cells. An important use of UV lamps has been as "germicidal" lamps, e.g., for destroying microorganisms in hospital operating rooms and in the air ducts of air conditioning systems (7).

In 1972, Bolon and Kunz (1) reported that UV light had the capability to depolymerize a variety of photoresist polymers. The polymer films were enclosed in a quartz tube that was evacuated and then backfilled with oxygen. The samples were irradiated with UV light from a medium-pressure mercury lamp that generated ozone. The polymer films of several thousand angstroms thickness were successfully depolymerized in less than one hour. The major products of depolymerization were found to be water and carbon dioxide. Subsequent to depolymerization, the substrates were examined by Auger electron spectroscopy (AES) and were found to be free of carbonaceous residues. Only inorganic residues, such as tin and chlorine, were found. When a Pyrex filter was placed between the UV light and the films, or when a nitrogen atmosphere was used instead of oxygen, the depolymerization was hindered. Thus, Bolon and Kunz recognized that oxygen and wavelengths shorter than 300 nm played a role in the depolymerization.

In 1974, Sowell et al. (2) described UV cleaning of adsorbed hydrocarbons from glass and gold surfaces, in air and in a vacuum system. A clean glass surface was obtained after fifteen hours of exposure to the UV radiation in air. In a vacuum system at 10^{-4} torr of oxygen, clean gold surfaces were produced after about two hours of UV exposure. During cleaning, the partial pressure of O_2 decreased, while that of CO_2 and H_2O increased. The UV also desorbed gases from the vacuum chamber walls. In air, gold surfaces which had been contaminated by adsorbed hydrocarbons could be cleaned by "several hours of exposure to the UV radiation." Sowell et al. also noted that storing clean surfaces under UV radiation maintained the surface cleanliness indefinitely.

During the period 1974 - 1976, Vig et al. (3)-(5) described a series of experiments aimed at determining the optimum conditions for producing clean surfaces by UV irradiation. The variables of cleaning by UV light were defined, and it was shown that, under the proper conditions, UV/ozone cleaning has the capability of producing clean surfaces in less than one minute.

To study the variables of the UV cleaning procedure, Vig and LeBus (5) constructed the two UV cleaning boxes shown in Fig. 1. Both were made of aluminum and both contained low-pressure mercury discharge lamps and an aluminum stand with Alzak (8) reflectors. The two lamps produced nearly equal intensities of short-wavelength UV light, about 1.6 mW/cm^2 for a sample 1 cm from the tube. Both boxes contained room air (in a clean room) throughout these experiments. The boxes were completely enclosed to reduce recontaminations by air circulation.

The tube of the UV lamp (8) in box 1 consisted of 91 cm of "hairpin-bent" fused quartz tubing. The fused quartz transmits both the 253.7 nm and the 184.9 nm wavelengths. The lamp emitted about 0.1 mW/cm^2 of 184.8 nm radiation measured at 1 cm from the tube. The lamp in box 2 had two straight and parallel 46 cm long high-silica glass tubes made of Corning UV Glass No. 9823, which transmits at 253.7 nm but not at 184.9 nm. Since this lamp generated no measurable ozone, a separate Siemens-type ozone generator (9) was built into box 2. This ozone generator did not emit UV light. Ozone was produced by a "silent" discharge when high-voltage AC was applied across a discharge gap formed by two concentric glass tubes, each of which was wrapped in aluminum foil electrodes. The ozone-generating tubes were parallel to the UV tubes, and were spaced approximately 6 cm apart. UV box 1 was used to expose samples, simultaneously, to the 253.7 nm and 184.9 nm wavelengths and to the ozone generated by the 184.9 nm wavelength. UV box 2 permitted the options of exposing samples to 253.7 nm plus ozone, 253.7 nm only, or ozone only.

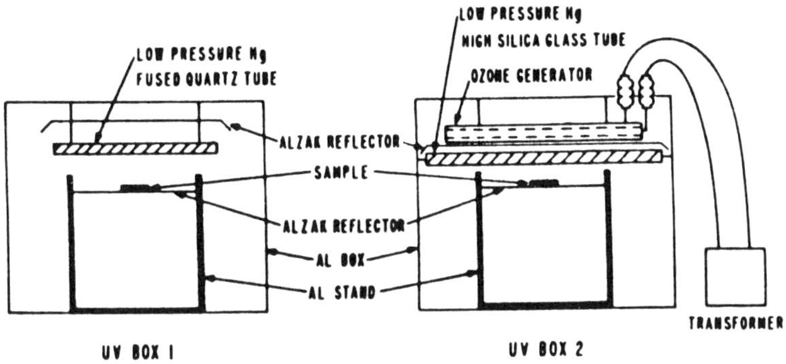

Figure 1. Apparatus for UV/ozone cleaning experiments.

Contact angle measurements, wettability tests, and Auger electron spectroscopy (AES) were used to evaluate the results of cleaning experiments. Most of the experiments were conducted on polished quartz wafers, the cleanliness of which could be evaluated by the "steam test," a highly sensitive wettability test (5)(11)(12). Contact angle measurements and the steam test can detect fractional monolayers of hydrophobic surface contamination.

Also tested was a "black-light," long-wavelength UV source that emitted wavelengths above 300 nm only. This UV source produced no noticeable cleaning, even after twenty-four hours of irradiation.

In the studies of Vig et al., it was found that samples could be cleaned consistently by UV/ozone only if gross contamination was first removed from the surfaces. The cleanliness of such UV/ozone-cleaned surfaces has been verified on numerous occasions, in the author's laboratory and elsewhere, by AES and electron spectroscopy for chemical analysis (ESCA) (1)(3)(4)(13)-(15). Figure 2 shows Auger spectra before and after UV/ozone cleaning (15). Ten minutes of UV/ozone cleaning reduced the surface contamination on an aluminum thin film to below the AES detectability level, about one percent of a monolayer. The effectiveness of UV/ozone cleaning has also been confirmed by ion scattering spectroscopy/secondary ion mass spectroscopy (ESS/SIMS) (16).

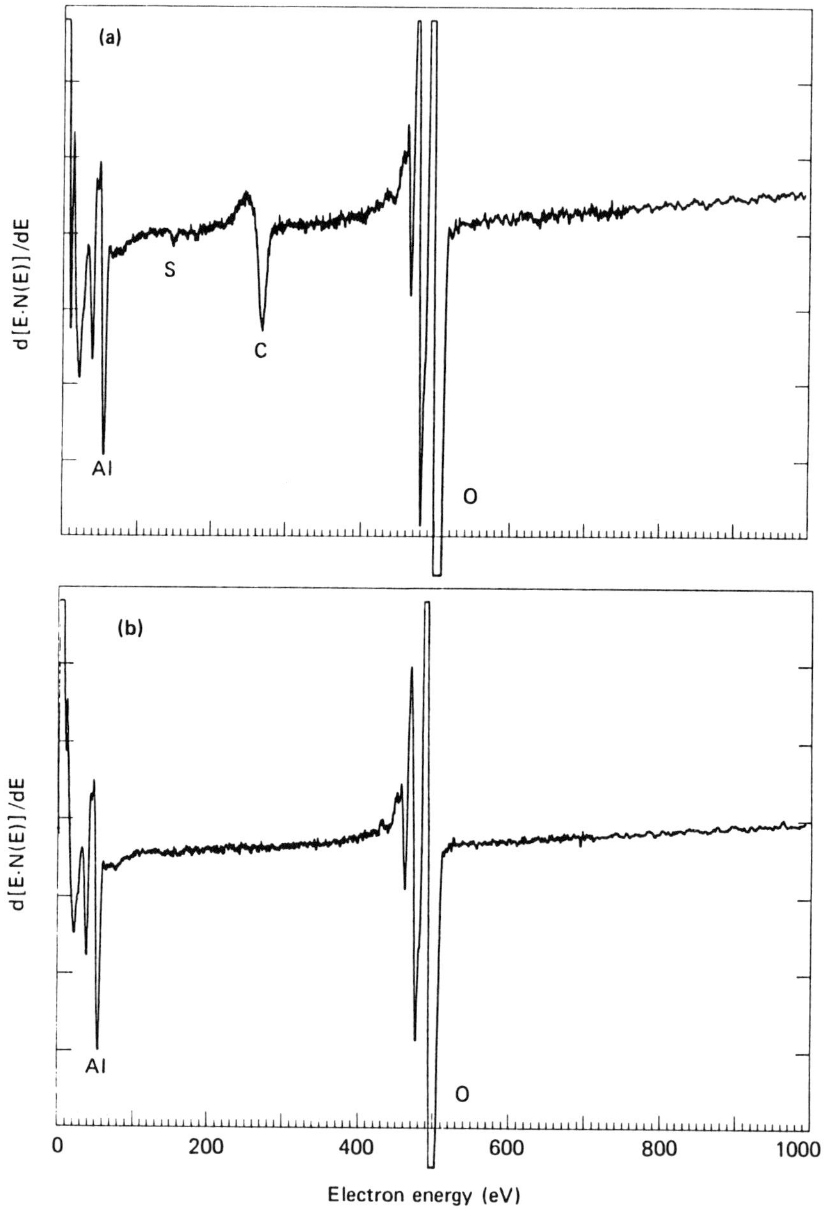

Figure 2. Auger spectra of evaporated aluminum film on silicon substrate: (a) before UV/ozone cleaning; (b) after UV/ozone cleaning.

A number of wafers of AT-cut quartz crystal were precleaned and exposed to the UV light in box 1 until clean surfaces were obtained. Each of the wafers was then thoroughly contaminated with human skin oil, which has been a difficult contaminant to remove. (The skin oil was applied by rubbing a clean wafer on the forehead of one of the researchers.) The wafers were precleaned again, groups of wafers were exposed to each of the four UV/ozone combinations mentioned earlier, and the time needed to attain a clean surface, as indicated by the steam test, was measured. In each UV box, the samples were placed within 5 mm of the UV source (where the temperature was about 70°C).

The wafers exposed to 253.7 nm + 184.9 nm + ozone in UV box 1 became clean in 20 seconds. The samples exposed to 253.7 nm + ozone in UV box 2 reached the clean condition in 90 seconds. Samples exposed to 253.7 nm without ozone and to ozone without UV light were cleaned within one hour and ten hours, respectively. The results are summarized in Table 1.

Table 1. Exposure Types vs. Cleaning Times

Exposure type	Time to reach clean condition
"Black light" (> 300 nm)	No cleaning
O_3, no UV	10 h
253.7 nm, no O_3	1 h
253.7 nm + O_3	90 s
253.7 nm + 184.9 nm + O_3	20 s

Although the 184.9 nm radiation is also absorbed by many organic molecules, it was not possible from these experiments to isolate the cleaning effect of the 184.9 nm radiation. The ozone concentrations had not been measured. As is discussed below, within each box the ozone concentrations vary with distance from the UV source. The UV/ozone cleaning method is now used in a variety of applications, in electronics, optics, and other fields.

3.0 VARIABLES OF UV/OZONE CLEANING

3.1 Wavelengths Emitted by the UV Sources

Since only the light that is absorbed can be effective in producing photochemical changes, the wavelengths emitted by the UV sources are important variables. The low-pressure mercury discharge tubes generate two wavelengths of interest: 184.9 nm and 253.7 nm. Whether or not these wavelengths are emitted depends upon the lamp envelopes. The emissions through the three main types of envelopes are summarized in Table 2. Pure quartz is highly transparent to both wavelengths.

Table 2. Principal Wavelengths of Low-Pressure Hg Discharge Lamps

	Lamp envelope*		
Wavelength (nm)	Fused quartz	High-silica glass	Glass
184.9	T	O	O
253.7	T	T	O
300.0	T	T	T

*T = transparent, O = opaque.

The 184.9 nm wavelength is important because it is absorbed by oxygen, thus leading to the generation of ozone (17), and it is also absorbed by many organic molecules. The 253.7 nm radiation is not absorbed by oxygen, therefore, it does not contribute to ozone generation, but is absorbed by most organic molecules (18)(19) and by ozone (17). The absorption by ozone is principally responsible for the destruction of ozone in the UV box. Therefore, when both wavelengths are present, ozone is continually being formed and destroyed. An intermediate product, both of the formation and of the destruction processes, is atomic oxygen, which is a very strong oxidizing agent. The absorption of either or both wavelengths by the organic and other contaminant molecules results in the dissociation or excitation of those molecules. The reaction of the atomic oxygen with excited or dissociated contaminant molecules is believed to be responsible for the cleaning action of UV/ozone, as is discussed below.

The absorption spectrum of oxygen is shown in Fig. 3 and that of ozone in Fig. 4. The effects of the principal wavelengths generated by low-pressure mercury discharge lamps are summarized in Table 3.

In the studies of Vig et al., wafers exposed to 253.7 nm + 184.9 nm + ozone became clean much faster than the samples exposed to 253.7 nm + ozone only, or to 253.7 nm without ozone, or to ozone without UV light, as is summarized in Table 1. Therefore, although both UV light without ozone and ozone without UV light can produce a slow cleaning effect in air, the combination of short-wavelength UV light and ozone, such as is obtained from a quartz UV lamp, produces a clean surface orders of magnitude faster.

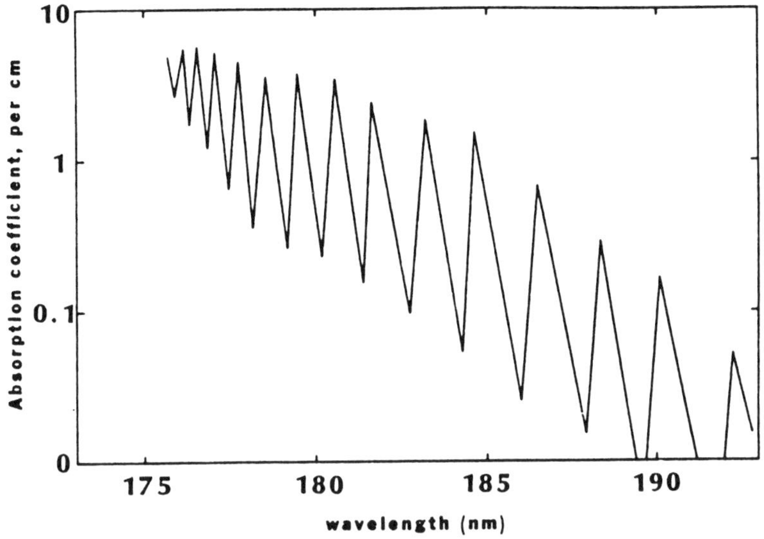

Figure 3. Absorption spectrum of oxygen.

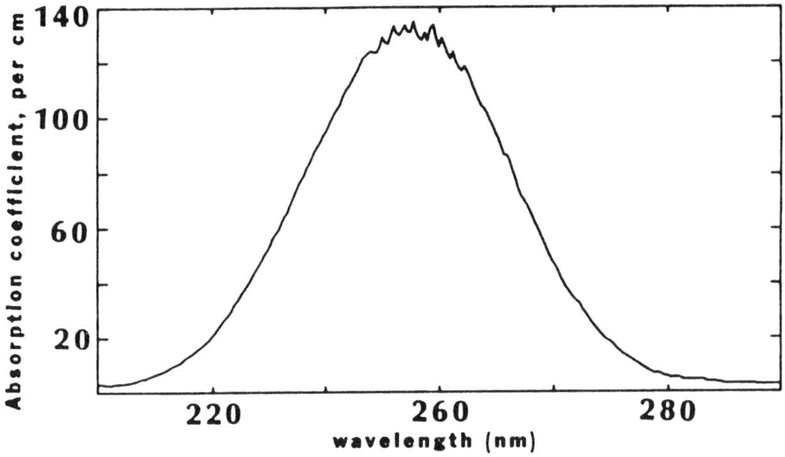

Figure 4. Absorption spectrum of ozone.

Table 3. Effects of the Principal Wavelengths Generated by Low-Pressure Hg Discharge Lamps

Wavelength (nm)	Effects
184.9	Absorbed by O_2 and organic molecules Creates atomic oxygen and ozone Breaks contaminant molecule bonds
253.7	Absorbed by organic molecules and O_3; not absorbed by O_2 Destroys ozone Breaks contaminant molecule bonds

3.2 Distance Between the Sample and UV Source

Another variable that can greatly affect the cleaning rate is the distance between the sample and the UV source. In Vig et al.'s experiment, the UV lamps were essentially plane sources. Therefore, one may conclude that the intensity of UV light reaching a sample would be nearly independent of distance. However, this is not so when ozone is present, because ozone has a broad absorption band (17)(20)(21) centered at about 260 nm, as is shown in Fig. 4. At 253.7 nm, the absorption coefficient is 130/cm·atm. The intensity I of the 253.7 nm radiation reaching a sample therefore decreases as

$$I = I_0 e^{-130pd}$$

where p is the average ozone pressure between the sample and the UV source in atmospheres at 0°C, and d is the distance to the sample in centimeters. When a quartz UV tube is used, both the ozone concentration and the UV radiation intensity decrease with distance from the UV source.

Two sets of identically precleaned samples were placed in UV box 2. One set was placed within 5 mm of the UV tube, the other was placed at the bottom of the box about 8 cm from the tube. With the ozone generator off, there was less than a thirty-percent difference in the time it took for the two sets of samples to attain a minimal (approximately 4°) contact angle, about 60 min vs. 75 min. When the experiment was repeated with the ozone generator on, the samples near the tube became clean nearly ten times faster (about 90 s vs. 13 min). Similarly, in UV box 1, samples placed within 5 mm of an ozone-producing UV tube were cleaned in 20 s vs. 20 - 30 min for samples placed near the bottom of the box at a distance of 13 cm. Therefore, to maximize the cleaning rate, the samples should be placed as close as practicable to the UV source.

3.3 Contaminants

Vig et al. tested the effectiveness of the UV/ozone cleaning procedure for a variety of contaminants. Among the contaminants were:

1. human skin oils (wiped from the forehead of one of the researchers)
2. contamination adsorbed during prolonged exposure to laboratory air
3. cutting oil (22)
4. beeswax and rosin mixture
5. lapping vehicle (23)
6. mechanical vacuum pump oil (24)
7. DC 704 silicone diffusion pump oil (25)
8. DC 705 silicone diffusion pump oil (25)
9. silicone vacuum grease (25)
10. acid (solder) flux (26)
11. rosin flux from a rosin core lead-tin solder
12. residues from cleaning solvents, including acetone, ethanol, methanol, isopropyl alcohol, trichloroethane, and trichlorotrifluoroethane.

After contamination the wafers were precleaned, then exposed to UV/ozone by placement within a few millimeters of the tube in UV box 1. After a 60 s exposure, the steam test and AES indicated that all traces of the contaminants had been removed.

Ion-implanted silicon wafers, each with approximately a 1 µm coating of exposed Kodak Micro Resist 747 (27), were placed within a few millimeters of the source in UV box 1. After an overnight (10 h) exposure to UV/ozone, all traces of the photoresist had been removed from the wafers, as confirmed by AES.

Films of carbon, vacuum-deposited onto quartz to make its surface conductive for study in an electron microscope, were also successfully removed by exposure to UV/ozone. Inorganic contaminants, such as dust particles, metals, and salts, cannot be removed by UV/ozone and should be eliminated in the precleaning procedure.

UV/ozone has also been used for waste-water treatment and for destruction of highly toxic compounds (28)-(31). Experimental work in connection with these applications has shown that UV/ozone can convert a wide variety of organic and some inorganic species to relatively harmless, mostly volatile products such as CO_2, CO, H_2O, and N_2. Compounds which have been destroyed successfully in water by UV/ozone include: ethanol, acetic acid, glycine, glycerol, palmitic acid; organic nitrogen, phosphorus and sulfur compounds; potassium cyanide; complexed Cd, Cu, Fe, and Ni cyanides; photographic wastes, medical wastes, secondary effluents; chlorinated organics and pesticides such as pentachlorophenol, dichlorobenzene, dichlorbutane, chloroform, malathion, Baygon, Vapam, and DDT. It has also been shown (32) that using the combination of UV and ozone is more effective than using either one alone in destroying microbial contaminants (*E. coli* and *streptococcus faecalis*) in water. UV/ozone has been used for the breakdown of PCBs (33). A combination of UV, ozone and hydrogen peroxide is used in a commercial water treatment method (34). The UV breaks the hydrogen peroxide into atomic oxygen and hydroxyl radicals. The hydroxyls assist with the breakdown of contaminant molecules. For example, benzene can be converted into carbon dioxide and water with this method.

Ozone, dissolved in fluorocarbon solvents, plus UV has been used in a chemical warfare agent decontamination system (35). The combined effect of ozone plus UV was found to be superior to either UV or ozone alone.

3.4 Precleaning

Contaminants, such as thick photoresist coatings and pure carbon films, can be removed with UV/ozone without any precleaning, but, in general, gross contamination cannot be removed without precleaning. For example, when a clean wafer of crystal quartz was coated thoroughly with human skin oils and placed in UV box 1 (Fig. 1) without any precleaning, even prolonged exposure to UV/ozone failed to produce a low-contact-angle surface, because human skin oils contain materials, such as inorganic salts, which cannot be removed by photosensitized oxidation.

The UV/ozone removed silicones from surfaces which had been precleaned, as described earlier, and also from surfaces which had simply been wiped with a cloth to leave a thin film. However, when the removal of a thick film was attempted, the UV/ozone removed most of the film upon prolonged exposure but it also left a hard, cracked residue on the surface, possibly because many chemicals respond to radiation in various ways, depending upon whether or not oxygen is present. For instance, in the presence of oxygen, many polymers degrade when irradiated; whereas, in the absence of oxygen (as would be the case for the bulk of a thick film) these same polymers crosslink. In the study of the radiation degradation of polymers in air, the "results obtained with thin films are often markedly different from those obtained using thick specimen..." (36).

For the UV/ozone cleaning procedure to perform reliably, the surfaces must be precleaned: first, to remove contaminants such as particles, metals, and salts that cannot be changed into volatile products by the oxidizing action of UV/ozone; and, second, to remove thick films the bulk of which could be transformed into a UV-resistant film by the crosslinking action of the UV light that penetrates the surface.

3.5 Substrate

The UV/ozone cleaning process has been used with success on a variety of surfaces, including glass, quartz, mica, sapphire, ceramics, metals, silicon, gallium arsenide, and a conductive polyimide cement. Quartz and sapphire are especially easy to clean with UV/ozone since these materials are transparent to short-wavelength UV.

For example, when a pile of thin quartz crystal plates, approximately two centimeters deep, was cleaned by UV/ozone, both sides of all the plates, even those at the bottom of the pile, were cleaned by the process. Since sapphire is even more transparent, it, too, could probably be cleaned the same way. When flat quartz plates were placed on top of each other so that there could have been little or no ozone circulation between the plates, it was possible to clean both sides of the plates by the UV/ozone cleaning method. It is interesting to note that Ref. 37 shows that photocatalytic oxidation of hydrocarbons, without the presence of gaseous oxygen, can occur on some oxide surfaces. This suggests that UV cleaning may also work on some surfaces in ultrahigh vacuum.

When white alumina ceramic substrates were cleaned by UV/ozone, the surfaces were cleaned properly. However, the sides facing the UV became yellow, probably due to the production of UV induced color centers. After a few minutes at high temperatures (>160°C), the white color returned.

Metal surfaces could be cleaned by UV/ozone without any problems, so long as the UV exposure was limited to the time required to produce a clean surface. (This time should be approximately one minute or less for surfaces which have been properly precleaned.) However, prolonged exposure of oxide-forming metals to UV light can produce rapid corrosion. Silver samples, for example, blackened within one hour in UV box 1 of Vig, et al. Experiments with sheets of Kovar, stainless steel (type 302), gold, silver, and copper showed that, upon extended UV irradiation, the Kovar, the stainless steel, and the gold appeared unchanged, whereas the silver and copper oxidized on both sides, but the oxide layers were darker on the sides facing away from the UV source. When electroless gold-plated nickel parts were stored under UV/ozone for several days, a powdery black coating gradually appeared on the parts. Apparently, nickel diffused to the surface through pinholes in the gold plating, and the oxidized nickel eventually covered the gold nearly completely. The corrosion was also observed in UV box 2, even when no ozone was being generated. The rates of corrosion increased substantially when a beaker of water was placed in the UV boxes to increase the humidity. Even Kovar showed signs of corrosion under such conditions.

The corrosion may possibly be explained as follows: as is known in the science of air-pollution control, in the presence of short wavelength UV light and impurities in the air, such as oxides of nitrogen and sulfur, combine with water vapor to form a corrosive mist of nitric and sulfuric acids. Therefore, the use of controlled atmospheres in the UV box may minimize the corrosion problem.

Since UV/ozone dissociates organic molecules, it may be a useful means of cleaning some organic materials, just as etching and electropolishing are sometimes useful for cleaning metals. The process has been used successfully to clean quartz resonators which have been bonded with silver-filled polyimide cement (38). Teflon (TFE) tape exposed to UV/ozone in UV box 1 for ten days experienced a weight loss of 2.5 percent (39). Also, the contact angles measured on clean quartz plates increased after a piece of Teflon was placed next to the plates in a UV box (40). Similarly, Viton shavings taken from an O-ring experienced a weight loss of 3.7 percent after 24 hours in UV box 1. At the end of the 24 hours, the Viton surfaces had become sticky. Semiconductor surfaces have been successfully UV/ozone-cleaned without adversely affecting the functioning of the devices. For example, after a 4 K static RAM silicon integrated circuit was exposed to UV/ozone for 120 min in a commercial UV/ozone cleaner, the device continued to function without any change in performance. This IC had been made using n-channel silicon gate technology, with 1 to 1.5 μm junction depths (41).

3.6 Rate Enhancement Techniques

UV/ozone cleaning "rate enhancement" techniques have been investigated by Zafonte and Chiu (42). Experiments on gas phase enhancement techniques included a comparison of the cleaning rates in dry air, dry oxygen, moist air, and moist oxygen. The moist air and moist oxygen consisted of gases that had been bubbled through water. Oxygen that had been bubbled through hydrogen peroxide was also tried. Experiments on liquid enhancement techniques consisted of a drop-wise addition either of distilled water or of hydrogen peroxide solutions of various concentrations to the sample surfaces. Most of the sample surfaces consisted of various types of photoresist on silicon wafers.

The gas-phase "enhancement" techniques resulted in negligible to slight increases in the rates of photoresist removal (3 - 20 Å /min without enhancement vs. 3 - 30 Å /min with enhancement). The water and hydrogen peroxide liquid-phase enhancement techniques both resulted in significant rate enhancements (100 - 200 Å /min) for resists that were not exposed to ion implantation. The heavily "ion implanted" resists (10^{15} to 10^{16} atoms/cm^2) were not significantly affected by UV/ozone, whether "enhanced" or not.

Photoresist removal rates of 800 to 900 Å /min for positive photoresists and 1500 to 1600 Å /min for negative photoresists (43) were reported by one manufacturer of UV/ozone cleaning equipment (43). The fast removal rate was achieved at 300°C by using a 253.7 nm source of UV, a silent discharge ozone generator, a heater built into the cleaning chamber, and oxygen from a gas cylinder to generate the ozone. A schematic drawing of this UV/ozone cleaner is shown in Fig. 5. The photoresist stripping rate vs. temperature for three different photoresists is shown in Fig. 6.

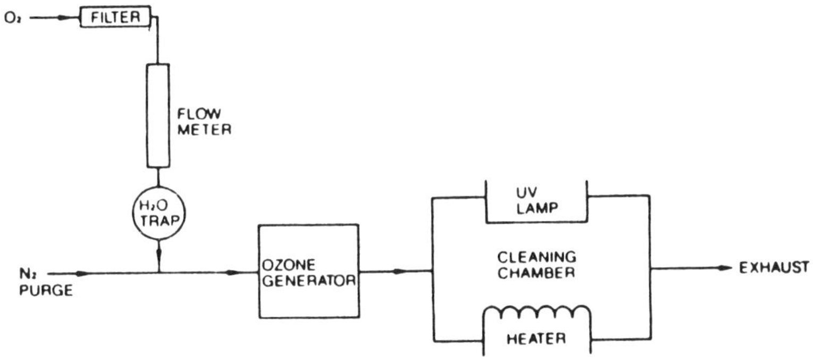

Figure 5. Schematic drawing of a UV/ozone cleaner that uses a silent-discharge ozone generator.

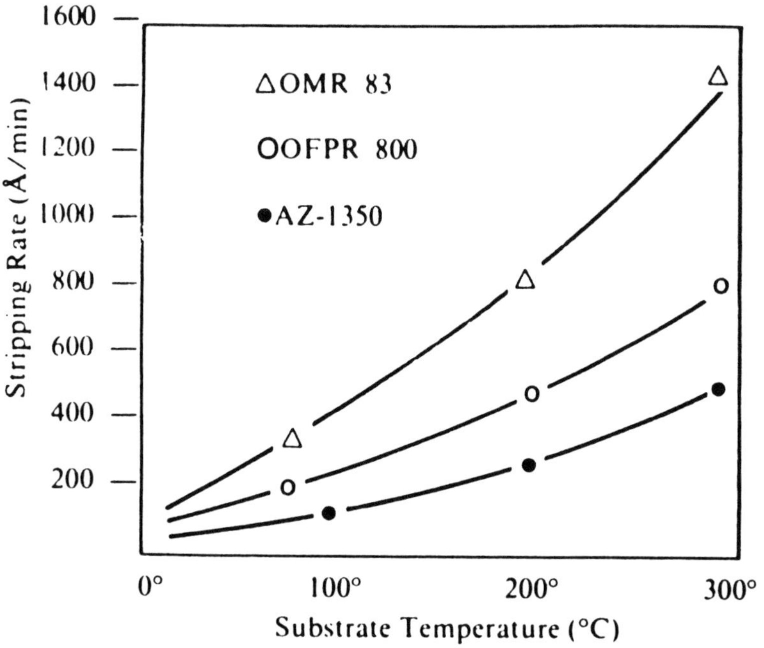

Figure 6. Photoresist stripping rate vs. substrate temperature for three types of photoresists.

4.0 MECHANISM OF UV/OZONE CLEANING

The available evidence indicates that UV/ozone cleaning is primarily the result of photosensitized oxidation processes, as is represented schematically in Fig. 7. The contaminant molecules are excited and/or dissociated by the absorption of short-wavelength UV light. Atomic oxygen and ozone are produced simultaneously when O_2 is dissociated by the absorption of UV with wavelengths less than 245.4 nm. Atomic oxygen is also produced when ozone is dissociated by the absorption of the UV and longer wavelengths of radiation (20)(21). The excited contaminant molecules and the free radicals produced by the dissociation of contaminant molecules react with atomic oxygen to form simpler, volatile molecules, such as CO_2, H_2O, and N_2.

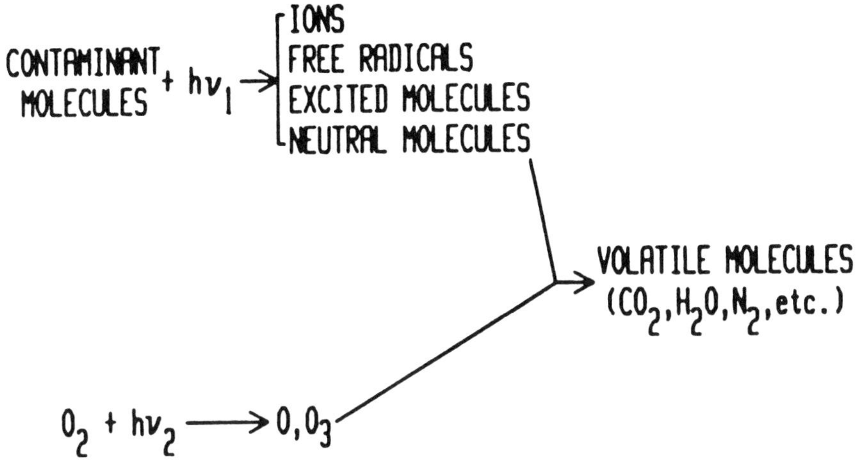

Figure 7. Simplified schematic representation of UV/ozone cleaning process.

The energy required to dissociate an O_2 molecule into two ground state O atoms corresponds to 245.4 nm. However, at and just below 245.4 nm the absorption of O_2 is very weak (17)(20)(21). The absorption coefficient increases rapidly below 200 nm with decreasing wavelengths, as is shown in Fig. 3. For producing O_3, a convenient wavelength is the 184.9 nm emitted by low-pressure Hg discharge lamps in fused quartz envelopes. Similarly, since most organic molecules have a strong absorption band between 200 nm and 300 nm, the 253.7 nm wavelength emitted by the same lamps is useful for exciting or dissociating contaminant molecules. The energy required to dissociate ozone corresponds to 1,140 nm; however, the absorption by ozone is relatively weak above 300 nm. The absorption reaches a maximum near the 253.7 nm wavelength, as is shown in Fig. 4. The actual photochemical processes occurring during UV/ozone cleaning are more complex than that shown in Fig. 7. For example, the rate of production of ozone by 184.9 nm photons is promoted by the presence of other molecules, such as N_2 and CO_2.

As was described previously, the combination of short-wavelength UV light and ozone produced clean surfaces about 200 to 2,000 times faster than UV light alone or ozone alone. Similarly, in their studies of wastewater treatment, Prengle et al. (28)-(31) found that UV enhances the reaction with ozone by a factor of 10^2 to 10^4, and the products of the reactions are materials such as CO_2, H_2O, and N_2. Increasing the temperature increased the reaction rates.

The physical and chemical mechanisms of hydrocarbon removal by UV/ozone cleaning was studied in an integrated processing facility with in situ analysis capabilities (44). On silicon wafers intentionally contaminated with heptanol, volatilization of the hydrocarbons appeared to proceed by means of oxidation of the long carbon chain at every carbon atom. In the final stage of the process, the carbon desorbed as CO_2 (but some carbon containing species remained on the surface). The desorbing gases were found to be mostly CO_2 and H_2O. UV/ozone cleaning at elevated (>60°C) temperature resulted not only in more complete hydrocarbon removal but also in the removal of surface OH-groups.

Mattox (45) found that mild heat increases the UV/ozone cleaning rates. Bolon and Kunz (1), on the other hand, found that the rate of ozone depolymerization of photoresists did not change significantly between 100°C and 300°C. The rate of destruction of microorganisms was similarly insensitive to a temperature increase from room temperature to 40°C (32). One manufacturer of UV/ozone cleaning equipment claims that the rate of photoresist stripping by UV/ozone increases several-fold as the temperature is raised from 20°C to 300°C (43).

5.0 UV/OZONE CLEANING IN VACUUM SYSTEMS

Sowell et al. (2) reported that, when 10^{-4} torr pressure of oxygen was present in a vacuum system, short-wavelength UV desorbed gases from the walls of the system. During UV irradiation, the partial pressure of oxygen decreased, while that of CO_2 and H_2O increased. Similar results were obtained by Hiroki et al. who found that upon turning on a short-wavelength UV lamp in a vacuum chamber, the outgassing of "H_2, CO, CO_2, CH_4, etc... increased, while H_2O and O_2 were slightly reduced" (46).

When UV/ozone cleaning of silicon surfaces in air was compared with UV/ozone cleaning in one to 20 torr of pure oxygen in a vacuum chamber, it was found that, whereas a clean surface could be produced in 7 seconds in air, it took about 50 minutes to produce the same cleanliness level in 5 torr of oxygen. It took 60 minutes in 20 torr of oxygen, and no cleaning effect was observed in 1 torr after 60 minutes of cleaning (47). (It should be noted, however, that the cleaning conditions differed. In the air cleaning, the UV source was 1 cm from the sample. In the vacuum chamber, the UV source was 6 cm from the sample and the UV passed through a quartz window before reaching the sample.)

A UV/ozone cleaning method that is suitable for use in an ultrahigh vacuum environment uses a low pressure Hg UV lamp and a separate ozone source (48). The ozone, generated in an oxygen glow discharge at liquid nitrogen temperatures, is admitted into the vacuum chamber through a valve. A slow cleaning action was observed at a 2×10^{-5} torr pressure. Using the same ozone source, cleaning was also observed without the UV light when the sample was heated to 500K (227°C). The ozone molecules that reach the sample surface decompose on the heated surface. The atomic oxygen created by the decomposition can react with the contaminant molecules.

One must exercise caution in using a mercury UV source in a vacuum system because, should the lamp envelope break or leak, mercury can enter, ruin the usefulness of the system and cause hazards due to its toxicity. Mercury has a high vapor pressure; its complete removal from a vacuum chamber is a difficult task. Other types of UV sources, such as xenon or deuterium lamps, may be safer to use in vacuum systems. The UV light can also be radiated into systems through sapphire or quartz windows, or through deep-UV fiber optic bundles. A small partial pressure of oxygen should be present during UV cleaning.

Caution must also be exercised when using UV/ozone in a cryopumped vacuum system, since cryopumped ozone is potentially explosive (49), particularly during regeneration of the cryopump. A convenient method of dealing with this potential hazard is to use two kinds of UV sources, one an ozone-generating source, the other an "ozone killer" source (50), as discussed in the next section.

Integrated processing systems, which incorporate UV/ozone cleaning, have been proposed (44)(51)-(53) and built (54)-(56) for processing devices in vacuum or in controlled atmospheres.

6.0 SAFETY CONSIDERATIONS

In constructing a UV/ozone cleaning facility, one must be aware of the safety hazards associated with exposure to short-wavelength UV light which can cause serious skin and eye injury within a short time. In the UV boxes used by Vig et al., switches are attached to the doors so that when the doors are opened the UV lamps are shut off automatically. If the application demands that the UV lamps be used without being completely enclosed (for example, as might be the case if a UV cleaning facility is incorporated into a wire bonder), then proper clothing and eye protection (e.g., UV safety glasses with side flaps) should be worn to prevent skin and eye damage.

Short-wavelength UV radiation is strongly absorbed by human cellular DNA. The absorption can lead to DNA-protein crosslinks, and can result in cancer, cell death, and cell mutation. It is now well-known that solar UV radiation is the prime causative factor in human skin cancer (57)(58), and is a significant risk factor in eye cancer (59). The 290 - 320 nm portion of solar UV radiation has been found to be the most effective wavelength region for causing skin cancer. Because the atmosphere filters out the shorter wavelengths, humans are not normally exposed to wavelengths as short as 254 nm. However, in a study of the effects of UV radiation on skin cancer rates, it was found that the 254 nm wavelength was many times more effective in causing cell mutations than were those above 300 nm wavelengths. Therefore, it is essential that personnel not be exposed to the short wavelengths needed for UV/ozone cleaning because even low doses of these wavelengths can cause significant damage to human cells. Safety exposure limits for ultraviolet germicidal irradiation have been set by the American Conference of Governmental and Industrial Hygienists (7)(60).

Another safety hazard is ozone, which is highly toxic. In setting up a UV cleaning facility, one must ensure that the ozone levels to which people are exposed do not exceed 0.1 ppm TWA and 0.3 ppm STEL (61). The time weighted average (TWA) exposure is a person's average airborne exposure in any 8 hour work shift of a 40 hour work week. The short term exposure limit (STEL) is a person's 15 minute time weighted average exposure which is not to be exceeded at any time during a work day. Ozone is a potential hazard in a cryopumped vacuum system because cryopumped ozone can become explosive under certain conditions (49).

One method of minimizing the hazards associated with ozone is to use two types of short-wavelength ultraviolet sources for UV/ozone cleaning (50): one, an ozone-generating UV lamp, e.g., a low-pressure mercury light in a fused quartz envelope, the other, a UV lamp that does not generate ozone but which emits one or more wavelengths that are strongly absorbed by ozone, e.g., a low-pressure mercury light in a high-silica glass tube which emits at 253.7 nm but not at 184.9 nm. Such a non-ozone generating UV source can be used as an "ozone killer." For example, in one cryopumped vacuum system, UV/ozone cleaning was performed in up to 20 torr of oxygen. After the cleaning was completed and the ozone-generating UV lamp was turned off, ten minutes of "ozone killer" UV light reduced the concentration of ozone to less than 0.01 ppm, a level that is safe for cryopumping (62). Therefore, with the "ozone killer" lamp, ozone concentrations were reduced by at least a factor of one hundred within ten minutes. Without the "ozone killer" lamp, the half-life of ozone is three days at 20°C (63).

The decomposition of ozone can also be greatly accelerated through the use of catalysts. For example, prior to 1980, in high-flying aircraft, ozone was found to be a causative factor for flight personnel and passengers experiencing headaches, eye, nose and throat irritations and chest pains. Passing the aircraft cabin air through a precious-metal catalytic converter reduced the ozone concentration from the 1 - 2 ppm level present in the troposphere to the low levels required for passenger comfort and safety (64).

7.0 CONSTRUCTION OF A UV/OZONE CLEANING FACILITY

The materials chosen for the construction of a UV/ozone cleaning facility should remain uncorroded by extended exposure to UV/ozone. Polished aluminum with a relatively thick anodized oxide layer, such as Alzak (8), is one such material. It is resistant to corrosion, has a high thermal conductivity, which helps to prevent heat buildup, and is also a good reflector of short-wavelength UV. Most other metals, including silver, are poor reflectors in this range.

Initially, Vig et al. used an ordinary shop-variety aluminum sheet for UV box construction, which was found not to be a good material because, in time, a thin coating of white powder (probably aluminum oxide particles) appeared at the bottom of the boxes. Even in a UV box made of standard Alzak, after a couple of years' usage, white spots appeared on the Alzak, probably due to pinholes in the anodization. The UV/ozone cleaning system should be inspected periodically for signs of corrosion to avoid the possibility of particles being generated. The use of "Class M" Alzak may also aid in avoiding particle generation, since this material has a much thicker oxide coating and is made for "exterior marine service," instead of the "mild interior service" specified for standard Alzak. Some commercially available UV/ozone cleaners are now constructed of stainless steel (65)(66). To date, no corrosion problems have been reported with such systems. The reflectance of stainless steel in the 200 to 250 nm range is about twenty percent (7).

Organic materials should not be present in the UV cleaning box. For example, the plastic insulation usually found on the leads of UV lamps must be replaced with inorganic insulation such as glass or ceramic. The box should be enclosed so as to minimize recontamination by circulating air, and to prevent accidental UV exposure and ozone escape.

The most widely available sources of short-wavelength UV light are the mercury arc lamps. Low-pressure mercury lamps in pure fused quartz envelopes operate near room temperature, emit approximately 90 percent at the 253.7 nm wavelength, and generate sufficient ozone for effective surface cleaning. Approximately five percent of the output of these lamps is at 184.9 nm. Medium- and high-pressure UV lamps (17) generally have a much higher output in the short-wavelength UV range. These lamps also emit a variety of additional wavelengths below 253.7 nm, which may enhance cleaning action. However, they operate at high temperatures (the envelopes are near red-hot), have a shorter lifetime, higher cost, and present a greater safety hazard. The mercury tubes can be fabricated in a variety of shapes to fit different applications. In addition to mercury arc lamps, microwave-powered mercury vapor UV lamps are also available (67).

Other available sources of short-wavelength UV include xenon lamps and deuterium lamps. These lamps must also be in an envelope transparent to short-wavelength UV, such as quartz or sapphire, if no separate ozone generator is to be used. In setting up a UV cleaning facility, one should choose a UV source which will generate enough UV/ozone to allow for rapid photosensitized oxidation of contaminants. However, too high an output at the ozone-generating wavelengths can be counterproductive because a high concentration of ozone can absorb most of the UV light before it reaches the parts to be cleaned. The parts should be placed as close to the UV source as possible to maximize the intensity reaching them. In the UV cleaning box 1 of Vig et al., the parts to be cleaned were placed on an Alzak stand the height of which can be adjusted to bring them close to the UV lamp. The parts to be cleaned can also be placed directly onto the tube if the box is built so that the tube is on the bottom of the box (68).

An alternative to using low-pressure mercury lamps in fused quartz envelopes is to use an arrangement similar to that of box 2, shown in Fig. 1. Such a UV/ozone cleaner, now also available commercially (43), uses silent-discharge-generated ozone and a UV source that generates the 253.7 nm wavelength, as is shown in Fig. 5. The manufacturer claims a cleaning rate that is much faster than that obtainable with UV/ozone cleaners that do not contain separate ozone generators. This cleaner also uses oxygen from a gas cylinder and a built-in sample heater that may further increase the cleaning rate.

8.0 APPLICATIONS

8.1 Cleaning of Silicon Surfaces

Photoresist removal (1)(6)(13)(43) and cleaning of silicon wafers for enhancing photoresist adhesion (69)-(71) and removing carbonaceous contamination have been primary applications of UV/ozone cleaning. The removal of carbonaceous contamination is important because, if carbon is not completely removed from the surface during the cleaning procedure, it can form silicon carbide on the surface at about 800°C that can be removed only by heating up to 1200°C (72).

Although wet-chemical cleaning has been widely used in the fabrication of semiconductor devices, as the device geometries have been reduced to submicron levels, the inherent shortcomings of wet-chemical cleaning methods have heightened interest in "dry" cleaning techniques (73). UV/ozone cleaning has been found to be a highly effective dry cleaning method for eliminating organic contaminants; it has also been found to lead to rapid oxidation of etched silicon surfaces (71)-(82). The oxide can be desorbed in vacuum at below 900°C to produce a contamination-free surface, as evidenced by Auger electron spectroscopy (71). UV irradiation using a high pressure mercury lamp and disilane gas at 20 torr and 730°C has also led to effective surface cleaning during silicon epitaxy (83).

When photochemical reactions (e.g., UV/ozone cleaning) were compared with plasma processes (e.g., plasma cleaning), the plasma processes were found to "cause harmful radiation damage. Moreover, because of the widely distributed electron energy in the plasma and the activation of a lot of reactions at the same time, the plasma process has poor controllability" (81).

A study of the surface chemistry of silicon wafers after various cleaning processes revealed that exposing anhydrous HF treated wafers to UV/ozone not only removed hydrocarbons and produced an oxide layer, but also removed the silicon fluoride species (76). When compared with wet cleaning techniques based on hydrogen peroxide, the UV/ozone was found to reduce the potential for contamination by the metallic impurities present in H_2O_2. "The cleanest silicon surface with respect to metallic and hydrocarbon impurities was achieved with a HF etch-H_2O rinse-UV/ozone oxidation process."

In another study, high resolution electron energy loss spectroscopy (HREELS) and other high sensitivity surface analytical techniques were used to investigate the mechanisms of hydrocarbon removal from Si wafers by UV/ozone and other cleaning techniques (82). A drop of a hydrocarbon (cyclohexane) was spun onto an HF-dip cleaned Si(100) surface. After the HF-dip cleaning, the surface was found to be hydrogen-passivated, i.e., saturated with SiH and SiH_2 groups. The adsorbed hydrocarbons did not replace the surface hydrogen, but adsorbed molecularly on top of the hydrogen. During the first 45 to 60 seconds of UV/ozone cleaning, at room temperature, very rapid oxidation of the Si surface occurred, and the hydrogen that saturated the surface after the HF dip was transformed into OH groups. Since hydrocarbons were still present at the end of this initial cleaning period, the oxidation apparently occurred underneath the contaminant layer. The UV/ozone treatment transformed the previously hydrophobic surface into a hydrophilic surface.

HREELS was also used to compare an RCA cleaned Si(100) surface with a UV/ozone cleaned one (82). The spectral signatures of the two surfaces were "very much alike". The authors conclude that "The UV-O_3 process is a gas-phase process that creates an oxide very similar to that after the standard RCA wet surface clean. The UV-O_3 process also removes hydrocarbons with similar efficiency as the RCA clean. Pending further study of metal removal, the UV-O_3 process seems a viable gas-phase replacement for the RCA clean" (82).

When several variations of HF treatments were examined for suitability as pretreatment for a silicon epitaxy process, the optimum treatment consisted of the steps of HF dipping, deionized water rinsing, nitrogen gas blowing for drying, and UV/ozone cleaning (77).

In the production of high-quality epitaxial films by molecular beam epitaxy (MBE), the cleaning of substrate surfaces is one of the most important steps. UV/ozone cleaning of silicon substrates in silicon MBE has been found to be effective in producing near defect-free MBE films (47)(81)(84)-(87). By using UV/ozone cleaning, the above 1200°C temperatures required for removing surface carbon in the conventional method can be lowered to well below 1000°C. The slip lines resulting from thermal stresses and thermal pits that are often produced by the high-temperature treatment are minimized in the lower temperature processing. Impurity redistribution in the substrate is also reduced.

Vacuum ultraviolet (VUV) light from a synchrotron source has been used in the low temperature cleaning of HF-passivated Si surfaces (88). In another study, VUV from a microwave-excited deuterium lamp was used in low-temperature (i.e., 650°C) silicon epitaxial growth (89). Organic contamination was effectively removed from Si, GaAs and MgF_2 substrates by

124 nm VUV radiation from a krypton source at 0.5 - 760 torr of air pressure (90). During Si molecular beam epitaxy, UV radiation from 193 nm ArF or 248 nm KrF lasers were found to enhance the interdiffusion of Si and B_2O_3, Sb incorporation, and the Hall mobility (91).

A two-step annealing method has been used to reduce the leakage currents in 64 Mbit silicon dynamic random-access memory chips (92). The first annealing step is at 300°C while the wafers are exposed to UV/ozone, and the second is at 800°C in dry oxygen.

In the processing of semiconductor wafers, a single UV/ozone exposure has been found to be capable both of "descumming" and of stabilizing (93). After developing and rinsing the photoresist pattern, the UV/ozone removes the thin layers of organic photoresist residue (scum) from the "clear" regions. The photoresist stabilization is believed to be due to crosslinking produced by the short-wavelength (deep) UV radiation (94). The stabilization rate is accelerated by increasing the temperature. For example, UV/ozone exposure times of 10 to 30 minutes from a 25 cm x 25 cm low-pressure mercury grid lamp at 100°C yields satisfactory results. The stabilized photoresist pattern exhibits *(i)* improved adhesion to the substrate, *(ii)* improved ability to maintain geometrical shape under thermal stress, and *(iii)* improved ability to withstand exposure to the etchants and solvents used to create the desired patterns in the circuit coatings (93).

UV/ozone cleaning has also been used in studies of the wetting of silicon and silica wafers (95)-(102). These studies included investigations of the evolution of tiny drops of polydimethylsiloxane and squalane on UV/ozone cleaned Si wafers, and the dynamics of ultra-thin wetting films under a controlled atmosphere. UV/ozone cleaning was also used in studies of surface-chemical reactions (103), and in the formation of diblock copolymer films on silicon wafers (104).

The use of UV/ozone treatment for the removal of contaminants from thin film transistors, and from substrates of complex composition or geometries has also been studied (105). Surfaces that were not directly irradiated by the UV became clean, but the required cleaning time was longer, in agreement with earlier results (3)-(6) on cleaning by UV/ozone vs. ozone alone. The chemistry of oxidized hydrocarbons on SiO_2 was found to differ from that on gold; UV/ozone was able to remove hydrocarbons from SiO_2 much faster than from gold. On thin film transistors, the UV/ozone cleaned the field oxide regions faster than the single component surfaces (105).

UV/ozone cleaning has been used in a variety of silicon processing studies. It was used as a precleaning step in investigations of: remote plasma cleaning using a hydrogen plasma (106), the breakup upon annealing of a thin oxide film between a polysilicon film and the silicon substrate (107), the effect of UV irradiation on minority-carrier recombination lifetime (108), and the chemical vapor deposition of titanium nitride onto silicon wafers (109).

8.2 Cleaning of Other Semiconductor Surfaces

UV/ozone cleaning has also been applied to the cleaning of gallium arsenide (GaAs) wafers (110)-(122), and to cleaning and "ozone etching" of indium phosphide (InP) substrates (122)-(126). In the growth of GaAs by molecular beam epitaxy and by chemical vapor deposition (CVD) substrate cleanliness is critically important. Contamination of the substrate/epitaxial layer interface leads to defects that reduce the yield of functional devices. Carbonaceous contamination is the primary problem. UV/ozone has been shown to be an effective means of removing carbonaceous contamination and, at the same time, producing an oxide-passivated surface (110). The carbon-free oxide can be removed by heating in ultrahigh vacuum prior to MBE layer growth.

The formation of a sacrificial oxide layer on GaAs is a well established step in the preparation of in situ cleaned substrates prior to MBE. A problem with oxides formed in air or in deionized (DI) water is that the oxide tends to be Ga rich, with As pile-up at the oxide/GaAs interface. When such an oxide is thermally desorbed, the stoichiometry of the surface is not preserved and the MBE layer is, thereby, degraded. In UV/ozone produced oxide layers, the As/Ga ratios and the As-oxide/Ga-oxide ratios are much closer to unity than for other oxidation methods (111)(118)(122). In the same amount of time, the UV/ozone also produces a much thicker oxide layer than air exposure. For example, in ten minutes, UV/ozone produces a 2.0 nm to 2.5 nm oxide layer. In air alone, 24 hours are required to produce the same thickness. A longer oxidation time increases the amount of adsorbed and absorbed carbonaceous contamination. The contamination rate of a UV/ozone produced oxide surface was found to be "at least an order of magnitude less" than that of a DI water produced oxide surface when observed for days in an x-ray photoelectron spectroscopy (XPS) system (111).

A problem with AlGaAs/GaAs heterostructure field effect transistors (HFETs) is sidegating, i.e., the electrical interaction between two closely spaced devices which were intended to be isolated from each other. Sidegating was traced to carbon contamination, presumably due to the adsorption of carbonaceous contamination from the atmosphere. When

UV/ozone cleaning was compared with other methods, "The carbon concentration of the interfacial region decreased by two orders of magnitude for the wafers exposed to ultraviolet radiation...A dramatic improvement in sidegating was observed for the wafers subjected to the ultraviolet-ozone cleaning procedure." (114). Similarly, another study found that "Ultraviolet/ozone cleaning of GaAs substrates prior to metalorganic molecular beam epitaxy at 500°C is shown to reduce the interfacial C and O concentrations by more than two orders of magnitude...UV/ozone cleaning...is a necessity for obtaining MESFET performance undegraded by parallel conduction from the substrate-epitaxial layer interface." (118).

In a study of light-enhanced oxidation of GaAs surfaces, it was found that photon energies higher than 4.1 eV (which is the energy needed to dissociate O_2^-) greatly enhanced the oxidation rate (117). The temperature at which the oxide desorbs from GaAs surfaces was found to be 638°C for UV/ozone produced oxide vs. 582°C for thermally produced oxide (120). When the native oxide and Fermi level of UV/ozone formed oxides on GaAs were investigated (127), it was found that the surface oxide consisted of a mixture of gallium and arsenic oxide phases which desorb at two different temperature ranges. Desorption of arsenic oxide phases and oxygen transfer from arsenic to gallium occurred at 250 - 500°C, and desorption of gallium oxide phases occurred at 550 - 600°C.

Oxide passivation with UV/ozone followed by thermal desorption also works well on InP. Epitaxial growth has successfully been carried out on InP surfaces so cleaned (122)(123). When the native oxides on InP surfaces were compared after solvent cleaning, etching with two different wet chemical etchants, and "ozone etching" with UV/ozone, the surface compositions were found to vary greatly with the surface treatment. The ozone-etched surface contained the most oxygen, and the In:P ratio increased as the surface treatment became more oxidizing (124). The oxides grown on InP can improve the electrical properties of InP interfaces (123).

8.3 Other Applications

The UV/ozone cleaning procedure is now used in numerous applications in addition to the cleaning of semiconductor surfaces. A major use is substrate cleaning prior to thin film deposition. The process is also being applied in a hermetic sealing method which relies on the adhesion between clean surfaces in an ultrahigh vacuum (14)(51)(128)(129). It has been shown that metal surfaces will weld together under near-zero forces if the surfaces are atomically clean. A gold gasket between gold metallized (UV/

ozone cleaned) aluminum oxide sealing surfaces is currently providing excellent hermetic seals in the production of a ceramic flatpack enclosed quartz resonator. It has also been shown (51)(128)(129) that it is feasible to achieve hermetic seals by pressing a clean aluminum gasket between two clean, unmetallized aluminum oxide ceramic surfaces.

The same adhesion phenomenon between UV/ozone cleaned gold surfaces has been applied to the construction of a novel surface contaminant detector (130)(131). The rate of decrease in the coefficient of adhesion between freshly cleaned gold contacts is used as a measure of the gaseous condensable contaminant level in the atmosphere.

The process has also been applied to improve the reliability of wire bonds, especially at reduced temperatures. For example, it has been shown (132)(133) that the thermocompression bonding process is highly temperature dependent when organic contaminants are present on the bonding surfaces. The temperature dependence can be greatly reduced by UV/ozone cleaning of the surfaces just prior to bonding, as is shown in Fig. 8. In a study of the effects of cleaning methods on gold ball bond shear strength, UV/ozone cleaning was found to be the most effective method of cleaning contaminants from gold surfaces (134). UV/ozone is also being used for cleaning alumina substrate surfaces during the processing of thin film hybrid circuits (135).

A number of cleaning methods were tested when the nonuniform appearance of thermal/flash protective electro-optic goggles was traced to organic contaminants on the electro-optic wafers. UV/ozone proved to be the most effective method for removing these contaminants, and thus it was chosen for use in the production of the goggles (136).

Other applications have been: photoresist removal (1)(6)(13)(43), the cleaning of vacuum chamber walls (2), photomasks (69), lenses (69), mirrors (69), solar panels (69), sapphire (69) (before the deposition of HgCdTe) and other fine linewidth devices (69)(70)(137), inertial guidance subcomponents (glass, chromium-oxide surfaced-gas bearings, and beryllium) (69)(138), the cleaning of stainless steel for studying a milk-stainless steel interface (139), the cleaning of amorphous alloy Metglas 2826 (140) and of sintered beryllium oxide (141), the cleaning of adsorbed species originating from epoxy adhesives (15), the removal of organic materials deposited during the deposition of antireflective silica coatings (142), the cleaning of surfaces prior to the deposition monolayer films (143)-(145), in a study of the frictional behavior of thin film magnetic disks (146), in friction studies in ultrahigh vacuum (147)(148), in studies of the spreading of liquid droplets (149), the cleaning of an x-ray grating which was carbon contaminated during synchrotron radiation (150), in the preparation of high temperature superconducting films (151), and in the fabrication of liquid crystal displays (152). Surface cleaning of niobium superconducting cavities with UV/ozone was found to result in RF performance that was superior to the performance of cavities cleaned by chemical or thermal methods (153). Since short-wavelength UV can generate radicals and ions, a side benefit of UV/ozone cleaning of insulator surfaces can be the neutralization of static charges (154).

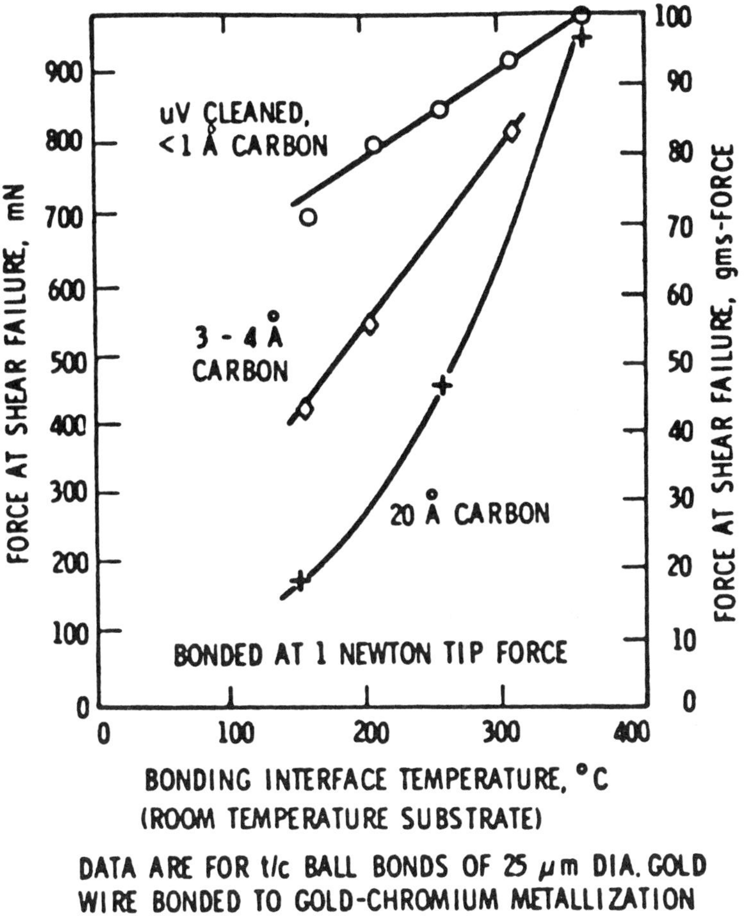

Figure 8. Effect of UV/ozone cleaning on gold-to-gold thermocompression bonding.

9.0 EFFECTS OTHER THAN CLEANING

Short-wavelength UV, ozone, and the combination of the two can have effects other than surface cleaning. The more significant of these effects are discussed below.

9.1 Oxidation

Ozone's oxidation power is second only to that of fluorine. Ozone can oxidize most inorganic compounds to their final oxidative state (63). For most substrates, UV/ozone cleaning, for the minimum time necessary to obtain a clean surface, will not cause a significant amount of oxidation. However, extended storage under UV/ozone may be detrimental for some oxidizable surfaces. In some cases, the enhanced oxide formation may be beneficial. For example, whereas the "native" oxide on GaAs is only about 3 nm thick, UV/ozone produces an oxide layer that is 10 - 30 nm thick (112), i.e., UV/ozone can produce a clean, oxide passivated surface. Similarly, the native UV/ozone-produced oxide layer at the interface of $HgCdTe/SiO_2$ has been found to enhance the interface properties (155). Solar radiation and atmospheric ozone have been found to markedly enhance the sulfidation of copper (156). Extended exposure to UV/ozone has been found to significantly increase the oxide layer thickness on aluminum surfaces (157). Whereas the oxide thickness on air-exposed aluminum surfaces is normally limited to about 50 Å, UV/ozone exposure increased the oxide layer thickness significantly beyond the "normal" 50 Å limit, as shown in Table 4.

Table 4. UV/Ozone Exposure vs. Oxide Thickness on Aluminum

Substrate treatment	Oxide thickness (Å)
Evaporate 1 μm of aluminum	47
10-minute UV/ozone cleaning	90
60-minute UV/ozone cleaning	200

When the oxidation of silicon surfaces was studied by Auger electron spectroscopy, it was found (72) that "an etched silicon surface can be exposed to air for about 1 hour without showing the silicon oxide signal at 82 eV. Under the UV/ozone treatment a rapid oxidation takes place. The peak characteristic of silicon oxide appears after one minute of irradiation. The increase of the intensity of this peak gives evidence for a thickening of the layer." Ten minutes of UV/ozone cleaning increased the oxide thickness on oxidized silicon substrates from 0.9 nm to 1.2 nm (85).

The ozone produced by a UV/ozone generator was found to enhance the growth rate of SiO_2 during the thermal oxidation of Si at 800°C (75). After a 140 minute oxidation period, the SiO_2 film thickness was 290 Å with ozone vs. 148 Å without ozone. The oxide growth rate enhancement decreased at higher temperatures and longer exposure times (i.e., with thicker films), presumably because "during the initial stage of silicon oxidation, the reaction at the silicon surface may be the controlling factor, whereas in the later stage, the diffusion of reactant through the oxide layer becomes important."

9.2 UV-Enhanced Outgassing

Short-wavelength UV has been found to enhance the outgassing of glasses (158). The UV light produced the evolution of significant quantities of hydrogen, water, carbon dioxide, and carbon monoxide. The hydrogen evolution was proportional to the amount of radiation incident on the samples. For UV-opaque glasses, the evolution occurred from the side exposed to the UV; for high-transmission samples, the gas evolved from both sides.

9.3 Other Surface/Interface Effects

Energetic radiation such as UV and gamma radiation has been reported to produce dehydration and the formation of free radicals on silica surfaces (159). However, dehydrated (or siloxinated) silica surfaces are hydrophobic (160)(161), whereas UV/ozone-cleaned silica (quartz) surfaces exhibit a very low (less than 4°) water contact angle, thus indicating that the UV/ozone does not dehydrate the surfaces, nor does it modify surface silanol groups the way high temperature vacuum baking does (162). UV/ozone has been shown to convert hydrophobic surfaces into hydrophilic ones. Short-wavelength UV has also been found to produce a bleaching effect in Si-Si_3 interfaces with thin oxides (163), and has also been found to produce yellowing (color centers) during the cleaning of aluminum oxide ceramics (39). The yellowing can be readily bleached by heating the sample to above 160°C.

9.4 Etching

Short-wavelength (193 nm) UV laser irradiation of biological and polymeric materials has been shown to be capable of etching the materials with great precision, via "ablative photodecomposition," and without significant heating of the samples. Linewidths 5 μm wide have been etched onto a plastic film to demonstrate the capability of this technique (164). Oxygen does not appear to have the same significance in this process as it does in UV/ozone cleaning. The etch depth vs. fluence in vacuum and in air were found to be the same (165).

In a study of the photodegradation of polyimide films, it was found that "the complete photooxidation process requires photolysis with light below 300 nm to produce both chain scission and photooxidative ablation efficiently," in the presence of oxygen (166).

UV light of wavelengths less than 200 nm has been proposed for selectively removing biological materials, e.g., skin lesions and decayed teeth (167). UV/ozone has been found to etch Teflon (39)(40), and Viton (39), and will likely etch other organic materials as well (168)(169). The susceptibility of polymers to degradation by ozone can be reduced by various additives and through the elimination of "the offending double bonds from the backbone structure of the polymers" (170). Vacuum ultraviolet radiation has been used to form images in polymer films (171)(172). Photoetching of polymer films with soft x-rays has also been studied (173). The etching of polymethyl mathacrylate (PMMA) by UV and VUV has also been investigated (174).

10.0 SUMMARY AND CONCLUSIONS

The UV/ozone cleaning procedure has been shown to be a highly effective method of removing a variety of contaminants from silicon, and compound semiconductor wafers, as well as from many other types of surfaces. It is a simple-to-use dry process that is inexpensive to set up and operate. It can produce clean surfaces at room temperature, either in a room atmosphere or in a controlled atmosphere. "The UV-O_3 process is a gas-phase process that creates an oxide very similar to that after the standard RCA wet surface clean. The UV-O_3 process also removes hydrocarbons with similar efficiency as the RCA clean" (82).

In combination with a dry method for removing inorganic contamination, such as cleaning with UV-excited high purity chlorine gas (175), the method may meet the requirements for the all-dry cleaning methods that will be necessary for processing of future generations of semiconductor devices. When compared with plasma cleaning, UV/ozone cleaning produces less radiation damage and is more controllable (81).

The variables of the UV cleaning procedure are the contaminants initially present, the precleaning procedure, the wavelengths and intensity emitted by the UV source, the atmosphere between the source and sample, the distance between the source and sample, and the time of exposure. For surfaces that are properly precleaned and placed within a few millimeters of an ozone-producing UV source, the process can yield a clean surface in less than one minute. The combination of short-wavelength UV plus ozone produces a clean surface substantially faster than short-wavelength UV without ozone or ozone without UV light. Clean surfaces will remain clean indefinitely during storage under UV/ozone, but prolonged exposure of oxide-forming metals to UV/ozone in room air can produce rapid corrosion.

The cleaning mechanism seems to be a photosensitized oxidation process in which the contaminant molecules are excited and/or dissociated by the absorption of short-wavelength UV. Simultaneously, atomic oxygen is generated when molecular oxygen is dissociated and when ozone is dissociated by the absorption of short and long wavelengths of radiation. The products of the excitation of contaminant molecules react with atomic oxygen to form simpler molecules, such as CO_2 and H_2O, which desorb from the surfaces.

REFERENCES

1. Bolon, D. A. and Kunz, C. O., *J. of Polymer Engin. Sci.* 12:109-111 (1972)
2. Sowell, R. R., Cuthrell, R. E., Mattox, D. M., and Bland, R. D., *J. Vac. Sci. Technol.* 11:474-475 (1974)
3. Vig, J. R., Cook, C. F. Jr., Schwidtal, K., LeBus, J. W., and Hafner, E., in: *Proc. 28th Ann. Symp. on Frequency Control,* Philadelphia, PA, AD 011113, pp. 96-108 (1974)
4. Vig, J. R., LeBus, J. W., and Filler, R. L., in: *Proc. 29th Ann. Symp. on Frequency Control,* Philadelphia, PA, AD A017466, pp. 220-229 (1975)
5. Vig, J. R., *IEEE Transactions on Parts, Hybrids, and Packaging,* PHP-12(4):365-370 (December 1976)

6. Vig, J. R., in: *Treatise on Clean Surface Technology, Vol. 1* (K. L. Mittal, ed.), pp. 1-26, Plenum Press, NY (1987)
7. Sylvania Engineering Bulletin 0-342, "Germicidal and Short-Wave Ultraviolet Radiation," GTE Products Corporation, Sylvania Lighting Center, Danvers, Massachusetts 01923 (1981)
8. Alzak is an aluminum reflector material with a corrosion-resistant oxide coating. The Alzak process is licensed to several manufacturers by the Aluminum Co. of America, Pittsburgh, PA 15219
9. Model No. R-52 Mineralight Lamp, UVP, Inc., San Gabriel, CA 91778.
10. See, e.g., *Encyclopaedic Dictionary of Physics,* 5:275, Pergamon Press, New York (1962)
11. Schrader, M. E., in: *Surface Contamination: Its Genesis, Detection and Control,* (K. L. Mittal, ed.), 2:541-555, Plenum Press, New York (1979)
12. Bryson, C. E., and Sharpen, L. J., in: *Surface Contamination: Its Genesis, Detection and Control,* (K. L. Mittal, ed.), 2:687-696, Plenum Press, New York (1979)
13. Holloway, P. H., and Bushmire, D. W., in: *Proc. 12th Ann. Reliability Phys. Symp.,* pp. 180-186, IEEE, Piscataway, NJ (1974)
14. Peters, R. D., *in: Proc. 30th Ann. Symp. on Frequency Control,* pp. 224-231, Philadelphia, PA, AD A046089 (1976)
15. Benson, R. C., Nall, B. H., Satkiewitz, F. G., and Charles, H. K. Jr., *Appl. Surf. Sci.* 21:219-229 (1985)
16. Braun, W. L., *Appl. Surf. Sci.* 6:39-46 (1980)
17. Calvert, J. G., and Pitts, J. N. Jr., *Photochemistry*, pp. 205-209, 687-705, John Wiley & Sons, New York (1966)
18. Fikhtengolts, V. S., Zolotareva, R. V., and L'vov, Yu A., *Ultraviolet Spectrum of Elastomers and Rubber Chemicals,* Plenum Press Data Div., New York (1966)
19. Lang, L., *Absorption Spectra in the Ultraviolet and Visible Region,* Academic Press, New York (1965)
20. McNesby, J. R., and Okabe, H., in: *Advances in Photochemistry,* (W. A. Noyes, Jr., G. S. Hammond, and J. N. Pitts, eds.), 3:166-174, Interscience Publishers, New York (1964)
21. Volman, D. H., in: *Advances in Photochemistry,* (W. A. Noyes, Jr., G. S. Hammond, and J. N. Pitts, eds.), 1:43-82, Interscience Publishers, New York (1963)
22. P. R. Hoffman Co., Carlisle, PA. 17013.

23. John Crane Lapping Vehicle 3M, Crane Packing Co., Morton Grove, IL 60053.
24. Welch Duo-Seal, Sargent-Welch Scientific Co., Skokie, IL 60076.
25. Dow Corning Corp., Midland, MI 48640.
26. Dutch Boy No. 205, National Lead Co., New York, NY 10006.
27. Eastman Kodak Co., Rochester, NY 14650.
28. Prengle, H. W. Jr., Mauk, C. E., Legan, R. W., and Hewes, C. G., *Hydrocarbon Processing* 54:82-87 (October 1975)
29. Prengle, H. W. Jr., Mauk, C. E., and Payne, J. E., *Forum on Ozone Disinfection (1976);* International Ozone Institute, Warren Bldg., Suite 206, 14805 Detroit Ave., Lakewood, OH 44107.
30. Prengle, H. W. Jr., and Mauk, C. E., *Workshop on Ozone/Chlorine Dioxide Oxidation Products of Organic Materials,* EPA/International Ozone Institute (November 1976)
31. Prengle, H. W. Jr., in: *Proc. International Ozone Symp.,* Warren Bldg., Suite 206, 14805 Detroit Avenue, Lakewood, OH 44107 (1978)
32. Zeff, J. D., Barton, R. R., Smiley, B., and Alhadeff, E., "UV-Ozone Water Oxidation/Sterilization Process," US Army Medical Research and Development Command, Final Report, Contract No. DADA 17073-C-3138, AD A0044205 (September 1974)
33. Raloff, J., *Science News* 132:154-155 (September 5, 1987)
34. Wiegner, K. K., *Forbes* p. 298 (July 22, 1991)
35. Snelson, A., Clark, D., and Brabets, R., *Proc. US Army Chemical Research, Development and Engineering Center Scientific Conf.* Chemical Defense Research, Aberdeen, Maryland, Vol. 1, AD-B113 947 (18-21 November 1986)
36. Boenig, H. V., *Structure and Properties of Polymers,* p. 246, Wiley, New York (1973)
37. Filimonov, V. N., *Elementary Photoprocesses in Molecules,* (B. S. Neporent, ed.), pp. 248-259, Consultants Bureau, New York (1968)
38. Filler, R. L., Frank, J. M., Peters, R. D., and Vig, J. R. *Proc. 32nd Ann. Symp. on Frequency Control,* Philadelphia, PA, pp. 290-298 (1978)
39. LeBus, J. W., and Vig, J. R., U.S. Army Electronics Technology and Devices Lab., Fort Monmouth, NJ 07703, unpublished information (1976)
40. Kusters, J., Hewlett Packard Co., Santa Clara, CA 95050, personal communication (1977)

41. Lasky, E., Aerofeed Inc., Chalfont, PA, personal communication (1978)
42. Zafonte, L., and Chiu, R., "Technical Report on UV-Ozone Resist Strip Feasibility Study," UVP, Inc., 5100 Walnut Grove Avenue, San Gabriel, CA 91778, September 1983; presented at the SPIE Santa Clara Conf. Microlithography in March 1984.
43. Application Note, "Photoresist Stripping With the UV-1 Dry Stripper," March Instruments Inc., Concord, CA 94520.
44. Kasi, S. R. and Liehr, M., *Appl. Phys. Lett.* 57(20):2095-2097 (12 November 1990)
45. Mattox, D. M., *Thin Solid Films* 53:81-96 (1978)
46. Hiroki, S., Abe, T., Murakami, Y., Kinoshita, S., Naganuma, T., and Adachi, N., *J. Vac. Soc. Japan* 31(10):850-853 (1989)
47. Kaneko, T., Suemitsu, M., and Miyamoto, N., *Japanese J. of Appl. Phys.* 28(12):2425-2429 (1989)
48. Lenssinck, J. M., Hoeven, A. J., van Loenen, E. J., and Dijkkamp, D., *J. Vac. Sci. Technol.* B9(4):1963-1966 (Jul/Aug 1991)
49. Chen, C. W., and Struss, R. G., *Cryogenics* 9:131-132 (April 1969)
50. Vig, J. R. and LeBus, J. W., "Method of Cleaning Surfaces by Irradiation With Ultraviolet Light," U.S. Pat. No. 4,028,135, issued June 7, 1977
51. Hafner, E., and Vig, J. R., "Method of Processing Quartz Crystal Resonators," U.S. Pat. No. 3,914,836, issued October 28, 1975.
52. Deal, B. E., McNeilly, M. A., Kao, D. B., and deLarios, J. M., *Solid State Technology* 73-77 (July 1990)
53. Offenberg, M., Liehr, M., and Rugloff, G. W., *J. Vac. Sci. Technol.* A9(3):10581065 (May/Jun 1991)
54. Ney, R. J. and Hafner, E., *Proc. 33rd Ann. Symp. on Frequency Control,* pp. 368- 373, AD-A213544 (1979)
55. Frank, J. M., *Proc. 35th Ann. Symp. on Frequency Control,* pp. 40-47, AD-A110870 (1981)
56. Liehr, M., *J. Vac. Sci. Technol. A,* 8(4):1939-1946 (May/June 1990)
57. Peak, M. J., Peak, J. G., and Jones, C. A., *Photochemistry and Photobiology* 42:141-146 (1985)
58. Kubitschak, H. R., Baker, K. S., and Peak, J. M., "Enhancement of Mutagenesis and Human Skin Cancer Rates Resulting From Increased Fluences of Solar Ultra-violet Radiation," to be published in *Photochemistry and Photobiology*

59. Tucker, M. A., Shields, J. A., Hartge, P., Augsburger, J., Hoover, R. N., and Fraumeni, J. F. Jr., *New England J. Medicine* 313:789-792 (1985)

60. American Conf. Governmental and Industrial Hygienists, Threshold Limit Values and Biological Exposure Indices for 1988-1989, ACGIH, Cincinnati, Ohio.

61. "Air Contaminants - Permissible Exposure Limits" (Title 29 Code of Federal Regulations Part 1910.1000), OSHA 3112, U. S. Department of Labor, Occupational Safety and Health Administration (1989)

62. Ehlers, D. A., "Ozone Generation and Decomposition by UV in the ERADCOM QXFF," Report No. PT81-004, General Electric Neutron Devices Dept., P.O. Box 2908, Largo, FL 34924 (January 26, 1981)

63. *Matheson Gas Data Book,* Published by Matheson Gas Products Co., East Rutherford, NJ, 6th Edition, pp. 574-577 (1980)

64. Bonacci, J. C., Egbert, W., Collins, M. F., and Heck, R. M., *International Precious Metals Institute Proceedings,* 1982; reprint and additional literature on DEOXO Catalytic Ozone Converters is available from Engelhard Corp., Specialty Chemicals Div., 2655 U.S. Rt. 22, Union, NJ 07083.

65. UVOCS Div., Aerofeed Inc. P.O. Box 303, Chalfont, PA 18914.

66. UVP, Inc., 5100 Walnut Grove Ave., San Gabriel, CA 91778.

67. Petelin, A. N., and Ury, M. G., in: *VLSI Electronics: Microstructure Science,* Vol. 8, (N. G. Einspruch, and D. M. Brown, eds.), Academic Press (1984)

68. Peters, R. D., General Electric Neutron Devices Dept., P.O. Box 2908, Largo, FL 34924, personal communication (1976)

69. Lasky, E., UVOCS Div., Aerofeed Inc., Chalfont, PA 18914, personal communication (1983)

70. Ruzyllo, J., Durnako, G. T., and Hoff, A. M., *J. Electrochemic. Soc.* 134(8):2052-2055 (August 1987)

71. Krusor, B. S., Biegelsen, D. K., Yingling, R. D., and Abelson, J. R., *J. Vac. Sci. Technol.* B7(1):129-130 (January/February 1989)

72. Baunack, S. and Zehe, A., *Phys. Stat. Sol.* 115(1):223-228 (16 September 1989)

73. Ruzyllo, J., *Microcontamination* 6(2):39-43 (March 1988)

74. Ruzyllo, J., Hoff, A. M., Frystak, D. C., and Hossain, S. D., *J. Electrochem. Soc.* 136 (5):1474-1476 (1989)

75. Chao, S. C., Pitchai, R., and Lee, Y. H., *J. Electrochem. Soc.* 136(9):2751-2752 (September 1989)

76. Zazzera, L. A. and Moulder, J. F., *J. Electrochem. Soc.* 136(2):484-491 (1989)

77. Suemitsu, M., Kaneko, T., and Miyamoto, N., *Japanese J. Appl. Phys.* 28(12):2421- 2424 (1989)

78. Hossain, S. D., Pantano, C. G., and Ruzyllo, J., *J. Electrochem. Soc.* 137(10):3287-3291 (1990)

79. Kao, D. B., deLarios, J. M., Helms, C. R., and Deal, B. E., *Proc. 27th Ann. IEEE Reliability Phys. Symp.*, pp. 9-16 (1989)

80. Baumgartner, H., Fuenzalida, V., and Eisele, I., *Appl. Phys.* A43:223-226 (1987)

81. Nara, Y., Yamazaki, T., Sugii, T., Sugino, R., Ito, T., and Ishikawa, H., *SPIE* 797:90-97 (1987)

82. Liehr, M., and Thiry, P. A., *J. Electron Spectroscopy and Related Phenomena* 54-55:1013-1032 (15 December 1990)

83. Yamazaki, T., Sugino, R., Ito, T., and Ishikawa, H., *Extended Abstr. 18th (1986 International) Conf. Solid State Devices and Materials,* pp. 213-216, Tokyo (1986)

84. Burger, W. R. and Reif, R., *J. Appl. Phys.* 62(10):4255-4268 (15 November 1987)

85. Tabe, M., *Appl. Phys. Lett.* 45:1073-1075 (1984)

86. Thornton, M. C. and Williams, R. H., *Physica Scripta* 41(6):1047-1052 (1990)

87. Sherman, A, *J. Vac. Sci. Technol.* B8(4):656-657 (July/August 1990)

88. Takakuwa, Y., Nogawa, M., Niwano, M., Katakura, H., Matsuyoshi, S., Ishida, H., Kato, H., and Miyamoto, N., "Low-Temperature Cleaning of HF-Passivated Si(111) Surface with VUV Light," Research Institute of Electrical Communication, Tohoku University, Sendai 980, pp. L1274-L1277 (1989)

89. Gonohe, N., Shimizu, S., Tamagawa, K., Hayashi, T., and Yamakawa, H., *Japanese J. Appl. Phys.* 26(7):L1189-L1192 (July 1987)

90. Dushenkov, S. D., Valiev, K. A., and Velikov, I. V., "Solid Surface Cleaning From Organic Substances by 124 nm Irradiation," Institute of Physics and Technology, USSR Academy of Sciences, Moscow, USSR, to be published (1991)

91. Rhee, S. S. and Wang, K. L., *Proc. Sec. International Symp. on Silicon Molecular Beam Epitaxy,* pp. 484-491 (1987)

92. Watson, G., *IEEE Spectrum*, p. 30 (January 1991)

93. Gardner, W. L., Engelhard Millis Corp., Millis, MA, 02054, personal communication (November 1985)

94. Matthews, J. C., and Wilmott, J. I. Jr., "Stabilization of Single Layer and Multilayer Resist Patterns to Aluminum Etching Environments," SPIE Conf., Optical Microlithography III, Santa Clara, CA, March 14-15, 1985; reprints available from Semiconductor Systems Corp., 7600 Standish Place, Rockville, MD 20855

95. Heslot, F., Cazabat, A. M., and Levinson, P., *The American Physical Soc.* 62(11):1286-1289 (13 March 1989)

96. Heslot, F., Cazabat, A. M., and Fraysse, N., *J. Phys.: Condens. Matter* 1:5793-5798 (1989)

97. Heslot, F., Fraysse, N., and Cazabat, A. M., *Nature* 338:640-642 (20 April 1989)

98. Heslot, F., Cazabat, A. M., Levinson, P., and Fraysse, N., *Physical Review Letters* 65(5):599-602 (30 July 1990)

99. Newcombe, G. and Ralston, J., *Langmuir* 8(1):190-196 (1992)

100. Silberzan, P. and Léger, L., *Macromolecules* 25(4):1267-1271 (1992)

101. Silberzan, P. and Léger, L., Étalement de Microgouttes de Polymères sur Surfaces Solides de Haute Énergie, *C. R. Acad. Sci. Paris* 312(II):1089-1094 (1991)

102. Redon, C., Brochard-Wyart, F., Hervet, H., and Rondelez, F., *Journal of Colloid and Interface Science,* 149(2):580-591 (15 March 1992)

103. Van Velzen, P. N. T., Ponjeé, J. J., and Benninghoven, A., *Appl. Surf. Sci.* 37:147-159 (1989)

104. Coulon, G., Collin, B., Ausserre, D., Chatenay, D., and Russel, T. P., *J. Phys. France* 51:2801-2811 (1990)

105. McIntyre, N. S., Davidson, R. D., Walzak, T. L., Williston, R., Westcott, M., and Pekarsky, A., *J. Vac. Sci. Technol.* A 9(3):1355-1359, (May/June 1991)

106. Cho, J., Schneider, T. P., VanderWeide, J., Jeon, H., and Nemanich, R. J., *Appl. Phys. Lett.* 59(16):1995-1997 (14 October 1991)

107. Ajuria, S. A. and Reif, R., *J. Appl. Phys.* 69:662-667 (15 January 1991)

108. Katayama, K., Kirino, Y., Iba, K., and Shimura, F., *Japanese Journal of Applied Physics* 30(11B):L1907-L1910 (November 1991)

109. Sherman, A., *J. Electrochem. Soc.* 137(6):1892-1897 (June 1990)

110. McClintock, J. A., Wilson, R. A., and Byer, N. E., *J. Vac. Sci. Technol.* 20:241-242 (February 1982)

111. Ingrey, S., Lau, W. M., and McIntyre, N. S., *J. Vac. Sci. Technol.* A4(3):984-988 (May/June 1986)
112. McClintock, J. A., Martin Marietta Laboratories, Baltimore, MD 21227, personal communication (1981)
113. Solomon, J. S., and Smith, S. R., *Mat. Res. Soc. Symp. Proc.* 54:449-454 (1986)
114. Gray, M. L., Reynolds, C. L., and Parsey, J. M., Jr., *J. Appl. Phys.* 68(1):169-175 (1 July 1990)
115. Gray, M. L., Yoder, J. D., and Brotman, A. D., *J. Appl. Phys.* 69:830-835 (15 January 1991)
116. Gray, M. L., Yoder, J. D., Brotman, A. D., Chandra, A., and Parsey, J. M., *J. Vac. Sci. Technol.* B9(4):1930-1933 (July/August 1991)
117. Yu, C. F., Schmidt, M. T., Podlesnik, D. V., Yang, E. S., and Osgood, R. M., Jr., *J. Vac. Sci. Technol.* A6(3):754-756 (May/June 1988)
118. Pearton, S. J., Ren, F., Abernathy, C. R., Hobson, W. S., and Luftman, H. S., *Appl. Phys. Lett.* 58(13):1416-1418 (1 April 1990)
119. Kopf, R. F., Kinsella, A. P., and Ebert, C. W., *J. Vac. Sci. Technol.* B9(1):132-135 (January/February 1991)
120. SpringThorpe, A. J., Ingrey, S. J., Emmerstorfer, B., Mandeville, P., and Moore, W. T., *Appl. Phys. Lett.* 50(2):77-79 (12 January 1987)
121. Flinn, B. J. and McIntyre, N. S., *Surface and Interface Analysis,* 15:19-26 (1990)
122. Lau, W. M., McIntyre, N. S., and Ingrey, S., in: *Semiconductor-Based Heterostructures: Interfacial Structure and Stability,* (M. L. Green, J. E. E. Baglin, C. Y. Chin, H. W. Deckman, W. Mayo, and D. Narasinham, eds.) The Metallurgical Society, Warrendale, PA, pp. 95-101 (1986)
123. Hollinger, G., Gallet, D., Gendry, M., Besland, M. P., and Joseph, J., *Appl. Phys. Lett.* 59(13):1617-1619 (23 September 1991)
124. Hoekje, S. J., and Hoflund, G. B., *Thin Solid Films* 197:367-380 (1991)
125. Ingrey, S., *J. Vac. Sci. Technol.* A10(4):829-836 (July/August 1992)
126. Gallet, D., Gendry, M., Hollinger, G., Overs, A., Jacob, G., Boudart, B., Gauneau, M., L'Haridon, H., Lecrosnier, D., *Journal of Electronic Materials* 20(12):963-965 (1991)
127. Lau, W. M., Sodhi, R. N. S., Jin, S., and Ingrey, S., *J. Vac. Sci. Technol. A,* 8(3):1899-1906 (May/June 1990)
128. Vig, J. R., and Hafner, E., "Packaging Precision Quartz Crystal Resonators," Technical Report ECOM-4134, US Army Electronics Command, Ft. Monmouth, NJ, AD 763215 (July 1973)

129. Wilcox, P. D., Snow, G. S., Hafner, E., and Vig, J. R., *Proc. 29th Ann. Symp. on Frequency Control,* Philadelphia, PA, pp. 202-210 (1975) See Ref. No. 4 for availability information.

130. Cuthrell, R. E. and Tipping, D. W., *Rev. Sci. Instrum.* 47:595-599 (1976)

131. Cuthrell, R. E., in: *Surface Contamination: Its Genesis, Detection and Control,* (K. L. Mittal, ed.) 2,:831-841, Plenum Press, New York (1979)

132. Jellison, J. L., *IEEE Trans. Parts, Hybrids, Packaging,* PHP-11, pp. 206-211 (1975)

133. Jellison, J. L., in: *Surface Contamination: Its Genesis, Detection and Control,* (K. L. Mittal, ed.), 2:899-923, Plenum Press, New York (1979)

134. Weiner, J. A., Clatterbaugh, G. V., Charles, H. K. Jr., and Romensko, B. M., *Proc. 33rd Electronic Components Conf.,* pp. 208-220 (1983)

135. Tramposch, R., *Circuits Manufacturing* 23:30-40 (March 1983)

136. Wagner, J. A., in: *Surface Contamination: Its Genesis, Detection and Control,* (K. L. Mittal, ed.), 2:769-783, Plenum Press, New York (1979)

137. Smith, H. I., Massachusetts Institute of Technology, unpublished class notes on "Cleaning of Oxides," and personal communications (1982)

138. Stemniski, J. R. and King, R. L. Jr., "Ultraviolet Cleaning: Alternative to Solvent Cleaning, Adhesives for Industry," pp. 212-228, Technology Conferences, El Segundo, CA (1980)

139. Almas, K. A., and Lund, B., *Surface Technology* 23:29-39 (1984)

140. Pregger, B. A. and Kramer, J. J., *IEEE Transactions on Magnetics* 25(5):3333-3335 (September 1989)

141. Musket, R. G., *Appl. Surf. Sci.* 37:55-62 (1989)

142. Musket, R. G. and Thomas, I. M., *J. Appl. Phys.* 66(10):5115-5118 (15 November 1989)

143. Silberzan, P., Léger, L., Ausserré, D., and Benattar, J. J., *Langmuir* 7(8):1647-1651 (1991)

144. McGarvey, C. E. and Holden, D. A., *Langmuir* 6(6):1123-1132 (1990)

145. Benattar, J. J., Daillant, J., Bosio, L., and Lé ger, L., *Colloque de Physique* C7(10):39-66 (October 1989)

146. Yang, M., Ganapathi, S. K., Balanson, R. D., and Talke, F. E., *5th Joint MMM-Intermag Conf., IEEE Trans. on Magnetics* (June 1991)

147. DeKoven, B. M. and Meyers, G. F., *J. Vac. Sci. Technol.* A9(4):2570-2577 (July/August 1991)

148. DeKoven, B. M. and Mitchell, G. E., *Applied Surface Science* 52:215-226 (1991)

149. Ondarçuhu, T. and Veyssié, M., *J. Phys. II* 1:75-85 (January 1991)

150. Harada, T., Yamaguchi, S., Itou, M., Mitani, S., Maezawa, H., Mikuni, A., Okamoto, W., and Yamaoka, H., *Appl. Optics* 30(10):1165-1168 (1 April 1991)

151. Nakayama, Y., Tsukada, I., and Uchinokura, K., *J. Appl. Phys.* 70:4371-4377 (15 October 1991)

152. Lasky, E., UVOCS, Inc., Montgomeryville, PA, private communication, August 1992.

153. Boussoukaya, M., *Nuclear Instruments and Methods in Phys. Re.*, North-Holland, Amsterdam, A245:13-19 (1986)

154. Baird, D. H., "Surface Charge Stability on Fused Silica," Final Tech. Report, TR 76-807.1, AD A037463 (December 1976)

155. Janousek, B. K., and Carscallen, R. C., *J. Vac. Sci. Technol.* pp. 195-198 (January/February 1985)

156. Graedel, T. E., Franey, J. P., and Kammlott, G. W., *Science* 224:599-601 (1984)

157. Clatterbaugh, G. V., Weiner, J. A., and Charles, H. K. Jr., *Proc. 34th Electronic Components Conf.*, pp. 21-30 (1984)

158. Altemose, V. O., in: *Methods of Experimental Physics,* (G. L. Weissler, and R. W. Carlson, eds.), Academic Press, New York, 14:329-333 (1979)

159. Tagieva, M. M. and Kiseler, V. F., *Russian J. Phys. Chem.* 35:680-681 (1961)

160. Hair, M. L., *Proc. 27th Ann. Symp. on Frequency Control,* AD 771042, pp. 73-78, (1973)

161. White, M. L., *Proc. 27th Ann. Symp. on Frequency Control,* AD 771042, pp. 79-88, 1973; also, in: *The Detection and Control of Organic Contaminants on Surfaces, Clean Surfaces: Their Preparation and Characterization for Interfacial Studies,* (G. Goldfinger, ed.), pp. 361-373, Marcel Dekker, Inc., New York (1970)

162. Lamb, R. N., and Furlong, D. N., *J. Chem. Soc., Faraday Trans.* 1(78):61-73 (1982)

163. Caplan, P. J., Poindexter, E. H., and Morrisson, S. R., *J. Appl. Phys.* 53:541-545 (1982)

164. Srinivasan, R., Conf. on Lasers and Electrooptics, as reported in: *Science News* 123:396 (June 18, 1983)

165. Srinivasan, R. and Braren, B., *J. Polymer Sci.: Polymer Chem. Ed.* 22:2601- 2609 (1984)

166. Hoyle, C. E. and Anzures, E. T., "Photodegradation of Polyimides.III. The Effect of Chemical Composition, Radiation Source, Atmosphere, and Processing," Office of Naval Research, AD-A236 197 (May 1991)

167. Blum, S. E., Srinivasan, R., and Wynne, J. J., "Far Ultraviolet Surgical and Dental Procedures," U.S. Pat. No. 4,784,135, issued November 15, 1988.

168. Alberts, G. S., "Process for Etching Organic Coating Layers," U.S. Pat. No. 3,767,490, issued October 23, 1973.

169. Wright, A. N., "Removal of Organic Polymeric Films from a Substrate," U.S. Pat. No. 3,664,899, issued May 23, 1972.

170. Robinson, L., *IEEE Electrical Insulation Mag.* 1:20-22 (1985)

171. Valiev, K. A., Velikov, L. V., Dushenkov, S. D., and Prokhorov, A. M., *Sov. Phys. Dokl.* 30(3):239-241 (March 1985)

172. Valiev, K. A., Velikov, L. V., Dushenkov, S. D., Mitrofanov, A. V., and Prokhorov, A. M., *Sov. Tech. Phys. Lett.* 8:15-16 (January 1982)

173. Valiev, K. A., Velikov, L. V., and Dushenkov, S. D., *Phys. Chem. Mech. Surfaces*, 4:1423- 1435 (1986)

174. Ueno, N., et al., *Polymers in Microlithography-Materials and Processes,* (E. Reichmanis, S. A. MacDonald, and T. Iwayanagi, eds.), pp. 425-436, American Chemical Society, ACS Symposium Series 412 (1989)

175. Ito, T., Sugino, R., Watanabe, S., Nara, Y., and Sato, Y., in: *Proceedings of the First International Symp. on Cleaning Technol. in Semiconductor Device Manufacturing,* 90-9:114-120, The Electrochemical Society, Inc., Pennington, NJ (1990)

7

Vapor Phase Wafer Cleaning Technology

Bruce E. Deal and C. Robert Helms

1.0 INTRODUCTION AND BACKGROUND

1.1 General

Cleaning of silicon wafers has been an integral part of semiconductor device fabrication since the 1950s, when silicon replaced germanium as the preferential semiconductor material. Now, as in the past, types of impurities encountered include oxides, heavy metals, alkali and other light elements, organics, and particles and residues of all kinds. In the proceedings of the first major symposium devoted to cleaning of electronic device components and materials in 1958 (1), two conclusions are apparent. First, similar procedures were used for cleaning silicon and germanium wafers and for tube materials. Second, until recently, the same types of aqueous batch cleaning processes have been used continuously during device fabrication (2), even though device complexity has increased from single transistors on 15 mm diameter wafers to complex integrated circuits containing more than one million components on 200 mm diameter wafers. Furthermore, as feature size decreased from mils to angstroms, and process complexity increased from the use of a few masking steps to twenty or more, and as process steps now number in the several hundred compared to a little more than a dozen in 1960, many of the process procedures used to fabricate devices have remained relatively unchanged in principal. The same is also true for the materials used in the device structures.

The above-mentioned conservative nature of semiconductor process and device engineers has continued to be apparent in the cleaning area as was demonstrated by a survey report a few years ago. In this report, it was concluded that "only one-third of the industry is working on new cleaning or monitoring methods." The "remainder are content in their ways" (3). This finding was even more surprising when it is considered that at that time, newer methods of masking lithography, etching, metal deposition, oxidation, and other processes were being actively pursued and implemented.

Fortunately, more recently, newer methods of wafer cleaning have been developed and are starting to be implemented. One of these, vapor phase cleaning, is the subject of this chapter. After a brief review of trends in wafer cleaning since 1960 and the conclusion that many of the so-called "improved" methods also have had shortcomings, vapor phase cleaning technology is discussed. First, historical developments of vapor cleaning are summarized, with discussions of the various types of systems and processes used. Next, the subject of vapor phase etching of oxides and dielectrics is reviewed, including a discussion of mechanisms involved. A critical aspect of wafer cleaning is the removal of trace impurities from the wafer surface and both analysis techniques (briefly summarized) as well as removal procedures using vapor phase technology are discussed. Perhaps the most important, and certainly the subject of greatest interest to the device engineer, is the effectiveness of the cleaning procedures with respect to device applications. Such results are presented in some detail for particular areas. Finally, recent developments involving advanced processing of semiconductor device wafers are reviewed. It is observed that cleaning and related technologies may actually "drive" the overall direction of device manufacturing. This is already resulting in the development and evaluation of various types of in situ, sequential processing involving cluster tool systems, also referred to as integrated processing. This concept is discussed in detail. The chapter ends with a summary of possible future trends in vapor cleaning of silicon wafers, including device requirements, possible new vapor chemistries, and other related topics.

It should be noted that most of the cleaning technology discussed in this chapter deals with silicon wafers. This is because a large percentage of today's semiconductor devices are silicon, and extensive cleaning efforts on other materials cannot be justified at present. On the other hand, cleaning problems associated with compound semiconductors are even more complex than silicon and will require even more effort to solve them. Undoubtedly, many of the advantages of vapor cleaning discussed in this chapter can also apply to such materials as gallium arsenide. Consequently, as this newer technology matures, it will very likely be used to advance the state-of-the-art processing for compound semiconductors.

1.2 Aqueous Cleaning Processes

It was mentioned above that the original wafer cleaning methods employed in the late 1950s and early 1960s were a carry-over from tube material cleaning procedures. In the late 1960s, the invention of integrated circuits and MOS device structures required improved procedures for cleaning wafers. These have been reported by Kern and Puotinen (4)-(7) and others (8)-(11), but still resemble the same type of aqueous processes used earlier. Typical sequences included solutions of acids (H_2SO_4, HCl) or bases (NH_4OH) with hydrogen peroxide, followed by appropriate deionized water rinses. In addition, an aqueous HF treatment was frequently used in various parts of the sequence. Kern has recently reviewed these developments (6) and they are also discussed in Ch. 3 of this volume. Some of the shortcomings of aqueous cleaning technology include:

1. Ineffective in cleaning small geometries
2. Contributes additional contamination and particles
3. High cost of chemicals
4. Environmental problems

Attempts have been made to overcome some of these problems. Approaches have included spray cleaning (12), displacement processing (13), jet aerosol cleaning (14), megasonics (15), recirculation of chemicals (16), and others (17). Along with the cleaning solutions themselves, severe problems have also been attributed to the deionized water rinse and drying steps. More often than not, the wafers were cleaner before these steps than after. Even with considerable efforts to provide satisfactory aqueous cleaning procedures, it was not believed that such cleaning technology would keep up with the requirements of the rapidly advancing device complexity.

1.3 "Dry" Cleaning Processes

Another approach to wafer cleaning involved the so-called "dry" processing. As early as 1960, when epitaxial technology was developed, it was found that a gas-phase in situ hydrochloric acid predeposition etch improved the epi-film quality (18)-(21). Soon, process engineers were experimenting with various gases, such as chlorine (22), hydrogen fluoride-hydrogen iodide (23), and other halides (24)(25). Because of the high temperatures required, it wasn't long before some sort of excitation was employed, such as reactive ions, ultraviolet light, plasmas, electron cyclotron resonance, and RF sputter or microwave radiation, in order to lower the process temperature. While these processes have demonstrated some value in wafer cleaning, it is now apparent that various problems are associated with them (26)-(28). The main difficulty has to do with the adverse radiation effects on the device structures themselves (29)-(31). In addition, various contamination problems are observed, both from bombardment of materials in the systems, and due to the formation of undesirable reaction byproducts from the complex chemistries involved. Some of these byproducts, such as polymers, are extremely difficult to remove from the wafer surfaces.

1.4 Other Types of Cleaning Processes

One other type of non-aqueous cleaning procedure that has been reported by several investigators is so-called ice, jet, or snow scrubbing (32)(33). In this method, jets of solid H_2O or CO_2 ice particles are directed at the wafer surface at high velocity. Some success has been reported in removing various organic films, but problems of non-uniformity and surface damage have prevented this procedure from moving from the laboratory to production.

An important type of dry cleaning procedure used to clean silicon wafers has involved ultraviolet light with ozone (UV/O_3) (34)(35). As in the case with aqueous cleaning procedures, various "dry" cleaning techniques including UV/O_3 and other related technologies are discussed more completely in Chs. 5 and 6 of this volume.

2.0 VAPOR CLEANING

2.1 Historical

The concept of vapor phase or gas processes for removing oxides has been known for a number of years. The original paper describing vapor HF removal of silicon oxide was written by Holmes and Snell in 1966 (36). A schematic diagram of the apparatus used is shown in Fig. 1. In their experiments, the authors suspended an oxidized silicon wafer above an aqueous solution of hydrofluoric acid. The solution was agitated by dry argon which also helped to transport the HF/H_2O to the wafer. Oxide etch rates were determined as a function of HF concentration and temperature. The authors observed that "the rate of attack of the oxide by the vapor close above the etching bath was comparable with that in the bath." Two other key results from this early work were that the reaction rate followed the familiar "volcano plot" peaking in their experiments at 20° to 30°C and falling to zero rapidly at higher (~40°C) temperatures. This plot is reproduced in Fig. 2. They also suggested the importance of surface condensation which, as will be shown below, is a key factor in oxide etching. These and other results of Holmes and Snell provide insight into etch mechanisms only now being completely understood (37)-(39). Those authors made use of information concerning vapor pressures of HF and H_2O in aqueous HF solutions at various temperatures reported in the 1940s by Brosheer and co-workers (40), by Munter and co-workers (41)(42), and by others (43)-(45). These data are still valid today.

Subsequent to the work of Holmes and Snell, Beyer and Kastl (46) also investigated a process based on vapor phase HF/H_2O mixtures for etching SiO_2 as did Blackwood, Biggerstaff, Clements, and Cleavelin (47). These investigators also observed the results to be similar to those obtained in HF/H_2O liquid phase processing.

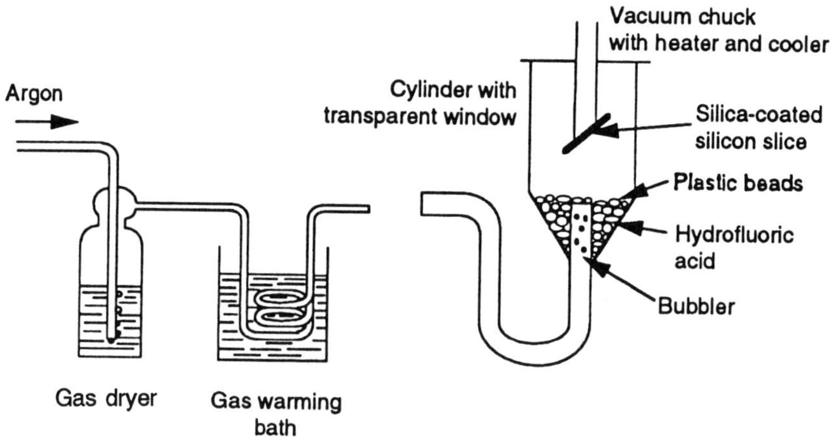

Figure 1. Equipment for experiments on etching of silica with a controlled flow of hydrofluoric acid vapor; after Holmes and Snell (36).

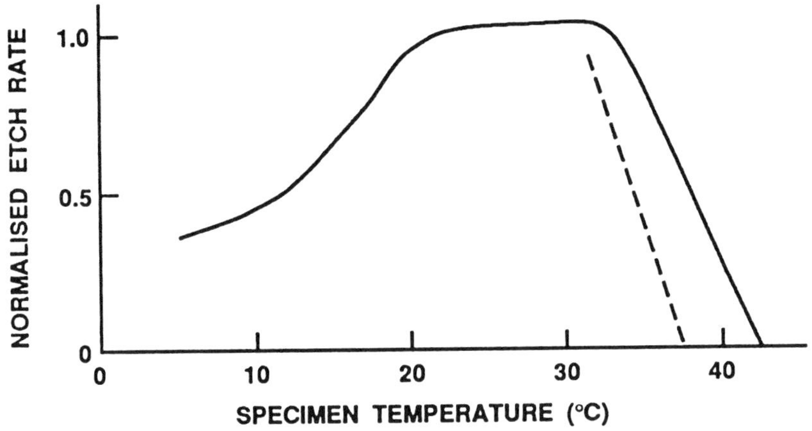

Figure 2. Etch rate of silica by hydrofluoric acid vapor at 24°C as a function of specimen temperature; after Holmes and Snell (36). Full line - 40% acid; broken line - dilute acid.

As noted later in this chapter, the mechanism for HF etching of SiO_2 involves an initial requirement for water availability before etching begins, even though water is a product of the reaction. An interesting application related to this effect was reported in 1977 by Bersin and Reichelderfer (48). In this case, the oxide etching was carried out selectively under a negative photoresist layer which contained enough water to initiate the reaction. In areas with no resist, the oxide etching did not occur.

About the same time as Holmes' work with vapor HF etching of silicon oxide, the use of HCl was reported as a pre-epitaxial silicon clean of the silicon substrate (18)(19). It was shown that by exposing the silicon substrate to HCl vapor at 1150° to 1250°C immediately prior to epitaxial deposition (in situ), low defect epi material could be deposited. The trend today towards low temperature (<1000°C) silicon epi deposition has caused process engineers to develop other in situ cleans such as vapor HF, but HCl is still used commercially in most production applications.

As mentioned earlier, a number of other gases have been evaluated during the past twenty-five years for cleaning or etching silicon as well as silicon dioxide surfaces. These have included halogen fluorides, such as ClF_3, BrF_3, and IF_3 (49), and other fluorides including NF_3, BF_3, PF_3, and PF_5 (50). Some of these procedures involved excitation of some kind, such as plasma; others did not. As of this date, few, if any, have been used for high volume device manufacturing applications.

In the employment of aqueous cleaning processes, various types of drying techniques have been used (51)-(54). These normally follow some sort of deionized water rinse, either a dip or spray. The drying steps consisted of blowing nitrogen, spinning, heating (in a hot inert ambient, vacuum, or infrared radiation), or immersing in a volatile organic liquid or vapor, such as isopropanol (53). Most of these drying techniques would be considered *vapor* or *dry* in nature. Certainly they could be compatible with vapor type cleaning processes. The subject of drying silicon surfaces in general is not completely understood or modeled from a mechanism standpoint, and is discussed more completely in a later section.

2.2 Advantages of Vapor Cleaning

During the development of improved wafer cleaning processes over a number of years, distinct advantages of vapor or gas phase cleaning have been demonstrated. Some of these are indicated here.

Reduced Contamination. One of the problems associated with conventional aqueous cleaning processes has been the redepositing of contaminants of various types back on the wafers. This has been observed in both the cleaning solution itself, as well as in the deionized water rinsing and drying steps. In the case of vapor processes, this self-contamination has not been observed, and very good results are typically obtained.

Improved Process Uniformity. Typically, as smaller and controlled amounts of reactants are used in a semiconductor fabrication process, better control is achieved for both the specific reaction involved and the overall process. This improvement has been noted in various process steps, and a good example is ion implantation versus furnace pre-deposition. Similarly, gas phase reactions of vapor etching have also resulted in superior control of etch selectivity, uniformity, and repeatability.

Reduced Chemical Usage and Disposal. An important consideration in today's manufacturing processes is that of environmental effects. Even though electronics manufacturing has been referred to as the "clean" industry, nevertheless, severe contamination problems have occurred in certain areas such as "Silicon Valley," California. Mainly affected have been ground water supplies through faulty storage tanks. Since gas or vapor phase cleaning typically uses less than 1/100 of a given chemical than does aqueous cleaning, considerable reduction of environmental problems as mentioned above will occur. Thus, the cost of the manufacturing is reduced as well as environmental contamination effects. An example of how vapor phase cleaning could replace several liquid cleaning, rinsing, and drying steps in a typical wafer processing sequence (MOS gate oxidation) is shown in Fig. 3.

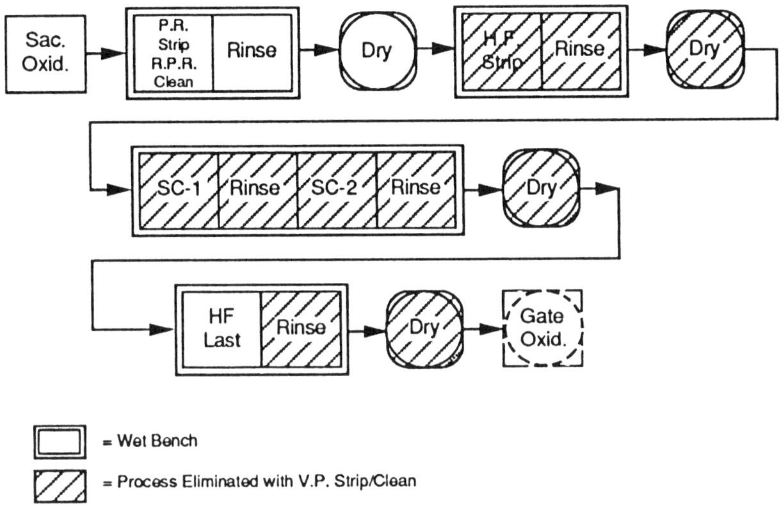

Figure 3. Potential replacement of individual processes in conventional aqueous cleaning sequence by vapor phase cleaning technology. *(Courtesy of Advantage Production Technology).*

Improved Safety Considerations. Along the lines of environmental improvements discussed above, vapor phase processing also provides a much safer working condition than encountered in typical *wet-bench* cleaning facilities. In addition, device processing may be carried out much more efficiently. Vapor cleaning is more adaptable to computer controlled processing and results are significantly improved because of this.

More Versatile Process Variables. Gas or vapor type processing allows much more versatility in the range of process variables permitted than do liquid or aqueous systems. For instance, it is much easier to vary either temperature or pressure in vapor systems. Thus, relative compositions of gas mixtures can be varied over a wide range, and resulting chemistries can be adjusted for particular applications. Furthermore, the volatility of certain reaction products can be maximized. Also, at reduced pressures it is possible for the reactants to penetrate narrow openings and to effect removal of unwanted materials from these openings.

New Chemistries Possible. Just about any liquid chemistry can be converted to gas or vapor phase. In addition, many additional gas species are available for improved chemical cleaning applications. It is anticipated that such new chemistries will be employed for the more efficient removal of all types of contaminants.

Sequential, In Situ Processing Possible. Perhaps one of the greatest advantages of vapor phase wafer cleaning will be its suitability for so-called integrated or cluster-type processing. Such capability will permit various pre-process cleaning steps for the more critical applications. This subject is discussed more completely in Sec. 7.

2.3 Current Vapor Cleaning Systems

At the present time, at least two systems employing vapor phase chemistries are being used for cleaning silicon wafers. These are briefly described below in chronological order.

Anhydrous Hydrogen-Fluoride/Water-Vapor. The industry's first commercial vapor phase etching tool was FSI International's Excalibur* system (55)-(57). Designed jointly with Texas Instruments for the etching of SiO_2, the system was introduced commercially in 1987. Operating at ambient temperature and pressure, the single wafer system provides a controlled mixture of anhydrous HF and H_2O vapor to the process chamber using programmable mass flow controllers. The N_2 carrier gas, anhydrous HF and the H_2O vapor ratios can be varied to satisfy selectivity requirements for etching dissimilar oxides simultaneously. Figure 4 provides a schematic diagram of the key components.

The initial systems used in production were etch-only systems that proved effective for thin etches, especially native oxide removal before CVD processes. An integrated rinse feature, added later, provided improved performance by reducing particles on the wafer and removing metallic and dopant residues resulting from various etches required for pre-CVD pre-gate and pre-contact processes (58).

Subsequent product improvements have included the addition of anhydrous HCl for enhanced metallic contamination removal, ozone gas for light organic cleaning and re-oxidation of silicon, a nitrogen ambient wafer staging area and loadlock capability for downstream processes.

Applications for the FSI system have included silicide cleaning, pre-gate cleans including the etching of the sacrificial oxide before gate

---

*Excalibur is a registrered trademark of FSI International.

oxidation, and pre-metal contact cleans, especially where CVD technology was chosen over PVD for first level metal contacts. Additional applications include pre-epi and pre-polysilicon deposition cleans and bond pad etching.

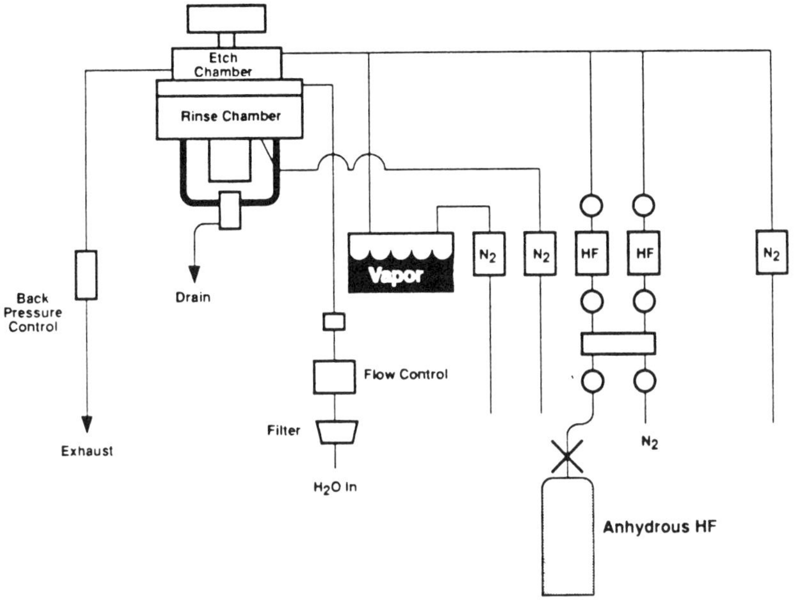

Figure 4. Schematic of Excalibur integrated vapor etch/rinse system. *(Courtesy of FSI International).*

Controlled etch rates can vary from 1 to 300 Å/second with total one sigma uniformity of ±2%, within wafer, wafer to wafer and batch to batch on a 200 Å etch of thermal oxide. The system is particle neutral at sizes equal to or greater than 0.2 μm. Oxide etch rates as a function of anhydrous HF concentration for various total flow rates are shown in Fig. 5.

Vapor Phase Reactants. Another type of wafer cleaning system involves *vapor phase* chemistries and was originally based on the vaporization of azeotropic mixtures of hydrogen fluoride and/or hydrogen chloride and water vapor (59)(60). This system is manufactured by Advantage Production Technology, Inc.; a schematic of the initial model (EDGE-2000) is shown in Fig. 6. The chamber of this system is about the size of a basketball and is constructed of a specially formulated silicon carbide material by Norton Co. (61). It consists of two hemispheres and a center ring. It will accommodate one wafer, up to 200 mm in diameter, positioned vertically. Gases are admitted through one side and evacuated from the other. In the first model, wafers were maintained at room temperature (22° - 25°C) but in later versions, higher wafer temperatures were employed (see below).

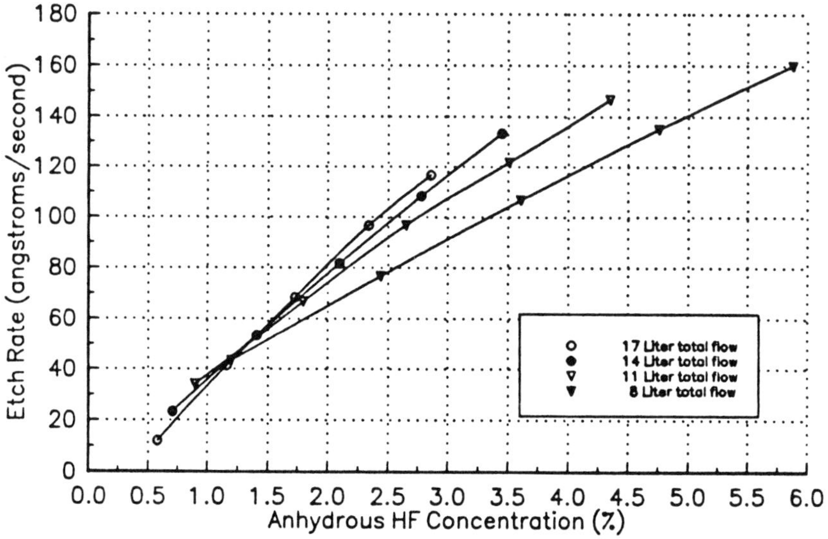

Figure 5. Thermal oxide etch rate versus anhydrous HF concentration at 25°C for various total flow rates in the Excalibur wafer cleaning system. *(Courtesy of FSI International).*

Vaporizers, shown in Fig. 6, can contain solutions of HF/H_2O, HCl/H_2O, H_2O, or various solvents such as CH_3OH. The solutions are heated to appropriate temperatures to provide various vapor pressures of the reactants. Vapors of these reactants are transported to the chamber by inert gases such as nitrogen or argon, whose flow rates are controlled by mass flow controllers. All materials in contact with the acids (HF, HCl, etc.) are silicon carbide or Teflon* based.

Also shown in Fig. 6 is a vacuum pump which permits the system to be evacuated to the low mtorr range. Thus, vapor phase cleaning and etching processes can be carried out at reduced pressure which assists in vaporization of reactants, reaction products, and contaminants. A typical cleaning sequence is indicated in Fig. 7, where a pressure-time cycle is shown. At the start of the cycle, the wafer is automatically inserted into the

*Teflon is a registered trademark of Dupont.

chamber, the door is closed, and the chamber evacuated to about 1 torr for a few seconds (I). Reactant gases are then admitted to the chamber to pressures ranging from 100 to 400 torr for times of 10 to 60 seconds (II). The system is then again evacuated to the low mtorr range for a few seconds (III), after which N_2 or Ar is readmitted to the chamber. As noted in a following section, the final evacuation or desorb step dries the wafer and no further drying is necessary. This overall cycle may be varied and is programmed and controlled by computer.

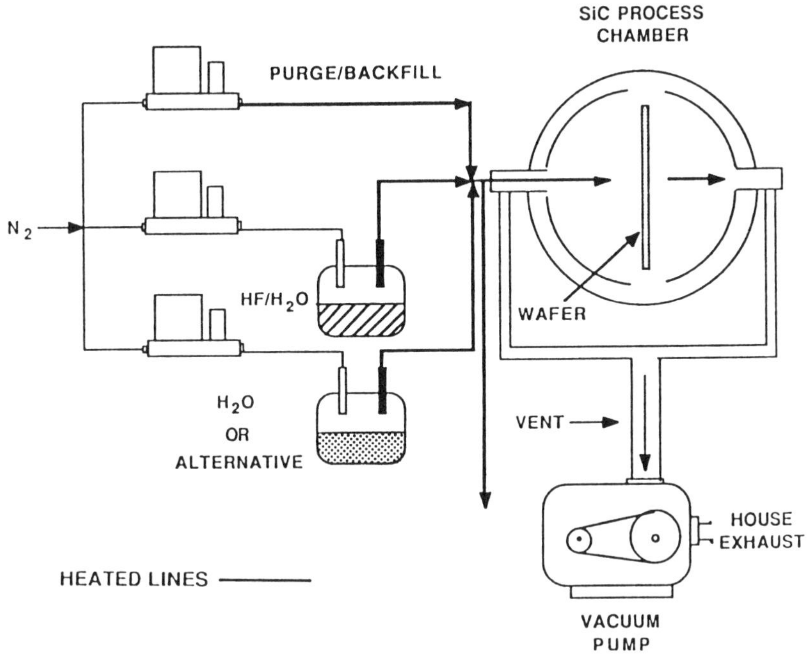

Figure 6. Schematic of Advantage Production Technology's EDGE-2000 vapor phase wafer cleaning/etching system; after Deal et al. (60). Note: Technology of Advantage presented by Genus, Inc. in 1992.

A more advanced model of the Advantage vapor phase cleaning system employs wafer heating and ozone activation (model EDGE-2002). A cross section of the process chamber is shown in Fig. 8. In this system, the silicon carbide chamber has been modified and consists of a single hemisphere with a flat quartz window across the diameter. This window or plate is covered by a special transparent material that resists attack of the quartz by hydrofluoric acid. The wafer can be exposed to IR (or UV) for heating up to 400°C. In addition, a corona discharge ozone source is connected to the inlet lines so that ozone may be admitted to the chamber before or during the clean/etch treatment cycle. It has been found that pre-treatment of oxide surfaces using ozone at elevated temperatures is a very effective treatment for removing deposited hydrocarbons, and much more uniform etching of silicon oxides can be achieved (62).

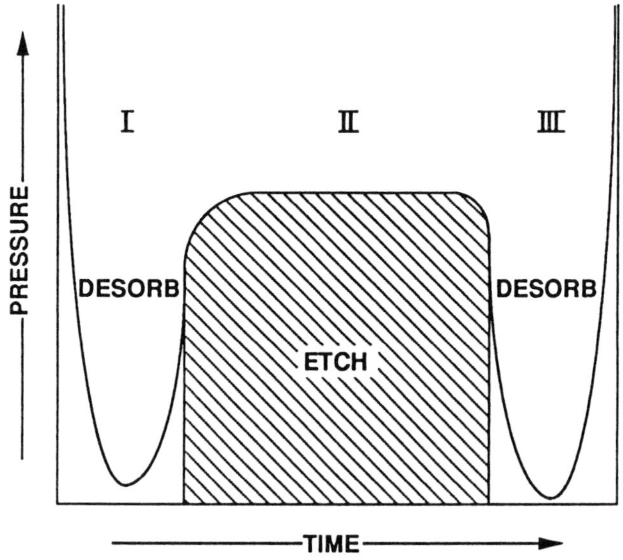

Figure 7. Pressure versus time characteristic of EDGE-2000 vapor phase oxide etching process; after Deal et al. (60).

Vapor phase cleaning and etching has been shown to eliminate many of the particle contamination problems discussed above which have been associated with aqueous-type cleaning processes. In general few, if any, particles are added during vapor phase processes, and often particles are actually removed. It is anticipated that in situ particle monitoring can be employed in more advanced systems (described later). It is also possible to employ other in situ monitoring tools, such as residual gas analyzers (RGA), ellipsometers, and in-line particle counters. Other experiments involving vapor phase HF processing, along with appropriate equipment, have been reported by van der Heide et al. (63), Onishi, Oki, and co-workers (54)(64)(65), Wong et al. (66)(67), and by Izumi and co-workers (68).

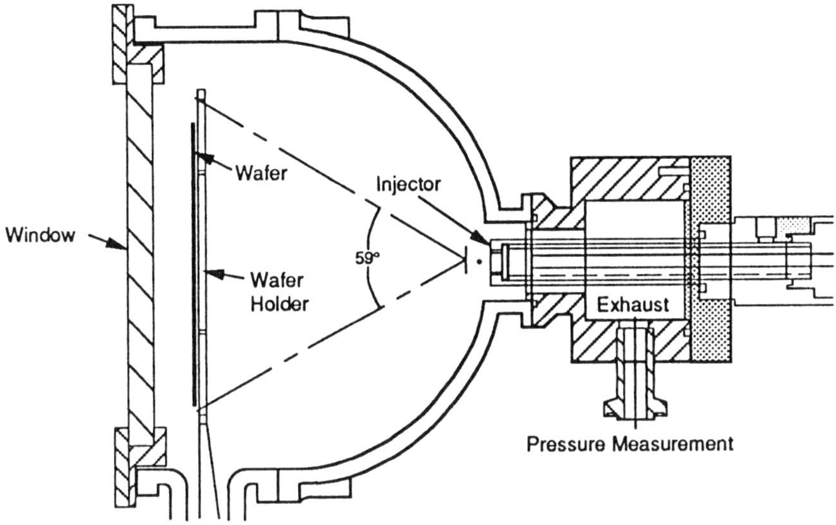

Figure 8. Modified process chamber of EDGE-2002 vapor cleaning system which permits UV/IR radiation of wafer through passivated quartz window at left. *(Courtesy of Advantage Production Technology).*

Activated Chlorine. While the use of activated chlorine gas for cleaning silicon wafers has not been developed to the point of commercial application, it is probably the most likely to be used in wafer production after HF vapor cleaning and etching. It was mentioned earlier that vapor HCl has been successfully used for a number of years as a pre-epitaxial silicon deposition step for removing native oxides and other contamination. It has since been reported by researchers at Fujitsu and others that the use of activated chlorine (Cl_2) gas can be an effective method for removing metal impurities from the silicon surface (69)-(74).

Most of the work in this area has involved ultraviolet activation of chlorine gas (70)-(74). The mechanism involves etching and removing a thin layer of silicon which either vaporizes the metal atoms as MCl_x along with the silicon, or by lifting the metal on top of the evaporating silicon chloride. Typically, UV light in the 180 to 350 nm wavelength range is used at temperatures from 100 to 400°C. In Fig. 9, the relationship of silicon removal rate to substrate temperature for Cl_2 with and without UV radiation is shown (75). The main challenge in establishing a production-worthy process is the control of the silicon etch rate so that a minimum but uniform layer is removed.

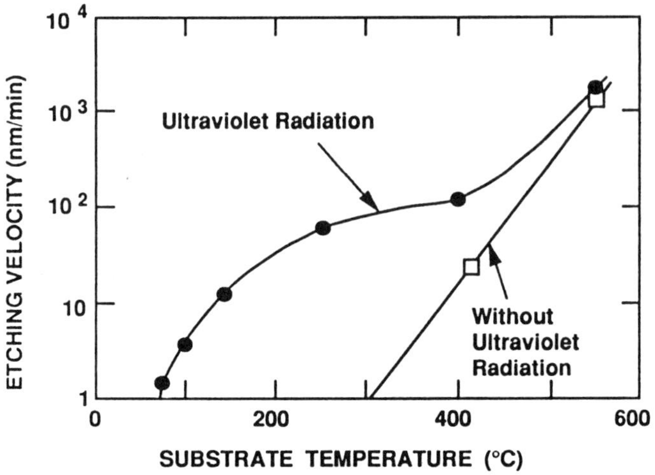

Figure 9. Photochemical etching of silicon wafer by chlorine gas with and without ultraviolet radiation; after Ito et al. (75).

A second type of Cl_2-activated etching of silicon has been reported using a downstream microwave discharge system (76). The authors investigated effects of various process variables, such as silicon orientations, doping concentration, and substrate temperature. As was the case for UV activation, the main problem in this type of cleaning process is the ability to uniformly etch the silicon. This will be a requirement for submicrometer device fabrication, but hopefully these uniformity problems will be solved. The current effectiveness of activated chlorine treatment for metal impurity removal is discussed in Sec. 5.5.

3.0 OXIDE ETCHING

3.1 Thermal Oxides

One of the more critical steps in semiconductor device fabrication is the complete or partial thickness removal of thermal oxide layers from the wafer surface. This removal or etching must be accomplished uniformly across the wafer or in selected regions. Typically, oxides are etched in hydrofluoric acid aqueous solutions. It is therefore appropriate to first mention aqueous etching processes. These etch processes are typically carried out by immersing a batch of wafers in a bath containing the etch solution. Appropriate temperature control and stirring are employed to maintain constant and uniform etch rate conditions. The overall chemistry may be expressed by one of the following reactions:

Eq. (1) $$4HF + SiO_2 \rightarrow SiF_4 + 2H_2O$$

Eq. (2) $$6HF + SiO_2 \rightarrow H_2SiF_6 + 2H_2O$$

In any such reactions, various intermediate species are produced, and Eqs. (1) or (2) are no exception. These intermediates can cause problems either during the oxide etching process or afterwards, and the subject is discussed in the following section which deals with oxide etching mechanisms.

Unfortunately, very few data concerning aqueous HF-H_2O etching exist in the literature, since most of these procedures involve so-called buffered oxide etch solutions. In this case, mixtures of ammonium fluoride and hydrofluoric acid (NH_4F + HF + H_2O) are generally used, which tend to prevent depletion of the fluoride ions—thus leading to stable etch characteristics (77). Other additives, such as glycerol are sometimes used as well.

Some early references by Mai and Looney (78), Judge (79), and Harrap (80), which provide thermal oxide etch rate data in HF-H_2O mixtures, are available, and these have been used to prepare the plot shown in Fig. 10. This plot will be used later to compare vapor phase HF etching of thermal oxides with aqueous etching. Etch rates are normally obtained by plotting oxide removed versus etch time for a particular set of process conditions. The etch rate is determined from the slope of the curve.

In vapor phase HF etching of oxides, an interesting phenomenon has been observed. While the process occurs as indicated in Eqs. (1) or (2), etching does not begin without the presence of condensed water on the oxide surface. The mechanism is discussed in detail in the next main section. Needless to say, the nature of the etch process is very dependent on the chemistry of the oxide surface and the effect on the water condensation. Variations in this chemistry will result in differences in delay time, that is, the on-set of etching. Once etching of the oxide does begin, it tends to be linear with time. Etch rate values are functions of HF concentration in the condensed aqueous layer, which in turn are dependent on temperature, pressure, and other process variables. A typical example of oxide etch rate plots is presented in Fig. 11. Normally, these types of measurements of oxide thickness removed during a particular etch time are carried out using suitable mapping equipment such as a Prometrix model SM 200/e, or a Gaertner model L115B. The variation of delay or off-set times with HF concentration in the vapor phase can be observed in Fig. 11, along with etch rates themselves. An estimation of HF concentration in the condensed aqueous layer for particular HF:H_2O vapor ratios can be made by comparing etch rates with those of Fig. 10 for aqueous solutions. For instance, a 4.7 Å/sec etch rate obtained by vapor etching with a 2.5 l/min azeotropic HF source and 12.5 l/min H_2O source corresponds to about a 5 wt% HF aqueous solution.

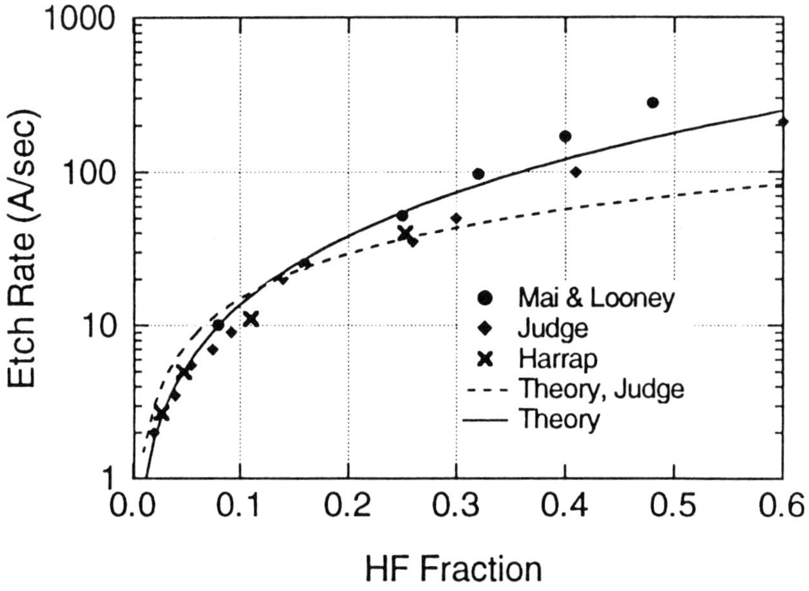

Figure 10. Thermal oxide etch rate as a function of HF concentration (wt.%) in aqueous HF solutions at 25°C. The data are from Refs. 78, 79, and 80. The dashed line is from Judge; the solid line is an improved fit discussed in Sec. 4.2.

In partially etching blanket oxide layers or selectively etching oxides, it is generally very important to etch as uniformly as possible. Good etch uniformity is important across the wafer as well as from wafer to wafer (often called repeatability). Typical data reported for the Advantage EDGE-2000 system are summarized in Table 1. Note that oxide etching uniformity and repeatability values are for clean oxides.

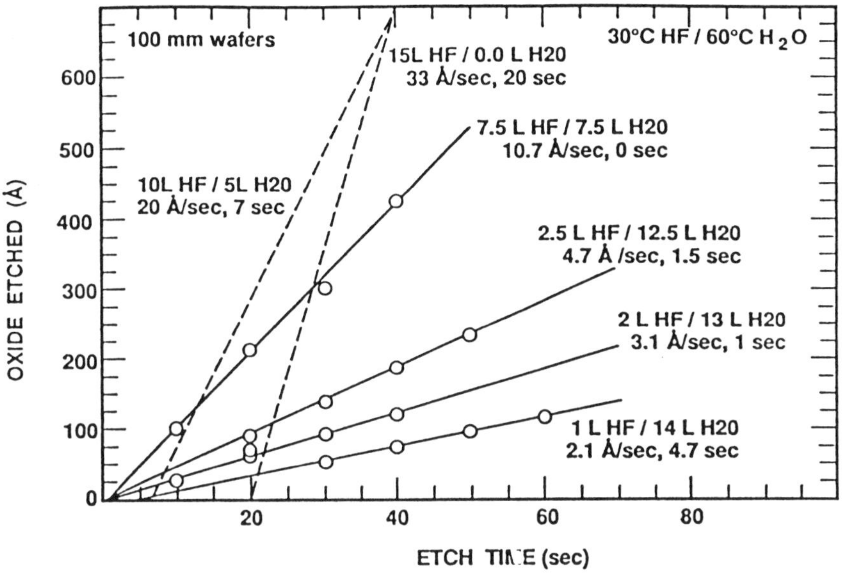

Figure 11. The amount of thermal oxide removed versus etch time for varying ratios of azeotropic HF/H_2O solutions (38.4 wt.% HF) to water vapor. The wafer temperature is 25°C, the azeotropic HF/H_2O vaporizer temperature is 30°C, and the water vaporizer temperature is 60°C. The total flow rate is 15 l/min. The etch rate equals the slope of plots (first number, second line); the delay time equals the time intercept (second number, second line); after Deal et al. (60).

Table 1. Typical Properties: Vapor Phase Oxide Etch Process (150 mm wafer, 22°C)

■ **ETCH RATES - THERMAL OXIDES**		
	AZEO HF	**30 - 70 Å/sec**
	HF - H_2O	**2 - 35**
	HF - HCl	**5 - 25**
	HF - IPA	**0.2 - 5**
■ **ETCH UNIFORMITY - THERMAL OXIDES**		
	≤2.0% (one sigma)	
■ **ETCH REPEATABILITY - THERMAL OXIDES**		
	≤3.0% (one sigma)	
■ **PARTICLE ADDITION (bare wafer)**		
	≥0.4 μm	**0 - 2**
	≥0.2 μm	**3 - 10**

The effect of organic impurities on the oxide surfaces or even the surface chemistry itself on oxide etch characteristics, especially for vapor etch systems, has already been mentioned. Data have been obtained which demonstrate this effect (62) and are shown in Fig. 12. This evaluation was carried out in the Advantage System 2002 which includes wafer heating and ozone capabilities. A single lot of twenty-four oxidized silicon wafers were selected and every other wafer was given an ozone pre-etch treatment at 200°C. All wafers were then vapor etched in HF-H_2O to remove 500 Å oxide. A distinct improvement in oxide etch uniformity was observed for the ozone pre-treated wafers, confirming that even for relatively clean oxides, enough organic contamination is present on the oxide surface to affect etching uniformity. On the other hand, etch repeatability from wafer to wafer is not affected appreciably.

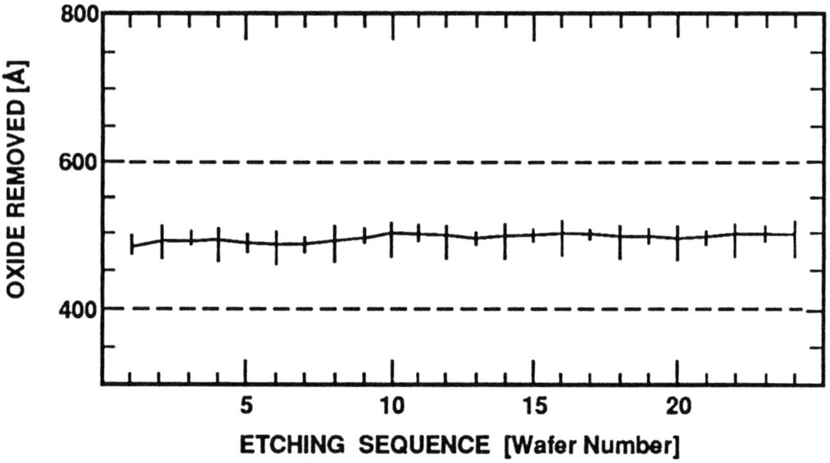

	Uniformity (1 sigma)	Repeatability (1 sigma)
Pre-etch Treat.	**0.90%**	**1.04% (1.31%)**
No Pre-etch Treat.	**2.17%**	**0.92%**

Note: Odd wafer numbers have ozone+heat pre-treatment

Figure 12. Uniformity of thermal oxide etching in vapor phase cleaning system with and without an in situ ozone/heat preclean; after Nobinger et al. (62).

As mentioned earlier, other process variables which can significantly affect vapor phase oxide etch characteristics include wafer temperature, chamber pressure, and any variable affecting composition of the condensed HF aqueous layer, such as gas flow rates, vaporizer composition and temperature, and type of solvent. The latter effect on etch rates is indicated in the first part of Table 1 where several solvents are listed.

3.2 Native/Chemical Oxides

Perhaps a great majority of applications involving vapor phase HF etching of oxides relate to complete removal of a thick (or thin) oxide. After this removal, and after various cleaning treatments of "stripped" silicon surfaces, a thin (5 - 15 Å) oxide layer generally remains. Thus, the important subject of so-called "native" or "chemical" oxides arises, including native oxide regrowth (81)-(87). Many device structures are critically affected by the presence or absence of this thin layer of questionable composition. These structures include interconnect contacts, poly-Si bipolar emitters, epitaxial silicon layers, and even MOS gate oxides. Practical aspects of pre-cleaning effects on device properties are discussed in a later section.

As mentioned above, any subsequent treatment of an HF-etched silicon surface will normally result in the formation of a very thin native or chemical oxide. This treatment can include cleaning solutions, DI water, process gases, or even exposure to room ambient. These layers vary in composition, depending on the treatment, but quite often contain silicon oxide, organics, various anions such as sulfates, and other impurities. Thus it has been very difficult to characterize them and to obtain reproducible evaluations in different laboratories. Various techniques have been used to determine their thickness and composition, such as XPS, Auger, SIMS, and ellipsometry. Kinetics of film growth have been investigated and typical results are presented in Fig. 13. In this figure, it can be noted that after a conventional aqueous HF etch and DI water rinse, the apparent native oxide thickness is 7 - 8 Å. For a vapor phase HF treatment, however, the thickness is about 4 Å for this particular substrate, as measured by an ellipsometer. If both wafers are exposed to room ambient, a steady thickness increase is observed. On the other hand, the increase is much less for exposure to either charcoal filtered air or argon. This has led to the conclusion that much of the so-called re-oxidation is really due to a condensation of organic and other impurities on the surface (62)(88).

One of the device fabrication requirements in removing native/ chemical oxides from contact openings is that attack of the surrounding dielectric should be minimal. Ideally it should not be etched at all. It has been found by groups at the Hashimoto Chemical Industries and Tohoku University that extremely dry anhydrous HF can remove native oxide layers without etching adjacent thermal oxides (89). In their work, a vapor phase HF critical concentration was found below which no etching was observed. This critical concentration depended on residual H_2O present, temperature, and flow rate, as well as the nature of the SiO_2 being etched. They suggested that, indeed, water was necessary for etching to proceed and that hydrated surfaces, such as chemical native oxides would therefore have lower HF partial pressure thresholds; whereas drier surfaces, such as thermal oxides, would have higher thresholds. Once the reaction is initiated at sufficiently high HF pressures, enough water will be produced to allow the reaction to continue. Data which demonstrate such selective oxide etching are presented in Fig. 14. It has also been observed that a similar selective etch process can be achieved using the vapor HF-H_2O reaction. This is accomplished for those conditions which produce a delay time of 10 seconds or more before etching of the thermal oxide begins. Since the condensed aqueous HF solution has not yet formed, the gaseous HF will etch the native oxide but not the thermal oxide. This phenomenon is illustrated in Fig. 15, where an 8 Å native oxide is removed before etching of the thermal oxide commences.

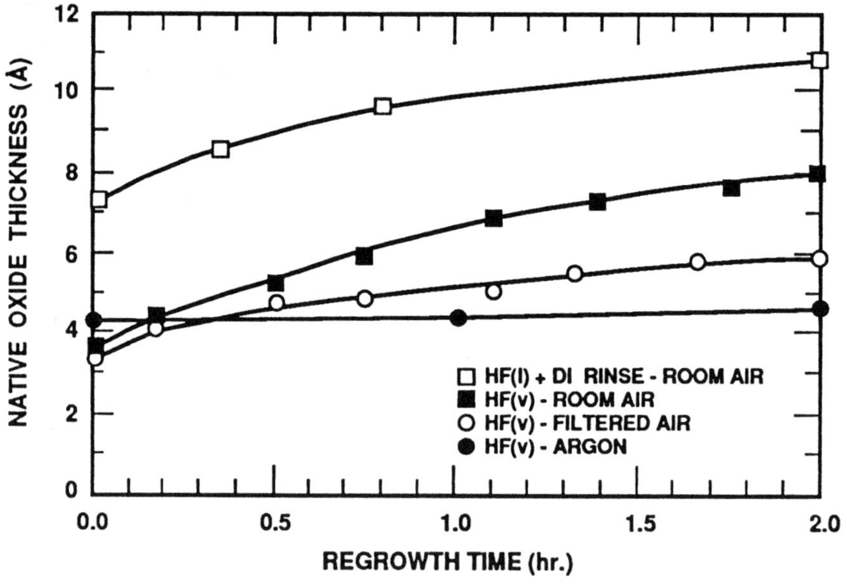

Figure 13. Native oxide regrowth at 25°C in various ambients following vapor and liquid HF strip of thermal oxide.

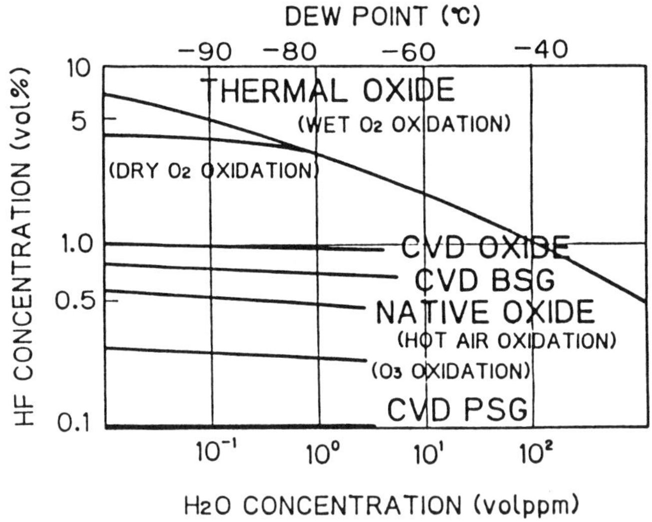

Figure 14. Relationship of HF critical concentration (for oxide etching) and moisture level at 25°C for various dielectric films. Native oxides were prepared by hot air and ozone oxidation; after Miki et al. (89). *(Copyright 1990, IEEE).*

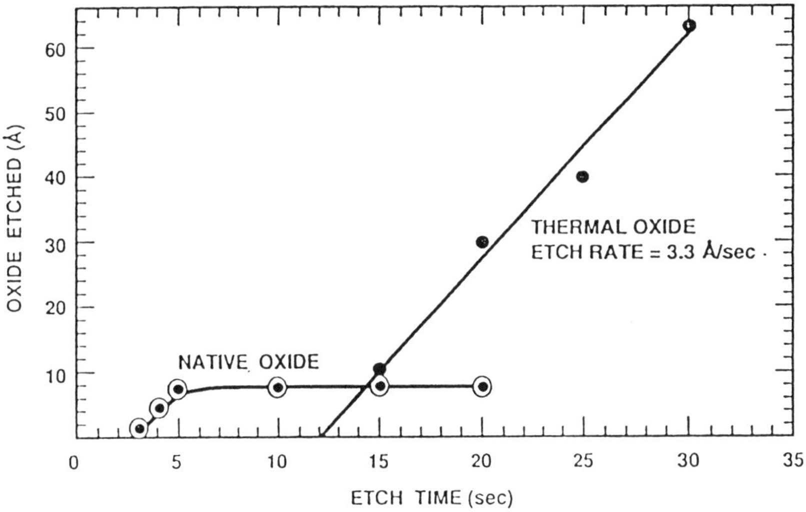

Figure 15. Selective etching of native oxides in presence of thermal oxides using vapor phase HF processes.

3.3 Deposited Oxides

Various types of dielectric films, usually based on SiO_2, are used in semiconductor device fabrication (18)(90)(91). These can be deposited by a variety of methods and may contain dopants, such as phosphorus and boron, in amounts up to 10%. As a result, etch rates can vary considerably. For a given etch process, aqueous or vapor, each film must be characterized. In addition, when two or more dielectrics are present on the same wafer, differential etch rates are often observed. Many times, etch rate differences can be used to advantage for fabricating a particular device, while in other cases severe problems result. In Table 2, examples of relative etch amounts for different dielectric films are shown for a given etch time in a vapor phase HF/H_2O system.

Similar data have been repeated for other vapor etch systems; an example is shown in Fig. 16. Note that etch rate increases with increasing dopant concentration and/or lower density deposited films.

Table 2. Relative Etch Selectivity of Various Thin Films (45°C HF Vaporizer; 12 sec. etch time)

Oxides	Thickness Etched	Selectivity (Compared to Thermal)
Thermal	240 Å	1.0
LTO	360 Å	1.5
TEOS	550 Å	2.1
PSG (4% P)	776 Å	3.2
BPSG (4% B, 7%P)	1288 Å	5.4

4.0 MECHANISM OF OXIDE ETCHING

4.1 Background

In our work on oxide cleaning and etching (37)-(39) we have taken the lead from the Holmes & Snell (36) work in light of subsequent data to develop a quantitative model of vapor phase HF/H_2O etching of SiO_2. The basis of the model is the initial assumption that etching will only proceed if a condensed HF-containing liquid layer is present on the surface. There is a significant amount of evidence in support of this assumption. An example is shown in a comparison of vapor phase etch rates illustrated in Figs. 5 and 11 with what is obtained in aqueous solutions as shown in Fig. 10. In both cases the ranges of etch rates are similar, suggesting similar mechanisms. In another example the thresholds in etching observed (see Figs. 2 and 14) for the vapor phase can be quite naturally related to the requirement for condensation for etching to occur. A major part of the model is therefore determining those conditions of temperature, HF, and H_2O partial pressures that lead to condensation. Proceeding on the assumption that etching then occurs in the same way as for liquid solutions, a steady state model allows the calculation of the HF concentration in the condensed liquid film. A comparison of that value to the liquid case then permits the vapor phase etch rate to be determined. This model has provided excellent agreement with experiment including the observations of etching selectivity, pressure and temperature thresholds, and induction times reported by numerous investigators.

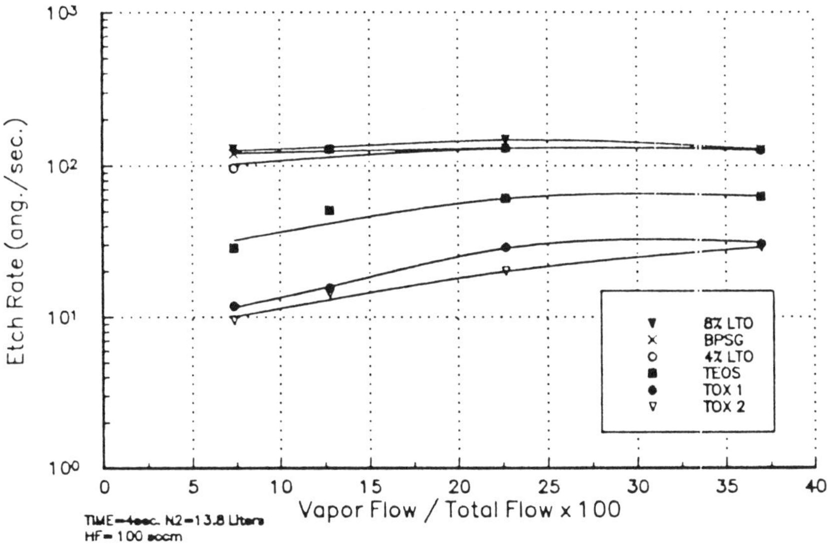

Figure 16. Etch rates at 25°C versus H_2/N_2 vapor flow for various dielectric films in Excalibur vapor cleaning system *(Courtesy of FSI International).*

A schematic of how this process works is shown in Fig. 17. Initially an SiO_2 surface is exposed to a vapor phase mixture of H_2O and HF (17a). If the partial pressures are sufficient, then a condensed film of HF and H_2O will form and continue to grow on the SiO_2 surface (17b). Etching proceeds with the formation of H_2SiF_6 and additional water (17c). Although the partial pressure of SiF_4 over dilute H_2SiF_6/HF/H_2O solutions is relatively low, some SiF_4 will evolve during the etching process. When the etching is complete and the HF and H_2O reactants are switched off, the liquid H_2SiF_6/HF/H_2O film can be evaporated (17d) or rinsed off.

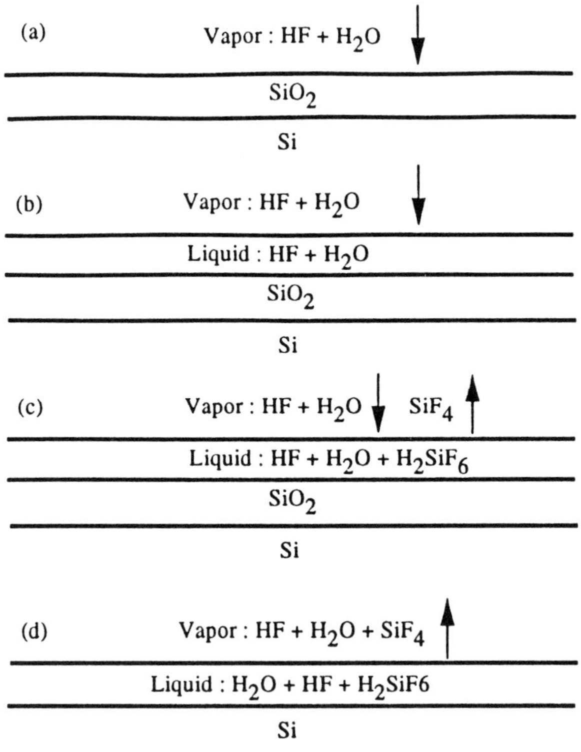

Figure 17. Schematic diagram of the various steps occurring during the etching of SiO_2.

4.2 Important Aqueous Chemistry

The chemistry of vapor phase HF or HF/H_2O etching or cleaning of SiO_2 on Si is closely related to the aqueous analog and is reviewed to provide a complete discussion of the vapor phase case. The chemistry of these processes has been discussed by Judge (79) and more recently by the Hashimoto/Tohoku University groups (92). In addition to a review of these works, additional information related to the effect of cleaning chemistry on surface roughness has been forthcoming (93).

The Chemistry of HF Etching of SiO_2. The overall chemistry of the etching process is described by Eqs. (1) and (2). However the properties of HF/H_2O mixtures are complicated due to the incomplete ionization of the HF and the formation of more complex species such as HF_2^-. Previous studies (79) have established the equilibrium for these species via

Eq. (3) $$[H^+][F^-] = K_1[HF]$$

and

Eq. (4) $$[HF][F^-] = K_2[HF_2^-]$$

giving values for K_1 and K_2 of 1.3×10^{-3} and 0.104 at 25°C, respectively. The corresponding concentration of the various species as a function of total HF concentration is shown in Fig. 18. Note the pH corresponds to the log$[H^+]$ which can be obtained from the left hand axis. From the perspective of SiO_2 etching, the question is, Which of these species are active? Two species H^+ and F^- are clearly inactive. We conclude this since other acids with high $[H^+]$ don't etch SiO_2 and pure NH_4F with a high $[F^-]$ also doesn't etch SiO_2. That leaves the HF and HF_2^- species. Tests of the activity of these species have been made by adjusting the ion concentrations in HF solutions using either HCl or NH_4F. For high HCl concentrations very little F^- of HF_2^- are present so the etching occurs primarily due to the neutral HF. For high NH_4F concentrations very little H^+ or neutral HF is present so the etching occurs primarily due to the HF_2^-. This is shown in Fig. 19 where the HF/HF_2^- concentration ratio is plotted versus HF concentration for the case of pure HF/H_2O and cases of HCl and NH_4F additions. Indeed, significant etch rates are found for both cases, indicating that there is significant activity for both HF_2^- and neutral HF. Judge (79) deduced a fit to much of his data as

Eq. (5) $$\text{Etch Rate (25°C)} = 2.5[HF] + 9.7[HF_2^-] - 0.14$$

indicating that the HF_2^- is considerably more active for SiO_2 etching than HF. This explains the high etch rates observed for buffered HF solutions with high F^- and therefore HF_2^- concentrations. Although the above relationship works well for some conditions it is not accurate for high HF concentration cases of interest for vapor phase etching. This is shown in Fig. 10 where Eq. (5) is shown as the dotted line. The fit is clearly poor, especially for the high concentration limit, and in addition, a negative term in Eq. (5) is not physically meaningful. A better fit can be obtained if a term in $[HF]^2$ or $[HF_2^-]^2$ is added. The solid line fit of Fig. 10 corresponds to:

Eq. (6) $$\text{Etch Rate (25°C)} = [HF] + 7[HF_2^-] + 0.3[HF]^2$$

For a specific set of conditions the dominant reactant can be determined by calculating the relative magnitude in the above equation. This is also illustrated in Fig. 19.

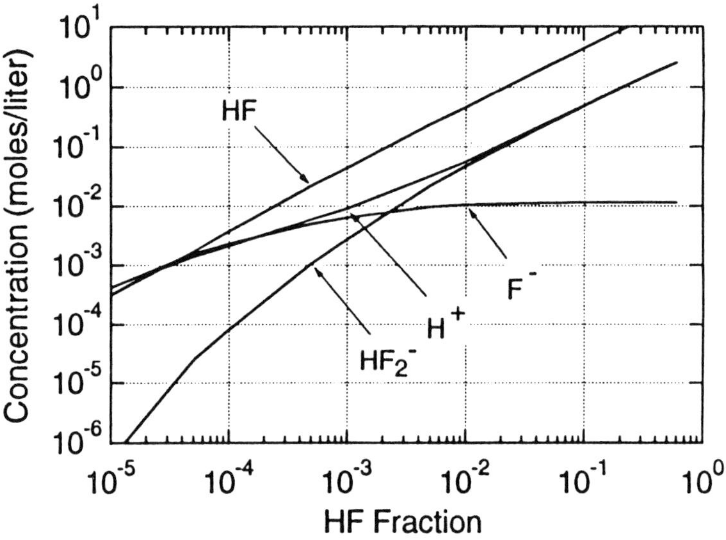

Figure 18. The concentration of various species is shown as a function of HF weight fraction from the model of Ref. 79. Note: the pH is the exponent of the H^+ concentration.

Surface Microroughness, Hydrogen Termination, and Electrical Implications. These considerations may be thought to be of only academic interest. However, recent findings show that the pH of HF solutions effects the promotion of hydrogen-terminated surfaces and surface microroughness. It is now well appreciated that HF-last processing can lead to Si surfaces that are atomically clean except for a monolayer of hydrogen that terminates the Si dangling bonds (93)-(100). Surprising as it may seem, the surface concentration of fluorine is not seen to increase above 0.1 monolayer. Haring & Liehr (101) have shown that even Si surfaces fluorinated in UHV are unstable in the presence of H_2O giving a surface where the Si is primarily bonded to oxygen and the remaining fluorine appears as OF radicals. Thus in an aqueous H_2O environment, fluorine terminated Si is clearly unstable. Both the low pH (HF dominated etching) and high pH (HF_2^- dominated etching) cases appear capable of producing these hydrogen-terminated surfaces. However, results from Higashi et al. (93)(100) show other differences between these two cases related to surface microroughness. This is shown in Fig. 20 where STM images of a low pH etched surface is shown in Fig. 20a and a high pH etched surface in Fig. 20b. The high pH case has led to a surface where the crystallinity is clearly evident, whereas the low pH case has produced a hydrogen-terminated surface which is considerably rougher. It is important to note that process induced surface microroughness has been correlated with poor interface and dielectric properties in MOS devices (102)-(110). In addition, cleaning induced microroughness has been attributed to less than ideal MOS properties for deposited SiO_2 (111)-(113).

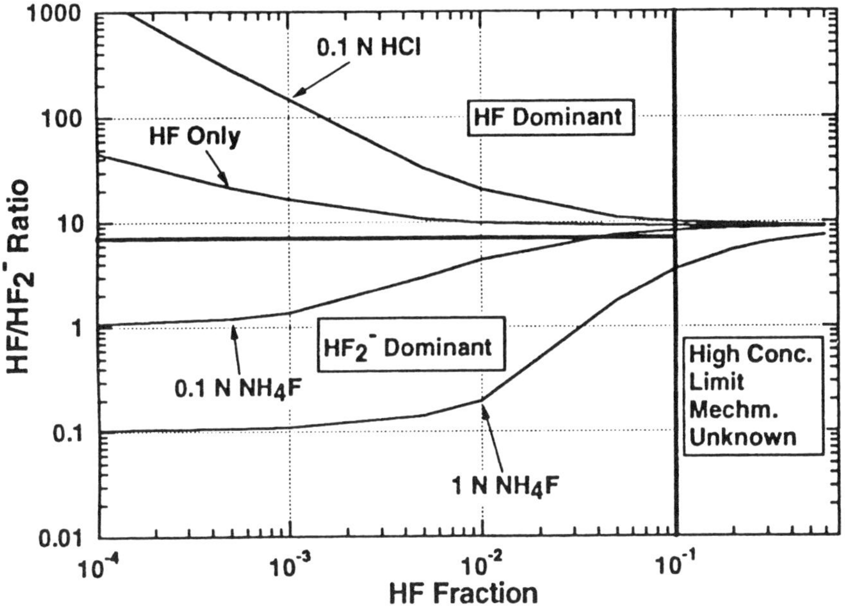

Figure 19. The HF/HF_2^- ratio plotted as a function of HF fraction for HF only, 0.1 normal HCl, 0.1 normal NH_4F, and 1 normal NH_4F. SiO_2 etching will be dominated by HF for high ratios whereas it will be dominated by HF_2^- for low ratios, corresponding to NH_4F additions. For large HF fractions as encountered in vapor phase etching the important chemical mechanisms for SiO_2 etching are unknown.

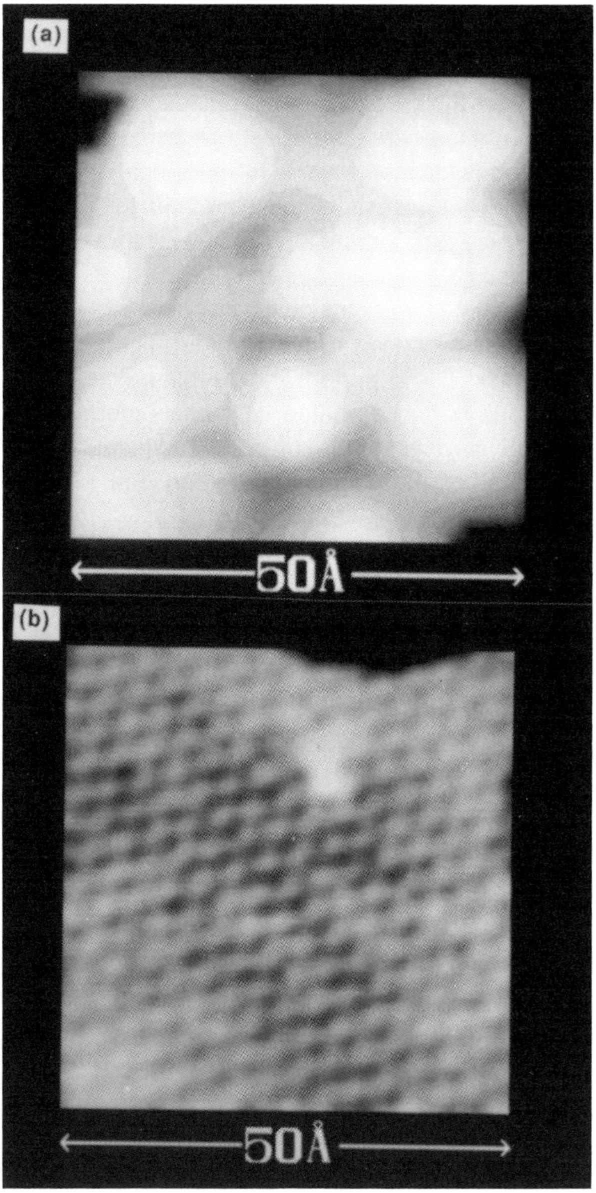

Figure 20. Scanning Tunneling Microscroscope (STM) images of Si(111) surfaces from Higashi et al. (93). Figure 20a shows a relatively rough surface morphology from low pH etching, whereas Fig. 20b shows a smooth morphology where atomic-scale features are resolved after a high pH (NH_4F) etch.

Currently there is no vapor phase analog to NH_4F buffered HF, since NH_4F is not volatile and the use of NH_3 with HF in the vapor phase will lead to NH_4F residues. Even though better device performance is expected for these HF/NH_4F-last surfaces, excellent device properties for vapor phase cleaned surfaces have been obtained , especially for HF/HCl-last surfaces, as discussed below.

4.3 Vapor Phase Mechanisms

For vapor phase etching we are interested in the comparison between measured liquid phase etch rates at a given HF concentration and the "vapor" phase etch rate at an identical calculated value of HF concentration in the condensed phase. We have shown that these etch rates are identical, validating the condensed film model (37)-(39). In addition, for vapor phase etching, the decomposition of aqueous H_2SiF_6 via

Eq. (7) $$H_2SiF_6 \rightarrow 2HF + SiF_4$$

with the desorption of SiF_4 is a necessary step, unless a subsequent water rinse is employed. Undesired secondary reactions can cause residue formation not encountered in aqueous processing, as discussed below.

Vapor Pressures of Aqueous HF/H_2O Mixtures. In our previous work we have used the equilibrium partial pressures of HF and H_2O over aqueous solutions to determine conditions for condensation and HF concentrations which can then be used to determine etch rates. In this approach we assume that the system is in steady state with a flux of reactants into the surface (at sufficiently low pressures) given by

Eq. (8) $$\Phi = P / (2\pi mkT)^{1/2}$$

where P is the pressure, m the mass, and kT the energy associated with temperature T. The flux of reactants leaving the surface is given by the appropriate reactant vapor pressures for the liquid phase. This is only valid if the sticking probabilities for the HF and H_2O are nearly equal; this is indeed the case, as shown below. Comparing masses we find that $m_{HF} = 1.05\ m_{H_2O}$ so assuming that the fluxes of each are proportional to their partial pressures leads to minimal error. For these conditions the net steady state flux into the condensed layer will be

Eq. (9) $$\Phi_{H_2O} = P_{H_2O} - P_0^{H_2O}\,(T,[HF])$$

and

Eq. (10) $$\Phi_{HF} = P_{HF} - P_0^{HF}\,(T,[HF])$$

where the Φ's are the fluxes, P is pressure, T surface temperature and [HF] is the HF concentration in the condensed layer. P_{H_2O} and P_{HF} are the vapor phase partial pressures right above the surface and $P_0^{H_2O}$ and P_0^{HF} are the equilibrium vapor pressures at sample temperature T and condensed layer composition [HF]. The HF concentration in steady state is then determined by the ratio of the net HF flux to the total reactant flux, so that

Eq. (11) $$[HF] = \frac{P_{HF} - P_0^{HF}\,(T,[HF])}{P_{HF} - P_0^{HF}\,(T,[HF]) + P_{H_2O} - P_0^{H_2O}\,(T,[HF])}$$

For a given set of experimental conditions this expression can be solved, providing values for the equilibrium vapor pressures are available. Fortunately there are extensive data in the literature on vapor/liquid equilibria for HF/H_2O mixtures. The most complete reference on HF/H_2O vapor/liquid equilibria appears to be that of Munter, Aepli and Kossatz (42); plots based on their data are shown in Fig. 21 where the equilibrium vapor pressures are plotted versus liquid composition for a series of temperatures including the values for pure HF. Additional recent data on the high HF concentration limit are also shown (114). Without further analysis we can make a few obvious statements from these curves. First, in the low HF concentration limit the vapor pressure of HF is quite small so that the HF fraction in the liquid phase will be much larger than in the vapor phase. The converse is true for the high HF concentration limit. This in part explains the "anhydrous" etching results, since any water present or created by the reaction can be retained in the liquid due to the low H_2O vapor pressure for high HF concentrations. If the sticking probabilities are equal, that point where the vapor phase and liquid phase partial pressures are equal will be the azeotrope; this occurs at approximately 39 wt.% HF (41), which is near the equal partial pressure points. Therefore, the assumption of equal sticking probabilities is justified. In order to assist in analyzing etching behavior, we have modeled the data used to prepare Fig. 21 using a near regular solution approach, resulting in the curves presented. The model has been refit from our original work (37)-(39) to recent data for the high HF partial pressure limit and all subsequent curves will reflect this fit (114).

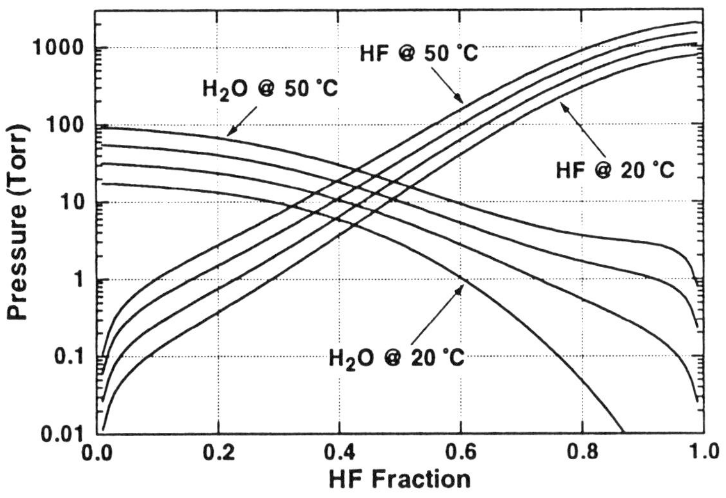

Figure 21. Fit to data of Refs. 42 and 114 showing the vapor pressures of both HF and H_2O as a function of HF weight fraction for 20°, 30°, 40°, and 50°C.

At a given temperature a condensed phase will form only if the combination of vapor phase partial pressures is high enough so that the flux of molecules impinging on the surface is greater than that leaving by evaporation. This manifests itself in Eq. (11) where both the numerator and denominator must be greater than zero. This is shown in Fig. 22 for 20°C where condensation will occur on the high pressure side of the line (upper right) and no condensation is expected on the low pressure side of the line (lower left).

So far we have neglected the problem of nucleating a condensed film. For condensation on clean SiO_2 surfaces this is undoubtedly justified due to the hydrophilic nature of such surfaces. This analysis cannot be applied in other cases and substantially higher pressures for condensation could be required and non-uniformities could also be expected for more hydrophobic surfaces.

Calculation of Etch Rate. Having now established the conditions that lead to condensation and therefore etching, we turn our attention to determining the etch rate which can be compared to the liquid etch rates of Fig. 10. This will be affected by the adsorption/desorption process, as well as the consumption of HF during etching and presence of reaction products. For the moment we will ignore these last factors so that we can consider the situation of condensation on an inert surface. Using the data of Fig. 21 via

Eq. (11), the HF concentration as a function of partial pressure can be determined. The etch rate can then be calculated using the value of [HF] and Eq. (6). The predicted etch rate is shown in Fig. 23 at 25°C for various H_2O partial pressures. Note again the upper left region of this figure corresponds to conditions leading to no condensation and therefore no etching. This also assumes a negligible pressure drop across any boundary layer or pressure gradient due to vapor phase diffusion limitations.

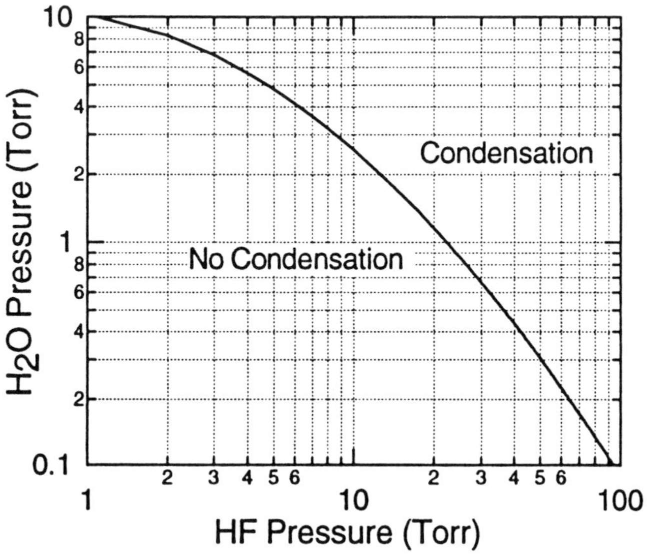

Figure 22. Partial pressure diagram indicating what ranges correspond to a stable condensed phase at 20°C.

The comparison of this theory to experiment is difficult since the experimental data reported seldom give values for the reactant partial pressures. Experimental etch rates have ranged from a few Å/sec for conditions where the HF concentration was clearly dilute to hundreds of Å/sec for cases which were reported to be anhydrous (see Figs. 5 and 14). From Fig. 23 we can see qualitatively that these values are in the expected range. In addition, vapor pressure data for the etch results of Deal et al. (59)(60) are available. Fig. 24 shows these experimental vapor phase etch rates as a function of calculated HF concentration in the condensed film. For comparison, the data of Fig. 10 for the liquid case are also shown. The agreement is reasonably good but shows higher vapor phase etch rates at low concentration and lower vapor phase etch rates at higher concentration. The latter effect may be due to the comparison to liquid rates for stirred conditions as well as the buildup of reaction products. The deviation at low concentration is likely due to the inaccuracy of the vapor pressure model for low HF concentrations where no data were actually available.

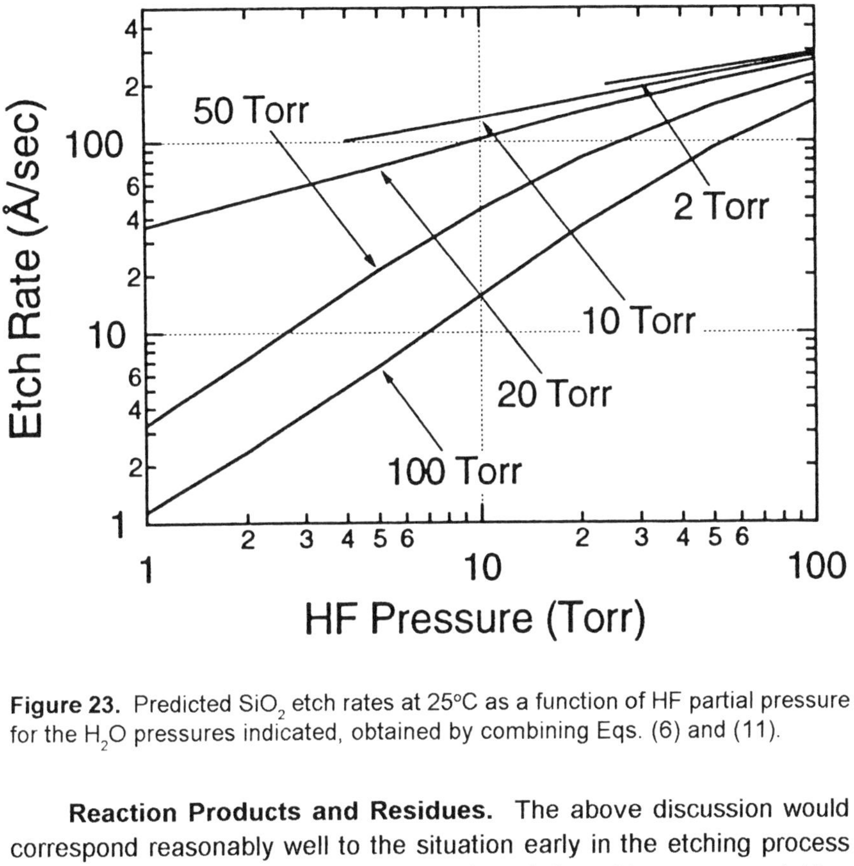

Figure 23. Predicted SiO_2 etch rates at 25°C as a function of HF partial pressure for the H_2O pressures indicated, obtained by combining Eqs. (6) and (11).

Reaction Products and Residues. The above discussion would correspond reasonably well to the situation early in the etching process where the concentration of reaction products is low. However, as etching proceeds, the reaction products may tend to build up in the condensed layer. If we consider reaction (2), we see six molecules of HF are consumed to form one molecule of H_2SiF_6 plus two molecules of H_2O. Therefore the case of "anhydrous" etching would also correspond to a case with a high concentration of reaction products, including H_2O. This points out two disadvantages of vapor phase oxide etching and cleaning: the lack of a nearly infinite source of solvent and the need to remove reaction products by evaporation rather than dissolution. This has been dealt with in some cases with a final water rinse (58), but the addition of a wet step at the end of the process defeats much of the purpose of vapor phase etching and cleaning in the first place. Two vapor phase methods have shown promise for providing residue-free etching and cleaning, one employing HF and H_2O at elevated temperatures (64)(65) and one employing HF with methanol as the condensable solvent medium (68).

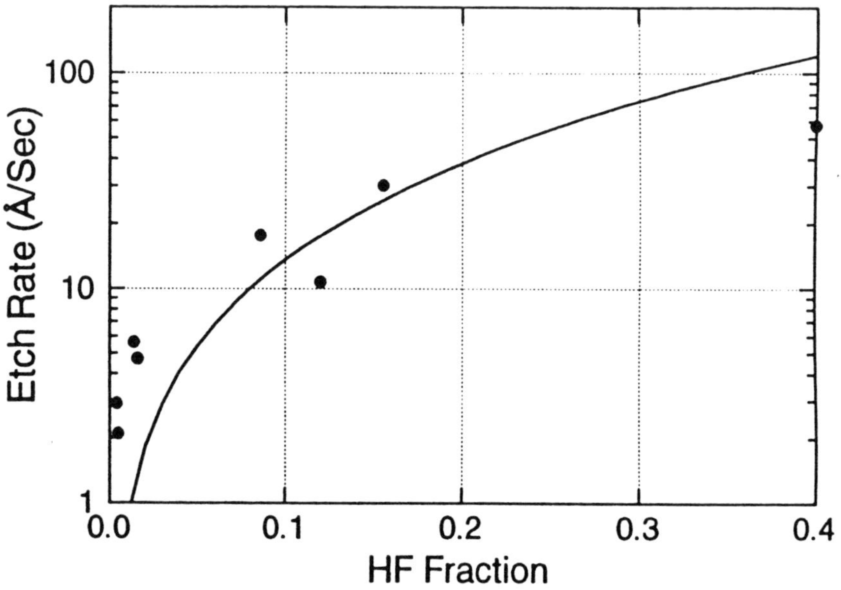

Figure 24. Vapor phase etch rates from Refs. 59 and 60 versus equivalent liquid HF fraction calculated from partial pressures using Eq. (11) shown as the points. Actual liquid phase etch rate from Eq. (6) is also shown as the line.

The residues that can form during etching and cleaning occur by the competition between reaction product desorption, represented by Eq. (7), and other reactions which tend to form silicic acids. These can be represented approximately as

Eq. (12) $$H_2SiF_6 + H_2O \rightarrow H_2SiO_3 + 6HF$$

The equilibrium suggests that conditions that favor rapid desorption of SiF_4, high HF pressures, and low water concentrations would lead to less H_2SiO_3 being formed. Onishi et al. (64)(65) have shown that increasing the reaction temperature by tens of degrees gives much lower residue concentrations.

This is likely related to the greater volatility of SiF_4 at elevated temperatures (44). Alcohols in place of H_2O as the solvent have also shown improvements in residue formation (68). In this case, reaction (12) is driven to the left by the tendency of H_2O to desorb, leading to lower H_2O concentrations.

There are other implications of the potentially high reaction product limit associated with vapor phase HF etching and cleaning. First, regarding etching itself, Thomsen (43) has shown that H_2SiF_6 itself is capable of etching SiO_2 via

Eq. (13) $$5H_2SiF_6 + SiO_2 \rightarrow 3H_2(SiF_6SiF_4) + 2H_2O$$

In addition, the presence of high concentrations ($\geq$10 mol%) of H_2SiF_6 in HF/H_2O solutions significantly reduces the vapor pressures of both HF and H_2O above such solutions. This is an additional mechanism which will stabilize a condensed liquid phase in the high HF concentration "anhydrous" situations, but may also lead to residue formation.

4.4 Summary

From the above discussion it is clear that the etching of SiO_2 from vapor phase HF/H_2O mixtures occurs after an initiation step by condensation of the HF and H_2O. The subsequent etching occurs by identical mechanisms compared to what would be obtained for liquid phase etching. The major differences between liquid and vapor phase etching are the initiation step, the possible presence of high concentrations of reaction products in the condensed phase, and the need to desorb or rinse off the reaction products after the etch.

Initially it might seem that the "anhydrous" etching results are counter to this model. However, small amounts of water present in the ambient, on the surface, or produced as a consequence of an initial surface reaction, especially in the high HF partial pressure limit are sufficient to produce an HF-rich condensed phase. In addition, the even further reduced H_2O vapor pressure in the presence of high H_2SiF_6 concentrations will lead to a condensed phase for "anhydrous" conditions. The model for the low HF concentration limit explains the observed thresholds for etching since, at a temperature which is too high or an HF and/or H_2O pressure which is too low, a condensed phase will not be supported and therefore no etching will occur.

Although we have not specifically discussed mechanisms for the selective etching that has been observed, these effects can be understood qualitatively within the same frame work. The key factor that seems to lead to differences in selectivity is the water content of the oxide being etched. For low water content oxides or oxide surfaces, higher HF/H_2O pressures will be necessary to initiate the reaction. This is the case for thermal oxides as discussed above. Native, chemical oxides inherently contain more water so that lower partial pressures will initiate condensation. This effect has been used to provide selective cleaning of native oxides while thermal oxides under the same conditions don't etch at all (58)(60).

5.0 IMPURITY REMOVAL

5.1 Types of Contamination

As indicated in the Introduction, several types or classes of contaminants are associated with semiconductor device fabrication. These include particles and residues, heavy metals and ionic impurities, organic and related species, and oxides. These contaminants can come from many sources: equipment, chemicals, ambient, humans, and even the process reactions themselves. A brief summary of analysis and detection methods is presented, followed by a discussion of specific origins and removal methods of impurities using current vapor phase cleaning techniques. These techniques include anhydrous and vapor HF chemistries, vapor HCl and activated Cl_2, and $UV/IR/O_3$ combinations. Finally, a short discussion of mechanisms of vapor phase impurity removal will be presented, including an appropriate comparison with aqueous cleaning techniques.

5.2 Evaluation Techniques

A more complete discussion of surface contamination analysis involving device wafers appears elsewhere in this volume (Ch. 12). However, a short summary is appropriate here. Typically, contamination levels which affect device properties have steadily decreased as device feature sizes have shrunk to sub-micrometer dimensions. Thus, critical surface concentrations of impurities, which were originally greater than 10^{12} cm^{-2}, are now rapidly approaching 10^8 cm^{-2}. As often happens, analytical capabilities can barely keep up with detection requirements.

Three general types of analysis procedures are used for surface contaminants. These are so-called "beam" techniques, chemical analyses, and device electrical properties. The latter obviously provide the most sensitive indication of effects of the various levels and types of contaminants. The beam techniques include TXRF (total reflection x-ray fluorescence), SIMS (secondary ion mass spectrometry), XPS (x-ray photoelectron spectroscopy), AES (Auger electron spectroscopy), STM (scanning tunneling microscopy), and AFM (atomic force microscopy), among others (115)-(119). In addition, various optical methods, such as ellipsometry and reflectivity, are used to measure thickness of oxide layers and other surface properties (82)(120)(121). These methods range in detection limit from below 10^{10} cm^{-2} to 100%, in depth resolution from a few angstroms to 100 Å or more, and in spot size from 100 Å to greater than 1 mm. It is thus obvious that different techniques (or combinations of techniques) will be required for different applications. For instance, SIMS or TXRF would be used for the detection of heavy metals, while Auger or XPS might be used to characterize chemical bonds. Also, correlations of results with electrical characteristics will generally be required for final interpretation.

Chemical analyses are being developed either alone or in combination with other methods. For instance, a current method of impurity analysis that shows great promise and appears capable of achieving 10^8 cm^{-2} detection levels involves vapor phase decomposition followed by atomic absorption spectroscopy (VPD/AAS) (122). This technique consists of dissolving the native oxide layer in a drop of HF solution which rolls over the surface, pipetting up the HF, and analyzing the drop using atomic absorption. An alternative procedure is to evaporate the drop and use TXRF on the surface. Detection levels of 10^8 cm^{-2} have been reported (122) for this method. Other chemical techniques involve rinsing the silicon surface and carrying out appropriate chemical evaluations of the solvent.

Finally, as indicated above, the most sensitive and probably the most relevant analysis technique involves electrical measurements of device related properties. These may be basic properties, such as carrier lifetime, junction I-V characteristics, or surface recombination velocity (123). Or they may involve capacitance-voltage (C-V) analysis, for measurement of ion drift, oxide charges, or charge trapping (124). In addition, such device properties as gate oxide integrity, tunneling oxide endurance, bipolar current gain and speed, and many others can be evaluated as a function of processing variables.

5.3 Particles and Residues

One of the critical problems associated with conventional aqueous cleaning of silicon wafers during device fabrication has been that of particles or residue contamination (13)(125)-(127). Particles and residues can cause severe yield and reliability problems, especially as device feature sizes move to sub-micrometer dimensions. They may be due to the environment, various process equipment, process chemicals and gases, or the actual process cleaning reactions, as well as rinsing and drying. It is now established that many of these problems, especially those involving particles, can be eliminated by the use of vapor cleaning processes. This is especially apparent for HF-last processing where the oxide-free hydrophobic surface is critically sensitive to particles and contamination in aqueous cleaning procedures. Part of this improvement arises because deionized water rinsing, either spray or dip, is no longer needed. A typical example of reduced particle contamination in vapor cleaning compared to aqueous processing is shown in Fig. 25 (66).

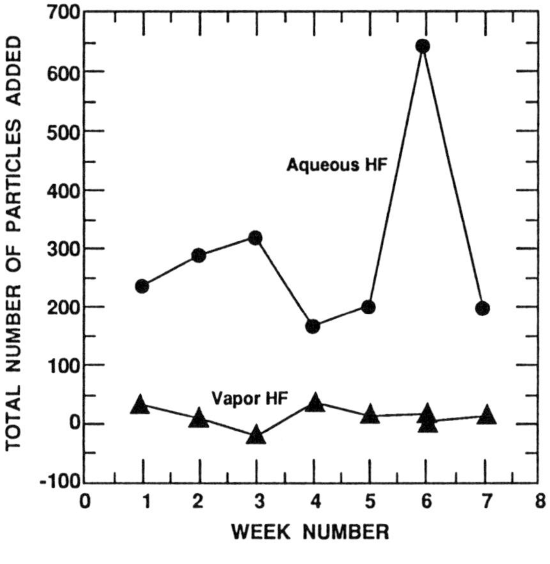

Figure 25. Particle trends for aqueous and vapor HF native oxide deglaze; after Wong et al. (66). *(Reprinted by permission of The Electrochemical Society, Inc.)*

The subject of so-called residues is more complex. In the case of oxide etching in HF solutions, for instance, aqueous processes appear to currently have an advantage due to the solubility of typical reaction products and by-products of the reaction in water rinsing solutions. There appear to be two possible solutions to the problem with respect to vapor phase processing. One is the use of a post-vapor etch rinse of some sort. This approach is discussed in (58), and appears to be a satisfactory interim solution to the problem. As integrated processing is developed (see Sec. 10), however, a more basic solution may be required. This will probably involve process chemistry modifications, which may also help to minimize particle contamination as well. For instance, control of pH and composition of the condensed layer on the wafer may provide considerable advantages (64)(65). More than one component or multiple cleaning steps may also be necessary (for instance, UV/O_3 + $HF/HCl/H_2O$).

5.4 Organic Contaminants

For many years, concern about carbon-containing impurities on device wafers has been expressed. However, much of this concern was based on intuitive thinking and speculation, rather than real experimental verification. Even so, various methods were developed for organic impurity removal, most involving UV/ozone chemistry (34)(35). More recently, as vapor phase cleaning technology was being developed, the sensitivity of silicon and oxide surfaces to ambient contaminants, especially hydrocarbons, became much more obvious. One such investigation was especially significant since it demonstrated that common laboratory air and materials could produce considerable organic contamination (88). Furthermore, it was proposed that a good percentage of the so-called "native oxide regrowth" observed using ellipsometric measurement techniques is not due to oxide growth at all, but merely a build-up of these organics and other impurities. (See previous discussions on organic effects on oxide etching in vapor phase systems—Sec. 3.1 and Fig. 12).

The use of UV/ozone chemistries or even heat at lower pressure has proven successful in removing unwanted organic impurities from wafer surfaces. Investigations involving these types of pre-cleaning treatments have been carried out in molecular beam epitaxy investigations, or even by conventional low temperature epitaxial processing (128)-(130). Some of these investigations have involved activated "dry" processes, but as was noted earlier, radiation-producing reactions such as plasma and sputtering tend to cause undesirable effects on device properties.

While millions, and probably billions, of dollars have been spent on minimizing particles in wafer fabrication environments, very little effort or cost has been expended in removing chemical impurities. Therefore, it is expected that either considerable attention must be paid to this particular problem, or new types of wafer fab facilities must be developed. Extensive filtering systems or integrated processing technology (discussed in Sec. 7.0) may be the answer.

5.5 Metallic Contaminants

It has long been known that heavy metal impurities can result in severe yield and reliability problems in silicon devices. These impurities are typically Fe, Cu, Zn, Ni, Cr, and even Au. They can originate at any process step, even with the starting silicon material. They will affect a variety of device characteristics, including junction leakage, surface and bulk recombination, emitter to collector shorting, and even gate oxide integrity. In addition to heavy metals, light elements such as C, Al, S, and even the alkali metals (Na, K, Li) can result in catastrophic failures in devices.

It is essential that all metallic impurities be analyzed and controlled to suitable minimum concentrations. This control may be difficult and depends on the location of the metal species in or on the device wafer, as well as its chemical nature. For instance, it may be located on top of an oxide layer, within the oxide, at the oxide-silicon interface, or even in the silicon itself. It also may be in elemental form, or as a compound such as an oxide or carbide. Thus, these properties of the metal impurities can determine the nature of the optimum removal process itself, whether it be gettering, aqueous cleaning or vapor type cleaning. It is possible that combinations of these may be appropriate.

Of all the types of impurities discussed above, vapor phase removal of metallic impurities has proven the most difficult. Part of this difficulty has been due to the high temperatures required for vaporization of metallic compounds such as oxides or halides. Also, the location of the metal impurities has posed some problems. On the other hand, aqueous cleaning followed by a water rinse has proven more satisfactory since most of the metal compounds are water soluble and can be rinsed away. Even so, it is believed that overall possibilities for vapor phase metal removal remain good, especially when considering related particle and organic impurity cleaning processes. A more detailed discussion on mechanisms and future possibilities involving vapor cleaning techniques follows.

Some investigations are now being reported concerning vapor phase removal of metallic impurities from silicon wafers. These involve HF chemistries (55)(59), combinations of HF and UV/ozone (131), dry HF-H_2O combinations (58), other gas chemistries (132), and UV excited Cl_2 processes (54). Data showing the effectiveness of HF/H_2O cleaning for metal impurity removal from silicon surfaces, with and without a post-clean DI water rinse, are shown in Fig. 26 (58). In the UV/Cl_2 case, photoexcited chlorine species are used to etch a thin silicon layer, thus removing metal impurities incorporated on or in the silicon. Mechanism of metal removal may be by volatilization of the metal halide or by liftoff. A critical part of this process involves the uniformity and control of etching the silicon layer. Data demonstrating this silicon etching are shown in Fig. 27, as reported by Ito (133).

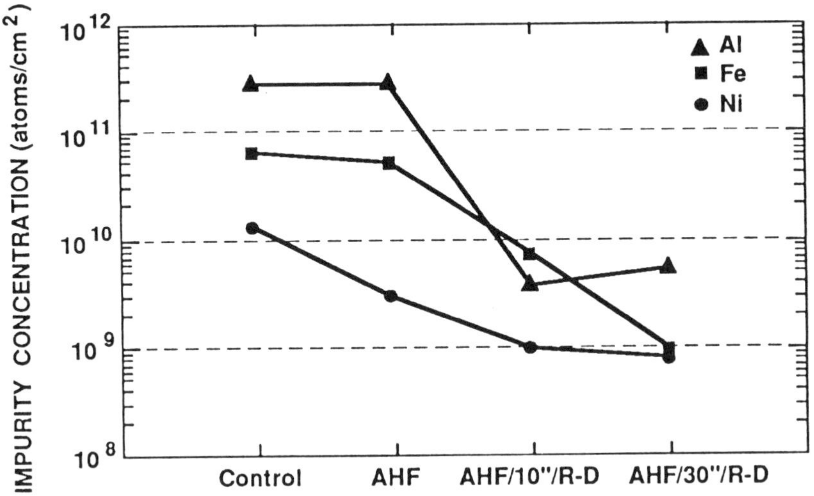

Figure 26. Metallic impurity removal by integrated etch rinse processing; after Syverson (58).

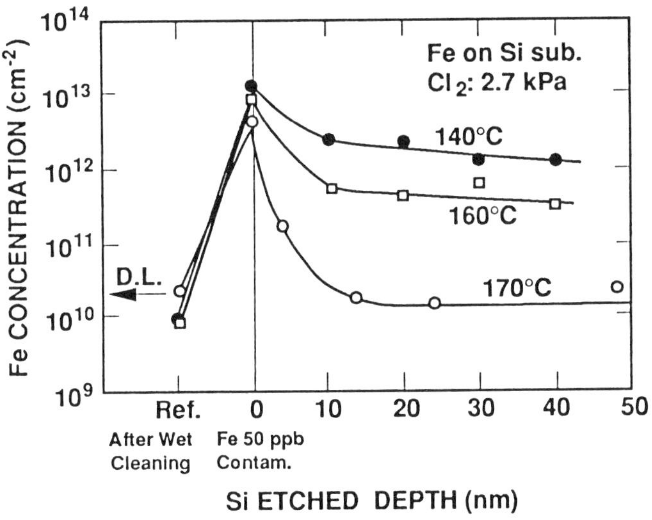

Figure 27. Iron concentration on the silicon surface before and after UV-excited dry cleaning; after Ito (133).

5.6 Mechanisms of Metal Impurity Removal

Vapor phase metal removal requires that the metal be in a form that can be volatilized, which typically requires elevated temperatures as well as the formation of a more volatile compound. For the purposes of this discussion we will assume that temperatures above 400°C are unacceptable for typical cleaning processes. Even at this temperature, however some metals may be volatile enough so that they can simply be thermally desorbed. This is illustrated in Fig. 28 where the vapor pressures of volatile metals are plotted versus temperature from the tabulations of Kubaschewski and Alcock (134). The right axis is the equivalent flux, assuming the sticking probabilities are unity for equilibrium vaporization. A vapor pressure of 10^{-6} torr corresponds to a equivalent flux of approximately one monolayer/sec, sufficient for removal of the metal in reasonable times. From this curve, therefore, we might expect all these metals to be volatile enough for simple thermal cycling to be sufficient to remove them. However, the analysis assumes that the metals are not chemically bonded to Si, oxygen, or other species. This may not be the case for Li, Ca, Mg, and Sr, which form both stable oxides as well as silicides. Na and K oxides are not stable in contact with Si, but with stable silicides this approach may not be effective with them either. Of the volatile metals, Zn, Cd, and Pb have the best chance for this process to work, since they don't have stable silicides and their oxides are unstable with respect to Si. We note at this point that this analysis (and that to follow) is rather speculative, and we know of no experimental data to test these ideas at this time.

A method that has been examined experimentally is the use of chlorine (normally excited by UV radiation) to produce volatile species of the metals which can be desorbed. A plot for some metal chlorides similar to that of Fig. 28 for the pure metals is shown in Fig. 29. The experimental results obtained on many of these metals can therefore be understood based on simple evaporation of volatile chlorides. However, Ito (133) has observed significant cleaning efficiencies at 170°C, below what would be necessary for this mechanism alone (except for possibly Fe). It seems likely that the UV radiation or the joint evaporation of $SiCl_x$ may well catalyze the metal chloride removal; it is also possible that some volatile metal-Si-Cl complex may form as well.

A difficulty with this method appears to be related to the rapid reaction and desorption rate of Si in the form of $SiCl_2$ or $SiCl_4$ under conditions that lead to efficient metal removal. The tendency for this to occur is also illustrated in Fig. 29 where the $SiCl_2$ and $SiCl_4$ vapor pressures are shown. Methods to retard this Si etching and the surface roughness it can cause are an active area of research today.

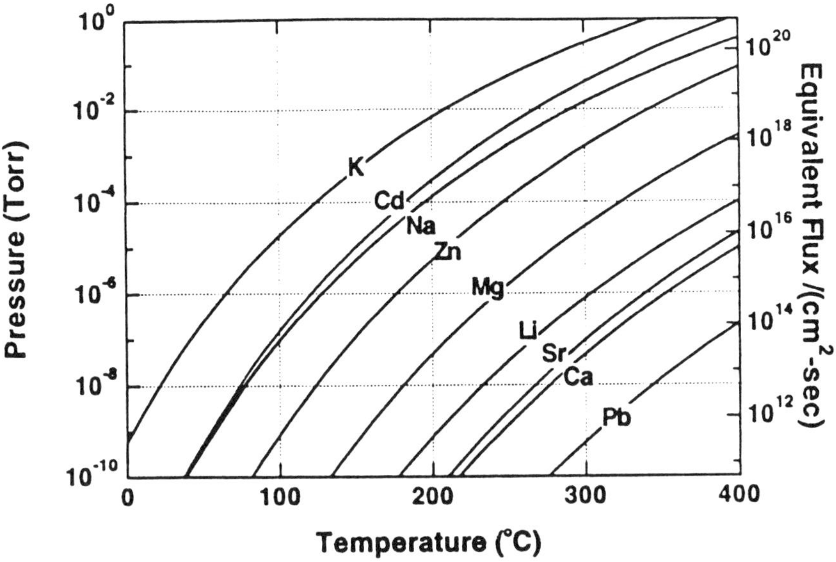

Figure 28. Vapor pressures for "high" vapor pressure metals as a function of temperature; the right hand axis would correspond to the equilivalent flux leaving the surface assuming a unity equilibrium sticking coefficient and molecular flow.

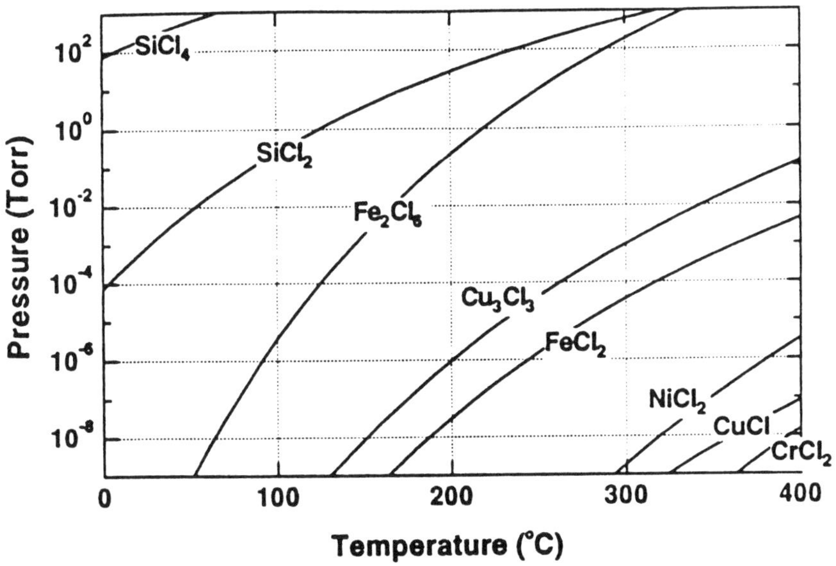

Figure 29. Vapor pressures for some important metal chlorides showing their volatility at temperatures used in UV enhanced vapor phase metal cleaning.

6.0 DEVICE APPLICATIONS

6.1 General Effects of Impurities on Device Properties

The fact that impurities can adversely affect both yield and reliability of semiconductor devices has been known from the days of the first silicon devices. Even though tolerance to impurity concentrations was much greater in the late 1950s, considerably higher levels of contaminants were also present. Over the years, as manufacturing and process procedures improved and contamination levels decreased, so did device feature sizes, and device purity requirements correspondingly increased. Thus the process engineer has continually been on the edge of his or her ability to maintain an impurity level low enough to achieve satisfactory device performance.

As was indicated above, several types of impurities must be controlled. Effects of characteristics generally fall into three main categories. These are: (1) junction characteristics, (2) contact or interface properties, and (3) MOS gate oxide properties. There are obviously some overlaps, such as MOS gate oxide - semiconductor interfaces. General considerations and application examples regarding these three categories are presented below with respect to vapor phase cleaning.

6.2 Junction Characteristics

A reasonable number of papers have been published relating impurity effects to junction characteristics (135)(136) but most of these deal with liquid precleans or gettering for the removal of contaminants from the junction region. It is anticipated that vapor cleaning technology will be successful in minimizing contaminant effects on junctions, but for now not much information is available.

6.3 Contact/Interface Properties

A number of device structures involve contacts or interfaces of some kind. These include: *(i)* metal-silicon or metal-metal contacts, *(ii)* epitaxial silicon-substrate silicon interfaces, *(iii)* poly-silicon-substrate silicon interfaces, and *(iv)* oxide- or dielectric-substrate silicon interface. Some of these, *(i)* and *(ii)*, require a minimum of resistance and thus as thin an oxide interface layer as possible. On the other hand, *(iii)* can involve poly-Si bipolar emitter structures and for this application, a thin, uniform oxide (about 10 Å) is often preferred. In the case of *(iv)*, the resistance is not a factor, but rather minimum structural defects which lead to interface traps are desired for optimum device performance. In all cases, proper cleaning of the surface prior to film deposition or oxide formation is necessary to remove all unwanted impurities including native or chemical oxides.

Various results have substantiated that vapor phase cleaning can result in improved surface and interface characteristics. Often this requires only vapor hydrofluoric acid (57). In the case of pre-epi-Si cleaning, several examples have been reported which demonstrate the effectiveness of vapor HF preclean (137)(138). Because a final water rinse is often not necessary, less oxygen, as well as carbon and other impurities, are found after vapor cleaning at the epi-silicon interface. The same is true for various structures involving metal contact depositions on silicon. Along related lines, it has been determined that the presence of any water on a doped poly-silicon surface will prevent satisfactory adherence of metal silicide films. The use of vapor phase HF cleaning prior to silicide deposition completely eliminates this problem. Yield data tabulated in Table 3 demonstrate the effectiveness of a vapor phase HF pre-clean for tungsten silicide adherence.

Table 3. Effect of Pre-Clean Process on Tungsten Silicide Lifting

■ VAPOR PHASE HF CLEAN

	SILICIDE LIFTING / DELAMINATION	
WAFER NO.	NO. LIFTED AREAS	NO. DIE INSPECTED
1	1	33
3	0	33
5	1	33
7	1	33
9	0	33
11	0	33
TOTALS	3	198
PERCENT FAILED	1.5	

■ STANDARD AQUEOUS CLEAN

	SILICIDE LIFTING / DELAMINATION	
WAFER NO.	NO. LIFTED AREAS	NO. DIE INSPECTED
2	3	33
4	21	33
6	12	33
8	16	33
10	6	33
12	26	33
TOTALS	84	198
PERCENT FAILED	42	

In the case of poly-Si emitters, more uniform, better controlled device characteristics have been obtained for vapor HF precleans (139). Figure 30 indicates results obtained where vapor HF was compared with other types of cleans, with and without integrated in situ processing. The results demonstrate that optimum results will require a thin, controlled oxide film to optimize both current gain and contact resistance. Finally, the interface between a gate oxide and the substrate must be properly cleaned to remove metals, ionics, carbon, and other impurities which contribute to the formation of interface charges (124)(140).

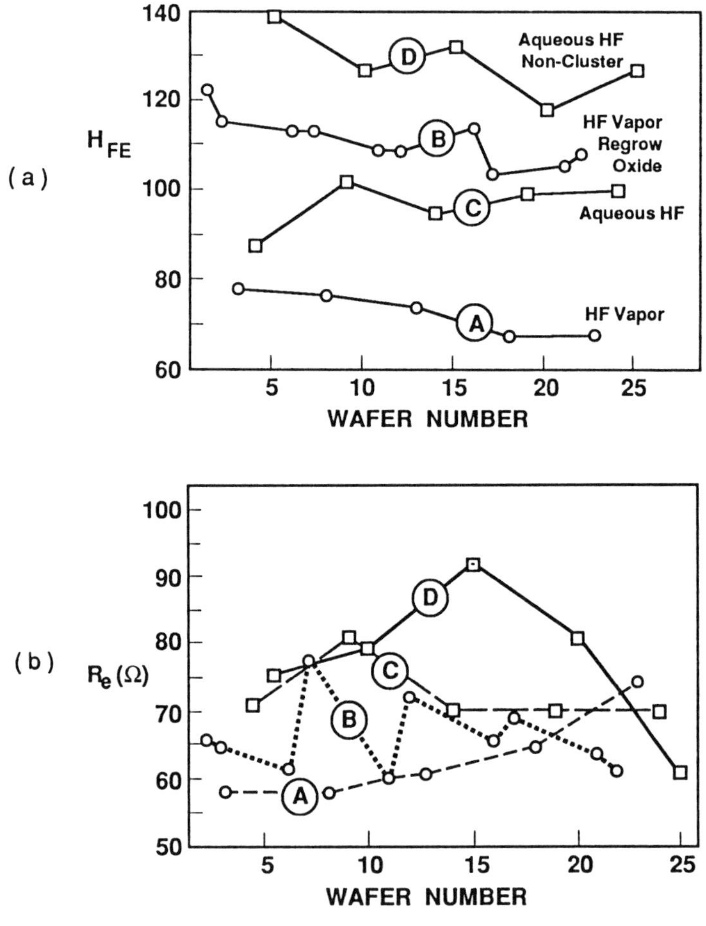

Figure 30. (a) Effect of various HF pre-treatments on current gain (H_{fe}) of poly-Si emitter transistors. (b) Effect on emitter resistance ($R_{e.}$); after deBoer and van der Linden (139). *(Reprinted by permission of The Electrochemical Society, Inc.)*

6.4 Gate Oxide Properties

Integrity of the thin gate oxide is perhaps the most critical property of MOS devices, and it is the most dependent on contaminants such as heavy metals or particles. These contaminants are generally present on or in the silicon prior to gate oxidation and become incorporated in the oxide as it is formed. The oxide integrity normally is reflected by low dielectric breakdown or even direct electrical shorts. Various tests are employed to evaluate gate oxide integrity. These include ramped voltage or field breakdown (V_{BD} or E_{BD}), charge-to-breakdown (Q_{BD}) or time-dependent dielectric breakdown (TDDB). Comparisons of vapor phase versus conventional liquid gate oxidation pre-cleans are now being reported which indicate improved oxide integrity for vapor HF pre-cleans (66)(67)(141). It is believed that vapor processes result in a reduction of particles and other impurities on the surface which helps to account for this improvement. A typical comparison of TDDB obtained for vapor and liquid pre-cleans is presented in Fig. 31.

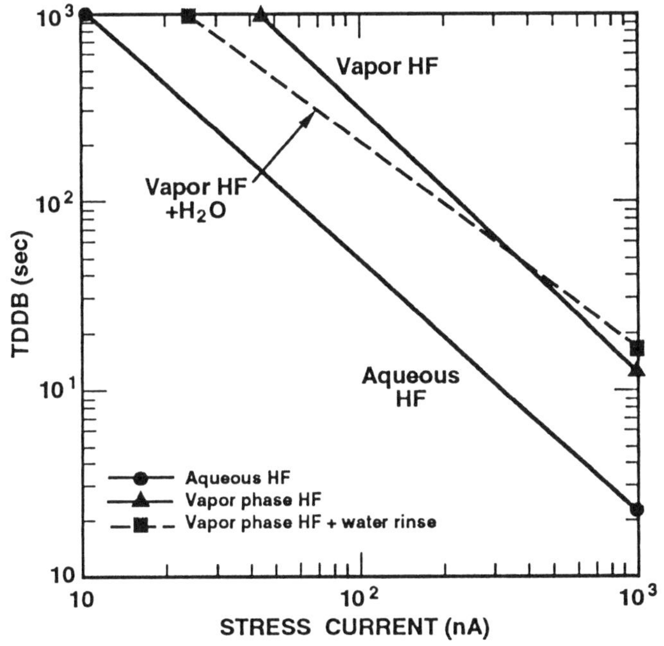

Figure 31. Effect of gate oxidation pre-clean on TDDB (time-dependent dielectric breakdown) using constant current stress; after Wong et al. (66). *(Reprinted by permission of The Electrochemical Society, Inc.)*

It has recently been noted (108)-(110) that silicon surface roughness is also partially responsible for low dielectric breakdown in gate oxides. This is undoubtedly due to higher electric fields at non-uniform areas of the oxide, which are caused by the rough silicon surface during oxidation. However, contaminants undoubtedly contribute as well. A primary cause for rough silicon has been found to be due to etching of the silicon by the ammonium peroxide liquid clean treatment. The effect is reduced by decreasing the ammonium hydroxide concentration. Typical results are shown in Fig. 32. The roughness effect is not observed after vapor phase cleaning treatments, unless they follow directly the liquid ammonium peroxide step.

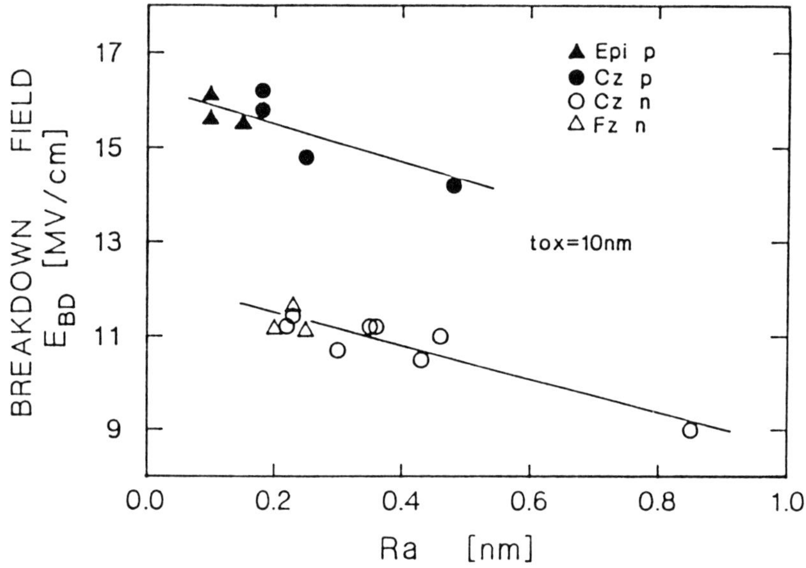

Figure 32. Surface microroughness dependence of gate oxide breakdown field; after Ohmi (109).

Another problem observed with gate oxides caused by surface contamination is excess carrier trapping in the oxide. This is especially critical with respect to memory-type devices where so-called "tunneling oxides" are employed. It is also important for devices exposed to ionizing radiation. The latter involves the same effects mentioned earlier where radiation-produced defects result from "dry" cleaning processes. Not only have vapor cleaning processes been found to incorporate fewer impurities in the oxides which can lead to charge or carrier trapping, but fluorine species left on the surface after vapor HF cleaning appear to provide trap- or radiati‹ n-resistant oxides (142)-(144). Somehow, the fluorine species neutralize the oxide traps and significantly retard charge trapping.

One other effect of contamination on gate oxide technology is that of oxidation kinetics. It has been determined that a preoxidation clean involving ammonium peroxide (NH_4OH + H_2O_2) can retard the rate of oxidation (145). Later investigations (146)-(148) indicated that this retardation was due to aluminum impurities in the cleaning solution, probably from the hydrogen peroxide. In addition, for very thin oxides (< 100 Å) the effect is reversed, and the oxidation rate is faster for aluminum-contaminated ammonium peroxide treatments. A kinetic plot is shown in Fig. 33 which demonstrates the effect of NH_4OH + H_2O_2 versus HF pre-oxidation cleaning treatment. The solution to the problem obviously involves the use of higher purity chemicals.

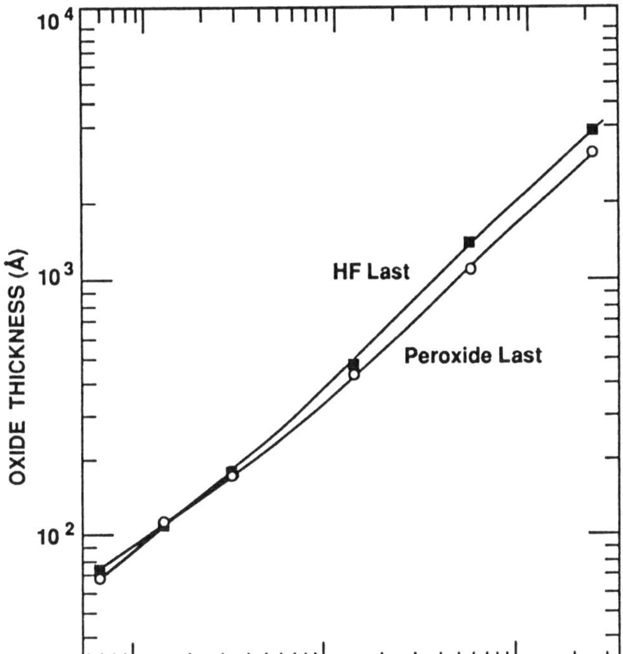

Figure 33. Best fit of present oxidation model to reverse-RCA and HF-cleaned surfaces for 800°C dry O_2 oxidation of (100) silicon; after deLarios (148).

7.0 INTEGRATED PROCESSING

7.1 Concept

Integrated processing is the term or expression normally used to describe the combination of two or more sequential processes in semiconductor device fabrication whereby these processes are carried out in situ in a controlled ambient. Often such processes involve a film deposition onto a silicon surface, and the resulting interface is critically dependent on the cleanliness of that surface. A typical example might involve a silicon cleaning treatment followed by the deposition of a contact metal such as aluminum. A more complex sequence would be the preparation of a MOS gate structure, where the integrated processing sequence would involve *(i)* sacrificial oxide strip, *(ii)* gate oxide pre-clean, *(iii)* gate oxidation, and *(iv)* poly-Si gate deposition. Thus the cleanliness of two critical interfaces (SiO_2-Si and poly-Si-SiO_2) is preserved by controlling the ambient and surface cleanliness throughout the multi-process sequence.

The cluster-tool configuration, used to carry out integrated processing, typically consists of a central automated handling system which transports device wafers from a loading station to and from various process modules. When the process is complete, the wafers are transported to an unloading station. The wafers are exposed during transport (and in the handler) to only vacuum or an inert gas such as argon or nitrogen. Vacuum level is usually maintained at least to 10^{-5} torr. While concepts for cluster tools are still developing, configurations are generally radial or linear. Schematic examples of these two types are shown in Fig. 34.

Several reviews which discuss the current status and future potential of integrated processing and cluster tool concepts are available (150)-(160). Certain conclusions may be drawn at the present time. First, it is apparent that only vapor or gas type processes will be practical in such systems. Thus, conventional liquid-type wafer cleaning processes will not be acceptable, providing considerable support for the concept of vapor or gas phase cleaning technology. Also, single wafer processing will be required which will impact such factors as throughput. This will be discussed below. Finally, it is not clear at this time how far the idea of integrated device processing can be taken. In other words, How soon, if ever, will the technology allow the often-spoken "sand-to-device-in-a-box" concept? It is much more likely that small numbers of processes (two to five) will be combined into small cluster groups, and wafers will be transferred from one group to the next by some appropriate method such as SMIF (161)(162). In any case, it is anticipated that the days of mega-dollar clean rooms may be numbered.

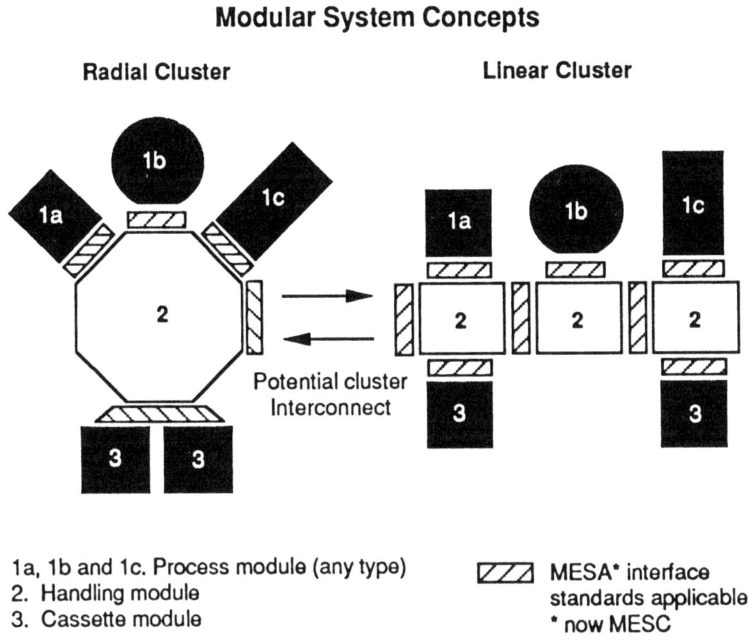

Figure 34. Schematic of radial and linear cluster configurations for integrated processing; after Van Leeuwen (149).

7.2 Advantages and Disadvantages of Integrated Processing

Several of the reasons for developing integrated or cluster tool processing for sub-micrometer device fabrication have already been given above, but a short summary of potential advantages and disadvantages (challenges) is appropriate. First, and probably most important, is the fact that tomorrow's device structures probably cannot be produced without reducing contamination levels several orders of magnitude from today's capabilities. This can only be accomplished by better controlling the ambient to which wafers are exposed during fabrication. Along these same lines, integrated processing, even on a limited basis, will allow more efficient and flexible processing. It should also result in fewer process steps as well, with much greater process control through computerized monitoring and programming of the process sequences. Finally, greater cooperation and planning between equipment manufacturers and device engineers will be required which will undoubtedly lead to improved overall results.

Integrated processing will also lead to additional challenges. It should be expected that any new complex equipment, as these systems are, will pose reliability questions which must be resolved. Since this technology utilizes single wafer processing, then thruput will be a consideration. Another challenge in this type of concept will be the matter of compatibility, both from a standpoint of module alignment and space considerations, but also questions of interference among the various reaction gases and ambients. For instance, one process may operate at semi-high vacuum and can't tolerate water vapor, while the next module may involve a water-saturated cleaning treatment. Finally, Is vapor-phase, integrated processing adaptable to all types of processes required for device fabrication? What about lithography? All of the questions and challenges must be considered and satisfactorily solved.

7.3 Requirements/Considerations of Integrated Processing

In addition to the discussion above related to advantages and challenges of integrated wafer processing, it is worthwhile to summarize the main requirements and other considerations of this technology with emphasis on wafer cleaning. First, the requirement for vapor cleaning processes in integrated processing has already been emphasized. In addition, the single wafer concept is already well established in vapor cleaning technology, and is rapidly becoming common in other processes. The same is true for the use of computer-controlled, automated systems of all types.

An area that at present is somewhat lacking in vapor cleaning systems and will be very important in cluster tools is that of in situ monitoring and control. It will be very desirable, for instance, to be able to measure oxide thickness as it is being etched back in HF/H_2O chemistries. Similarly, determining complete removal of an oxide to a hydrophobic surface would be most important. Monitoring particle concentration on the wafer, as well as in the ambient in situ, will provide a lot better control of the process. Of course, temperature and pressure monitoring are already being accomplished in the currently available cleaning systems. Likewise, gas reactant composition is being monitored and controlled by a wide variety of RGAs (residual gas analyzer). The ultimate, however, will be to have the capability to measure surface impurity levels as the wafers are being cleaned. This will require analytical beam techniques, such as TXRF (total reflection x-ray fluorescence spectroscopy), to be incorporated in the reaction module (116). Hopefully most of these capabilities will be possible in the not-too-distant future.

7.4 Applications Involving Vapor Cleaning

Investigations are already being reported on the use of integrated processing for selected critical sequences in device fabrication. Even though these tend to be limited to a few steps, they show the great potential for this kind of technology. As might be expected, all of them include a pre-process cleaning step, and this cleaning involves vapor or dry processing. Following is a summary of reported results.

MOS Structure. This process sequence includes gate oxide pre-clean (may include sacrificial oxide strip), gate oxidation, and finally poly-Si gate deposition. As indicated earlier, this sequence includes two of the most sensitive interfaces from the standpoint of contamination effects. Results reported involving part or all of the above sequence indicate significant improvements in gate oxide quality of MOS structures (163)-(166).

Metal or Poly-Si Contacts. Another natural application for vapor cleaning/integrated processing involves metal or poly-Si contacts to silicon. This can be accomplished rather simply by combining a pre-clean with the metal deposition. As is the case for MOS structures above, initial reported results appear very promising.

Epitaxial Silicon Deposition. It was mentioned above that epitaxial silicon deposition following a pre-clean involving gaseous HCl represents the first example of integrated processing (18). Also, MBE (molecular beam epitaxy) scientists have used various vapor or gas type precleans for the preparation of exotic structures––both silicon and compound semiconductor-based (137). More recently, investigators have demonstrated the use of in situ vapor cleaning processes in combination with lower temperature (750°C or thereabouts) silicon epitaxial deposition (137)(138). The results indicate that such integrated processing will be required for contamination-free epi deposits, especially as process temperatures drop even further.

Bipolar Poly-Si Emitter Structures. The poly-Si emitter structure in bipolar device technology has shown great promise but has been very difficult to prepare. In this technology, the poly-Si is deposited onto a single crystal region. An n-type dopant, such as arsenic, implanted into the poly-Si is diffused into a shallow, p-type base region to form the emitter. The presence of (or lack of) a thin oxide layer between the poly-Si and single crystal silicon regions will affect both current gain and emitter resistance. Results now demonstrate that integrated processing permits in situ removal of the initial thin oxide by vapor HF, regrowing a controlled 10 Å oxide, and then depositing the poly-Si. The resulting bipolar device exhibits tighter distribution of both emitter resistance and current gain (139)(167). Results have been plotted in Fig. 35. Thus, device characteristics can be tailored by controlling the thickness of the regrown oxide interface layer. Other similar results have been reported (168)(169).

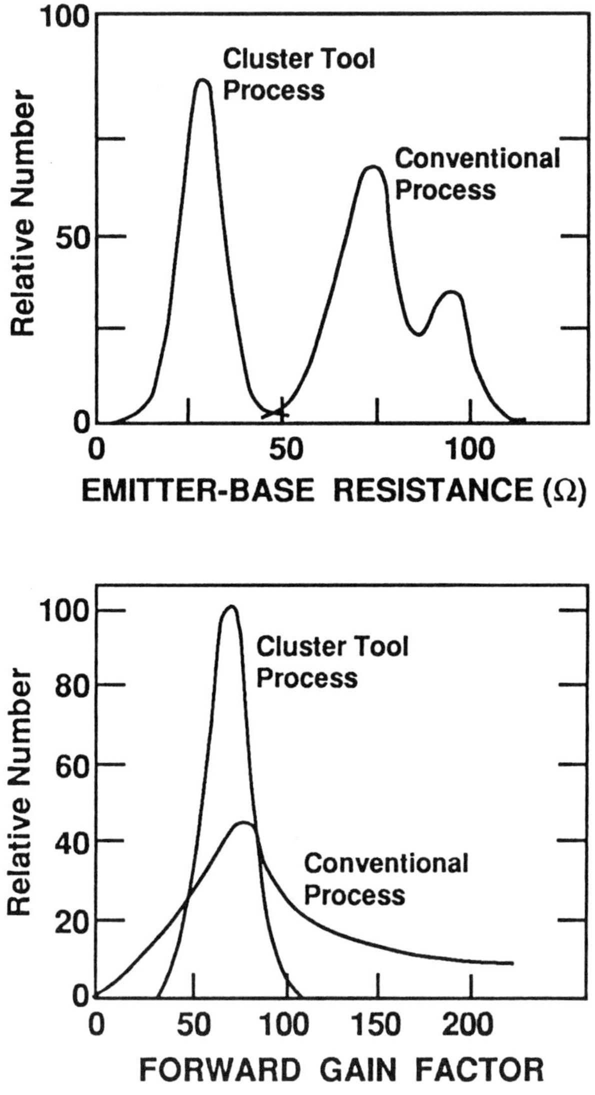

Figure 35. Distribution of emitter-base resistance and gain comparison between cluster tool process and conventional process; after Werkhoven et al. (167).

Sidewall or Trench Passivation. One other application of vapor cleaning combined with integrated processing involves the passivation of sidewall or trench structures. In this situation, polymeric and other residues remaining from reactive ion etching are removed from the sidewalls by vapor cleaning, which is followed by suitable thermal oxidation or CVD deposition of various dielectric films. As with the other examples described above, improved results are obtained by this combination (109)(110).

8.0 CONCLUSIONS AND SUMMARY

In this chapter, we have reviewed vapor phase wafer cleaning technology. After summarizing historical developments and practices concerning silicon wafer cleaning which primarily involved aqueous technology, the concept of vapor phase wafer cleaning technology has been presented and discussed in detail. This discussion includes historical trends which justify the incorporation of vapor cleaning processes in current or future advanced device fabrication facilities. One of these justifications is the requirement for the cleaning reagent to penetrate deep trenches or grooves in the 0.5 µm range or below. Another is the critical requirement to simplify process and facility complexity and cost. Still another relates to environmental concerns. It should be recognized, however, that as of this date (1992), vapor phase cleaning only exists in current production lines on a very limited basis. On the other hand, it is evident that this concept is overwhelmingly endorsed by most engineers looking toward the future.

One of the important aspects of current and future device fabrication involves the uniform removal or etching of oxide layers. Examples have been presented showing how such etching can be carried out, both in blanket form across a wafer as well as in selected regions. The concept of "native oxide" formation and regrowth has been reviewed. In addition, we have discussed differences in etch rates and reaction mechanisms for different types of oxides and dielectric films. Moreover, a detailed discussion of mechanisms involved in vapor phase etching, and cleaning in general, has been presented, which should help considerably in the development of more advanced cleaning processes in the future.

Finally, we have discussed the importance and requirements for reducing the levels of various types of contaminants using vapor phase cleaning. These impurity types include organics, oxides, particles and residues of all kinds, and different kinds of metals. It is believed that the difficulty of accomplishing such minimization of these contaminants by vapor cleaning increases in the order listed above. In other words, a form of vapor cleaning, ultraviolet light plus ozone, has been used for some time to remove exposed resist from wafer surfaces. On the other hand, at least some of the metals may be very difficult to remove completely using known vapor cleaning techniques without subsequent water rinsing.

We strongly believe, however, that for device complexities and feature sizes of the future (1 Gbit and 0.2 µm), vapor cleaning will be an absolute requirement. Many engineers also feel that an HF-last process with no subsequent rinse will provide the most optimum surface for enhanced device performance and reliability. This will best be accomplished using vapor technology. It is also believed that some form of integrated, in situ processing involving so-called "cluster tools" will solve many of the current technical, financial, and environmental problems that seriously limit future advances in integrated circuit technology.

ACKNOWLEDGMENTS

The authors wish to thank T. Ohmi of Tohoku University, Sendai, Japan, for providing Figs. 14 and 30, D. B. Deal, M. Kohl, and D. Syverson of FSI International for providing Figs. 4, 5, 16, and 25, and G. Higashi of AT&T Bell Laboratories for providing Fig. 20. Helpful discussions with J. M. deLarios, D. B. Kao, M. A. McNeilly, and G. L. Nobinger of Advantage Production Technology, and M. Wong and M. Moslehi of Texas Instruments are gratefully acknowledged. One of us (CRH) would also like to acknowledge financial support from Advantage Production Technology, Texas Instruments, SEMATECH through Advantage, and Stanford's Center for Integrated Systems.

REFERENCES

1. *Symposium on Cleaning of Electronic Device Components and Material,* 246, ASTM, Philadelphia (1958)
2. Faust, J. W., Jr., in: *Symposium on Cleaning of Electronic Device Components and Materials,* 246:66, ASTM, Philadelphia (1958)
3. Khilnani, A., in: *Particles on Surfaces I, Detection, Adhesion, and Removal,* (Mittal, ed.) p. 17, Plenum Press, Thornwood (1986)
4. Kern, W. and Puotinen, D. A., *RCA Review* 31(6):187 (1970)

5. Kern, W., *RCA Engineer* 28(4):99 (1983)
6. Kern, W., *J. Electrochem. Soc.*137:1887 (1990)
7. Kern, W., *Semiconductor International* 7(4):94 (1984)
8. Amick, J. A., *Solid State Technology,*19(11):47 (1976)
9. Henderson, R. C., *J. Electrochem. Soc.*119:772 (1972)
10. Tolliver, D., *Solid State Technology,*18(11):33 (1975)
11. Burggraaf, P., *Semiconductor International,*13(7):58 (1990)
12. Burkman, D., *Semiconductor International,* 4(7):103 (1981)
13. Walter, A. E. and McConnell, C. F., *Microcontamination,* 8(1):35 (1990)
14. Jurcik, B. J. Jr., Brock, J. R. and Tractenberg, I., *J. Electrochem. Soc.* 138:2141 (1991)
15. Shwartzman, S., Mayer, A. and Kern, W., *RCA Review,* 46(3):81 (1985)
16. Skidmore, K., *Semiconductor International,*11(8):64 (1988)
17. Skidmore, K., *Semiconductor International,*10(9):80 (1987)
18. Bean, K. E., in: *Semiconductor Materials and Process Technology Handbook for VLSI and ULSI,* (McGuire, ed.) p. 80, Noyes Publications, Park Ridge (1988)
19. Lang, G. A. and Stavish, T., *RCA Review,* 24:488 (1963)
20. Shepherd, W. H., *J. Electrochem. Soc.*112:988 (1965)
21. Chu, T. L., Gruber, G. A. and Stickler, R., *J. Electrochem. Soc.*113:156 (1966)
22. Dismukes, J. P. and Ulmer, R., *J. Electrochem. Soc.*118:634 (1971)
23. Dismukes, J. P. and Levin, E. R., *Proc. Am. Inst. Chem. Eng.,* p. 135 (1970)
24. Reisman, A. and Berkenblit, M., *J. Electrochem. Soc.* 112:812 (1965)
25. Gregor, L. V., Balk, P. and Campagna, F. J., *IBM J. Res.* 9:365 (1965)
26. Ruzyllo, J., *Microcontamination,* 6(3):39 (1988)
27. Ruzyllo, J., *Solid State Technology,* 33(3):S1 (1990)
28. Ruzyllo, J., Frystak, D. C. and Bowling, R. A., in: *Proc. IEEE International Electron Device Meeting,* San Francisco, p. 409, (1990)
29. Ruzyllo, J., Hoff, A. M., Frystak, D. C. and Hossain, S. D., *J. Electrochem. Soc.*136(5):1474 (1989)
30. Moghadam, F. K. and Mu, X.-C., *IEEE Transactions on Electron Devices,* 36(9):1602 (1989)

31. Fonash, S. J., *J. Electrochem. Soc.* 137(12):3885 (1990)
32. Ohmori, T., Fukumoto, T., Kato, T., Tada, M. and Kawaguchi, T., in: *First Intern. Symposium on Cleaning Technology in Semiconductor Device Manufacturing,* (Ruzyllo and Novak, ed.) 90-9:182, The Electrochemical Society, Inc., (1989)
33. Sherman, R. and Whitlock, W., *J. Vac. Sci. Technol. B* 8(3):563 (1990)
34. Tabe, M., *Appl. Phys Lett.* 45(10):1073 (1984)
35. Vig, J. R., *J. Vac. Sci. Technol. A* 3(3):1027 (1985)
36. Holmes, P. J. and Snell, J. E., *Microelectronics and Reliability,* 5:337 (1966)
37. Helms, C. R., Deal, B. E. and McNeilly, M. A., in: *1991 Proceedings,* p. 822, The Institute of Environmental Sciences, Mount Prospect, ILL (1991)
38. Helms, C. R. and Deal, B. E., in: *Fall Meeting Extended Abstracts,* 91-2, p. 807, The Electrochemical Society, Phoenix (1991)
39. Helms, C. R. and Deal, B. E., "Mechanisms of the HF/H_2O Vapor Phase Etching of SiO_2," to be published *J. Vac. Sci. Technol.*
40. Brosheer, J. C., Lenfesty, F. A. and Elmore, K. L., *Industrial & Eng. Chem.* 39:423 (1947)
41. Munter, P. A., Aepli, O. T. and Kossatz, R. A., *Industrial and Engineering Chemistry,* 39:427 (1947)
42. Munter, P. A., Aepli, O. T. and Kossatz, R. A., *Industrial and Engineering Chemistry,* 41(2):1504 (1949)
43. Thomsen, S. M., *J. Am. Chem. Soc.*74:1690 (1952)
44. Whynes, A. L., *Trans. Instn Chem. Engrs.* 34:117 (1956)
45. Illarionov, V. V., Smirnova, Z. G. and Knyazeva, K. P., *Zhurnal Prikladnoi Khimii.*36(2):237 (1963)
46. Beyer, K. D. and Kastl, R. H., U. S. Patent 4,264,374 (April 28, 1981)
47. Blackwood, R. D., Biggerstaff, R. L., Clements, D. and Cleavelin, R., U. S. Patent 4,749,440 (June 7, 1988)
48. Bersin, R. L. and Reichelderfer, R. F., *Solid State Technology* 20(4):78 (1977)
49. Saito, Y., Yamaoka, O. and Yoshida, A., *Appl. Phys. Lett.* 56(12):1119 (1990)
50. Ibbotson, D. E., Mucha, J. A., Flamm, D. L. and Cook, J. M., *J. Appl. Phys.* 56(10):2930 (1984)
51. Skidmore, K., *Semiconductor International* 12(7):80 (1989)

52. Mishima, H., Ohmi, T., Mizuniwa, T. and Abe, M., *IEEE Transactions on Semiconductor Manufacturing* 2(4):121 (1989)
53. Mishima, H., Yasui, T., Mizuniwa, T., Abe, M. and Ohmi, T., *IEEE Transactions on Semiconductor Manufacturing* 2(3):69 (1989)
54. Oki, I., Biwa, T., Kudo, J. and Ashida, T., in: *Fall Meeting Extended Abstracts,* 91-2:790, The Electrochemical Society, Inc., Phoenix, Arizona (1991)
55. Syverson, D. J. and Duranko, G. T., *Solid State Technology* 31(10):101 (1988)
56. Cleavelin, D. C. R. and Duranko, G. T., *Semiconductor International* 10(11)(1987)
57. Novak, R. E., *Solid State Technology* 31(3):39 (1988)
58. Syverson, D., in: *1991 Proceedings,* p. 829, Institute of Environmental Sciences, Mount Prospect, IL (1991)
59. Deal, B. E., McNeilly, M. A., Kao, D. B. and deLarios, J. M., in: *Proceedings of the First International Symposium on Cleaning Technology in Semiconductor Device Manufacturing,* (Ruzyllo and Novak, ed.) 90-9:121, The Electrochemical Society, Inc., (1989)
60. Deal, B. E., McNeilly, M. A., Kao, D. B. and deLarios, J. M., *Solid State Technology* 33(7):73 (1990)
61. Norton Co., 1. N. B. S., Worcester, MA 01615, U. S. Patent 4,761,134 (Aug. 2, 1988)
62. Nobinger, G. L., Moskowitz, D. J. and Krusell, W. C., "Vapor Phase Technology: Sensitivities and Uniform Etching of SiO_2", to be published.
63. van der Heide, P. A. M., Baan Hofman, M. J. and Ronde, H. J., *J. Vac. Sci. Technol.* A7(3):1719 (1989)
64. Onishi, S., Matsuda, K. and Sakiyama, K., in: *Spring Meeting Extended Abstracts,* 90-1:519, The Electrochemical Society, Washington D.C. (1990)
65. Onishi, S., Matsuda, K. and Sakiyama, K., in: *Third International Symposium on Ultra Large Scale Integration Science and Technology,* (Andrews and Celler, ed.) 91-11:226, The Electrochemical Society, Inc., (1991)
66. Wong, M., Moslehi, M. M. and Reed, D. W., *J. Electrochem. Soc.* 138(6):1799 (1991)
67. Wong, M., Liu, D. K. Y., Moslehi, M. M. and Reed, D. W., *IEEE Electron Device Letters* 12(8):425 (1991)

68. Izumi, A., Matsuka, T., Takeuchi, T. and Yamano, A., in: *Extended Abstracts of the 1991 International Conference on Solid State Devices and Materials,* p. 135, Japan Society of Applied Physics, Yokohama (1991)

69. Ehrlich, D. J., Osgood, R. M., Jr. and Deutsch, T. F., *Appl. Phys Lett.* 38(12):1018 (1981)

70. Okano, H., Horiike, Y. and Sekine, M., in: *Spring Meeting Extended Abstracts,* 83-1:673, The Electrochemical Society, Inc., San Francisco (1983)

71. Horioka, K., Okano, H. and Horiike, Y., in: *16th International Conference on Solid State Devices and Materials Final Program,* p. 50, Japan Society of Applied Physics, Kobe (1984)

72. Sugino, R., Nara, Y., Yamazaki, T., Watanabe, S. and Ito, T., in: *19th Conference on Solid State Devices and Materials,* p. 207, Japan Society of Physics, Tokyo (1987)

73. Sato, Y., Sugino, R., Okuno, M. and Ito, T., in: *22nd (1990 Intl.) Conference on Solid State Devices and Materials,* p. 1103, Japan Society of Physics, Sendai (1990)

74. Ito, T., Sugino, R., Watanabe, S., Nara, Y. and Sato, Y., in: *First International Symposium on Cleaning Technology in Semiconductor Device Manufacturing,* (Ruzyllo and Novak, ed.) 90-9:114, The Electrochemical Society, (1989)

75. Ito, T. and Sugino, R., in: *Semiconductor World,* p. 120, (March, 1989)

76. Ogryzlo, E. A., Ibbotson, D. E., Flamm, D. L. and Mucha, J. A., *J. Appl. Phys* 67(6):3115 (1990)

77. Kern, W. and Deckert, C. A., in: *Thin Film Processes,* (Vossen and Kern, ed.) p. 401, Academic Press, New York (1978)

78. Mai, C. C. and Looney, J. C., *SCP and Solid State Technology* 9(1):19 (1966)

79. Judge, J. S., *J. Electrochem. Soc* 118:1772 (1971)

80. Harrap, V., in: *Semiconductor Si,* (Huff and Burgess, ed.) p. 354, The Electrochemical Society, (1973)

81. Deal, B. E. and Kao, D.-B., in: *Tungsten and Other Refractory Metals for VLSI Applications II,* (Broadbent, ed.) p. 27, Materials Research Society (1986)

82. Archer, R. J., *J. Electrochem. Soc.* 104:619 (1957)

83. Raider, S. I., Flitsch, R. and Palmer, M. J., *J. Electrochem. Soc.* 122:413 (1975)

84. Hattori, T., Takase, K., Yamagishi, H., Sugino, R., Nara, Y. and Ito, T., *Jpn. J. Appl. Phys.* 28(2):296 (1989)

85. Morita, M., Ohmi, T., Hasegawa, E., Kawakami, M. and Ohwada, M., *J. Appl. Phys.* 68(3):1272 (1990)

86. Graf, D., Grundner, M., Schulz, R. and Muhlhoff, L., *J. Appl. Phys.* 68(10):5155 (1990)

87. Ting, W., Hwang, H., Lee, J. and Kwong, D. L., *Appl. Phys. Lett.* 57(26):2808 (1990)

88. Olsen, J. E. and Shimura, F., *J. Vac Sci. Technol.* A7(6):3275 (1989)

89. Miki, N., Kikuyama, H., Kawanabe, I., Miyashita, M. and Ohmi, T., *IEEE Transactions on Electron Devices* 37(1):107 (1990)

90. Kern, W., Schnable, G. L. and Fisher, A. W., *RCA Review* 37(3):3 (1976)

91. Bean, K. E., *Thin Solid Films* 83:173 (1981)

92. Kikyuama, H., Miki, N., Saka, K., Takano, J., Kawanabe, I., Miyashita, M. and Ohmi, T., *IEEE Transactions on Semiconductor Manufacturing* 4(1):26 (1991)

93. Higashi, G. S., Becker, R. S., Chabal, Y. J. and Becker, A. J., *Appl. Phys. Lett* 58(15):1656 (1991)

94. Ubara, H., Imura, T. and Hiraki, A., *Solid State Communications* 50(7):673 (1984)

95. Grundner, M. and Jacob, H., *Appl. Phys.A* 39:73 (1986)

96. Yablonovitch, E., Allara, D. L., Chang, C. C., Gmitter, T. and Bright, T. B., *Physical Review Letters.*57(2):249 (1986)

97. Burrows, V. A., Chabal, Y. J., Higashi, G. S., Raghavachari, K. and Christman, S. B., *Appl. Phys. Lett.* 53(11):998 (1988)

98. Graf, D., Grundner, M. and Schulz, R., *J. Vac. Sci. Technol* A7(3):808 (1989)

99. Chabal, Y. J., Higashi, G. S., Raghavachari, K. and Burrows, V. A., *J. Vac. Sci. Technol.* A 7(3):2104 (1989)

100. Higashi, G. S., Chabal, Y. J., Trucks, G. W. and Raghavachari, K., *Appl. Phys. Lett.* 56(7):1990 (1990)

101. Haring, R. A. and Liehr, M., "Reactivity of a Fluorine Terminated Si Surface," to be published, *J. Vac. Sci. Technol.*

102. Helms, C. R., Johnson, N. M., Schwarz, S. A. and Spicer, W. E., *J. Appl. Phys.* 50(11):1979 (1979)

103. Hahn, P. O. and Henzler, M., *J. Appl. Phys.* 52(6):4122 (1981)

104. Hahn, P. O., Yokohama, S. and Henzler, M., *Surface Science* 142:545 (1984)

105. Hahn, P. O. and Henzler, M., *J. Vac. Sci. Technol.* A2(2):574 (1984)

106. Hahn, P. O., Grundner, M., Schnegg, A. and Jacob, H., *Applied Surface Science* 39:436 (1989)

107. Hahn, P. O., Grundner, M., Schnegg, A. and Jacob, H., *Semiconductor Silicon 1990,* (Huff, Barraclough and Chikawa, ed.) 90-7:296, Electrochemical Society, Inc., (1990)

108. Heyns, M. M., *Microcontamination* 9(4):29 (1991)

109. Ohmi, T., in: *Extended Abstracts of the 1991 International Conference on Solid State Devices and Materials,* p. 481, Japan Society of Applied Physics, Yokohama (1991)

110. Offenberg, M., Liehr, M. and Rubloff, G. W., *J. Vac. Sci. Technol.* A9(3):1058 (1991)

111. Batey, J., Tierney, E. and Nguyen, T. N., *IEEE Electron Device Letters* EDL-8(4):148 (1987)

112. Stasiak, J., Batey, J., Tierney, E. and Li, J., *IEEE Electron Device Letters* 10(6):245 (1989)

113. Batey, J., Tierney, E., Stasiak, J. and Nguyen, T. N., *Applied Surface Science* 39:1 (1989)

114. Miki, N., Maeno, M., Maruhasi, K. and Ohmi, T., *J. Electrochem. Soc.* 137(3):787 (1990)

115. Meieran, E. S., Flinn, P. A. and Carruthers, J. R., *Proc. of the IEEE* 75:908 (1987)

116. Hockett, R. S. and Katz, W., *J. Electrochem. Soc.* 136(11):3481 (1989)

117. Davis, L. E., Katz, W., Eberle, W. J. and Zazzera, L. A. *Microcontamination Conf. Proc.* (1989)

118. Phillips, B. F., Burkman, D. C., Schmidt, W. R. and Peterson, C. A., *J. Vac. Soc. Technol.* Al(2): 646 (1983)

119. Pool, R., *Science* 247:634 (1990)

120. Grosse, P., Harbecke, B., Heinz, B., Meyer, R. and Offenberg, M., *Appl. Physics* A39:257 (1986)

121. Aspnes, D. E., in: ACS Symposium Series 295, *Microelectronic Processing: Inorganic Materials Characterization,* (Casper, ed.) p. 192, American Chemical Society, Washington DC (1986)

122. Shimazaki, A., in: *Defects in Silicon II,* (Bullis, Gosele and Shimura, ed.) 91-9:47, The Electrochemical Society, Inc., (1991)

123. Hahn, S., Eichinger, P., Park, J. G., Kwack, Y. S., Cho, K. C. and Choi, S. P., in: *Semicon/Korea Technical Proceedings,* p. 60, Semiconductor Equipment and Materials International, Seoul (1991)

124. Deal, B. E., in: *Proc. First Electronic Materials and Processing Congress,* p. 41, ASM International, Chicago (1988)

125. Menon, V. B., Clayton, A. C., Michaels, L. D. and Donovan, R. P., *Solid State Technology* 32(10):S9 (1989)

126. Bowling, R. A. and Larrabee, G. B., *J. Electrochem. Soc.* 136(2):487 (1989)

127. Onishi, S., Matsuda, K. and Sakiyama, K., A in: *22nd International Conference on Solid State Devices and Materials,* p. 1127, Japan Society of Physics, Sendai (1990)

128. Kasi, S. R., Liehr, M., Thiry, P. A., Dallporta, H. and Offenberg, M., *Appl. Phys. Lett.* 58(1):106 (1991)

129. Kasi, S. R. and Kiehr, M., *Appl. Phys. Lett.* 57(20):2095 (1990)

130. Kaneko, T., Suemitsu, M. and Miyamoto, N., *Jpn. J. Appl. Phys.* 28:2425 (1989)

131. Zazzera, L. A. and Moulder, J. F., *J. Electrochem Soc.* 136(2):484 (1989)

132. Gluck, R. M., in: *Electrochemical Society Fall Meeting Extended Abstracts,* 91-2:759, Electrochemical Society, Phoenix (1991)

133. Ito, T., in: *1991 Proceedings,* p. 808, Institute of Environmental Sciences, Mount Prospect, Ill (1991)

134. Kubaschewski, O. and Alcock, C. B. *Materials Science and Technology.* (Raynor, ed.), Fifth Ed., Pergamon Press, Oxford (1983)

135. Ohsawa, A., Honda, K., Takizawa, R., Nakanishi, T., Aoki, M. and Toyokura, N., in: *Sixth International Symposium on Silicon Materials Science and Technology,* (Huff, Barraclough and Chikawa ed.) 90-7:601, The Electrochemical Society (1990)

136. Tamura, M., Isomae, S., Ando, T., Ohyu, K., Yamagishi, H. and Hashimoto, A., in: *Defects in Silicon II,* (Bullis, Gosele and Shimura, ed.) 91-9:3, The Electrochemical Society, Inc., (1991)

137. Iyer, S. S., Arienzo, M. and Fresart, E. D., *Appl. Phys. Lett.* 57:895 (1990)

138. Galewski, C., Lou, J.-C. and Oldham, W. G., *IEEE Transactions on Semiconductor Manufacturing* 3(3):93 (1990)

139. de Boer, W. B. and van der Linden, R. H. J., in: *Fall Meeting Extended Abstracts,* 91-2:808, The Electrochemical Society, Phoenix (1991)

140. Deal, B. E., *J. Electrochem. Soc.* 127(4):979 (1980)
141. Kao, D. B., Cairns, B. R. and Deal, B. E., in: *Fall Meeting Extended Abstracts,* 91-2:802, The Electrochemical Society, Phoenix (1991)
142. Lo, G. Q., Ting, W., Kwong, D.-L., Kuehne, J. and Magee, C. W., *IEEE Electron Device Letters* 11(11):511 (1990)
143. Nishioka, Y., E. F. da Silva, J., Wang, Y. and Ma, T. P., *IEEE Electron Device Lett* 9:38 (1988)
144. Morita, M., Kubo, T., Ishihara, T. and Hirose, M., *Appl. Phys. Lett.* 45(12):1312 (1984)
145. Schwettmann, F. N., Chiang, K. L. and Brown, W. A., in: *Spring Meeting Extended Abstracts,* 78-1:688, The Electrochemical Society, Seattle (1978)
146. deLarios, J. M., Kao, D. B., Helms, C. R. and Deal, B. E., *Appl. Phys. Lett.* 54(8):715 (1989)
147. deLarios, J. M., "Effect of Aqueous Chemical Cleaning on the Si and Silicon Dioxide Surface and Silicon Oxidation Kinetics," Ph.D Dissertation, Stanford University (1989)
148. deLarios, J. M., Kao, D. B., Deal, B. E. and Helms, C. R., *J. Electrochem. Soc.* 138:2353 (1991)
149. Van Leeuwen, C., *Semiconductor International* 13(1):68 (1990)
150. Newboe, B., *Semiconductor International* 13(8):82 (1990)
151. Shankar, K., *Solid State Technology* 33(10):43 (1990)
152. Burggraaf, P., *Semiconductor International* 13(9):56 (1990)
153. McNab, T. K. P., *Semiconductor International* 13(9):58 (1990)
154. Bergendahl, A. S., Horak, D. V., Bakeman, P. E. and Miller, D. J., *Semiconductor International* 13(10):94 (1990)
155. McNab, T. K. P., *Semiconductor International* 13(11):86 (1990)
156. Bader, M. E., Hall, R. P. and Strasser, G., *Solid State Technology* 33(5):149 (1990)
157. Korolkoff, N. O., *Solid State Technology* 33(8):73 (1990)
158. Ohmi, T. and Shibata, T., *Microelectronic Engineering* 10:177 (1991)
159. Ohmi, T., *Microcontamination* 8(6):27 (1990)
160. Ohmi, T. and Shibata, T., *Microcontamination* 8(7):25 (1990)
161. Parikh, M. and Bonora, A. C., *Semiconductor International* 8(5):222 (1985)
162. Harada, H. and Suzuki, Y., *Solid State Technology* 29(12):61 (1986)

163. Liehr, M. and Kasi, S. R., in: *Extended Abstracts of the 1991 International Conference on Solid State Devices and Materials,* p. 484, Japan Society of Applied Physics, Yokohama (1991)
164. Offenberg, M., Liehr, M., Rubloff, G. W. and Holloway, K., *Appl. Phys. Lett.* 57:1254 (1990)
165. Lucovsky, G., Kim, S. S., Fitch, J. T., Wang, C., Rudder, R. A., Fountain, G. G., Hattangady, S. V. and Markunas, R. J., *J. Vac. Sci. Technol.* A9(3):1066 (1991)
166. Pan, P., Berry, W., Kermani, A. and Liao, J., *Solid State Technology* 33(1):37 (1990)
167. Werkhoven, C. J., Westendorp, J. E. M., Huusen, F. and Granneman, E. H. A., *Semiconductor International* 14(6):228 (1991)
168. Zhou, Z.-H., Yu, F. and Reif, R., *J. Vac. Sci. Technol.* B9(2):374 (1991)
169. Kermani, A., Johnsgard, K. E. and Wong, F., *Solid State Technology* 34(5):71 (1991)

8

Remote Plasma Processing for Silicon Wafer Cleaning

Ronald A. Rudder, Raymond E. Thomas, and Robert J. Nemanich

1.0 INTRODUCTION

The increasing complexity of silicon-based microelectronics is requiring greater numbers of processing steps: epitaxial growths, gate oxidations, field oxidations, planarizations, metal depositions, reactive ion etching, photoresist patternings, resist removals, etc. Between many of these processing steps, residues from the previous process must be removed to insure that the subsequent step provides adequate electrical and mechanical performance. This may require that cleaning procedures between steps remove impurities on surfaces to the sub-ppm level. These cleaning procedures must be reproducible and must not deteriorate the underlying material, or else product yield will suffer. In addition to contamination control during processing, thermal budgets for the overall manufacturing process are being reduced as device dimensions shrink. Of particular concern with plasma based processes is ion-induced damage of the substrate. As processes are performed at lower and lower temperatures, plasma-induced damage to substrates will not be annealed during wafer processing. One suitable candidate for low-temperature cleaning is remote plasma cleaning. By generating an active plasma in a spatial region remote from the substrate surface, active chemical and physical species can diffuse or drift to the substrate surface. Typically, the electric field strength in the region of the substrate is weak, avoiding ion acceleration into the substrate. Traditionally, the separation of the plasma generation region from the substrate has been sufficient to thermalize high energy ions that might travel to the substrate. More recently, plasma sources such as electron cyclotron resonance (ECR) sources and distributed electron cyclotron resonance (DECR) sources, which operate at low pressures, have been applied to low temperature cleaning. These ECR sources offer control over incident ion energy, and do not depend as critically as earlier remote plasma sources did on the separation between the plasma generation region and the substrate.

Remote plasma processing, in general, has been developed for a variety of semiconductor processing steps. An important application of remote plasmas is deposition of materials including SiO_2 (1)-(8), Si_3N_4 (9)-(11), amorphous Si (12)(13), microcrystalline-Si (14)-(16), epitaxial Si (17)-(26), epitaxial Ge (27)-(32), epitaxial SiGe alloys (33), epitaxial GaN (34), and dilute-C SiC alloys (35)(36). During development of these deposition technologies, researchers have from necessity developed in situ cleaning techniques, especially for the removal of native oxides from semiconductor surfaces. The removal of native oxides has permitted both homoepitaxial and heteroepitaxial growths on Si(100) at extremely low substrate temperatures. Oxide removal has allowed formation of low defect density SiO_2/Si interfaces between deposited SiO_2 films and in situ cleaned Si(100) wafers. In some cases a single chamber can be used for both the cleaning cycle and the deposition cycle.

Remote plasma cleaning with an external RF coil (electrodeless) was first used for oxide removal from III-V (1) and Ge surfaces (27). External fixtures, (inductors) served to couple RF power to the plasma gas through a dielectric medium. External coupling has the advantage of eliminating wafer contamination from the electrode material. Native oxides of Ge and GaAs are chemically unstable and electrically trapping. Researchers seeking to passivate or gate III-V or Ge surfaces needed to eliminate the native oxide on those surfaces prior to insulator deposition. In situ or in vacuo techniques were not available on the early systems to ascertain the efficiency of the cleaning process. However, in situ NH_3 plasma treatments of InGaAs surfaces prior to $SiO_2/Si_3N_4/SiO_2$ insulator stack deposition dramatically improved the transconductance in FET channels (1). The vast improvement in device performance convinced early researchers in the field that in situ cleaning would be a critical enabling technology for device fabrication. However, controlling and developing the cleaning process would be far more difficult without direct observation of the surface chemistry or structure.

The decision to implement in situ analytical tools to ascertain the state of the surface represented a marked departure from more traditional plasma processing. Traditional plasma processing has used immersion plasma systems that have a vacuum quality which is too poor to permit preservation of the plasma-treated surface after processing. Necessarily, the development of plasma processing in these immersion plasma systems concentrated on understanding plasma chemistry. Resultant surface contamination, combined with ion damage to the surface, has prevented the use of surface electron spectroscopic and diffraction techniques. In contrast, remote plasma processing coupled with UHV practice and high-purity gas delivery permits surface analytical tools to study plasma processing on crystalline semiconductor surfaces. While operating pressures of the remote plasma techniques preclude real time electron diffraction or electron spectroscopic techniques, rapid recovery to ultrahigh vacuum conditions allows step-wise determination of the efficacy of various components of the cleaning process. Reflection high energy electron diffraction (RHEED) was the first in situ tool adapted to a remote RF plasma system to assess remote plasma processing of semiconductor surfaces. Surface diffraction studies were performed on Ge surfaces after in situ cleaning in conjunction with the development of remote plasma processing for epitaxial Ge deposition (27)-(32) and for insulator-gating of Ge (37)-(39). In both cases, removal of oxides from Ge surfaces was found to be quite straightforward. Germanium substrates showed only diffuse electron scattering from the native oxide before hydrogen-plasma treatments, but exhibited integral and half-order diffraction streaks after a 10 - 20 sec remote hydrogen-plasma treatment. On these surfaces, homoepitaxial Ge layers and pseudomorphic Si layers were deposited by plasma-enhanced CVD. Following successful germanium oxide reduction, workers began to address silicon wafer cleaning for homoepitaxial growth, heteroepitaxial Ge growth, and gate-dielectric deposition. This work relied heavily on in situ RHEED to ascertain oxide reduction and subsequent epitaxial growth. In addition to RHEED diffraction, a variety of surface analytical techniques have been subsequently employed either in situ to the remote plasma chamber or in vacuo through transfer to adjacent surface analytical units (13)(23)(40)-(42). The analytical units have employed x-ray photoelectron spectroscopy (XPS), Auger electron spectroscopy (AES), ultraviolet photoelectron spectroscopy (UPS), low energy electron diffraction (LEED), and reflection high energy electron diffraction (RHEED) as tools to examine plasma processing for oxygen and carbon removal.

Complementing the in situ techniques, studies have begun employing ex situ techniques to examine structural effects of the cleaning process. Raman spectroscopy has been used to examine Si(100) surfaces for Si-H bonding following extensive hydrogen dosing (43). Cross-sectional transmission electron microscopy (XTEM) has been used to examine roughening of the silicon wafer surface and H-induced dislocations (44). A number of studies have used functional definitions to define a successful cleaning process. In several cases, the quality of epitaxy for homoepitaxial Si (17)(19)(22)(23) and heteroepitaxial Ge (30) has been used to arrive at a critical assessment of the cleaning technique employed. Secondary ion mass spectroscopy (SIMS) coupled with depth profiling has also been used to evaluate residual interfacial impurities in homoepitaxial Si layers (20)(45). Residual impurities at the SiO_2/Si interface can appear as interface state traps in capacitance-voltage (CV) measurements on metal-oxide on semiconductor (MOS) capacitors. Fountain et al. have fabricated MOS test structures for electrical evaluation by depositing SiO_2 films on remote plasma cleaned Si(100) (46). Cleaning is perhaps more critical for a deposited SiO_2 layer on Si where the electrical interface created is nearer to the wafer surface than for a thermally grown SiO_2 layer where the electrical interface created is beneath the original wafer surface. Breakdown statistics have also been used on MOS capacitors to evaluate effects of the remote plasma cleaning process (47)(48). Ultimately, the performance of MOS device structures verifies the success of cleaning for this electronic application where trapping surface states at levels of one in 10^5 are critical. These ex situ assessments (as in the previous in situ analysis) show that remote plasma processing is indeed a viable low-temperature plasma-assisted cleaning process.

2.0 PLASMA CLEANING CRITERIA

Due to the ever increasing number of Si processing steps, a cleaning process suitable at one point in wafer processing may not suffice at another. It is likely that individual processing steps will each leave specific residues on the surface which may require tailored cleaning chemistries. The discussion of all possible cleaning chemistries and requirements is beyond the scope of this review. Nonetheless, two particular examples (given below) illustrate vividly that the cleaning process is not independent of the processing environment, but must be an integral part of the subsequent processing. The cleaning process must focus on preparing the silicon surface chemically and structurally for the next step. The next two examples demonstrate the necessity for both determination and control of the structural and chemical perfection of the surface.

2.1 Cleaning for Low-Temperature Epitaxial Growth

The necessity for low-temperature epitaxial Si technology in future Si microelectronics is mandated by a low thermal budget required to maintain doping profiles for smaller and shallower channel sections. Cleaning of silicon surfaces for epitaxial growth requires that the surfaces be both structurally well ordered and chemically devoid of atmospheric impurities such as nitrogen, carbon, and oxygen. However, not all surface adsorbates are necessarily detrimental. For example, with hydride-based chemical vapor deposition of silicon, the presence of hydrogen passivation of the Si(100) surface may not preclude epitaxy. While physical sputtering processes can be used to remove impurities from the silicon surface at low temperature, such processes require subsequent high temperature anneals to recrystallize the surface which is made amorphous by the ion bombardment (49)(50). Low-temperature cleaning of Si(100) surfaces by remote plasma processes for subsequent epitaxial growth at low temperature has been well documented (20)(23)(51)(52). Typically, these remote plasma cleans have been carried out on Si(100) wafers loaded immediately after a modified RCA cleaning treatment. Following the ex situ clean and wafer transfer, in air, silicon surfaces show small amounts of carbon and oxygen contamination. Nitrogen has been occasionally observed (20). Subjecting the wafer to remote hydrogen plasma treatments removes carbon and oxygen impurities. Remote hydrogen plasma treatments are very efficient in removing carbon-containing contaminants, but they do not seem to be as proficient in removing oxygen contaminants (20)(23)(53). Although carbon contamination is reduced below the sensitivity of XPS, early work showed that residual oxygen concentrations were reduced to levels comparable to the noise in the AES or XPS techniques. More recent results have demonstrated the proficiency of the remote hydrogen plasma process to eliminate both carbon and oxygen from the Si(100) surface (24). Electron diffraction of these surfaces following hydrogen plasma treatments at temperatures greater than 300°C show 1/2-order reconstruction streaks in RHEED and 2x1, 1x2 spots in LEED. UPS measurements show the presence of hydrogen bonding on the silicon surface, suggesting that the diffraction patterns are from 2x1:H and 1x1:H terminated sites (44). The hydrogen plasma cleaned surface is structurally well ordered, is void of substantial quantities of oxygen and carbon, and is frequently hydrogen-terminated. After remote plasma cleaning, ultra-low temperature silicon homoepitaxy at 150°C has been successfully demonstrated on these surfaces (19)-(26).

2.2 Cleaning for Interface Formation with Deposited-SiO_2 on Silicon

It would seem that the cleaning requirements for interface formation with a deposited SiO_2 layer on silicon would require exacting control of chemical impurities through elimination of nitrogen, carbon, and metals on the Si(100) surface without regard to oxygen contamination. In reality, the deposition of SiO_2 for high quality SiO_2/Si interface formation places very severe demands on native oxide removal. Unlike thermal oxide growth where the SiO_2/Si interface is buried beneath the original wafer surface, a deposited oxide layer creates an SiO_2/Si interface that forms at or near the interface of the original wafer surface. Thus, residual impurities and structural imperfections on the Si surface may produce electrical traps at the SiO_2/Si electrical interface. Any native oxide remaining on the surface is incorporated into the electrically active interfacial region as the SiO_2 is deposited. Native oxides and suboxides are typically not well behaved electrically and, as a consequence, must be removed thoroughly from the silicon surface prior to SiO_2 deposition. Structural imperfections near the surface must not be induced by the cleaning. Such imperfections can also degrade the electrical performance of the device. Imperfections can result in electrically active interface states, reduced mobility in the conduction channel, and can limit the gate voltage modulation of the channel.

Several groups of researchers, by using remote hydrogen plasma cleaning followed by in situ SiO_2 deposition, have produced SiO_2/Si interfaces wherein the interface state trap density (D_{it}) is as low as 10^{10}/cm^2-eV (5)(17)(46). The performance of the corresponding metal-oxide-semiconductor (MOS) capacitor structures is comparable to state of the art thermal SiO_2/Si MOS structures where the SiO_2/Si interfaces are formed buried underneath the wafer surface. Channels beneath gate insulators deposited on remote plasma cleaned surfaces have shown peak effective mobility in the channel of 700 cm^2/V-sec (54). It has been reported that D_{it} for in situ cleaned, remote plasma deposited SiO_2 structures correlates with the observed surface reconstruction (46). Upon SiO_2 deposition, Si(100) 1x1:H terminated surfaces produced relatively high D_{it}, 7 - 8 x 10^{10}/cm^2-eV. Si(100) 2x1:H terminated surfaces, upon SiO_2 deposition, produced much lower D_{it}, 3 - 4 x 10^{10}/cm^2-eV. Consequently, MOS capacitors fabricated on carefully prepared Si(100) 2x1:H surfaces exhibit lower D_{it} due to fewer Si dangling bond sites at the SiO_2/Si interface Currently, D_{it} of 1 - 2 x $10^{10}/cm^2$-eV is achieved on Si(100) surfaces that have received remote hydrogen plasma cleans.

3.0 MECHANISMS

Given the success of remote plasma processing in cleaning silicon surfaces, it is illustrative to discuss mechanisms for oxygen and carbon removal before discussing experimental details. In designing a cleaning technique, reactions with both the surface contaminants and the substrate have to be taken into account. The majority of remote plasma cleaning work to date has revolved around hydrogen-based plasmas. Consequently, this section will briefly review relevant data on hydrogen interactions with Si(100) surfaces, with silicon oxides, and possible mechanisms for contamination removal by activated hydrogen.

Interactions of atomic hydrogen with the Si(100) surface are fairly complex and still not completely understood. A clean Si(100) surface will reconstruct to a rotated 2x1 structure. Exposure of the clean 2x1 surface can result in either a 1x1, 2x1, or 3x1 hydrogen terminated structure depending on the substrate temperature during exposure (55). Figure 1 shows diagrams of the three structures. Dosing the surface with atomic hydrogen at less than 100°C saturates the available bonds in the 1x1 configuration, while dosing at approximately 125°C results in the 3x1 structure. At temperatures above 325°C the surface remains in the 2x1 configuration, but hydrogen will bond to the single dangling bond on each silicon atom. Scanning tunneling microscope (STM) studies of the three configurations indicates that the 2x1 and 3x1 structures are more resistant to atomic hydrogen etching of the silicon than the 1x1 state (55). It should be noted that in typical UHV surface studies a silicon surface receives a maximum atomic hydrogen exposure of only a few tens of monolayers. In a typical remote cleaning system a silicon surface is exposed to hundreds of monolayers of atomic hydrogen per second.

At fluxes between those typically used in UHV dosing studies and fluxes produced in a cleaning system, modulated molecular beam spectroscopy has shown that atomic hydrogen etches the silicon surface at a substrate temperature ranging from room temperature to 700°C (56). Figure 2 presents a plot of silicon/hydrogen reaction probability as a function of substrate temperature (56). The etch rate decreases as the temperature is increased. This effect can be understood in terms of hydrogen desorption from the silicon (100) surface. Starting from the fully saturated dihydride surface, molecular hydrogen is seen to desorb from the surface at two distinct temperatures, 425°C and 540°C, for samples given a linear temperature ramp of approximately 5°C/sec (57)(58). A thermal desorption spectrum for hydrogen on Si(100) is shown in Fig. 3. The lower temperature desorption peak corresponds to loss of one hydrogen atom from each silicon atom and attendant restructuring to the 2x1 state. The 540°C peak corresponds to loss of the second hydrogen atom from the silicon atom which combines with another hydrogen on the surface and desorbs as H_2. Desorption of hydrogen from the surface partially explains the decrease in etch rate with increasing substrate temperature. At high temperature, hydrogen is not resident on the surface long enough to break Si bonds and create SiH_4. The decrease in etch rate with increasing substrate temperature for temperatures below the first desorption peak (425°C) can be explained on the basis of the high diffusion rate of atomic hydrogen into bulk silicon. The diffusion of hydrogen into the bulk of the silicon at higher temperatures decreases the surface concentration of hydrogen. At lower temperatures, less hydrogen diffuses into the bulk resulting in higher surface concentrations and correspondingly higher SiH_4 production and an increased etch rate. In general, a cleaning process involving atomic hydrogen will produce less substrate etching if the process occurs at higher temperatures, e.g., 200 - 500°C.

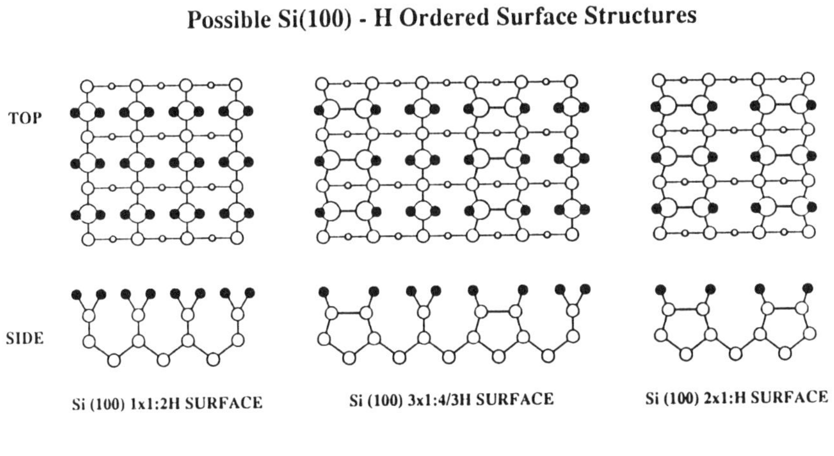

Figure 1. Schematic view of 1x1, 2x1, and 3x1 hydrogen terminated surface structures on a Si(100) surface.

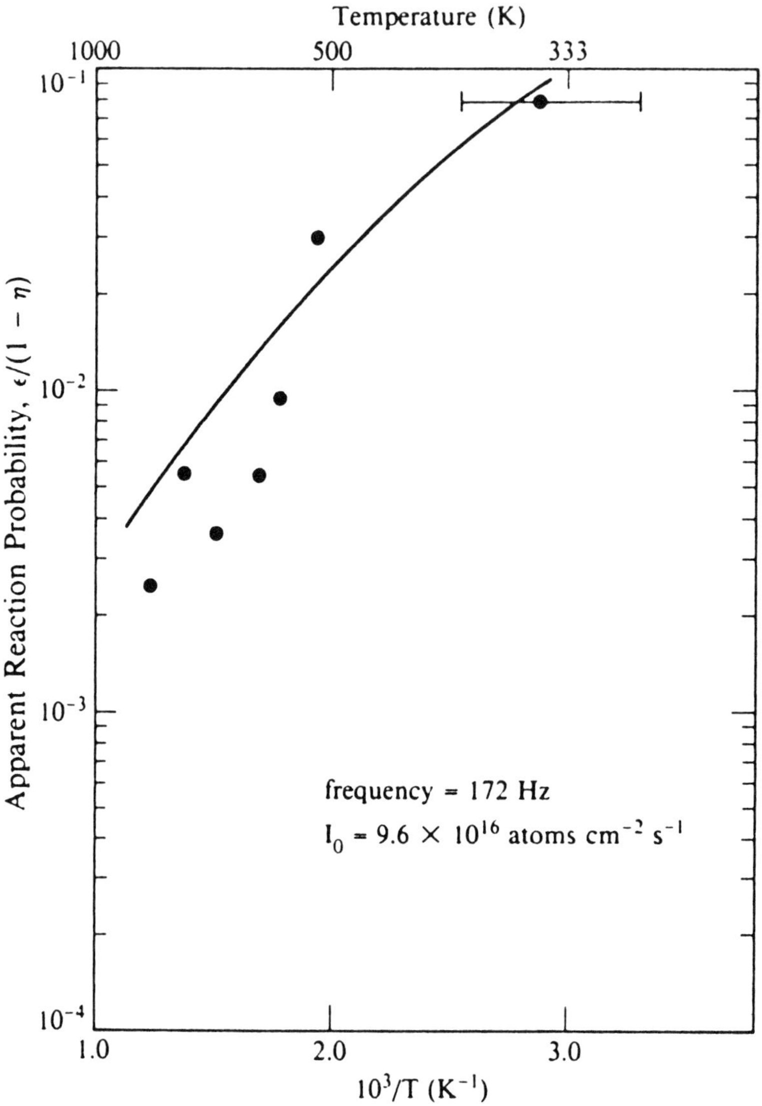

Figure 2. Results from modulated molecular beam study of interactions of atomic hydrogen with a Si(100) surface. The graph shows the apparent reaction probability for the production of silane as a function of substrate temperature (56). *(Reprinted with permission of* Surface Science.*)*

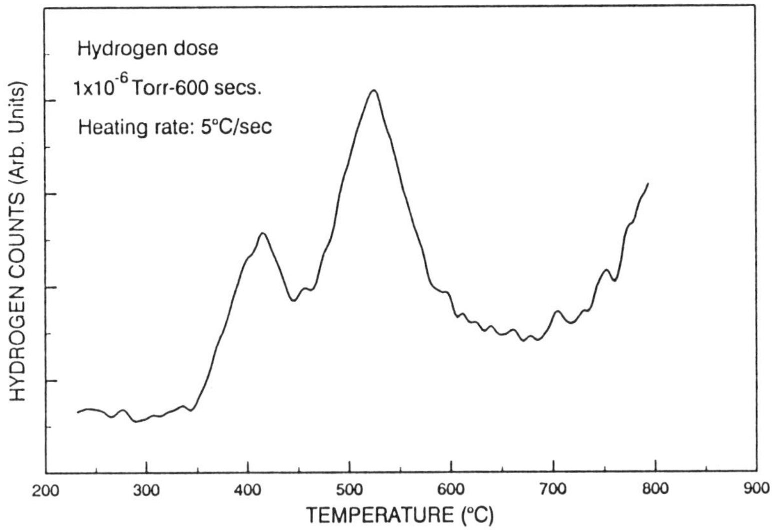

Figure 3. Thermal desorption spectrum of a hydrogen-dosed silicon surface. The surface was dosed at room temperature with atomic hydrogen and was in a 1x1 configuration after dosing (58).

We are also interested in understanding interactions of atomic hydrogen with surface species other than silicon. One difficulty in describing mechanisms for contamination removal from the surface is that the contamination is so diverse. Carbon can arrive at the surface in a variety of forms, such as hydrocarbons, halocarbons, inorganic carbon, etc. Oxygen on the surface can assume a variety of bonding states, and as the oxide thickness increases the structure of the oxide changes (59). Work has barely begun on the variety of contaminants such as metals that may also be present on the silicon surface. Possible mechanisms for carbon and oxygen reduction and removal through interactions with atomic H are discussed below.

Although the chemical forms of carbon adsorbed on silicon surfaces are not known in many cases, carbon removal is thought to occur through a process of hydrogenation followed by volatilization. Short-chain saturated hydrocarbons (C_nH_{2n+2}, $n \leq 3$) in general have very low sticking coefficients on the clean silicon surface (60). Unsaturated hydrocarbons, in contrast, are easily adsorbed on the surface. The sticking coefficient increases with both the degree of saturation and the chain length. Higher molecular weight hydrocarbons tend to decompose rather than desorb as the surface is heated. Again, the percentage desorbing intact decreases as the chain length and degree of saturation increases (60). One possible mechanism for hydrocarbon removal from the silicon surface would be atomic hydrogen attack of the double and triple bonds on adsorbed hydrocarbons. The saturated hydrocarbons would then be able to desorb from the surface. Another possibility for carbon removal from the silicon surface is gasification of solid carbon. Atomic hydrogen has the ability to etch graphite directly (61). It thus seems plausible that polymeric forms of carbon could easily be removed from the silicon surface through hydrogenation to CH_4. In spite of the ill-defined nature of carbon on the surface, the results of remote plasma cleaning show very efficient removal of carbon from the silicon surface by atomic hydrogen exposure.

Oxide removal from the surface occurs much more slowly than carbon removal. The Si-O bond strength and the O-H bond strength are comparable in magnitude (111 kcal/mol each). Consequently, a direct reduction process of Si-O would not be an energetically favored event. (In addition, significant kinetic barriers may exist to prevent hydrogen reduction of the oxide.) It is illustrative to compare remote hydrogen plasma cleaning processes for the reduction of germanium oxide to that for silicon oxide reduction. In contrast to silicon oxide removal, reduction of germanium oxides using remote hydrogen plasma exposures is easily accomplished. Germanium surfaces oxidized by plasma oxidations are easily dissolved producing clean, reconstructed Ge surfaces (29). The relative ease for Ge oxide reduction can partially be explained by the bond energies. The O-H bond strength of 111 kcal/mol is significantly stronger than the Ge-O bond strength of 86 kcal/mol. Thus, the reduction of germanium surface oxides with atomic H should proceed on energetic considerations, and, indeed as mentioned earlier, remote hydrogen plasma treatments do remove surface oxides from germanium. For the reduction of silicon oxide, the slower reaction rates of atomic H with Si-O can be partially rationalized by the comparable bond strengths between O-H and Si-O. Consequently, the reaction of atomic H with SiO is not as fast as the reaction of atomic H with GeO. The slower reaction rate places stringent requirements in practice on the process. The H-flux must be increased and the impurity flux minimized in order to accomplish oxide reduction without subsequent reoxidation. In any cleaning process, it must be remembered that the cleaning is a competition between reduction and reoxidation reactions at the surface.

4.0 PROCESSING EQUIPMENT

Currently, several types of remote plasma sources are employed for in situ surface cleaning. These sources include RF discharges, microwave discharges, and electron cyclotron resonance sources. Except for remote RF discharges, the development of many of these sources for cleaning has been paralleled by their development for dry etching processes. This chapter will focus mainly on a review of remote RF plasma sources and summarizes cleaning results from work with remote microwave sources and ECR.

The basic design of a RF remote plasma cleaning system is shown schematically in Fig. 4. An induction coil is typically wound around a quartz tube appended to one side of a metal vacuum cross designed for substrate manipulation, in situ characterization, and temperature control. Electrodeless designs have been chosen for cleaning processes to avoid potential electrode contamination problems associated with in vacuo capacitive plates. Frequently, the vehicle by which power is imported to the gas stream is through an external inductor or external capacitor. The use of these external devices allows sharp resonant peaks with high Q values to develop in the RF circuit. High Q values result in high fields and high circulating currents. The relative magnitudes of the fields determine whether coupling to the gas is either primarily inductive or capacitive. In using RF induction coils it should be realized that power coupling at low powers is principally capacitive (62)(63). At some critical power level and gas ionization, coupling with the induction coil can become principally inductive. Given the low powers typically employed in using remote RF plasmas for cleaning, the vast majority of work to date has probably been through capacitive coupling.

Typically, in a remote RF cleaning process, a process gas is introduced through the plasma tube where it is ionized and/or dissociated by the RF plasma. Operating pressures are generally in the range of 5 - 100 mtorr. Reactive species, such as atomic H generated by electron impact dissociation of H_2, diffuse from the plasma region to the substrate surface. Ions generated in the plasma-coil region, are "thermallized" during transport from this region to the substrate located downstream from the RF coil. An afterglow extending from the plasma coil towards the substrate is often observed under typical pressure and power conditions. It should be remembered that the emission is indicative of an electron population with significant energy so as to induce photon emission upon impact with ground state molecules or atoms. While this electron population can induce ionization, ions created outside the plasma coil region are not subjected to high **E**-fields and hence cannot gain appreciable energy.

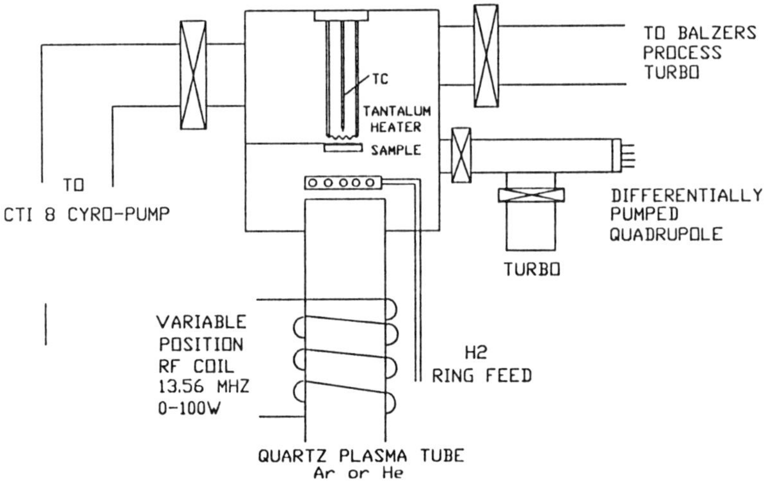

Figure 4. Schematic diagram of RF source for activated gases. In this system the RF coil is located approximately 45 cm from the sample. The hydrogen gas is injected approximately 5 cm from the sample via the ring feed (53).

Alternatively, an indirect plasma technique for activating the hydrogen gas has been applied (40)(64). A remote noble gas plasma dissociates molecular hydrogen downstream from the RF coil. The noble gas discharge serves as a vehicle to couple energy (through electrons, ions, metastables) into molecular hydrogen which is introduced downstream near the sample. This indirect technique minimizes interactions of atomic hydrogen with the reactor wall by minimizing the contact area of the atomic H with the wall. This technique, originally implemented on 50 mm diameter silica plasma tubes, becomes increasingly more difficult for larger diameter tubes. At larger diameters, the plug velocity of the noble gas required to prevent hydrogen back-diffusion would require an enormous flow rate of noble gas (approximately 10 L/min at 100 mm diameter and 50 mtorr).

A typical microwave source for the production of active gases is shown in Fig. 5. The geometry is similar in many ways to that of the RF source. A plasma is formed in the microwave cavity and active gases then flow into the sample chamber. The source typically operates at a pressure near 1 Torr. Although no provisions were made with the apparatus shown, the flux of charged particles reaching the sample could be controlled by substrate biasing (45). Microwave sources normally operate with resonant cavities which provide higher plasma densities than RF plasmas operating in a capacitively-coupled mode.

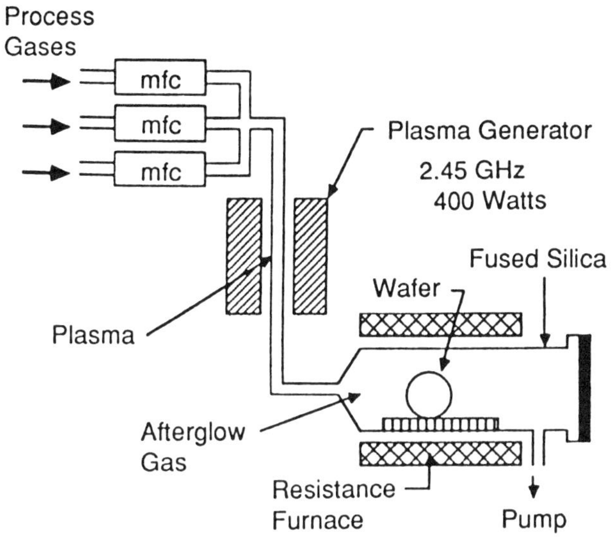

Figure 5. Schematic diagram of remote microwave source reactor (47). *(Reprinted with permission of The Electrochemical Society.)*

An electron cyclotron resonance (ECR) source for generating active gases is shown in Fig. 6 where the gas is activated by microwave excitation (2.45 GHz) in a resonant cavity. Activation efficiency is enhanced by magnetically confining the electrons in the plasma chamber. With electron cyclotron resonance, a discharge can be sustained at a pressure of approximately two orders of magnitude lower (2×10^{-4} Torr) than that used for remote RF geometries. Neutrals and charged particles flow out of the plasma chamber and toward the sample stage. Interactions between charged particles and the substrate can be controlled through bias voltages applied to the substrate.

Aside from considerations for generating activated gases, many of the system design constraints for achieving acceptable residual gas levels, sample heating, and ultrapure gas delivery are generic to all remote plasma configurations. Besides thorough baking and evacuation, it is expedient to plasma-condition the fused quartz tube walls. Impurities, such as carbon-containing compounds originating from the manufacture and handling of the quartz, can be leached from the near surface walls by hydrogen plasmas. Failure to plasma-condition the tube prior to wafer processing results in chemical vapor transport of these impurities to the silicon wafer surface. Fortunately, once these impurities are leached from the surface of the plasma tube, extensive plasma-conditioning is not required after each venting of the vacuum system. Heating of the samples has been accomplished with several types of in situ radiant heaters. Quartz-halogen lamps have been used to irradiate the backside of silicon wafers either directly or indirectly by heating some moderator in physical contact with the wafer. Alternatively, bare-tantalum resistive heaters have also been used in situ. Although tantalum contamination on the front side of wafers has not been observed, contamination of the back side of wafers has been found. The cleaning process demands delivery of high-purity gases to the processing chamber. Typically, 99.9999% pure gases are used for the systems. In-line purifiers have also been used to further reduce background water and oxygen. Contaminants such as H_2O and O_2 are easily plasma activated and can then oxidize the sample surfaces. Gas lines, like the system chamber, are baked and evacuated. In order to maintain the cleanliness of the cleaning chamber, some systems have incorporated turbomolecular pumps as the process pump to provide better isolation of the ultrahigh vacuum components from the roughing vacuums of the secondary process pumps (roots/mechanical exhaust pumps).

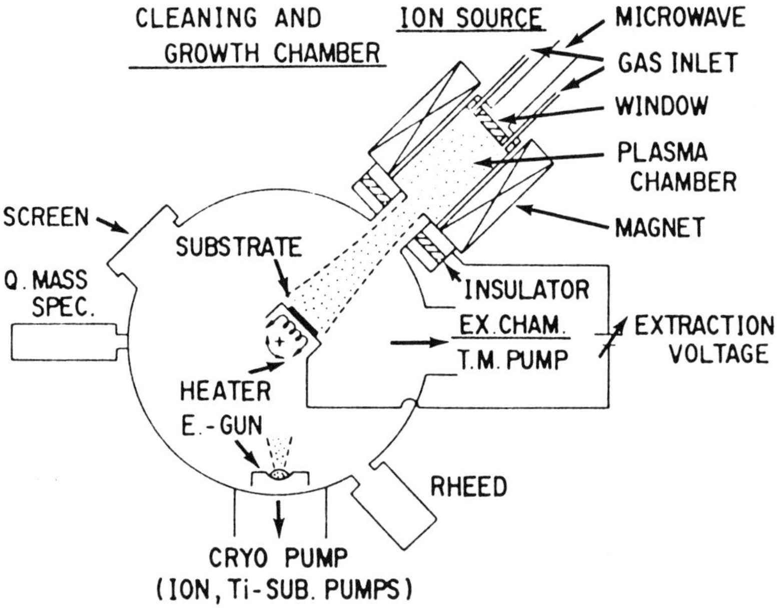

Figure 6. Schematic diagram of electron cyclotron resonance (ECR) system. In the illustrated reactor the substrate bias can be independently controlled (76). *(Reprinted with permission of American Institute of Physics.)*

Adaptation of ultrahigh vacuum practices with rigorous control of gas purity permits remote plasma processing systems to recover to UHV conditions rapidly from processing pressure. Consequently, in situ electron diffraction or electron spectroscopy techniques can be used to assess the cleaning process in a stepwise manner. While RHEED has been used successfully as an in situ tool, later practice has been to locate both electron diffraction and electron spectroscopic techniques in adjacent vacuum chambers connected through in vacuo transfer tubes. The schematic arrangement of one such integrated processing system is shown in Fig. 7. Transfers on integrated processing systems are maintained under ultrahigh vacuum conditions (<1 x 10^{-9} Torr) so as to preserve the surface quality before and after analysis. The vacuum interface to surface analytical equipment permits a broad range of analytical tools to be focused on the cleaning process. These include RHEED, LEED, XPS, AES, and UPS. These characterization tools have permitted thorough evaluation of the hydrogen cleaning process. As shown in Fig. 7, it has become preferred practice in research facilities to establish a dedicated cleaning facility in vacuo to other deposition capabilities rather than to integrate the cleaning process into a deposition system. Integrating a cleaning process into a deposition system can generate problems as contamination from wall deposits complicate cleaning procedures (42)(46)(64).

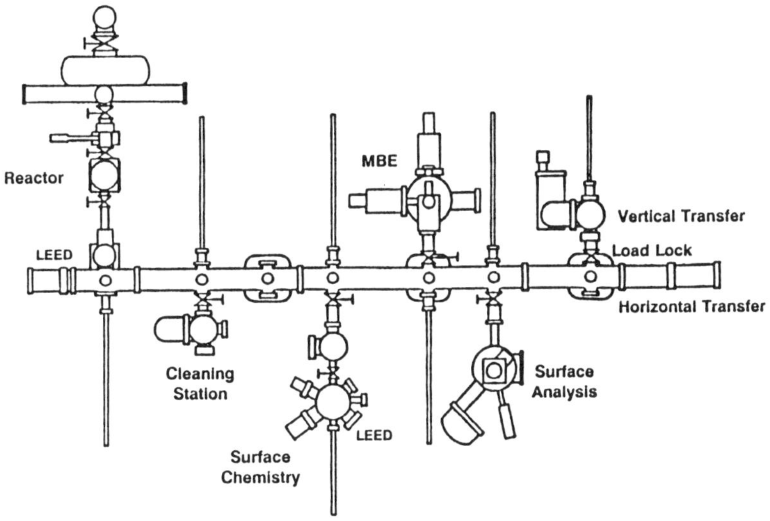

Figure 7. Example of an integrated processing facility combining a CVD reactor, MBE system, remote plasma cleaning system, and two analysis facilities. Sample transfers can be accomplished in an ambient pressure of approximately 5 x 10^{-10} Torr (41).

5.0 WAFER PROCESSING

Remote plasma processing for silicon wafer cleaning has involved both ex situ and in situ processing. As previously noted, the ex situ processing has involved wet chemistry to minimize contaminants on the silicon surface prior to introduction into the vacuum environment. The wet chemistry has usually involved a modified RCA treatment followed by a dilute HF immersion and deionized water rinse. Recently, UV-ozone treatments for wafer preparation have been developed. (A detailed review of this process is given in Ch. 6 of this book.) The in situ plasma treatments have involved predominately hydrogen plasmas, although there has been some work using HCl /Ar mixtures (48). Hydrogen plasma treatments of the Si(100) surfaces are performed usually at substrate temperatures less that 400°C; these temperatures are too low for hydrogen to completely desorb from the Si(100) surface. The cleaning process results in hydrogen termination of the Si dangling bonds at the surface. The hydrogen terminated silicon surface is in general much less reactive than the pure silicon terminated surface. Thus, hydrogen plasma treatments can provide surface passivation prior to transfer of the silicon wafer to other wafer processing stations (24).

Removal of the hydrogen from the silicon surface, when warranted, can easily be accomplished by an anneal at » 500°C. The efficacy of the remote plasma cleaning process to remove contaminants from the silicon surface is highly dependent on processing conditions: plasma source, system background pressure, chemical constituency of the residual chamber gasses, purity of the process gases, power levels applied to the discharge, etc. Consequently, details of the ex situ and in situ processing are discussed in more detail below.

5.1 Ex Situ Processing

The ex situ processing of silicon wafers has involved rigorous wet chemistry to minimize surface contamination. Modified RCA cleaning treatments have been used extensively. A typical process sequence for silicon wafers is given below:

1. ultrasonically degrease in tricholoroethylene, acetone, and methanol
2. rinse for 10 min in flowing deionized water
3. immerse in 5:1:1 H_2O:H_2O_2:HCl at 80°C for 10 min

4. rinse for 10 min in flowing deionized water
5. immerse in 50:1 H_2O:HF at 20°C for 30 sec
6. immerse in 5:1:1 H_2O:H_2O_2:NH_4OH at 80°C for 10 min
7. deionized water rinse
8. immerse in 50:1 H_2O:HF at 20°C for 10 sec
9. deionized water rinse for 30 sec

This process produces hydrogen-terminated Si(100) surfaces with minimal oxygen, carbon, or fluorine contamination. In some instances wafers loaded immediately into UHV systems show faint one-half order RHEED diffraction streaks. Samples exposed to ambient atmospheric conditions for prolonged periods of time prior to loading show substantially higher levels of contamination. Atmospheric exposure allows hydrocarbons and water to physisorb on the Si(100):H surfaces. A short vacuum anneal at <300°C can remove almost all of these contaminants.

As an alternative to the "modified RCA pretreatment," a combination of UV-ozone and HF based spin etch processes has yielded excellent results. The basic procedure is to use the UV-ozone exposure to remove hydrocarbon contaminants followed by a spin etch to remove the oxide. Typical UV-ozone pretreatment involves a ~ 5 min exposure to UV radiation from a low-pressure Hg lamp. The UV light causes O_2 in air to form O_3 and atomic O, resulting in hydrocarbon removal by oxidation. The process also results in oxide growth to a thickness of ~1.5 nm. The oxide is removed in the spin etch process where several drops of an HF-based etch are directed towards a spinning wafer. An unusual aspect of the treatment is that the HF is diluted with ethanol. A standard preparation involves HF/H_2O/ethanol at a ratio of 1:1:10 (65). The actual process and efficacy of other solvents have been studied in some detail (66). The result is a hydrogen passivated, hydrophobic surface which is less reactive to oxidation. In some cases, the whole process is repeated. It is likely that the final H-passivated surface is similar to that obtained from the standard RCA cleaning process followed by an HF etch, although infrared analysis of these surfaces indicates differences in surface Si-H bonding configurations. After direct loading of the wafer into a UHV analysis chamber, surface analysis of the UV-ozone/spin etch surface shows some (~5% of a monolayer) residual hydrocarbon and oxide. Here it is presumed that the hydrocarbons are physisorbed, and that some oxidation occurs during transfer and pump-down. The LEED of the surface shows a 1x1 pattern, but annealing in H_2 to ~300°C results in the appearance of weak half order spots characteristic of the 2x1 reconstruction. Angle resolved UV-photoemission spectroscopy (ARUPS) can serve as a probe for the electronic states of the surface. ARUPS spectra of the UV-ozone/spin etch surface shows disorganized electronic states with no evidence of sharp Si-H or dangling bond resonances (67).

5.2 In Situ Processing: Remote RF Sources

Wafers entering remote plasma systems following the aforementioned wet chemical treatments have only low levels of oxygen and carbon contaminations. Integral order diffraction streaks can be observed often upon loading from atmosphere. Short duration hydrogen plasma treatments result in the faint appearance of half-order streaks. However, the surface diffraction is only a qualitative measure of chemical cleanliness of the silicon surface. To supplement the in situ diffraction studies, surface analytical equipment, vacuum interfaced to the remote plasma systems, has been used to evaluate the cleaning processes. The combination of chemical analysis by various surface analytical techniques, structural determination through electron diffraction, and electronic determination through UV-photoemission, has allowed the interactions of the plasma generated hydrogen species with the sample surface to be evaluated. Perhaps not surprisingly, the understanding of the interactions of the remote plasma cleaning process with the silicon surface has relied heavily on the background work of surface scientists who, by using high temperature anneals to produce atomically clean surfaces and by employing hot-filaments for atomic hydrogen dosing, had previously characterized interactions of atomic H with Si(100) surfaces (57)(68). The capability of the remote plasma cleaning process to emulate the essentials of the hot filament dosing results, but at low temperatures and with short duration exposures to remote plasma-generated atomic hydrogen, is remarkable.

The removal of silicon surface contamination by exposure to an RF remote plasma has been demonstrated by a number of workers (24)(53). Figure 8 shows AES from silicon wafers loaded after the modified RCA clean and HF dip and after hydrogen plasma treatments as a function of increasing substrate temperatures (24). The results in Fig. 8 show more effective oxygen removal at temperatures greater than 150°C. It can be seen that the relatively small amounts of carbon and oxygen that existed on the surface after wet chemistry are removed by the remote hydrogen plasma treatment. Generally, it has been easier to remove carbon than oxygen from the silicon surface. The oxygen removal illustrated in Fig. 8 was accomplished by using a low-power hydrogen plasma treatment for a prolonged period of time. Oxygen removal has also been effected by a thermal desorption coupled with a hydrogen plasma treatment for carbon removal (53). Figure 9a shows XPS spectra from a silicon sample after carbon removal and then after oxygen desorption. Carbon removal was accomplished by a 5 min exposure to hydrogen plasma at a substrate temperature of 400°C. The oxygen was then desorbed, presumably as a suboxide, with an anneal at 700°C for 5 min. Figure 9b shows XPS spectra from a silicon sample where the processing times and temperatures for selected steps were reduced. For the carbon removal step the exposure time was reduced from 5 min to 30 sec. Carbon levels were further lowered below the XPS detection limit. The second step annealing temperature was reduced from 700°C to 620°C. At the lower annealing temperature no decrease was seen in the oxygen levels on the surface. The failure to remove oxygen is easily understood in light of thermal desorption results for oxygen on silicon (59). The desorption temperature for silicon oxides on the silicon surface tends to increase as the oxide thickness increases; the minimum desorption temperature was found to be approximately 700°C. The hydrogen plasma exposure to remove carbon, coupled with a brief thermal anneal, represents a modest thermal budget process for producing atomically clean surfaces.

Hydrogen plasma interactions with Si(100) surfaces have also been studied using low energy electron diffraction (LEED) and ultraviolet photoelectron spectroscopy (UPS) (44). The silicon surface after remote hydrogen plasma exposure exhibits different structures that are dependent upon the operating parameters of the process. By varying the process conditions, 1x1, 2x1, or 3x1 H-terminated, ordered surface structures can be obtained. A schematic of Si-H bonding characteristic of these surfaces was shown in Fig. 1. Typical LEED patterns associated with the Si(100)1x1:H, Si(100)2x1:H, and Si(100) 3x1:H reconstructions are shown in Fig. 10. It has been found that both the substrate temperature and plasma parameters play an important role in determining the surface structure. Processing regimes (pressure and temperature) for the different reconstructions are indicated in Fig. 11. One aspect that seems fairly consistent between different investigators is the temperatures where the Si(100)1x1:H and Si(100)2x1:H surfaces are obtained (1x1 - below 100°C and 2x1 - above 250°C). The electronic states on the different surfaces obtained by angle resolved UV-photoemission are summarized in Fig. 12. H-induced electronic states are observed at 4 - 6 eV below the Fermi energy level. The most well defined states were observed for the 2x1 surface. Broader H-induced states were identified in the 1x1 surface, while no obvious H-related features were observed after the ex situ wet chemical clean. We note that no states in the forbidden gap were observed for any of the H-terminated surfaces. The ARUPS measurements revealed that the hydrogen, presumably as H_2, had evolved from the surface after vacuum annealing at 500°C. Surface electronic states of the nonterminated surface were identified to be at 1.2 - 1.8 eV below the Fermi energy. Of particular note was the observation that the whole spectra shifted by ~0.5eV upon annealing, which was attributed to a change in surface Fermi energy. It was proposed that the H-terminated surfaces were unpinned (i.e., no band bending at the surface).

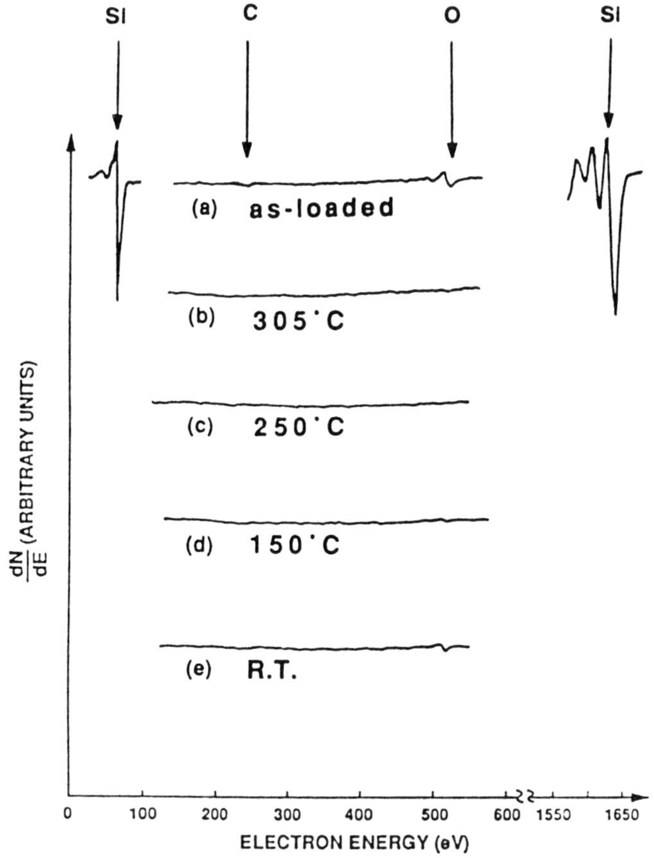

Figure 8. Auger electron spectroscopy analysis of a Si(100) surface after (a) a 40:1 H_2O:HF (49%) dip and remote hydrogen plasma clean at (b) 305°C (c) 250°C (d) 150°C and (e) room temperature for 45 min each (24). *(Reprinted with permission from Journal of Electronic Materials, Vol. 20, No. 3, p. 281, a publication of The Minerals, Metals & Materials Society, Warrendale, Pennsylvania 10586.)*

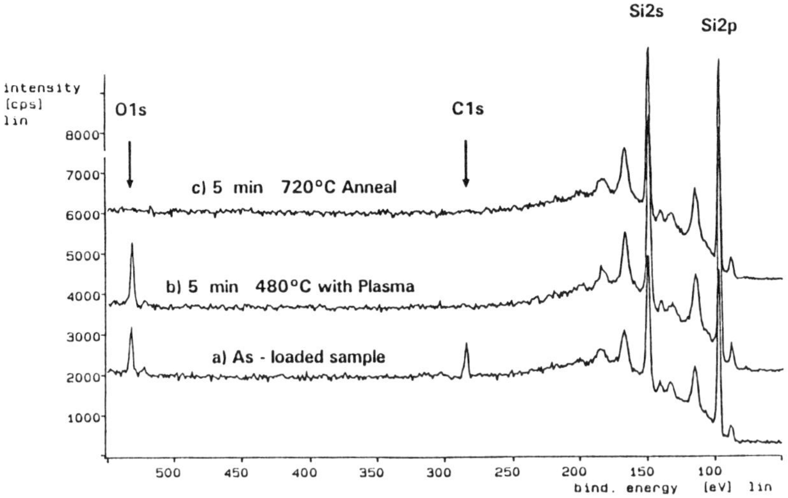

Figure 9a. XPS spectra from a Si(100) sample at selected points of a cleaning process which combines a remote hydrogen plasma exposure at 480°C for 5 min and a thermal anneal to desorb the residual oxide (53).

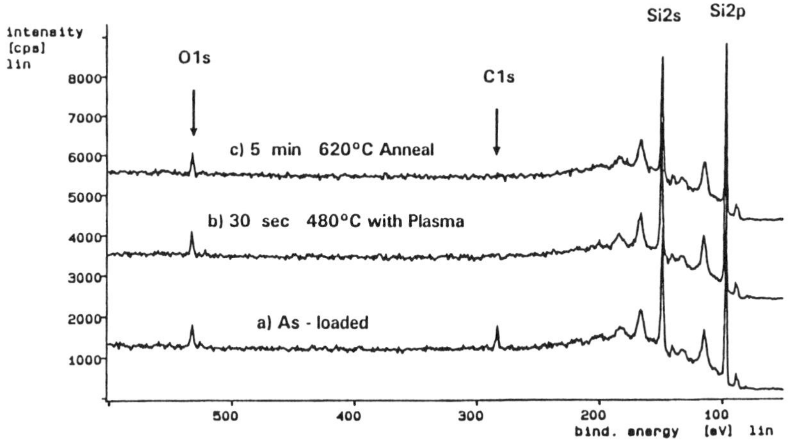

Figure 9b. A cleaning sequence similar to that shown in Fig. 9a except that the plasma exposure time has been reduced to 30 sec and the subsequent annealing temperature lowered to 620°C (53).

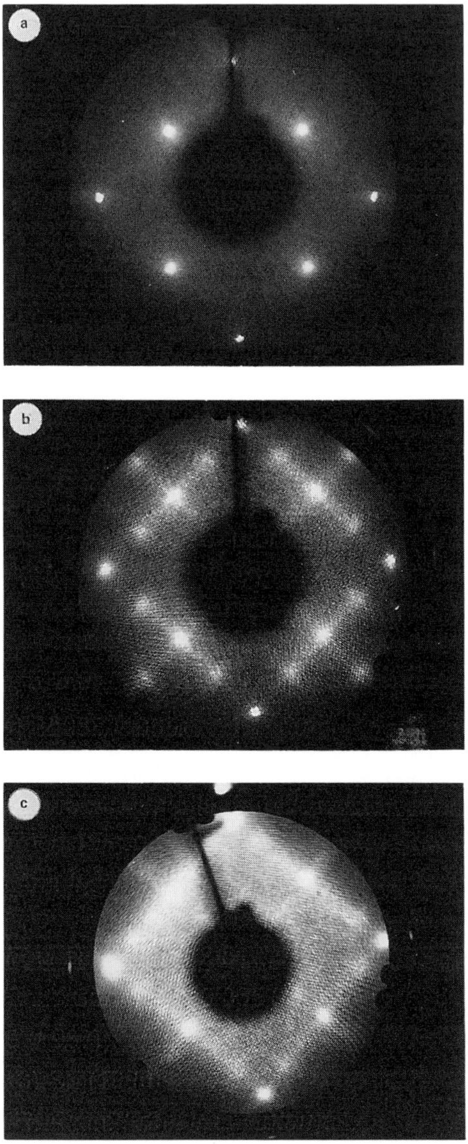

Figure 10. LEED patterns obtained with a beam energy of 60.8 eV: a) 1x1 after 100°C H-plasma cleaning that indicates a bulklike arrangement of surface atoms; b) 2x1 after 300°C H-plasma cleaning indicating a reconstructured surface; and c) 3x1 hydrogen terminated structure (67).

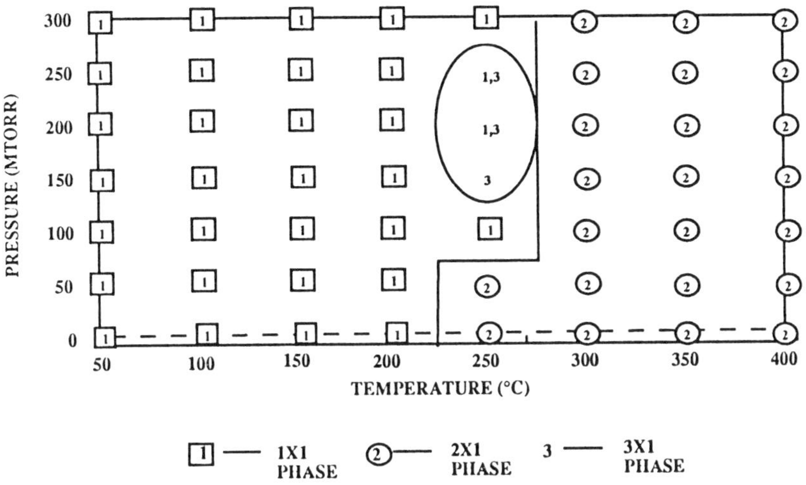

Figure 11. The temperature-pressure dependence of the surface structure of Si(100) for remote H-plasma cleaning (44).

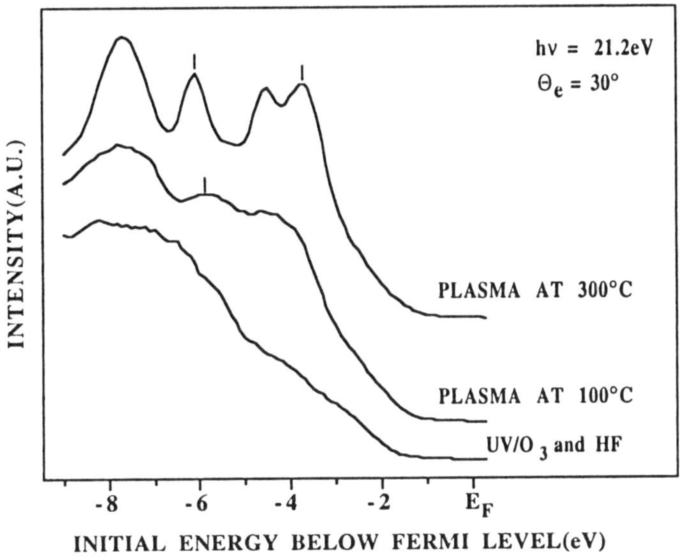

Figure 12. ARUPS spectra: a) after UV-ozone/HF treatment showing a broad spectra with no organized electronic states due to Si-H bonding; b) after 100°C H-plasma exposure which shows a broad feature at ~6 eV below E_f which is associated with the dihydride phase of Si-H bonding; and c) after 300°C H-plasma cleaning showing the monohydride Si-H features at 5.4 and 6.2 eV. Also, the feature at ~ 3.5 eV is attributed to the Si-Si surface dimer bonding (44).

Passivation of the surface dangling bonds with hydrogen results in surfaces which are more resistant to oxidation than the nonterminated 2x1 surface. Figures 13a and 13b illustrate the oxidation resistance and demonstrate the effectiveness of the hydrogen plasma treatments for removing oxygen from freshly loaded surfaces (22). Figure 13a shows the oxidation resistance of a silicon wafer following a modified RCA treatment. The sample following wet chemistry shows remarkably little oxygen on the surface and no AES detectable carbon. Upon exposure of the sample to the atmosphere for 10 hours, substantial carbon and oxygen appear on the surface. It is not clear whether the oxygen and carbon are physisorbed or chemisorbed on the surface. Figure 13b shows a sample which 15 min after wet chemical treatments was exposed to a hydrogen plasma (9 W, 250°C, 45 mtorr). The oxygen was removed from the silicon surface. Exposure of the sample to the atmosphere resulted in an uptake of oxygen similar to what was observed after wet chemistry, but showed significantly lower uptake of carbon. Resistance to oxidation has also been demonstrated by exposure to the atmosphere for times on the order of minutes. Upon re-introduction into the vacuum system, the surfaces still exhibited the reconstructed LEED patterns. H-terminated surfaces can thus survive a transfer in air between separate processing tools. For example, transfer has been accomplished between a remote RF cleaning tool and a plasma-assisted CVD tool and between a remote RF cleaning tool and an electron-beam MBE (44). Following the transfer, plasma-assisted CVD was used to obtain high-quality epitaxial growth at ~150°C (19). For MBE, growth was obtained at 600°C. In both cases, it was not necessary to expose the samples to temperatures greater than the growth temperature.

Surface analytical techniques have also been coupled with gas phase analysis to provide insight into the cleaning process. Specifically, both SiH_4 generation by surface etching and H-diffusion into the bulk have been identified. Both of these processes have the potential of limiting the effectiveness of the cleaning process. The etching process, which was reviewed in the mechanisms section, can result in surface roughness. The diffusion process can result in formation of electronic and structural defects. The effects of diffusion of plasma generated atomic hydrogen into a silicon substrate were explored in detail in experiments that were carried out with microwave excitation at relatively high processing pressures of ~2 Torr (69). Cross-sectional TEM micrographs from samples exposed to a remote hydrogen plasma showed extensive subsurface damage. The defects were identified as platelet structures oriented largely along (111) planes. Further-

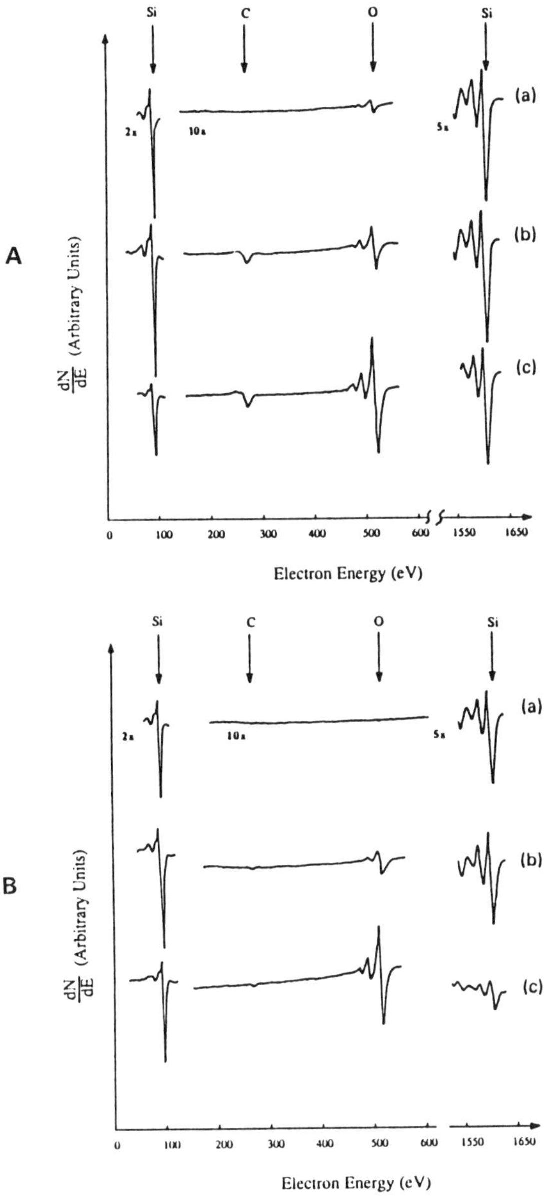

Figure 13. Auger spectra comparing uptake rates of carbon and oxygen during air exposure following (A) modified RCA clean with final HF dip and (B) modified RCA clean with final HF dip and remote hydrogen plasma clean (22). *(Reprinted with permission from Journal of Electronic Materials, Vol. 19, No. 10, p. 1029, a publication of The Minerals, Metals & Materials Society, Warrendale, PA 15086.)*

more, the hydrogen was found to form complexes with donor or acceptor atoms which resulted in passivation of the dopants. Recent work has shown that under some conditions of remote RF-plasma exposure and substrate temperature both surface roughening and subsurface diffusion occur (43). The plasma conditions which showed the most extensive surface roughness and subsurface diffusion were low pressure (<20 mtorr) and high RF-power. The results should be contrasted with optimal cleaning conditions in which smooth surfaces and no platelet defects were observed. Similarly, no defects by delineation of the substrate with a dilute Schimmel etch or by plan view TEM were observed for remote plasma cleaning at 12 W RF-power and a hydrogen pressure of 45 mtorr (23).

The application of these in situ and in vacuo techniques to study remote RF hydrogen plasma treatments has yielded considerable insights into the mechanisms of the cleaning process. Hydrogen plasma treatments are quite proficient in carbon removal from Si(100) surfaces. Oxygen reduction from the Si(100) surfaces requires stringent control and minimization of the native oxide thickness upon introduction from atmosphere into the vacuum chamber. The hydrogen plasma treatments produce a well-ordered hydrogen-passivated surface. Potential pitfalls are surface roughening and bulk-defect generation that can be minimized by appropriate choice of operating conditions.

5.3 In Situ Processing: ECR Sources

Besides hydrogen plasma cleaning techniques implemented with remote RF plasmas, other studies have been reported on characterization of ECR sources for hydrogen plasma cleaning of Si(100) surfaces (70)-(76). The ECR techniques, unlike the higher pressure RF inductive techniques, operate in a pressure range wherein ion bombardment of the silicon surfaces may provide a critical component to the cleaning process. Characterization of interactions between ECR generated atomic hydrogen and Si(100) surfaces has been accomplished with both in vacuo and in situ analytical techniques. Several studies have used XPS to evaluate the effectiveness of the low pressure ECR sources to remove oxygen and carbon (72)(73). Spectroscopic ellipsometry has also been used as a real-time monitor to indirectly ascertain oxide reduction and hydrogen termination (74)(75). Figure 14 shows XPS spectra from a silicon surface as loaded and for increasing exposure times to an ECR hydrogen plasma (72). After a 10 min exposure both the carbon and oxygen signals were dramatically

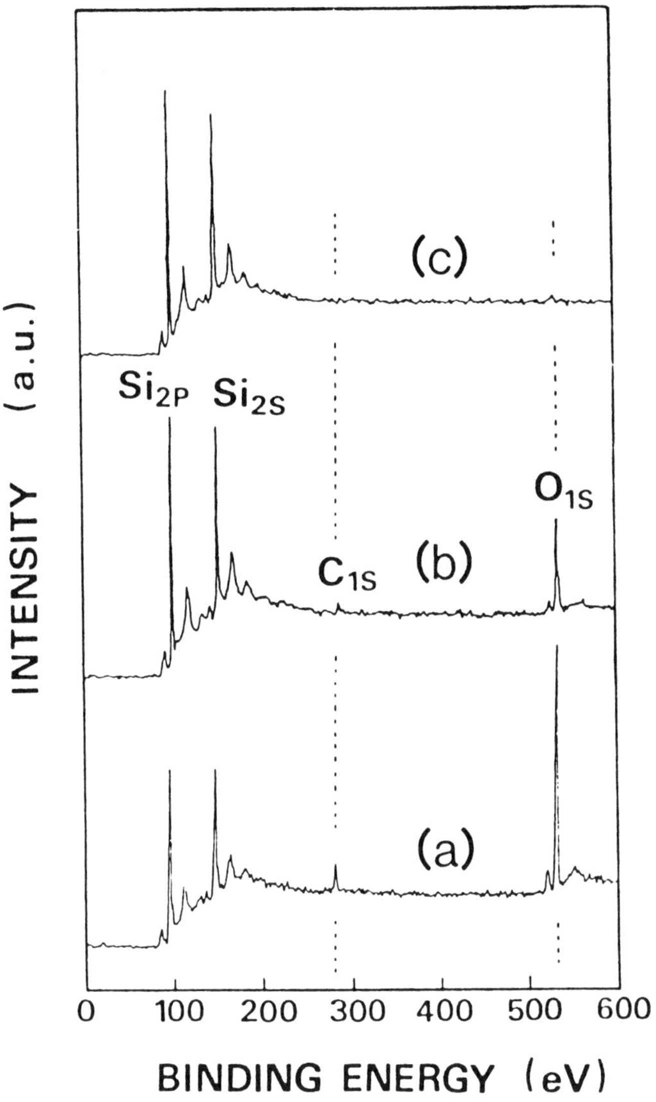

Figure 14. XPS spectra from Si(100) surface (a) as received and after ECR hydrogen plasma exposure for (b) 1 minute and (c) 10 min (72). *(Reprinted with permission of the American Institute of Physics.)*

reduced. Surface roughening was inferred from RHEED patterns from samples exposed to the plasma for times as short as 1 min. Transmission IR measurements showed the presence of both SiH_2 and SiH bonding in the Si crystal. Since this method of analysis is not sensitive to surface hydrogen bonding, hydrogen must have penetrated into the silicon bulk (72). Unlike the remote RF studies, surface cleaning has been achieved even under moderate background vacuum conditions (10^{-6} Torr), but only when a critical concentration of atomic hydrogen was generated (73). Figure 15 shows a plot of electron density versus pressure superimposed on a plot of residual oxide thickness as a function of pressure. At 10 mTorr the electron density reaches a maximum and the residual oxide thickness is reduced to zero; subcritical atomic hydrogen concentrations yield only asymptotic oxide reduction. At the critical concentration, SiO_2 is reduced below the detection capability of XPS. Unfortunately, exposures of the Si surface to these critical levels result in microscopic roughening, observed by spot-diffraction from RHEED. The low processing pressures of the ECR sources (10 mTorr) makes control of the ion energy a critical parameter in this process. It is expected that application of the ECR sources to systems with lower residual gas background levels would reduce the flux of atomic hydrogen needed for cleaning. These reduced fluxes would produce correspondingly less surface roughening and reduce H-incorporation.

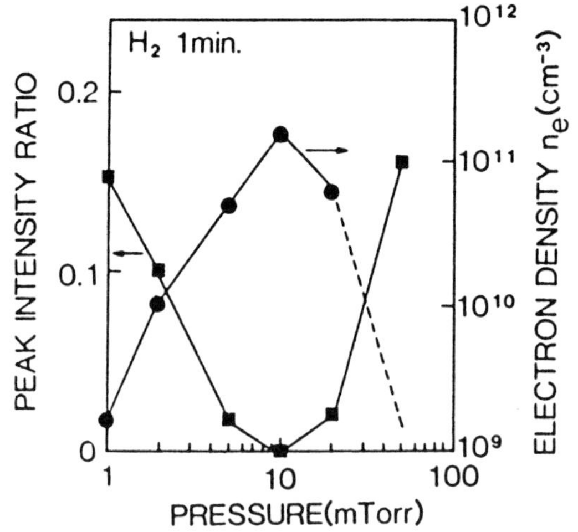

Figure 15. Pressure dependence of peak intensity ratio of SiO_2 peak to Si peak in Si 2p XPS spectra and electron density. Vertical value of 0.2 corresponds to a SiO_2 film thickness of about 8 Å (73). *(Reprinted with permission of the American Institute of Physics.)*

In addition to ECR cleaning with hydrogen plasmas, both H_2 and H_2 - SiH_4 plasmas have been used to clean Si surfaces at elevated temperatures (700°C) (76). The author proposed that solid SiO_2 and atomic H react to form gaseous SiH_x and O_2. The addition of SiH_4 to the plasma gas allows other reducing reactions to occur, as shown below:

Eq. (1) $SiO_2\ (s) + Si\ (s) \rightarrow 2\ SiO\ (g)$

Eq. (2) $SiO_2\ (s) + SiH_2\ (g) \rightarrow 2\ SiH\ (g) + O_2\ (g)$

The "s" and "g" denote solid and gas phases, respectively.

Equilibrium calculations predict that the concentration of reactive atomic H must be 10^6 - 10^7 times greater than the concentrations of O_2. These calculations nicely illustrate the critera of high-purity cleaning conditions. The ratio of 10^6 - 10^7 is an exceedingly high ratio to achieve between relatively short-lived atomic hydrogen and residual oxygen concentrations. For systems with base vacuum pressures of 10^{-8} Torr, this would require 10 mtorr of atomic H disregarding any impurity flux associated with plasma exposure to the walls. Water vapor, common to almost all non-UHV systems, would also contribute as a residual contaminant forcing even higher pressures of atomic hydrogen. The temperature dependence of the relative etch rates of silicon and SiO_2 on silicon (100) has also been studied recently (77). An ECR source for atomic hydrogen was used and etch rates for Si and SiO_2 were measured over the temperature range of 300 - 550°C. Figure 16 shows that at 300°C the etch rate for silica is much higher than that for SiO_2 but it decreases more rapidly as the temperature increases. At 400°C the Si and SiO_2 etch rates are almost equal and by 550°C the SiO_2 etch rate is approximately a factor of three larger than that of silicon.

5.4 In Situ Processing: Remote Microwave Sources

In addition to remote plasma cleaning by remote RF or ECR plasmas, studies have been made to apply remote microwave plasma sources for substrate cleaning. These systems have also been studied for ion-free dry etching (78)-(80) and for wafer cleaning prior to SiGe alloy deposition (81). The quality of the epitaxial material grown after the clean was used to assess the efficiency of the clean. In this work, mixtures of GeH_4 and H_2 were incident on a sample surface whose temperature was varied from 650 -

800°C. Although no plasma was involved in the cleaning process, (which was reported to occur through silicon oxide reduction by the formation of GeO and subsequent sublimination of the GeO), the authors did report that a remote H_2 plasma did not show a significant effect on the GeH_4-assisted cleaning process. At temperatures between 650°C and 800°C, hydrogen interaction with the Si surface would be limited due to rapid thermal desorption. Remote microwave plasmas can replace wet processing steps in the modified RCA clean with dry processing (47)(48)(82). Specifically, remote HCl/Ar, $NF_3/O_2/Ar$, and $HCl/Ar/O_2$ processes were evaluated (47)(48). While no in situ characterizations were employed, test MOS capacitors were fabricated on these processed surfaces. Breakdown statistics were used to assess the effect of the cleaning process. The primary problem associated with these reactive microwave discharges was silicon surface roughening from Cl or F. For metals removal, the $HCl/Ar/O_2$ mixtures with O_2 concentrations of 0.5 to 1.0% were unsuccessful in preventing silicon surface roughening. However, the use of a remote microwave O_2 discharge to form a thin protective oxide on the Si surface prior to HCl/Ar exposure did prevent surface roughening. SIMS data supported the results of metals removal from such a two-step approach.

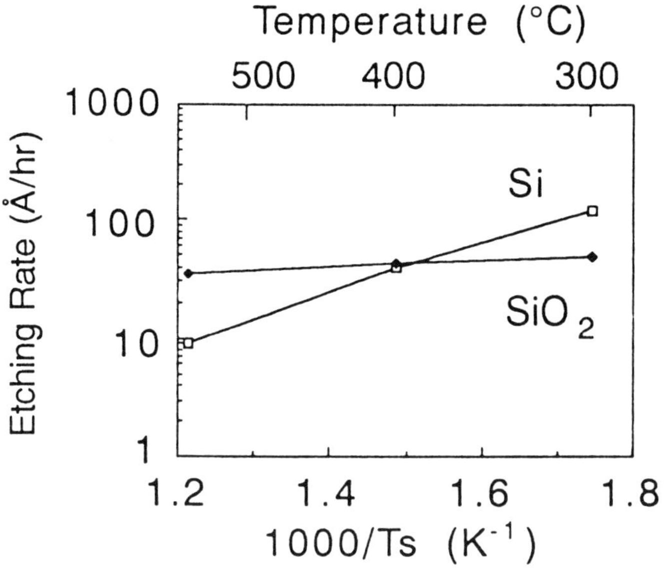

Figure 16. Temperature dependence of etch rates of Si and SiO_2 on Si(100) surface exposed to ECR hydrogen plasmas (77). *(Reprinted with permission of* Japanese Journal of Applied Physics.)

In order to minimize surface roughness induced by oxide removal a selective NF_3/H_2 process based on a $NF_3/H_2/Ar$ afterglow was studied and reported (82). By combining separate processes, a 5-step cleaning process was proposed for 1) removing resist, 2) residual oxide etching, 3) fine organics removal, 4) metals removals, and 5) residual oxide etching. The processes are all carried out in the microwave discharge afterglow by adjusting the gas composition. The gas mixtures proposed to accomplish these cleaning steps are as follows, 1) O_2, 2) NF_3 (1%) / H_2 (2%) / Ar (97%), 3) O_2, 4) HCl (10%) / Ar (90%), and 5) NF_3 (1%) / H_2 (2%) / Ar (97%).

6.0 CONCLUSIONS

Remote hydrogen plasma cleaning processes are capable of producing atomically clean or hydrogen passivated Si(100) surfaces. A variety of plasma sources have been utilized to create atomic hydrogen remote from the silicon surface. The interactions of the remote-plasma-generated atomic hydrogen with as-loaded Si(100) wafers have produced clean Si surfaces with monohydride and dihydride surface terminations. In addition to impurity atom removal, competing silicon etching has been minimized. Silicon etching leads rapidly to microscopic surface roughening. The interdiffusion of hydrogen into the bulk of the crystal is also an issue in remote hydrogen plasma processing. Effects of hydrogen interdiffusion include bulk-structural damage and dopant compensation. Remote hydrogen plasma processing is being optimized to minimize these problems by processing for shorter durations and at lower temperatures. Processing conditions are now such that combinations of hydrogen surface saturation under bulk-structural damage is not observed in XTEM. Conditions that do not permit substantial bulk diffusion, combined with annealing to desorb the impurities, may offer alternative processing paths. The continued development of remote processing will rely heavily on in situ or in vacuo processing. The development will undoubtedly progress as newer plasma sources are implemented. In particular, electron cyclotron resonance sources permit hydrogen cleaning under high atomic hydrogen flux and with higher residual gas pressure. Newer RF sources including helicon designs, quarter-wave cavities, and RF magnetically coupled sources will undoubtedly be investigated. With these high flux sources one will still have to address tradeoffs between interactions with reactive atomic hydrogen and the bulk substrate, versus interactions of the reactive hydrogen with surface impurity atoms.

ACKNOWLEDGMENTS

We would like to acknowledge many helpful discussions with R. J. Markunas, M. J. Mantini, G. G. Fountain, S. V. Hattangady, J. Posthill, T. P. Schneider, Jaewon Cho, and T. P. Humphreys. RAR and RET are indebted to R. C. Hendry, R. V. Durkee, and S. T. Ammons for the design and construction of the RF cleaning systems. We would also like to thank K. Barbour for cheerful and patient assistance with numerous revisions of the manuscript. RJN acknowledges the support of the National Science Foundation through grant CDR 8721505.

REFERENCES

1. Richard, P. D., Markunas, R. J., Lucovsky, G., Fountain, G. G., Mansour, A. N., and Tsu, D. V., *J. Vac. Sci. Technol.* A3:867-872 (1985)
2. Lucovsky, G. and Tsu, D. V., *J. Vac. Sci. Technol.* A5:2231-2238 (1987)
3. Kim, S. S., Tsu, D. V., and Lucovsky, G., *J. Vac. Sci. Technol.* A6:1740-1744 (1988)
4. Fountain, G. G., Hattangady, S. V., Rudder, R. A., Markunas, R. J., Lucovsky, G., Kim, S. S., and Tsu, D. V., *J. Vac. Sci. Technol.* A7:576-580 (1989)
5. Lucovsky, G., Kim, S. S., Tsu, D. V., Fountain, G. G., and Markunas, R. J., *J. Vac. Sci. Technol.* B7:861-869 (1989)
6. Kim, S. S., Stephens, D. J., Lucovsky, G., Fountain, G. G., Markunas, R. J., *J. Vac. Sci. Technol.* A8:2039-2045 (1990)
7. Lucovsky, G., Richard, P. D., Tsu, D. V., Lin, S. Y., and Markunas, R. J., *J. Vac. Sci. Tech.* A4:681-688 (1986)
8. Meiners, L. G., *J. Vac. Sci. Technol.* 21:655-658 (1982)
9. Tsu, D. V. and Lucovsky, G., *J. Vac. Sci. Technol.* A4:480-484 (1986)
10. Hattangady, S. V., Fountain, G. G., Rudder, R. A., and Markunas, R. J., *J. Vac. Sci. Technol.* A7:570-575 (1989)
11. Fountain, G. G., Hattangady, S. V., Rudder, R. A., Posthill, J. B., and Markunas, R. J., *MRS Symp. Proc.* 146:139-145 (1989)
12. Kim, S. S., Parsons, G. N., Fountain, G. G., and Lucovsky, G., *J. Non-Cryst. Solids*, 115:69-71 (1989)

13. Lucovsky, G., Kim, S. S., Tsu, D. V., Parsons, G. N., and Fitch, J. T., *J. Vac. Sci Technol.* A8:1947-1954 (1990)

14. Wang, C., Williams, M. J., and Lucovsky, G., *J. Vac. Sci. Technol.* A9:444-449 (1991)

15. Wang, C., Parsons, G. N., Kim, S. S., Buehler, E. C., Nemanich, R. J., and Lucovsky, G., *MRS Symp. Proc.* 192:535-540 (1990)

16. Wang, C., Parsons, G. N., Buehler, E. C., Nemanich, R. J., and Lucovsky, G., *MRS Symp. Proc.* 164:21-26 (1989)

17. Fountain, G. G., Rudder, R. A., Hattangady, S. V., Vitkavage, D. J., and Markunas, R. J., *AIP Conf. Proc.* 167:338-346 (1988)

18. Hattangady, S. V., Posthill, J. B., Fountain, G. G., Rudder, R. A., Mantini, M. J., and Markunas, R. J., *Appl. Phys. Lett.* 59:339-341 (1991)

19. Breaux, L., Anthony, B., Hsu, T., Banerjee, S., and Tasch, A., *Appl. Phys. Lett.* 55:1885-1887 (1989)

20. Anthony, B., Breaux, L., Hsu, T., Banerjee, S., and Tasch, A., *J. Vac. Sci. Technol.* B7:621-626 (1989)

21. Breaux, L., Anthony, B., Hsu, T., Banerjee, S., and Tasch, A., *Proceedings of the Industry-University Advanced Materials Conference*, Denver, CO, p. 47-59, Materials Research Society, Boston, MA (March 6-9, 1989)

22. Anthony, B., Hsu, T., Breaux, L., Qian, R., Banerjee, S., and Tasch, A., *J. Electron. Mater.* 19:1027-1032 (1990)

23. Hsu, T., Breaux, L., Anthony, B., Banerjee, S., and Tasch, A., *J. Electron. Mater.* 19:375-384 (1990)

24. Hsu, T., Anthony, B., Qian, R., Irby, J., Banerjee, S., and Tasch, A., Lin, S., Marcus, H., and Magee, C., *J. Electron. Mater.* 20:279-287 (1991)

25. Hsu, T., Anthony, B., Breaux, L., Qian, R., Banerjee, S., and Tasch, A., *J. Electron. Mater.* 19:1043-1050 (1990)

26. Anthony, B., Hsu, T., Qian, R., Irby, J., Banerjee, S., and Tasch, A., *J. Electron. Mater.* 20:309-313 (1991)

27. Rudder, R. A., Fountain, G. G., and Markunas, R. J., *J. Appl. Phys.* 60:3519-3522 (1986)

28. Rudder, R. A., Hattangady, S. V., Posthill, J. B., and Markunas, R. J., *MRS Symp. Proc.* 116:529-533 (1988)

29. Hattangady, S. V., Rudder, R. A., Fountain, G. G., Vitkavage, D. J., Markunas, R. J., *MRS Symp. Proc.* 102:319-322 (1988)

30. Rudder, R. A., Hattangady, S. V., V̂itkavage, D. J., Markunas, R. J., *MRS Symp. Proc.* 116:519-522 (1989)

31. Parikh, N. R., Hattangady, S. V., Posthill, J. B., King, M. L., Rudder, R. A., Vitkavage, D. J., *MRS Symp. Proc.* 102:275-278 (1988)

32. Posthill, J. B., Rudder, R. A., Hattangady, S. V., Fountain, G. G., Vitkavage, D. J., Markunas, R. J., Parikh, N. R., and Yu, N., *J. Vac. Sci. Technol.* A7:1130-1135 (1989)

33. Posthill, J. B., Malta, D. P., Hattangady, S. V., Parikh, N. R., Humphreys, T. P., Rudder, R. A., Fountain, G. G., Nemanich, R. J., and Markunas, R. J., *Proc. of the XIIth Int. Cong. for Elec. Micro.*, 4:646-647 (1990)

34. Humphreys, T. P., Sukow, C. A., Nemanich, R. J., Posthill, J. B., Rudder, R. A., Hattangady, S. V., and Markunas, R. J., *MRS Symp. Proc.* 162:531-536 (1990)

35. Posthill, J. B., Rudder, R. A., Hudson, G. C., Hattangady, S. V., Fountain, G. G., and Markunas, R. J., *Proc. of the 12th International Congress for Electron Microscopy*, 4:558-559 (1990)

36. Posthill, J. B., Rudder, R. A., Hattangady, S. V., Fountain, G. G., and Markunas, R. J., *Appl. Phys. Lett.* 56:734-736 (1990)

37. Fountain, G. G., Rudder, R. A., Hattangady, S. V., Vitkavage, D. J., Markunas, R. J., and Posthill, J. B., *Electron. Dev. Lett.* 24:1010-1011 (1988)

38. Vitkavage, D. J., Fountain, G. G., Rudder, R. A., Hattangady, S. V., and Markunas, R. J., *Appl. Phys. Lett.* 53:692-694 (1988)

39. Hattangady, S. V., Fountain, G. G., Rudder, R. A., Mantini, M. J., Vitkavage, D. J., Markunas, R. J., *Appl. Phys. Lett.* 57:581-583 (1990)

40. Rudder, R. A., Fountain, G. G., Hattangady, S. V., Posthill, J. B., and Markunas, R. J., *MRS Symp. Proc.* 165:151-159 (1990)

41. Rudder, R. A., Hendry, R. C., and Markunas, R. J., *J. Vac. Sci. Technol.* A7:802-807 (1989)

42. Lucovsky, G., Kim, S. S., Fitch, J. T., and Wang, C., Rudder, R. A., Fountain, G. G., Hattangady, S. V., and Markunas, R. J., *J. Vac. Sci. Technol.* A9:1066-1071 (1991)

43. Schneider, T. P., Bernard, B. L., Chen, Y. L., and Nemanich, R. J., *Mat. Res. Soc. Symp. Proc.* 259: in press (1992)

44. Schneider, T. P., Cho, J., Aldrich, D. A., Chen, Y. L., Maher, D., and Nemanich, R. J., *Proceedings of the 2nd Int. Symp. of Cleaning Technology in Semiconductor Device Manufacturing,* (J. Ruzyllo and R. E. Novak, eds.), 92-12:122-132, The Electrochemical Society Inc., Pennington, NJ (1992)

45. Shibata, T., Nanishi, Y., and Fujimoto, M., *Jap. Appl. Phys. (Let.)* 29:L1181-L1184 (1990)

46. Fountain, G. G., Rudder, R. A., Hattangady, S. V., Vitkavage, D. J., Markunas, R. J., and Posthill, J. B., *J. Appl. Phys.* 63:4744-4746 (1988)

47. Frystak, D. C. and Ruzyllo, J., *Proceedings of the First International Symposium on Cleaning Technology in Semiconductor Device Manufacturing, Semiconductor Cleaning Technology/1989* (J. Ruzyllo and R. E. Novak, eds.), 90-9:129-140 (1989)

48. Ruzyllo, J., Hoff, A. M., Frystak, D. C. and Hossain, S. D., *J. Electrochem. Soc.* 136:1474-1476 (1989)

49. Yew, T.-R., and Reif, R., *J. Appl. Phys.* 68:4681-4693 (1990)

50. DelFino, M., Saliman, S., and Hodul, D., *J. Appl. Phys.* 70:1712-1717 (1991)

51. Rudder, R. A., Hattangady, S. V., Posthill, J. B., Hudson, G. C., Mantini, M. J., and Markunas, R. J., *SPIE Conf. Proc.* 1188:125-133 (1989)

52. Anthony, B., Hsu, T., Breaux, L., Qian, R., Banerjee, S., and Tasch, A., *J. Electron Mater.* 19:1089-1094 (1990)

53. Thomas, R. E., Mantini, M. J., Rudder, R. A., Malta, D. P., Hattangady, S. V., and Markunas, R. J., "Carbon and Oxygen Removal from Silicon (100) Surfaces by Remote Plasma Cleaning Techniques," to be published *J. Vac. Sci. Tech.* A10: (Jul/Aug) (1992)

54. Fountain, G., "Oxide-Nitride-Oxide Structures by Remote Plasma Enhanced Chemical Vapor Deposition," presented at SRC Topical Research Conference on High Reliability Gate Dielectrics, Albuquerque, New Mexico (January 30-31, 1990)

55. Boland, J. J., *Surf. Sci.* 261:17-28 (1992)

56. Abrefah, J. and Olander, D. R., *Surf. Sci.* 209:291-313 (1989)

57. Gates, S. M., Kunz, R. R., and Greenlief, C. M., *Surf. Sci.* 207:364-379 (1989)

58. Thomas, R. E., Rudder, R. A., and Markunas, R. J., *MRS Symp. Proc.* 204:327-332 (1991)

59. Sun, Y. K., Bonser, D. J., and Engel, T., *Phys. Rev.* B43:14309-14312 (1991)

60. Bozack, M. J., Taylor, P. A., Choyke, W. J., and Yates, J. T. Jr., *Surf. Sci.* 177:L933-L937 (1986)

61. Balooch, M., *Jap. J. of Appl. Phys.* 16:1557-1561 (1977)

62. Amorium, J., Maciel, H. S., and Sudano, J. P., *J. Vac. Sci. Technol.* B9:362-365 (1991)

63. Mackinnon, K. A., *Philos. Mag.* 8:605-616 (1929)
64. Hattangady, S. V., Rudder, R. A., Mantini, M. J., Fountain, G. G., Posthill, J. B., and Markunas, R. J., *MRS Symp. Proc.* 165:221-226 (1990)
65. Grunthaner, P. J., Grunthaner, F. J., Futhauer, R. W., Lin, T. L., Hecht, M. H., Bell, L. O., Kaiser, W. J., Schowengerdt, F. D., and Mazur, J. H., *Thin Solid Films* 183:197-212 (1989)
66. Fenner, D. B., Biegelsen, D. K., and Brigans, R. D., *J. Appl. Phys.* 66:419-424 (1989)
67. Cho, J., Schneider, T. P., Vanderweide, J., Jeon, H., and Nemanich, R. J., *Appl. Phys. Lett.* 59:1995-1997 (1991)
68. Chabal, Y. J., and Raghavachari, K., *Phys. Rev. Lett.* 54:1055-1058 (1985)
69. Johnson, N. M., Ponce, F. A., Street, R. A., and Nemanich, R. J., *Phys. Rev.* B35:4166-4169 (1987)
70. Hirayama, H. and Tatsumi, T., *Appl. Phys. Lett.* 54:1561-1563 (1989)
71. Burke, R. R., Pelletier, J., Pomot, C., and Vallier, L., *J. Vac. Sci. Technol.* A8:2931-2938 (1990)
72. Ishii, M., Nakashima, K., Tajima, I., and Yamamoto, M., *Appl. Phys. Lett.* 58:1378-1380 (1991)
73. Nakashima, K., Ishii, M., Tajima, I., and Yamamoto, M., *Appl. Phys. Lett.* 58:2663-2665 (1991)
74. Raynaud, P., Booth, J. P., and Pomot, C., *Physica* B170:497-502 (1991)
75. Raynaud, P., Booth, J. P., and Pomot, C., *Appl. Surf. Sci.* 46:435-440 (1990)
76. Yamada, H., *J. Appl. Phys.* 65:775-781 (1989)
77. Kishimoto, A., Suemune, I., Hamaoku, K., Koui, T., Honda, Y., and Yamanishi, M., *Jap. J. Appl. Phys.* 29:2273-2276 (1990)
78. Lowenstein, L. M., *J. Vac. Sci. & Tech.* A7:686-690 (1889)
79. Lowenstein, L. M., *J. Vac. Sci. & Tech.* A6:1984-1988 (1988)
80. Lowenstein L. M. and Tipton, C. M., *J. Electrochem. Soc.* 138:1389-1394 (1991)
81. Moslehi M. M. and Davis, C. J., *J. Mater. Res.* 5:1159-1162 (1990)
82. Frystak, D. C. and Ruzllo, J., *Proc. Second International Symposium on Cleaning Technology in Semiconductor Device Manufacturing*, (J. Ruzyllo and R. E. Novak, eds.); 92-12:58-71, The Electrochemical Society Inc., Pennington, NJ (1992)

Part IV.

Analytical and Control Aspects

9

Measurement and Control of Particulate Contaminants

Venu B. Menon and Robert P. Donovan

1.0 INTRODUCTION

Wafer cleaning (including non-plasma film etching, e.g., oxide etching with HF, or nitride film etching with H_3PO_4) is the most frequently repeated operation in a semiconductor processing line. In a 4-megabit dynamic random access memory (DRAM) or equivalent manufacturing line, approximately 15% of the total steps involve some form of wafer cleaning. These cleaning operations, which may be a single process step, or a sequence of steps, are used to remove contaminants after etching or film deposition, or to prepare the wafer surface prior to etch or deposition.

There are many wet-chemical cleaning sequences that are routinely used for removing organic, inorganic, and ionic materials from wafer surfaces. These include the use of hydrofluoric acid to etch oxide films, sulfuric-acid/hydrogen-peroxide mixtures to remove heavy organics such as photoresist, hydrochloric-acid/hydrogen-peroxide/water mixtures to remove ionics, and ammonium-hydroxide/hydrogen-peroxide/water solutions to remove light organics and particles. Most of these chemical solutions are optimized to remove non-particulate contaminants, and it is not uncommon to find cleaning processes that *add* a substantial amount of particles to the wafer surface. This chapter describes the measurement and control of particles during cleaning, and recommends practices to minimize particle deposition on wafers.

1.1 Scope

Most wafer cleaning processes are liquid-based, hence the primary focus of this chapter will be limited to wet-chemical cleaning. The major components of a wet-chemical cleaning system are *(i)* the cleaning liquid, *(ii)* the distribution system, *(iii)* the process container (note: can be a drum, vessel, closed reactor, etc.), *(iv)* the rinse tank, and *(v)* the dryer. The major variables are chemical purity, process effects, and cleanliness of the rinse and dry processes.

The scope of this chapter is to:

1. Describe techniques for measuring particles in liquid chemicals, which are the raw materials for cleaning
2. Outline control strategies to reduce particle sources and prevent particle deposition during cleaning
3. Develop strategies to remove particles that are already deposited on wafers
4. Describe recommended practices for monitoring particles in wafer cleaning equipment

1.2 Chapter Organization

The chapter is divided into sections that deal with the scope described above. The sections are:

Particle measurement in liquids
Particle control in chemicals and distribution systems
Particle control during wafer cleaning
Post-processing particle removal technologies
Particle monitoring practices
Conclusions

2.0 PARTICLE MEASUREMENT IN LIQUIDS

The various chemicals used in wafer cleaning include deionized water, mineral acids (e.g., HF, H_2SO_4, H_3PO_4, HNO_3, HCl), alkalis (e.g., NH_4OH, KOH), solvents (e.g., isopropyl alcohol, n-methyl pyrrolidone), and oxidants (e.g., H_2O_2, $H_2S_2O_8$). Typically, one is interested in the particle number concentration, size distribution, and morphology/composition (when possible) at various points in the chemical distribution system.

The commercially available techniques for size and number concentration measurement fall into three broad categories:

1. optical light scattering from a sample of liquid
2. microscopy of particles collected on a filter membrane
3. optical light scattering from a silicon wafer surface

2.1 Optical Light Scattering

Liquid particle counters are the most popular instruments for measuring particles in DI water and chemicals. These particle counters, often referred to as optical particle counters (OPCs) work on the principle of light scattering. A sample of liquid is passed through a tube where it passes through a beam of laser light. Individual particles scatter light, which is detected by a photodetector. The intensity of scattered light is proportional to particle size, with smaller particles scattering less light. The detection sensitivity of these counters is typically improved by using low sample flow rates.

An excellent review of the performance and use of optical particle counters is found in Knollenberg et al. (1). Grant (2) also gives a clear description of the various particle counters that are commercially available:

> Two types of OPC sensors are commonly used to detect particles smaller than 0.5 microns in diameter in liquids: volumetric and in situ (see Fig. 1). Both have laser light sources. Volumetric sensors examine all of the fluid flowing through the sensor cell; in situ sensors examine only a small portion of the flow through the sensor cell by focusing light onto a small region in the center of the fluid flow path.
>
> In situ sensors offer the advantage of increased light intensity in the sensing volume as the laser beam is focused on a small volume. In addition, in situ sensors have decreased background noise because the amount of stray scattered light reaching the sensor collection optics is smaller than that in volumetric sensors. These advantages result in increased instrument sensitivity.

Commercial OPCs are available from Particle Measuring Systems Inc., Hiac-Royco, Horiba, Kowa, etc. State-of-the-art OPCs have best sensitivities around 0.1 microns for liquid chemicals, and around 0.07 microns for DI water. They can be used in-line for continuous monitoring, or off-line for periodic monitoring at different points in a chemical distribution system or process equipment.

Automated liquid particle counters are typically calibrated to one specific refractive index, and will miss particles that have low refractive index contrast with the bulk fluid. Also, these particle counters will size non-

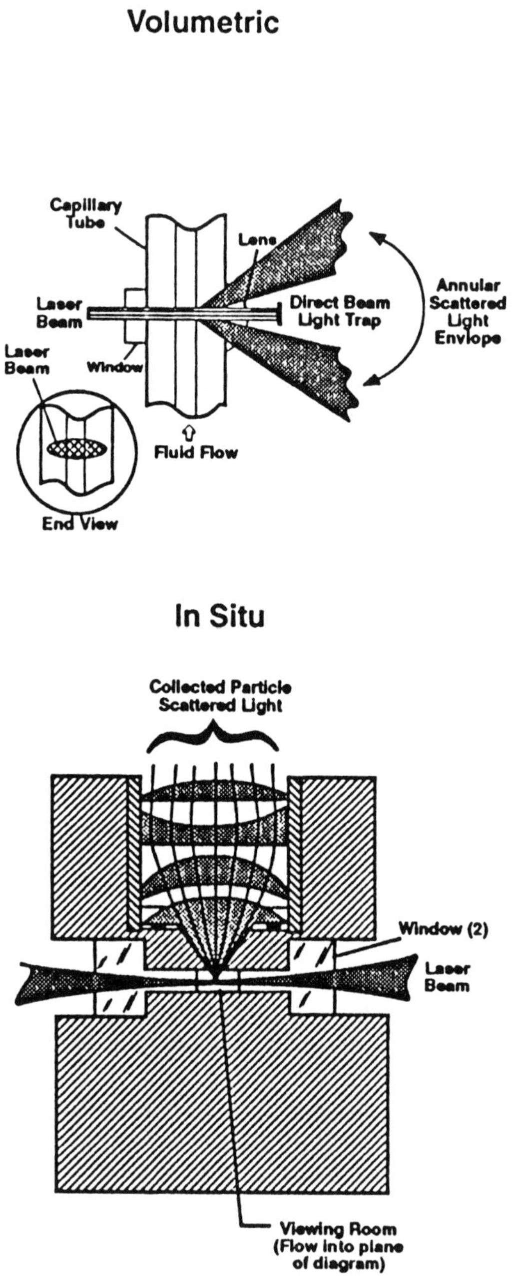

Figure 1. Liquid flow cell designs (1). *(Reprinted with permission from the* Proceedings of the Institute of Environmental Sciences, *pp. 757-758, San Diego, CA 1991)*

spherical particles to a mean scattering diameter, which could be very different from their true size and shape. In addition, bubbles within the fluid stream will scatter light and get counted as particles. This could be particularly problematic in ozonated DI water, hydrogen peroxide, or ammonium hydroxide solution, which tend to de-gas. For such chemicals, it is preferable to use a pressurized system, and/or use an oscilloscope in conjunction with an OPC, to manually separate spikes associated with bubbles (3).

Batchelder and Taubenblatt (4) have identified some of the deficiencies of current OPCs as *(i)* poor counting statistics; *(ii)* high initial, installation and maintenance costs; *(iii)* inability to identify particle composition; and *(iv)* poor reliability. They have developed a modified Nomarski style interferometer that not only detects single particles, but also provides, in real-time, some information about the composition of the detected particles. Using this technique they were able to identify air bubbles in water and separate their signal from that of polystyrene, aluminum oxide or silver particles (5).

2.2 Nonvolatile Residue Monitor

There are several nonvolatile colloidal species in DI water that plate onto the surface of wafers during the rinsing and drying steps of a cleaning process. These include colloidal silica, bacteria, particles, and certain ionic impurities. Blackford et al. (6)(8)(9) describe a nonvolatile residue monitor that quantitatively measures the amount of these nonvolatile species in water. Figure 2 depicts a schematic diagram of this monitor. A sample of water is atomized into an air stream that carries it through a diffusion dryer. The water is evaporated, leaving a particulate phase that represents the nonvolatiles. These particles are typically well below 0.1 microns in size, and conventional aerosol optical particle counters cannot be used to size and count them. Blackford et al. (6) recommend the use of a condensation nucleus counter (CNC) to count these particles. In a CNC, small particles are coated with a film of alcohol (e.g., butanol) that increases the effective size to around 10 microns, allowing them to be counted by standard optical techniques. The total nonvolatile residue is reported in parts per billion. This technique can also be used to measure the nonvolatile residue of water soluble solvents by injecting pulses of the solvent into the DI water passing through the monitor (7)(9).

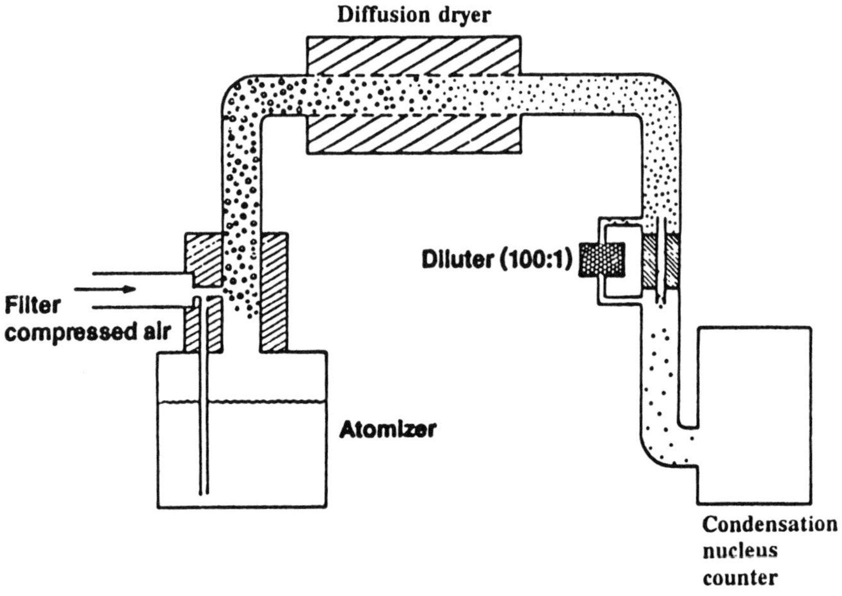

Figure 2. Nonvolatile residue monitor (6). *(Reprinted with permission from the* Journal of the Institute of Environmental Sciences, *pp. 43-47, July/Aug., 1987)*

2.3 Microscopy

To alleviate some of the problems associated with OPCs, some fabs use manual microscopic counting in addition to OPCs. Particles are collected on filter membranes, such as the Nucleopore® absolute membrane, and then manually counted using a microscope. This technique is used commonly only with DI water where proprietary staining procedures (e.g., the Nomura method) are used to increase the contrast between the particles and the background. The advantage of this method is that particles of different shapes and refractive indices can be differentiated and counted. The staining techniques are especially sensitive for visual classification of bacteria, PVC piping fragments, and stainless steel. Typical particle size sensitivities are 0.2 to 0.3 microns.

Microscopic counting is extremely laborious and dependent on trained operators. Particle counts have been known to be widely different depending on the operator.

2.4 Particles on Wafers

Many particle problems in liquid chemicals manifest themselves as defects on wafers. Hence, detection of particles or defects on wafers can be used as a means to quantify particles in liquids. One advantage is that it directly measures particles that actually are deposited onto wafers and thereby have real potential to cause yield loss (10). However, this approach is fraught with difficulties because the particles that are observed on wafers may be attributed to many potential sources such as the process equipment, chemical distribution system, cleanroom, wafer handling, etc. Isolating the liquid chemical as the source of particles and then quantifying the number concentration and size distribution in the liquids requires a systematic approach and often considerable expense of time and money. Silicon wafers are expensive; a 200 mm diameter wafer can cost over $100. We recommend use of silicon wafers for measuring particles in liquids only for first-pass trouble shooting. Once a particle problem has been identified, it is better to use liquid particle counters or microscopic methods to quantify the level of liquid-based contamination.

Most fabs use some form of routine wafer particle counting to detect early problems in process equipment. A particle-per-wafer-pass (PWP) method is used to make these measurements. A detailed description of the PWP methodology is available in Tullis (10). "PWP is the number of particles added to wafers in one pass through one or more pieces of process equipment" (10). PWP can also be used to evaluate subsets of a piece of equipment. Isolating the particle contributions of components of a wet bench using a PWP method assumes that the contributions are additive. This is not necessarily true for some chemical cleaning operations.

Particle counting is usually done on bare silicon wafers using a surface scanner, which works on the principle of light scattering. Such scanners are made by Tencor Instruments, Estek, Hitachi, Particle Measuring Systems, etc. Figure 3 shows the principle of operation (11). A beam of light from a laser source is used to illuminate the wafer surface. Any particles present on a smooth surface will reflect light in directions different from that of the surface reflection. The amount of light scattered (usually expressed as a scattering cross section) by a particle is a function of its size, shape and refractive index. State-of-the-art surface scanners have detection limits around 0.1 microns on planar surfaces, with higher sensitivities reported at slower detection speeds (12).

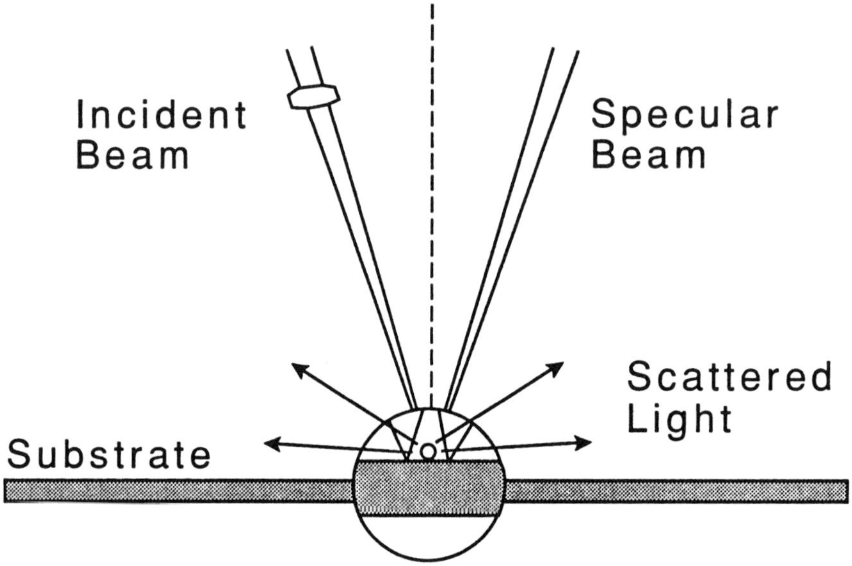

Figure 3. Principles of operation of a laser surface scanner (11). *(Source: Tencor Instruments Inc.)*

During a PWP measurement, relatively clean wafers are initially scanned through the surface particle counter to obtain a baseline particle count. These wafers are then passed through the process equipment as shown in Fig. 4, and the resulting counts measured. The difference between the pre- and post-counts is the number of particles added to (or removed from) the wafer by the process equipment. It is important to conduct the PWP measurement several times to obtain statistically meaningful results. A formal test method for measuring particles per wafer pass is in preparation by SEMATECH and Semiconductor Equipment and Materials International (SEMI) as a revision to document number E14-90.

3.0 PARTICLE CONTROL IN CHEMICALS

Particle control in the liquid chemicals used for wafer cleaning is achieved by using high quality chemicals, clean chemical distribution systems, and filters within the process equipment and in the distribution line.

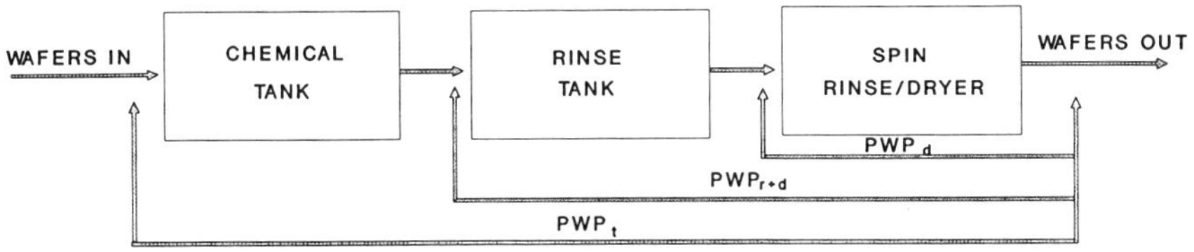

PWP_{dryer} = Particle per wafer pass for dryer

$PWP_{rinse\ tank} = PWP_{r+d} - PWP_{d}$

$PWP_{chemical\ tank} = PWP_{t} - PWP_{r+d}$

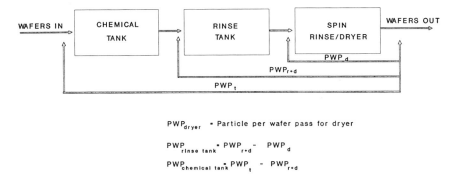

Figure 4. Simplistic diagram of particle-per-wafer-pass calculations for a wet bench.

3.1 Incoming Chemical Quality

Most inorganic chemicals used in semiconductor wafer cleaning are aqueous solutions. Typically the manufacturers of these chemicals mix water with concentrated chemicals to prepare the concentrations that are demanded by customers. Hence, the quality of the water used in chemical manufacturing is critical to the quality of the final product. Unfortunately, most chemical manufacturers do not have the resources to build the kind of state-of-the-art water purification systems that are used by the IC producers. In recent years, there is a trend for IC producers to buy concentrated chemicals and blend it on site with their own high-grade water, just prior to pumping it through their distribution systems. Table 1 shows the average particles/liter concentration in source drums reported by Rosenfeld et al. (13). Typical particle concentrations for bulk chemicals in 1989 were between 10 - 300 particles/mL (>0.5 µm). Menon (14) has used historical data to estimate the particle levels expected in 1993 to be 1 particle/mL, provided continuous improvements are made to chemical quality (Fig. 5). These requirements are well within the realm of continuous improvement and can be achieved by systematic reduction of particle sources during chemical production.

The materials of construction for reactors, pumps, transfer lines, absorbers, storage and shipping containers affect the particle levels for all chemicals produced (16). Until recently, the cost and chemical compatibility of the packaging materials have determined their choice as a packaging material. Table 2 provides a suggested list of materials of construction for chemical process systems (16). Glass bottles, which have traditionally been used for shipping chemicals, have been shown to add significantly more particles than fluoropolymers, such as PFA Teflon®. Table 3 shows a comparison of the particle performance of glass and polyethylene bottles for sulfuric acid. The polyethylene bottles consistently contributed fewer particles to the chemical. For polymeric bottles or drums, the process of molding the container contributes to its particle shedding behavior. Blow molding produces a smoother surface over rotary molding.

Table 1. Particle count profile of the source drums for some process chemicals (13). *(Reprinted with permission from* Proceedings of Microcontamination '90 Conference, *pp. 225-258, 1990)*

CHEMICAL	SOURCE	Particles per liter (at the source drum) ≥0.50 µm Avg.	≥1.0 µm Avg.
Ammonium Hydroxide	Chemical Supplier	298,700	2,140
Hydrochloric Acid	Chemical Supplier	95,670	1,090
Hydrofluoric Acid	Chemical Supplier	35,230	10,860
Sulfuric Acid	Reprocessor	20,660	7,090

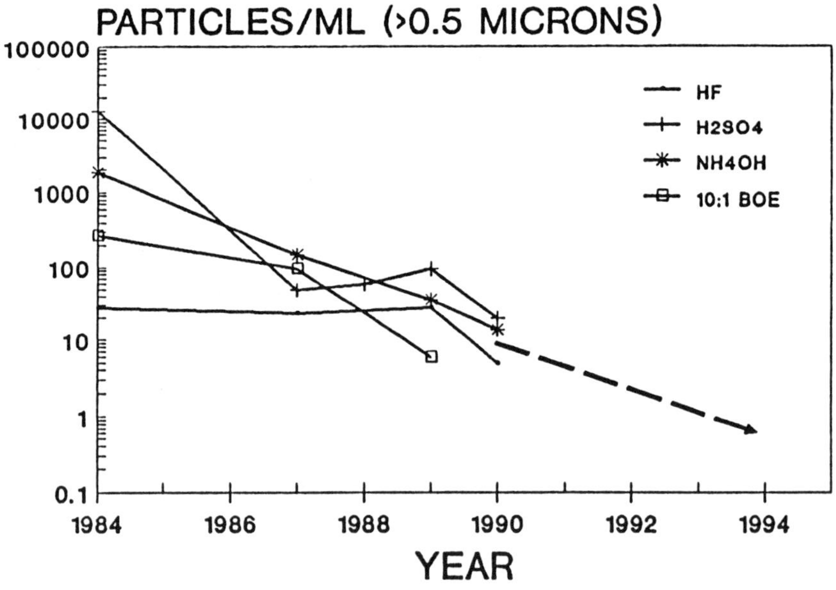

Figure 5. Chemical purity trends (14).

Table 2. Suggested materials for the construction of selected chemical process systems (16). *(Reprinted from* Particle Control for Semiconductor Mfg., *p. 185, by courtesy of Marcel Dekker, Inc., 1990)*

	Elastomers					Piping Materials				
Chemicals	Viton A	Ethylene propylene	Teflon	304 L Stainless Steel	316 Stainless Steel	Aluminum[a]	PVDF[b]	Poly-propylene	Teflon	ABS[c]
Acids										
Sulfuric Acid (Conc)	✶		✶				✶		✶	
Phosphoric Acid (Conc)	✶	✶	✶	✶	✶					
Nitric Acid (Conc)	✶		✶				✶		✶	
Acetic Acid (Glacial)		✶	✶		✶					
Hydrocholoric Acid (Conc)	✶		✶				✶			
Hydrofluoric Acid (49%)		✶	✶				✶	✶	✶	
Solvents, etc.										
Acetone		✶	✶	✶						
n-Butyl Acetate		✶	✶	✶						
Isopropyl Alcohol	✶	✶	✶	✶						
Methanol		✶	✶	✶						
Xylene	✶		✶	✶						
Deionized Water	✶		✶				✶		✶	✶
Hydrogen Peroxide	✶		✶		✶	✶		✶	✶	

✶ = Acceptable material
[a] Aluminum = Type 1060 for hydrogen peroxide service
[b] ABS = Dura Plus ABS Industrial; acrylonitrile butadiene styrene
[c] PVDF = Polyvinylidene fluoride

Table 3. Comparison of particle concentrations of sulfuric acid stored in polyethylene vs. glass bottles (16). *(Reprinted from* Particle Control for Semiconductor Mfg., *p. 199, by courtesy of Marcel Dekker, Inc., 1990)*

Sample	Poly Bottles (≥0.5 μm/mL)	Glass Bottles (≥0.5 μm/mL)	% Difference
1	4.92	22.04	448
2	54.32	99.64	183
3	57.96	80.04	138
4	13.82	42.76	309
5	50.72	93.24	184
6	23.00	45.48	198

Miki and Ohmi (17) have shown the benefits of the use of ultraclean, high surface finish stainless steel in the manufacture of HF. They report particle counts in 49% HF that are below 1000/liter (>0.5 μm). Their data also show that very few particles are added to the chemical during storage and transport, provided that the packaging container is selected carefully and cleaned thoroughly before filling. Polyethylene bottles, blow molded under cleanroom conditions, showed the lowest level of particulate contamination (17).

A recent advance in the quality of liquid chemicals was reported by Mathews (15) where ultrapure gases such as HCl and NH_3 were mixed with DI water in a reactor inside the semiconductor fab to produce the liquids required for wafer cleaning. This technique, referred to as "point of use chemical generation" appears to have the potential of substantially reducing metallics and particulate contamination in liquids. Matthews (15) reports particle levels (>0.5μm) and metallic contaminants below 0.1 ppb.

3.2 DI Water Quality

Depending on the size of the fab, water usage, principally in wafer cleaning tools, varies from less than 40 to greater than 200 liters/min. Because water is used in virtually every cleaning process, the impact of water quality on defect levels is considerable. Figure 6 shows a correlation of particles adhered to a hydrophilic wafer after immersion in water containing different levels of polystyrene latex spheres. In all cases, the number of particles added to a hydrophilic silicon wafer surface increased with particle concentration in water. Riley and Carbonell (18) show that the number of added particles does not necessarily increase with immersion time (see Ch. 4 by Donovan and Menon). This depends on the zeta potentials of the particles and wafers, and their wettability in water. The particle level of DI water entering the fab is around 100 - 1000 particles/liter for particles of size greater than 0.1 microns. This particle level is achieved through multiple pass filtration using coarse and fine filters, and the use of continuous or periodic ozonation (with ultraviolet light) to break down bacteria. The final filter in the finishing loop is usually a charged membrane filter.

Previous chapters have shown that many particles in DI water are colloidal species that are usually negatively charged, especially colloidal silica and lipopolysaccharides (19). The use of a positively charged modified membrane filter aids in removing these negatively charged particles. Because the nature of particle capture is related to the attraction of colloidal particles of charge opposite to that of the membrane medium (often termed "electrokinetic adsorption"), these filters are very efficient at removing particles much smaller than the pore size of the membrane. Companies such as Cuno and Pall Corporation make these charge modified filters.

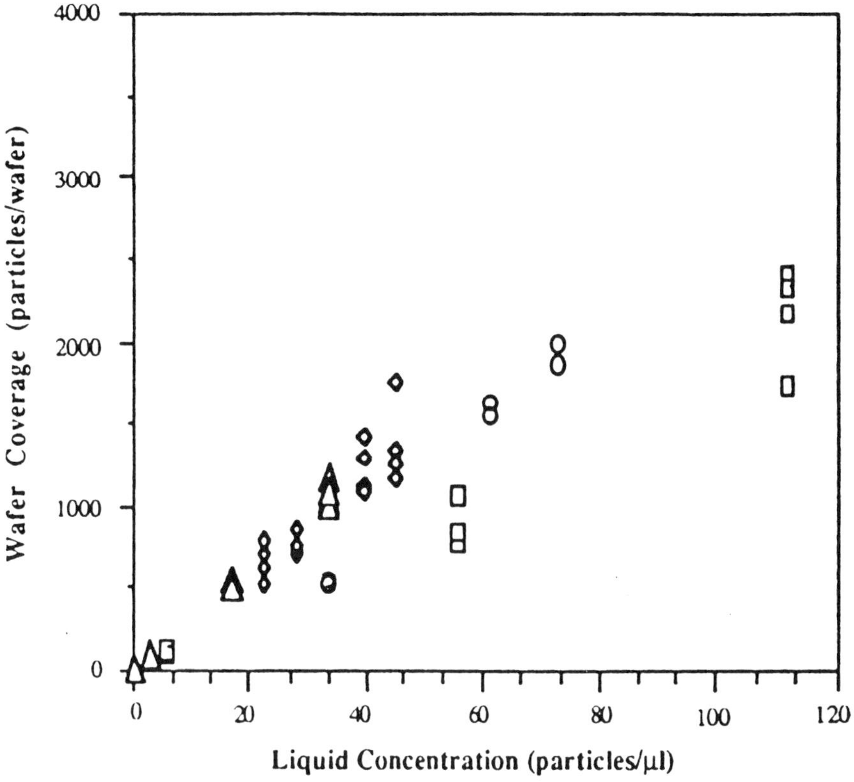

Figure 6. Comparison of wafer coverage (number of 0.487 μm PSL particles on a 100 mm silicon wafer) for various liquid bath concentrations of PSL (18). *(Reprinted with permission from the* Proceedings of the Institute of Environmental Sciences, *p. 225, New Orleans 1990)*

Recent research at the University of Arizona has shown some interesting properties of charge modified filters. Jan et al. (19) have extensively characterized the electrokinetic properties of these filters and found discrepancies between the claims of suppliers and the actual performance. Jan et al. measured the zeta potential, surface charge density, and saturation capacity of two different charge modified Nylon® membrane filters. They found that suppliers were claiming particle filtration at pH values of water ranging from 3 to 10, while their results showed that these filters became negatively charged (and hence ineffective) at pH values below 7.6. Figure 7 depicts the variation of zeta potential with pH for two different commercial filters, showing clearly the point where the zeta potential becomes zero (referred to as the isoelectric point). The wider the pH range over which the zeta potential is positive, the more efficient the filter. Nonetheless, these charged filters have been shown in field use to be very efficient in small particle removal from water.

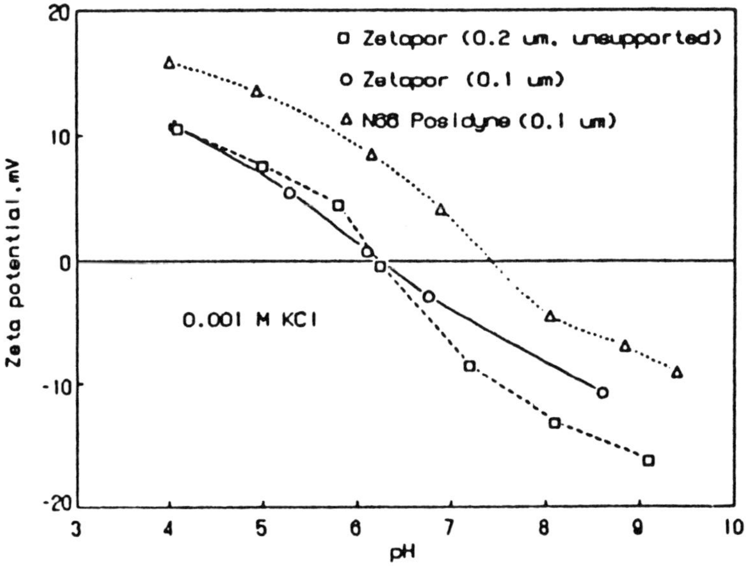

Figure 7. Comparison of zeta potential versus pH profiles for N66 Posidyne and Zetapor membranes in 0.001 M KCl solutions (19). *(Reprinted with permission from the* Proceedings of the Institute of Environmental Sciences, *p. 852, San Diego, CA, 1991)*

The sequence of UV light use, ozonization and filtration can also affect the final particle count (20)(21). The UV light breaks bacteria into smaller fragments, which manifest themselves as particles. One bacterium can generate several fragments. Shadman et al. (20) showed that the total organic carbon (TOC) level dropped more dramatically when the membrane filter was placed before the UV unit. The filter removes particles which are larger than a certain size, while the remaining organic particles are oxidized and broken down in the UV unit. In the reverse sequence, where the water is exposed to the UV unit first, the smaller fragments that are generated are more difficult to remove in the filter, thus resulting in a higher TOC level (Fig. 8). In spite of some of these research observations, most semiconductor factories still favor the use of a final filter before pumping water into the fab.

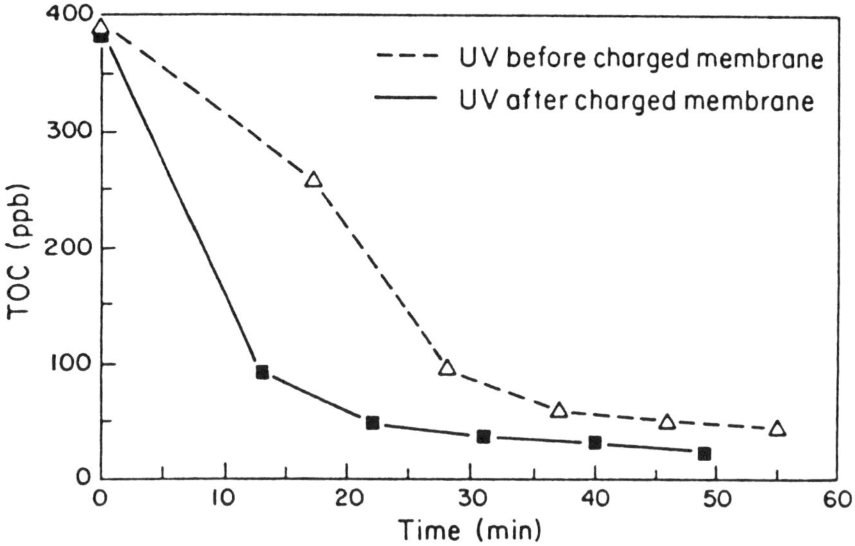

Figure 8. The effect on TOC of sequencing UV with a charged membrane filter; 15.2 L/min (20). *(Reprinted with permission from the* Proceedings of the Institute of Environmental Sciences, *p. 223, New Orleans, LA, 1990)*

Ozonized water has a high dissolved oxygen content, typically in the parts per million level. There are differing schools of thought on the use of ozonized water for wafer cleaning. Yabe et al. (22) prefer the use of water with low particle per billion (ppb) levels of dissolved oxygen, because it apparently inhibits the formation of a native oxide layer on wafers after a water rinse. Others prefer to use ozonized water because it provides for a thin native oxide layer which passivates the wafer surface, thus preventing subsequent contaminant deposition (23). Several suppliers now offer an ozonized rinse as a part of their wafer cleaning equipment.

The preferred material of construction for most surfaces that are in contact with DI water is PVDF, although some of the older fabs use PVC, or stainless steel. DI water piping must be free of additives or contaminants that can elute into the fluid, and it must be compatible with sterilizing chemicals such as hydrogen peroxide and hot water (22). PVC has been shown to shed particles, while stainless steel adds metallic contaminants into the water stream. Polyether-ether-ketone (PEEK) is a newer material that is recommended for DI water transport systems because of its smooth surface properties, lower leaching potential, and temperature insensitivity. PEEK, however is very expensive.

3.3 Chemical Distribution System

Chemical distribution systems consist of pumps, chemical storage tanks, and distribution piping that are required to automatically dispense chemicals into process equipment. These systems improve safety and reliability, eliminate the need to bring bottles into the cleanroom, incorporate multiple pre-filtering stations to reduce particles, and reduce the variability associated with manual operation (24). The chemical compatibility and cleanliness of storage drum and piping materials, the filtration scheme, and the particle generation characteristics of pumps and valves are important aspects in providing a low particle chemical to the point of use.

Present day IC production typically requires chemicals with particle levels of less than 500 particles/liter of size greater than 0.5 µm. The manufacture of 16 Mb DRAMs or other devices of equivalent complexity requires low levels of particles in the 0.1 to 0.2 micron size range. While chemical suppliers are working towards such levels of particulate contaminants in incoming chemicals, the purity at the point of use depends entirely on the filtration scheme in the chemical distribution system and at the recirculating process tank. Chemical distribution systems are sold in the U.S. by many manufacturers including FSI International and Systems Chemistry, Inc. Figure 9 depicts a schematic diagram of a typical bulk chemical distribution system (25). The chemical drums, storage tanks, pumps, and filter banks are usually located in an area isolated from the main cleanroom. Chemicals are pumped from the source drum into storage tanks, often called "day" tanks which may, or may not, have a nitrogen blanket above the liquid surface. From the storage tanks, the chemical is continuously recirculated through a bank of filters up to the point of connection to the process tools. Fluid flow within the distribution system is usually accomplished either with a pump or pressurized nitrogen. On demand, chemical is drawn from the central recirculation loop to fill a process tank. When chemical is not being called for by the process tools, it is being continuously circulated through the filters and the storage tank. Some facilities prefer to have just a local recirculation loop that keeps chemicals moving through the storage tank and the filter bank, as shown in Fig. 10.

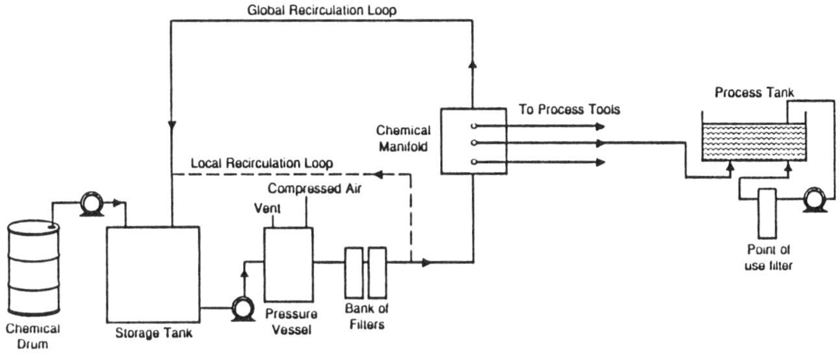

Figure 9. Schematic of a typical chemical distribution system (25). *(Reprinted from* Microcontamination, *©1990, Canon Communications Inc., Santa Monica, CA)*

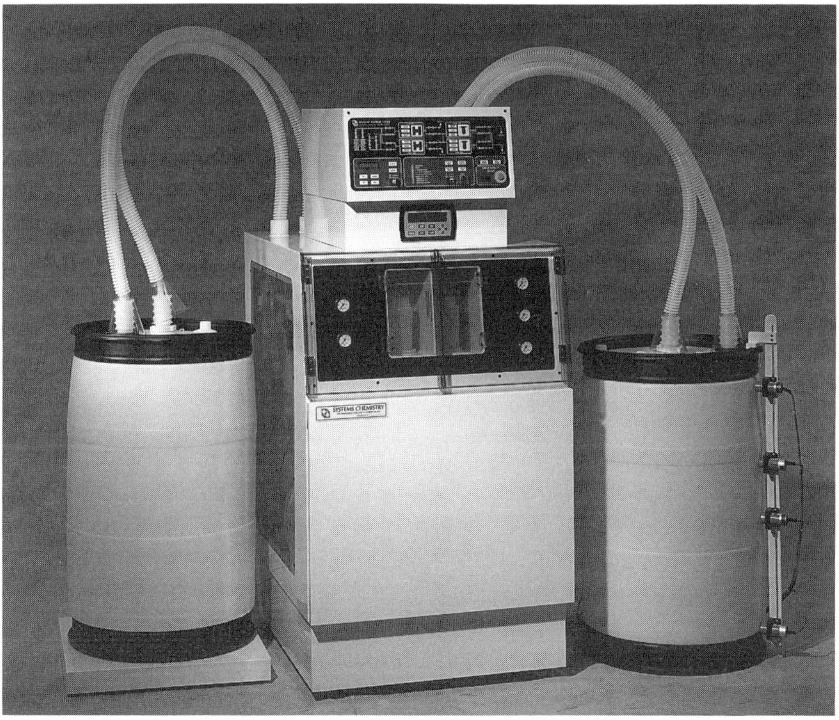

Figure 10. Photograph of a chemical distribution pumping and recirculating unit. *(Source: Systems Chemistry, Inc.)*

Filters capable of efficiently removing particles of sizes less than 0.1 μm are commercially available, however, there are compatibility issues that must be carefully examined before selecting a filter for a specific process. Fluorocarbon resins such as PFA, PVDF and PTFE are generally compatible with most inorganic acids and alkalis (26). Table 4 is a compatibility chart of filter materials with commonly used chemicals (27). Often, the membrane of a filter is compatible with the chemical being filtered, but the cartridge material, filter housing, or the O-ring is not. This can result in change in material composition, leading to gas permeation, polymer degradation, particle shedding, and overall loss of chemical quality. Gotlinsky warns that the material compatibility issues sometimes leads to the selection of Teflon® as the material of choice for all filters, while lower cost alternatives may be available (28). Hence, a systematic program for selection, maintenance and replacement of filters is essential to successfully controlling particles in a wet-chemical process tank.

Table 4. Chemical and material compatibility. Most of the concentrations are typical for those used in silicon wafer processing (27). *(Reprinted with permission from* Semiconductor International, *Oct., 1988)*

Chemical	Concentration	PP	PVDF	Nylon	PFA	Polysulfone
Acids						
Acetic	80%	L	R	NR	R	R
Hydrochloric	35%	L	R	NR	R	R
Hydrofluoric	60%	R	R	NR	R	L
Nitric	70%	NR	L	NR	R	L
Perchloric	50%	L	R	NR	R	NR
Phosphoric	80%	R	R	MR	R	R
Sulfuric	98%	NR	L	NR	R	NR
Bases						
Ammonium hydroxide	28%	R	R	R	R	R
Potassium hydroxide	50%	R	R	L	R	R
Sodium hydroxide	70%	L	R	L	R	R
Other Chemicals						
Acetone	100%	R	L	R	R	NR
Ammonium fluoride	25%	R	R	R	R	R
Ethyl Acetate	—	R	NR	R	R	NR
Hydrazine	—	L	NR	NR	R	NR
Hydrogen peroxide	90%	R	R	NR	R	R
Methanol	—	R	R	NR	R	R
2-Propanol	—	R	R	L	R	R
Xylene	—	NR	R	R	R	NR
Mixtures						
Aqua regia	—	NR	R	NR	R	NR
Buffered HF	—	R	R	NR	R	L

R = Recommended; L = Limited; NR= Not Recommended.
Based on ambient temperature exposure. *(Source: Millipore)*

For sulfuric acid, hydrogen peroxide, ammonium hydroxide, and hydrochloric acid, the particle counts after 6 - 10 minutes of operation of a chemical distribution system loop in a recycle mode is sufficient to bring the counts well below those observed in bottled chemicals. The particle concentrations at 0.3 micron size are typically 2 to 5 times that at 0.5 micron. Table 5 shows the performance of a chemical distribution system, reported by Gruver et al. (30). For all key chemicals, the final particle count at the point of use was below 500 particles/liter. On an average, the chemical distribution system with its filters can provide a 100 to 500-fold decrease in particle levels. This reduction does not include any point of use filtration. Rosenfeld and Menon (13) published recent results of another distribution system. Their results indicate a 35 to 300-fold decrease in particle levels from the drum to the point of use, depending on the chemical composition and type of filters being used.

Table 5. Particle comparisons of bottle, drum, and bulk delivery systems. Particle size >0.5 µm (30). *(Reprinted with permission from the* Proceedings of the Institute of Environmental Sciences, *p. 314, New Orleans, LA, 1991)*

	Particles / mL								
	10:1 BHF	Sulfuric	Nitric	KOH	CR/Phos	Phos	HCl	10:1 HF	NH_4OH
Bottle	6	96	46	33	326	39	66	28	36
Drum	132	11000	126	181	525	631	67	91	76
Bulk Delivery System	0.02	0.47	0.22	0.02	0.1	0.2	0.008	0	0.02

When designing a chemical distribution system, it is worthwhile to include sample ports at various points in the system to extract chemicals that can be analyzed for particles. Sample ports are recommended after the drum, the pumping unit, the storage tank, and at the point of connection to the process equipment. Also, building in some redundancy in pumping capability is useful as usage demands change, or if one pump is in repair.

3.4 Point Of Use Filtration

State-of-the-art fabs generally prefer to do all their DI water filtration at a central purification facility, and not at the point of use. Point-of-use water filters are breeding grounds for bacteria because of the low volumetric flow rate of water through each filter. Also, since most filters are not compatible with ozone, they cannot be used with ozonized water.

Most wafer cleaning and etching processes are conducted in an immersion tank or spray processor. These systems accommodate point-of-use recirculation filters quite easily. Until recently, recirculation tanks, with continuous filtration, were only available for processes that are conducted at temperatures below approximately 80°C owing to limitations in filter availability and pump materials of construction. With recent advances in polymeric materials technology, higher temperature recirculation tanks are now available for operating temperatures as high as 180°C. Nisso Engineering and Lufran, Inc. are two of the many companies that offer recirculation systems for sulfuric-acid/hydrogen-peroxide baths (piranha) that are maintained at 120 - 130°C, and for 85% phosphoric acid maintained at 180°C.

As with filters for chemical distribution systems, it is important to select a point-of-use filter that is compatible with the chemical being filtered. Compatibility charts are available from most filter manufacturers. Unlike DI water filters, bacterial growth has not been a problem with chemical filters. Before a filter is installed it is a good idea to ask the following questions:

1. Is the filter chemically compatible?
2. Does the filter need pre-wetting before it can be exposed to the chemical?
3. What are the ionic extractables from the filter, and will they detrimentally impact the wafers?
4. Will the filter handle the viscosity of the chemical without clogging and creating a high pressure drop, which may ruin the pumps?
5. How often does the filter have to be replaced?

Once a point-of-use filter is installed, it will shed particles for a short period of time. Also, particle levels in a chemical do not decrease instantaneously after a recirculation pump has started to run the chemicals through the filter. Figure 11 illustrates the effects of various flow rates on the removal of particles from recirculating acid etch baths (28). As the recirculation flow rate was increased, the particle removal efficiency also increased. This figure illustrates that the particle level can drop from, say, 100,000 per liter to 400 per liter in 4 minutes at a recirculation rate of 53 liters/min. For viscous materials, including some buffered HF solutions (31), it may take more than an hour for the particle level to attain its lowest steady state value. When chemicals are supplied manually in bottles to such a recirculating tank, it is necessary to provide adequate time for particle levels to stabilize.

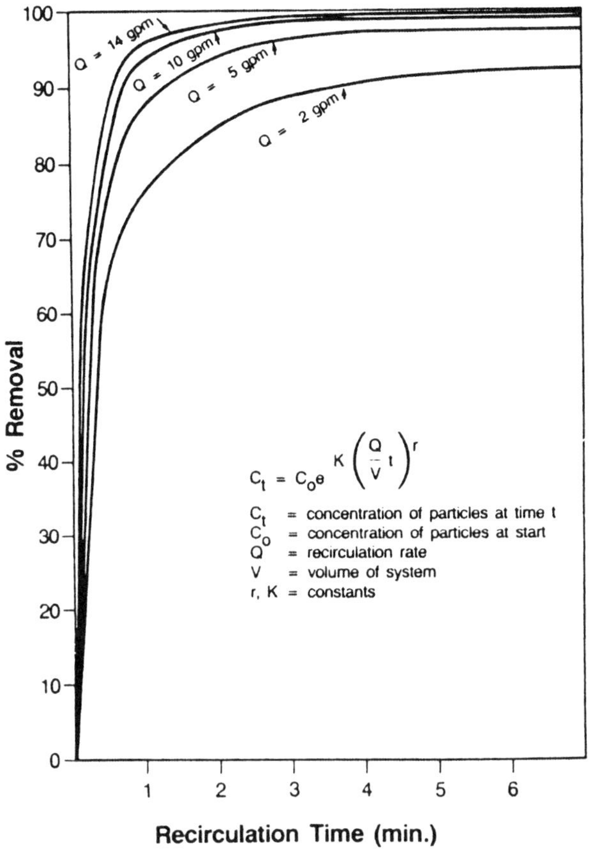

Figure 11. Particle removal vs. recirculation rates in acid etch baths (28). *(Reprinted with permission from the author.)*

3.5 Chemical Reprocessing

Reprocessing of used chemicals provides three benefits: *(i)* a reduction in waste disposal volumes and costs; *(ii)* a reduction in fresh chemicals costs; and *(iii)* improvements in chemical purity. Reprocessors are available for sulfuric acid, HF, isopropyl alcohol, and N-methyl pyrrolidone. Most of these reprocessors use distillation, in combination with filtration and ion exchange resins to remove contaminants from the chemical stream. The Athens, Inc. sulfuric acid reprocessor uses a vacuum distillation column to purify sulfuric acid after the water and residual hydrogen peroxide have been thermally eliminated. In addition, it generates peroxydisulfuric acid by oxidizing sulfuric acid in an electrochemical cell. This mixture of sulfuric acid and peroxydisulfuric acid is used for resist stripping in lieu of the standard "piranha" solution, which consists of sulfuric acid and hydrogen peroxide. Rosenfeld et al. (13) measured the mean average particle count in H_2SO_4 drums to be around 820/mL (>0.5 µm). The particle count exiting the reprocessor was less than 6/mL (>0.5 µm) and less than 0.007/mL (>1 µm). It is clear that reprocessing not only lowers operating cost by recycling used chemicals, but also increases chemical quality. The particle counts and metals levels are several-fold lower in the reprocessed material. A sulfuric acid reprocessor is also manufactured by Alameda Instruments. The acid output from this system is reported to contain less than 0.8 particles/mL (>0.5 µm) and less than 1.8/mL for particles greater than 0.3 µm (32). Most reprocessors can fit easily into a chemical distribution system.

Athens, Inc. recently introduced a HF reprocessor which uses beds of ion exchange resins to extract metallic contaminants and organics from diluted HF solutions (see Fig. 12). Impure HF is pumped through a coarse filter, an ion exchange system, a medium filter and into an ion-free HF tank. Then, the HF is recirculated through 0.1 micron filters to lower the final particle count. An on-line particle counter is used to determine number of recirculation loops required, and a conductivity sensor automatically measures the HF concentration and adds ultrapure water to make up for any losses. The particle levels on this system are reported to be around 10/mL for particles greater than 0.5 microns (33).

4.0 PARTICLE CONTROL DURING PROCESSING

4.1 Effect of Process Chemistry

Depending on the application, a variety of acids, alkalis, and solvents are used for cleaning wafers, the most common of which is the RCA clean. The RCA cleaning chemistry (34) is meant to remove organic residues, trace metals and ionic species, but not primarily particles. However, a 5:1:1 ratio of H_2O:NH_4OH:H_2O_2 (referred to as a standard clean 1, or SC-1), used primarily for removing light organics and many metals, also aids particle removal in a megasonic cleaner. A 5:1:1 ratio of H_2O:HCl:H_2O_2(SC-2) is used for removing metallic hydroxide contaminants and alkali metals, such as Na, Ca, Fe, Al, etc.

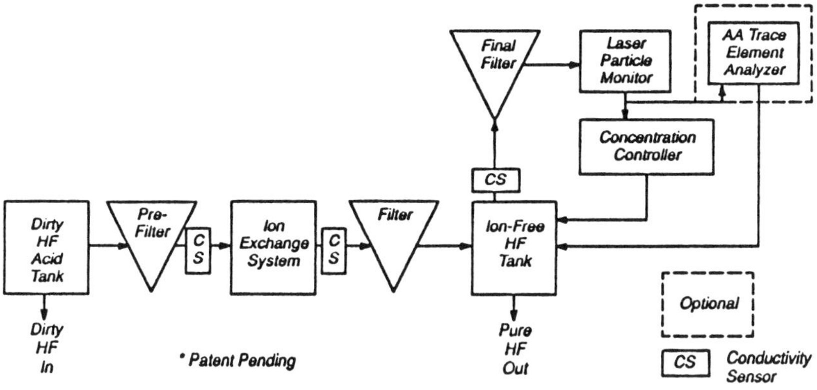

Figure 12. HF reprocessor block diagram (33). *(Reprinted with permission from* Proceedings of the Electrochemical Society, Inc., *P. 90, Fig.2, Vol. 90-9, 1990)*

Various compositions of buffered and non-buffered solutions of HF are used for stripping native oxide layers, and controlled oxide etching. Except for SC-1, these other cleaning chemistries do not remove particles from the wafer surface.

Menon et al. (31) conducted a study of particles added by various steps of an RCA clean. Fig. 13 shows the number of particles, in the size range of 0.2 - 0.5 micron, added during each step of an RCA cleaning sequence (SC-1, SC-2, buffered 10:1 HF), and for the entire clean (31). The SC-2 (standard clean 2) and buffered HF treatment added particles to the wafer, while the SC-1 (standard clean 1) treatment removed a few particles. The overall number of particles added to the wafer was very close to that added during the HF-last process. The addition of the particle contributions from the individual steps of the cleaning sequence does not equal the particles added for the entire clean. This implies that some of the particles added during one cleaning step are removed during the next step, possibly due to interfacial effects or chemical dissolution (18)(36).

Use of HF solutions has generally been known to add particles to wafers (36)(37). Stripping an oxide film from a silicon wafer renders the surface hydrophobic. When this wafer is immersed in the DI water rinse tank, the hydrophobicity causes organic particles in the DI water to be preferentially deposited on the wafer surface. Hence, particle levels on wet etched wafers generally correlate well with particle levels in the rinse tank, but not the HF bath (31). In the next section, various rinse tank configurations to minimize particle deposition are discussed.

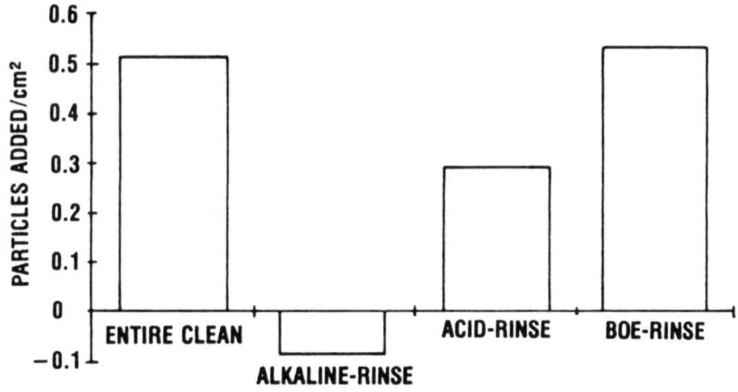

Figure 13. Particle deposition on wafers during a wet-chemical cleaning sequence (31).

The hydrophobic or hydrophilic nature of a surface can be quantified by the *contact angle* of a drop of water on that surface. A surface with a contact angle of zero degrees is completely hydrophilic, while a contact angle of 180° defines a completely hydrophobic surface. The contact angle of water with an HF-etched surface is approximately 66 to 85°, while that of buffered HF on a hydrophobic surface is around 70° (38)(39). The wettability of buffered HF is not significantly different from that of water. To improve the wettability of the etchant, Kikuyama et al. (38) recommend the use of hydrocarbon surfactants. Their studies have shown that the use of carefully selected surfactants can significantly reduce the particle contamination in wet etch processing. The application of surfactants in chemical solutions, especially in pre-gate oxide cleans, should be approached very carefully. Surfactant molecules are potential sources of carbonaceous residues. Aggregates of these molecules, called micelles, are in the submicron-size range and can represent an added source of particles. Also, many surfactants cause foaming, which can leave stains on the wafer surface and cause uneven etching.

The hydrophilic/hydrophobic nature of wafer surfaces after various chemical cleaning steps is noted in Table 6 (25). The SC-1 and SC-2 chemical treatments leave the wafer extremely hydrophilic. Generally, HCl solution has been easier to filter than NH_4OH, and particle specifications for bottled HCl have been lower than those for bottled NH_4OH. Yet, Fig. 13 shows that SC-2 adds particles to a wafer surface, while SC-1 actually removes a few particles. Both SC-1 and SC-2 leave the wafer hydrophilic, and the surface tensions of both chemical solutions are close (40), hence, the differences in particle removal behavior must be attributed to chemical interactions at the liquid-wafer interface. Kern (41) reports that SC-1 etches SiO_2 at a rate of about 0.5 Å/min, whereas SC-2 shows very little change in oxide thickness. The small amount of etching in the alkaline SC-1 solution probably undercuts oxide beneath particles, allowing them to be more easily dislodged from the wafer. Also, Niida et al. (42) have shown that the zeta potential of particles in alkaline solutions is significantly more negative than in acidic solutions, as was illustrated in Fig. 11 of Ch. 4 by Donovan and Menon. In SC-2 solutions, the zeta potential of most particles is positive. The zeta potential of the silicon surface is typically negative in most solutions (isoelectric points of silicon and silicon oxide surface are 1.5 - 4.0). Hence, particles tend to be attracted to the wafer surface in SC-2, while they are repelled in SC-1 solutions (42).

Table 6. Effect of cleaning chemistry on contact angle (25). *(Reprinted from* Microcontamination, *©1990, Canon Communications Inc., Santa Monica, CA)*

System	Contact Angle Degree ± Standard Deviation
Bare wafer after cleaning with:	
1. DI water	16 ± 1.3
2. $5H_2O{:}1H_2O_2{:}1NH_4OH$*	10 ± 0.8
3. $5H_2O{:}1H_2O_2{:}1HCl$	9 ± 0.5
4. choline-peroxide solution	16 ± 0.5
5. $10H_2O$:1 buffered HF solution	66 ± 2.1

* All cleans were followed by a DI water rinse which preceded the measurement of contact angle using a goniometer.

Mishima et al. (43) have investigated the effect of different concentrations of NH_4OH in an ammonium-hydroxide/water/hydrogen-peroxide mixture on particle removal. They recommend that the SC-1 solution should have lower NH_4OH content (0.5 to 0.05 times that conventionally used) to improve the particle removal capability by a factor of two, without any increase in surface roughness. Ohmi et al. (44) and Meuris et al. (45) have also studied the effect of ammonium hydroxide concentration on surface roughness and particle removal capability. Ohmi recommends a 0.05:1:5 SC-1 as the optimum mixture based on particle removal and etch rate studies. The optimum ammonium hydroxide concentration was correlated to the solution etch rate at which particle removal efficiency was highest. The roughness at 0.05:1:5 (0.2 nm Ra) was half that at 1:1:5. This surface roughness increase was considered to be the cause of a lowering of gate dielectric breakdown charge (Q_{bd}) (see Fig. 14). Meuris et al. (45) also showed that increase in surface roughness caused by an SC-1 solution can directly cause a decrease in capacitor yield. They recommend a 0.25:1:5 mixture of ammonium hydroxide, hydrogen peroxide and water for good particle removal without significant surface roughness. The differences in optimum ammonium hydroxide concentration between Ohmi's and Meuris et al's studies have been reported by Meuris to be due to a difference in solution temperatures. Ohmi's studies were conducted at temperatures around 80°C, while Meuris's work was at 70°C.

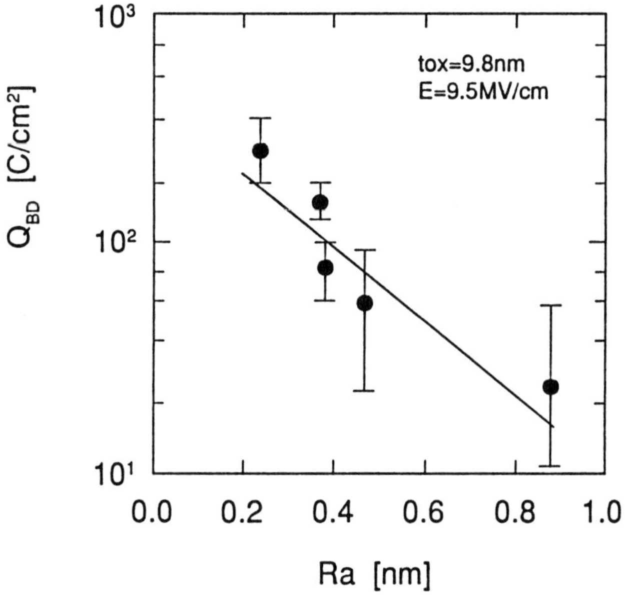

Figure 14. Effect of surface roughness on Q_{bd} (44).

In a recent study of wafer cleaning chemistries and their influence on oxidation-induced stacking faults, Hariri et al. (46) conclude that choline/surfactant solutions may be more effective than RCA solutions in removing heavy metals from the wafer surface. The impact of such a chemistry on removal of organic species and particles has not been evaluated.

To control particle deposition during cleaning, the following should be considered:

1. What is the contaminant to be removed?
2. What are the requirements for the process after the cleaning step, especially in terms of particle levels, surface roughness, and residual molecular species?
3. Is a native-oxide free surface absolutely required?

If metallic contaminants on wafers are not a problem, then the SC-2 cleaning operation can be excluded from the cleaning sequence. Historically, SC-2 has been used to remove alkali metals and metal hydroxides, and because the NH_4OH, and H_2O_2 used in SC-1 added certain metallic contaminants to the wafer. With the availability of purer grades of NH_4OH and non-stabilized H_2O_2, the need for SC-2 should be revisited. Elimination of SC-2, removes one potential source of particles.

If a native oxide layer on the wafer surface, such as one grown with H_2O_2, is acceptable for the next deposition operation, then the HF etch could be conducted before the SC-1 operation. HF-last operations are the biggest source of particles. There is no single cleaning sequence that works best for all applications. A designed experiment in which particle level on the wafer is an output variable, is recommended to optimize the cleaning process. For cleaning operations prior to the deposition of thermal oxides and nitrides, an HF, SC-1, SC-2 sequence has yielded good results with average particle counts below 10 per 150 mm wafer for particle sizes greater than 0.2 microns.

4.2 Process System Configuration

Wet-chemical cleaning systems are either immersion-based, or spray-based. Immersion wet stations consist of one or more chemical baths, rinse tanks, and a dryer. A wafer cassette is first immersed in the chemical bath for a specified length of time, then it is transported manually, or by a robot, to the rinse tank, and finally to a dryer. In a spray processor, a cassette of wafers is placed in a bowl, which is rotated at a certain speed while chemicals are sprayed on the wafers. After the chemical spray is stopped, the wafers are rinsed with a DI water spray, and then spun dry in a nitrogen ambient. There are several major differences between the two approaches to cleaning:

Immersion baths (Fig. 15) reuse chemicals for several batches of wafers, thus building up contamination with time (47). Recirculation with point-of-use filtration is essential to keep the particle level in the bath to an acceptable level. The section on point-of-use filtration discusses the options for immersion baths. For baths that use hydrogen peroxide, e.g., SC-1, SC-2, or piranha, the hydrogen peroxide decomposes rather rapidly. Hence, the bath needs to be replenished with fresh H_2O_2 before a new batch of wafers is processed. Typical bath life for SC-1, SC-2 mixtures operating at temperatures around 70°C is about half an hour. Thus, chemical usage costs can be very high. From a particle control viewpoint, the time allowed for recirculation filtration between wafer loads, and at the start of a fresh batch of chemicals is very important. If adequate time is not allowed for the particle level in the bath to decrease by filtration, gradual build-up of particles will rapidly cause the bath to be ineffective. Wet immersion stations are sold by Santa Clara Plastics, Submicron Systems, Semifab, Sankhyo Engineering, Dai Nippon Screen, etc.

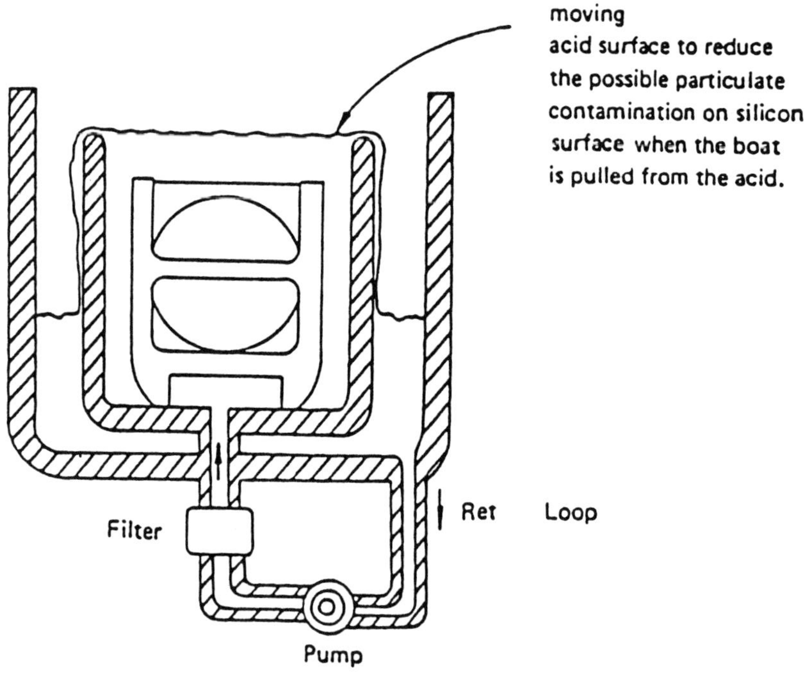

Figure 15. Schematic of recirculation immersion tank (47). *(Reprinted from* Particle Control for Semiconductor Mfg., *p. 408, Fig. 23-1, by courtesy of Marcel Dekker, Inc., 1990)*

A new form of immersion wet station is manufactured by CFM Technologies Inc. (48). In this system, a cassette of wafers is kept stationary in an enclosed chamber during the cleaning, rinsing and drying operations

(Fig. 16). After wafer loading, the chamber is closed and filled with hot cleaning solutions such as SC-1 or SC-2 until the vessel is completely filled. After cleaning, the solution is displaced by rinse water entering from the bottom, followed by IPA for drying. Each solution is introduced such that it displaces the previous solution in the chamber. This prevents the wafers from being exposed to the air/liquid interface during the cleaning sequence, thus eliminating particle problems due to wafer transport across an air/liquid interface. Figure 17 shows that particles were removed from wafers after an HF etch and a rinse-dry when wafers with high initial counts were used, while very few particles were added when wafers with low initial count were cleaned (48). McConnell noted an interesting effect when wafers were repeatedly exposed to the HF-rinse/dry process. The particle count after the first etch does not increase. Not until after the second and subsequent rinses does the particle count increase. Even after a long second rinse (which one would believe makes the wafer hydrophilic), the increased particle deposition did not disappear. Leaving the wafers overnight in a box, to allow for the growth of a native oxide, did not diminish the particle addition (see Fig. 18). This behavior of the wafer surface may be due to the stable nature of the hydrogen terminated silicon surface.

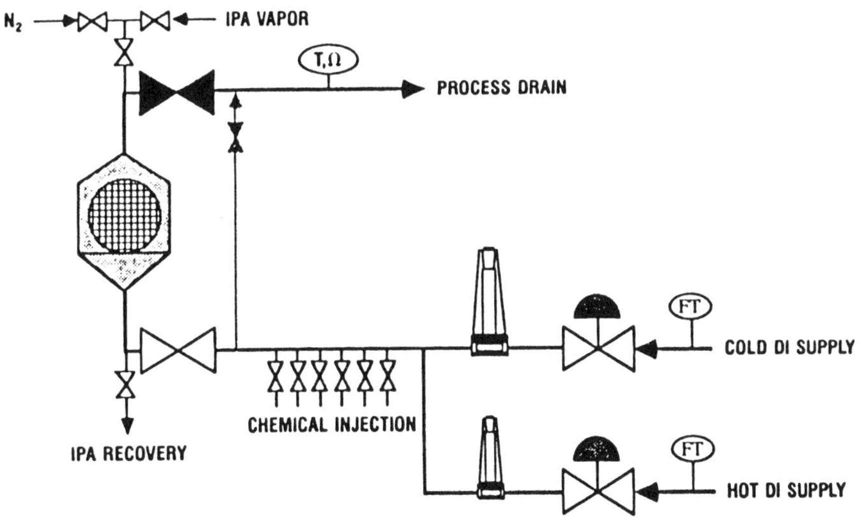

Figure 16. Schematic of the closed system apparatus (48). *(Reprinted with permission from* Microcontamination, *9(2):35-40, Fig. 1, 1991)*

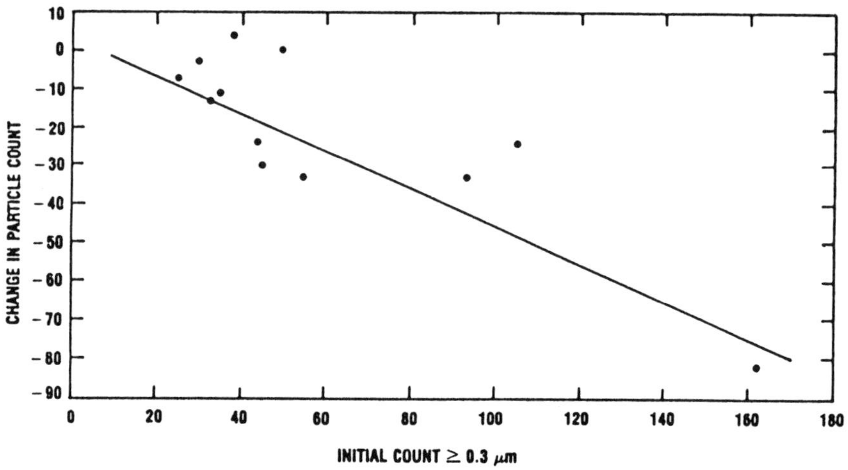

Figure 17. Particle removal from silicon with 40:1 H_2O HF treatment (48). *(Reprinted with permission from* Microcontamination, *9(2):35-40, Fig. 5, 1991)*

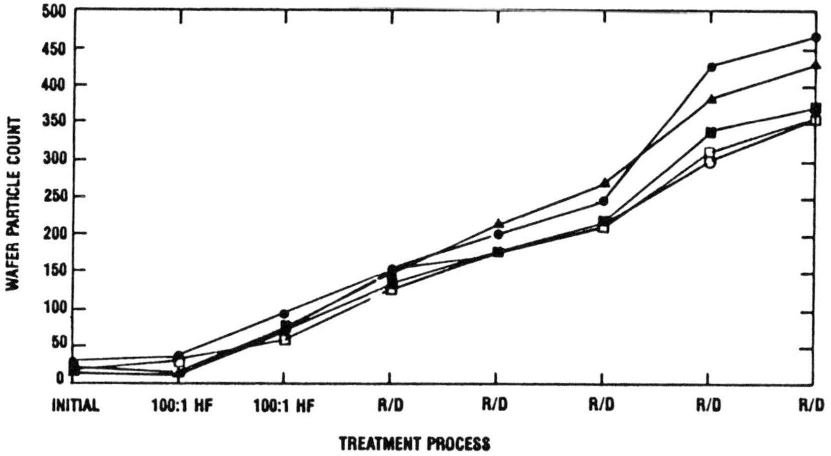

Figure 18. Example of direct-displacement technique, after wafers were held in storage box overnight (49). *(Reprinted from* Microcontamination, *©1990, Canon Communications Inc., Santa Monica, CA)*

Spray processors use fresh chemicals for every batch, thus providing purer chemicals for cleaning. Figure 19 depicts a schematic of a spray acid processor (37). Cassettes of wafers rotate past a central spray post which dispenses a fine mist of chemicals. Rinsing and drying are conducted in the same chamber. However, spray processors are maintenance intensive (37) because of a large number of rotating parts. In comparison to manual immersion wet stations, spray processors do not require operator intervention between the cleaning, rinsing and drying steps. This reduces human error and does not expose wafers to cleanroom contaminants. In the past, spray processors have added a larger number of particles to wafers than their equivalent immersion baths. This was due to particle generation from nozzles, seals, etc., and due to the phenomenon of *water spots*. Water spots are caused by inadequate rinsing and drying of the wafer leaving small circular spots of residues. Spray rinsing is prone to water spotting. Newer spray processors have attempted to eliminate the particle problem by using more effective spray posts, and chemically compatible nozzles and seals. Spray processors are made by FSI International (Fig. 20), Semitool Inc., Dai Nippon Screen, etc.

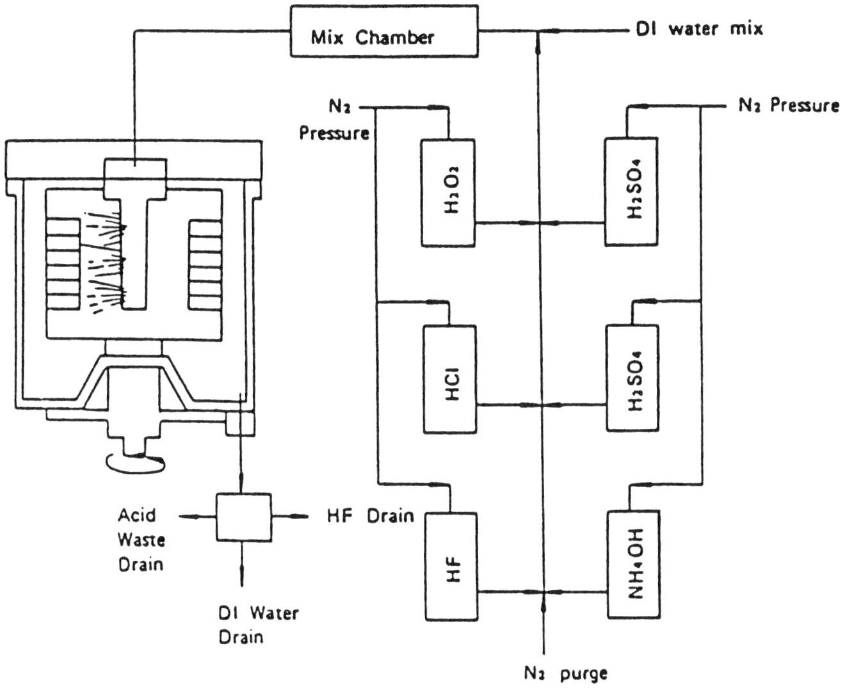

Figure 19. Spray acid processing system (37). *(Reprinted from* Particle Control for Semiconductor Mfg., *p. 203, by courtesy of Marcel Dekker, Inc., 1990)*

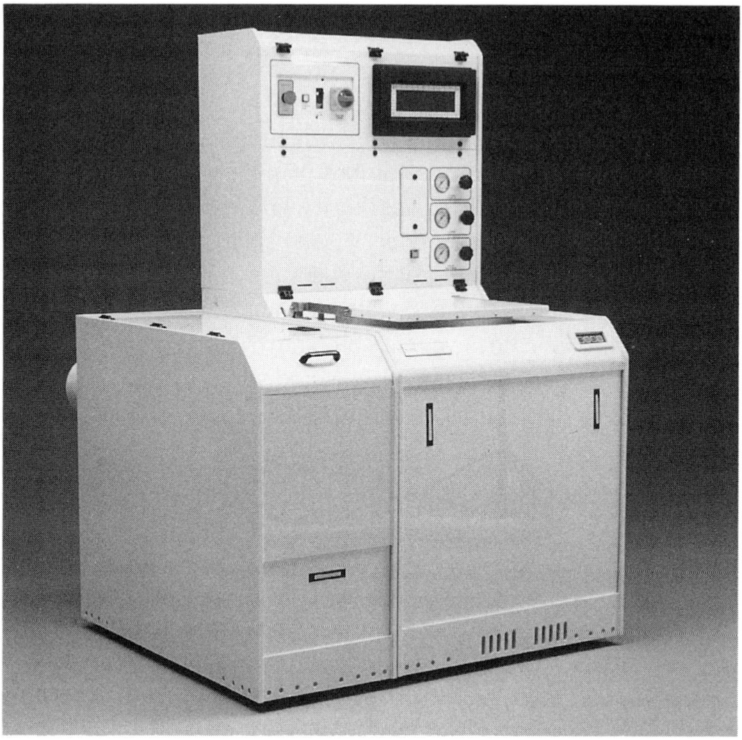

Figure 20. FSI Mercury OC Acid Processor. *(Source: FSI International)*

4.3 Rinsing and Drying

Major sources of particulate contamination during wet-chemical cleaning are often in the rinse tank and spin dryer, and not in the chemical bath. Careful selection of rinse tank and dryer configuration is therefore critical. The most commonly used rinse tank mode is the "quick dump with top spray." Wafers are exposed to a spray of DI water which fills the rinse tank. When the tank is full, the water is dumped from a valve at the bottom. This cycle is repeated several times. The advantage of this configuration is that it removes chemicals from the wafer surface very rapidly, periodically dumping the water containing the chemicals. However, spray nozzles tend to generate particles and grow bacteria. Also, quick dump rinse (QDR) tanks generally produce turbulent convective currents in the water, which increase particle mobility to the wafer surface. Hence, the QDR is a poor choice from a particle control point of view.

Particle control in the rinse tank becomes especially critical for wet etch processes using HF chemistry. Hydrophobic wafers exiting from the HF bath and entering the rinse tank are very susceptible to particle deposition. Figure 21 depicts a comparison of various rinse tank configurations for cleaning etched hydrophobic wafers (50). The cascade overflow configuration adds the least number of particles to the wafer surface. There is a steady flow of water from the bottom of the rinse tank to the top which accomplishes the rinse. There are no sprays. Rinse tanks with megasonic transducers also perform well.

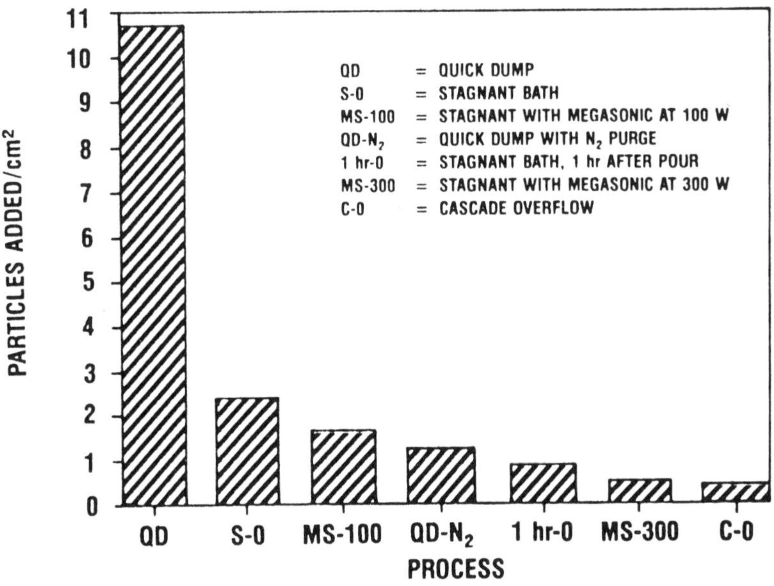

Figure 21. Effect of rinse tank configuration on particulate contamination on etched wafers (particle size: 0.2 - 0.5 µm; initial particle count <1.2/cm^2) (50). *(Reprinted from* Microcontamination, *©1989, Canon Communications Inc., Santa Monica, CA)*

At the end of a rinsing operation, a hydrophilic wafer exiting the rinse tank carries with it a thin film of water of about 20 microns in thickness. The boundary layer thickness during wafer pullout, on the other hand, is around 2500 microns (51). This implies the carryover film is entirely within the boundary layer. Transport of residual ions or particles is slower within the boundary layer than in the bulk rinsing medium. When the boundary layer is much larger than the carryover thickness, there will be ionic and particulate species that are not rinsed out even after a long rinsing period. Tonti (51) estimates that this boundary layer, where particles get trapped and subsequently deposit on wafers, can be reduced to below 1 micron by using megasonic or ultrasonic energy in the rinse tank. Hence, megasonic rinsing may have potential for effective rinsing especially for hydrophobic wafers.

Routine correlation of particle counts in the DI water with overall water quality is useful in determining the cause of particle excursions. For HF-etched wafers, variations in the TOC level in the rinse tank can cause major changes in the number of particles added to wafers.

It is common practice in the semiconductor industry to expose wafers leaving the rinse tank to another level of rinsing in the spin dryer. A spin dryer contains moving parts, spray nozzles, and various kinds of gaskets and seals. These components are sources of particles. Conducting a centrifugal rinse in the dryer results in the generation of micrometer and sub-micrometer size water droplets which may not be effectively removed in the dry cycle. A dry-only mode is, by far, a lower particle adder than a rinse-and-dry mode. As a general practice, it is recommended that the rinsing process be accomplished in the rinse tank, and not the spin dryer. An efficient dryer in the dry-only mode can add fewer than 1 particle ($\geq$0.2 μm) on a 150 mm diameter wafer.

Viscous chemicals, such as sulfuric acid and phosphoric acid, are not effectively removed from the wafer surface in a cold cascade overflow rinse. For such systems, the use of hot DI water or a good DI water spray has generally proven to be effective. While a hot DI cascade rinse is more expensive to install (due to heater costs) than a QDR with spray nozzles, it adds fewer particles.

Isopropyl alcohol (IPA) vapor dryers were introduced to achieve particle-free wafer drying. In IPA dryers, rinsed wafers are exposed to hot IPA vapors, whereby the water is displaced by the IPA, which then evaporates from the wafer surface (Fig. 22). These dryers have fewer moving parts and tend to have fewer particle problems than centrifugal spin dryers. Ohmi et al. (52) report that the water content in the IPA, the IPA heating system, and the IPA vapor velocity are major variables affecting the performance of a vapor dryer. So far, the number of particles by use of an IPA vapor dryer has not been reported to be significantly lower than that added by a well-functioning spin dryer (53). However, as particles smaller than 0.1 microns become yield detractors, IPA dryers will pose a distinct advantage over spin dryers because of the absence of mechanical motion and its associated vibration. IPA dryers are now becoming more common than spin dryers, especially in newer fabs.

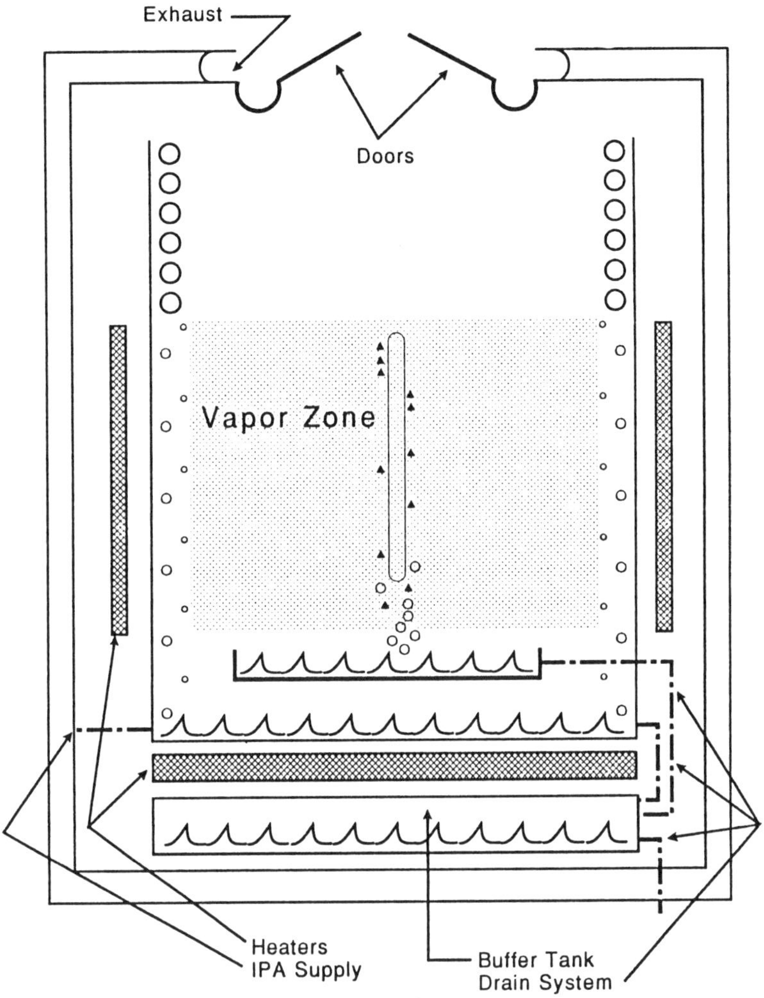

Figure 22. IPA Vapor Dryer (Process illustration - not to scale).

The CFM, Inc. system (48)(49) uses a different approach for IPA drying. This is referred to as IPA direct-displacement drying. After the wafer is cleaned and rinsed by the process described in a previous section, the drying process is initiated by shutting off the DI water supply and drain, to isolate the chamber. Subsequently a valve is opened to allow the rinse water to drain at a specific rate. At the same rate, hot filtered IPA is allowed to enter the chamber from the top to replace the void left by the descending water level. IPA forms a layer on top of the wafer surface. After the entire wafer load is exposed to IPA, it is drained, while a nitrogen purge removes any residual vapors. This approach is claimed to remove particles from the wafer surface more effectively than is possible with conventional IPA dryers.

A new drying technique, referred to as "Marangoni drying," has been proposed by researchers at Phillips Research Laboratories (53)(54). As the wafers are removed from the rinse tank at speeds of 1 - 2 mm/sec, the air/water/silicon interface is exposed to the vapors of a water-soluble organic liquid. This liquid is absorbed at the interface and induces a surface tension gradient within the curvature of the interface, causing the water meniscus to contract. This effect allows the water to sheet off a smooth hydrophilic wafer surface (see Fig. 23). The technique apparently leaves a hydrophilic wafer surface drier than a spin drying operation, as measured by the amount of water evaporated from the surface after the completion of the drying operation. Vapors of diacetone alcohol, 1-methoxy-2-propanol, or isopropyl alcohol (56) have been found to be suitable for Marangoni drying. This system is not commercially available at the present time.

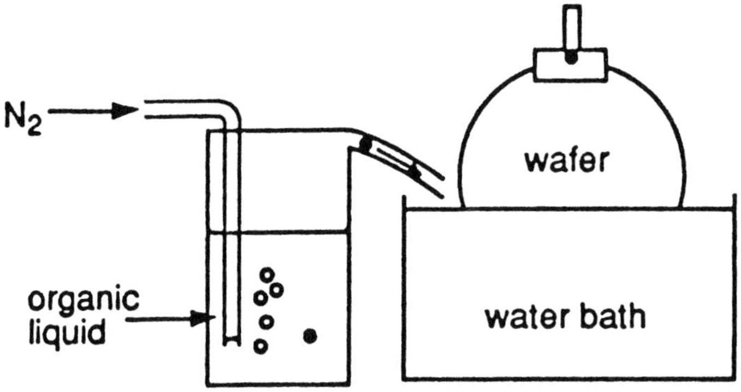

Figure 23. Marangoni Drying (54). *(Reprinted with permission from the author.)*

4.4 Gas/Vapor Phase Cleaning

It is not the intent of this chapter to review the technology of gas phase cleaning, but rather to discuss particulate contamination issues associated with this cleaning approach. The FSI Excalibur® uses anhydrous HF, while the Advantage Production Technology, Inc. (APTI) and Semitool units use HF/water-vapor mixtures to etch oxides. APTI was recently acquired by

Genus, Inc. For native oxide etching, data from all wafer cleaning systems (54)(57) reveal that they are essentially particle neutral. It is important to remember that the commercially available gas/vapor phase cleaning systems are not meant to remove particles. The best we can expect is to minimize particle addition, and in some cases, the removal of oxidic particles due to etching. Particle control in these tools is primarily accomplished through the use of ultrapure gases, point-of-use gas filtration, and the design of reactor chambers that minimizes flaking and debris formation. Minimizing particle generation due to the process itself is also important.

Both anhydrous HF gas and vapor phase HF-water mixtures react with surface oxides to form fluorides. These fluorides may manifest themselves as particulate residues on the wafer surface. Fluoride residues may be detrimental to the subsequent wafer processing operation. Their removal requires a DI water rinse or some type of vapor desorption technique.

5.0 POST-PROCESSING PARTICLE REMOVAL TECHNOLOGIES

Diligent process optimization notwithstanding, in several process steps, equipment adds particles to the wafer surface. There are certain applications like laser scribing and mechanical planarization which are known to be major particle adders. These particles have to be removed before subsequent processing can commence. As shown in Tables 7 and 8, several techniques are available for the removal of particles from wafer surfaces (58). No matter what the approach, the applied force must overcome the forces of adhesion to dislodge particles, or some fraction thereof (see Donovan and Menon, Ch. 4). That is:

$$F_{ad} / F_r < 1$$

where: F_{ad} = adhesive force between particles and surface
F_r = removal force applied to particle

Particle removal can be accomplished by decreasing F_{ad}, increasing F_r, or both. The force of adhesion due to van der Waals and capillary forces is proportional to particle diameter (d), while the removal forces (i.e., gravity, drag) are proportional to multiple powers of the particle diameter.

$$F_{ad} \propto d \qquad \text{(van der Waals, capillary, etc.)}$$

$$F_r \propto d^3 \qquad \text{(gravity)}$$

$$F_r \propto d^2 \qquad \text{(drag)}$$

Hence,

$$F_{ad}/F_r \propto d^{-2} \text{ or } d^{-1}$$

As the particle diameter decreases, the removal forces typically decrease much more than the adhesion force. This implies that smaller particles are harder to remove.

F_{ad} can be reduced by selecting a cleaning medium that lowers the Hamaker constant (and hence the van der Waals force of adhesion) of the system. The Hamaker constant of most systems in liquids is approximately 3 to 5 times lower than that in air or gases. This is one of the major reasons for the choice of liquids as cleaning solvents. Another approach is to select fluids that increase the electrostatic repulsion between particles and wafers. If the zeta potential of particles and the wafer surface is of the same sign, they will repel each other. Riley and Carbonell (18) and Ohmi et al. (59) have measured the zeta potential of various particles in liquids of different pH. The zeta potential is positive in acidic solutions and becomes negative in alkaline solutions. The silicon surface carries typically a negative zeta potential over a wide pH range. Hence, particle removal is favored in alkaline solutions. Based on these observations, it is not surprising that the SC-1 solution removes particles from wafers, while SC-2 adds particles!

Table 7. Mechanism of action of common wafer cleaning techniques (58). *(Reprinted from* Particle Control for Semiconductor Mfg., *p. 371, by courtesy of Marcel Dekker, Inc., 1990)*

Technique	Cleaning Action
RCA method and its variations	Chemical dissolution, electrostatic repulsion
Scrubbing	Mechanical force, often combined with hydrodynamic drag force
Pressurized fluid jets	Hydrodynamic and/or centrifugal drag force
Ultrasonics	Vibrational force
Megasonics	Vibrational force
UV/ozone	Chemical decomposition

Table 8. Advantages and disadvantages of common wafer cleaning techniques (58). *(Reprinted from* Particle Control for Semiconductor Mfg., *p. 372, by courtesy of Marcel Dekker, Inc., 1990)*

Cleaning Method	Advantages	Disadvantages
Wet chemical cleaning	• Removes metal ions and soluble impurities	• Not good for particle removal • Requires wafer rinsing and drying • Known to add particles to wafer
Scrubbing	• Removes large particles (>1μm) effectively • Suitable for cleaning hydrophobic wafer surfaces	• Unsuitable for removing submicrometer particles from patterned wafers • Requires regular maintenance • Could damage wafers
Pressurized fluid jets	• Removes small particles from patterned wafers	• Possibility of wafer damage • Static buildup
Ultrasonics	• Removes small particles effectively	• Wafer damage possible • Control of cavitation intensity difficult • Difficult to use reactive solutions
Megasonics	• Removes both soluble and insoluble contaminants • Removes small particles effectively • Can be used with chemical cleaning solutions • Less likelihood of wafer damage	• Contamination from chemical solutions possible • Wafer rinsing and drying required
Strippable polymer	• Capable of removing both soluble and insoluble impurities • Convenient for wafer storage and transport • Dry technique	• Possibility of depositing polymeric residue • Effectiveness not demonstrated adequately
UV/ozone	• Removes organic impurities • Dry technique	• Not proven for particle removal • Possibility of wafer damage

5.1 Brush Scrubbing

Brush scrubbers are commonly found in silicon wafer slicing and sawing applications. Large particles are hydrodynamically dislodged from the surface with a rotating brush made of a hydrophilic material (e.g., Nylon®), while water (with a surfactant such as Triton X-100®) or a solvent are sprayed on the wafer. The hydrophilic brush ensures that a cushioning film of scrubbing solution always remains between the wafer and the brush bristles. This is important because the contact of the bristles with the wafer surface could cause scratching and other damage. Brush scrubbers can clean one side (single sided) or both sides (double sided) of a wafer surface.

In addition to use in silicon slicing operations brush scrubbers can be used during device fabrication, for example, after chemical-mechanical planarization (CMP). CMP may be carried out at several points during the manufacture of a device to provide a smooth, flat surface for subsequent lithographic operations. Typically it involves the polishing of a thick insulator layer by use of a caustic slurry of colloidal particles. After the polish, the particles from the slurry that have deposited on the wafer surface can be removed by brush scrubbing with or without a surfactant solution.

The design and maintenance of brush scrubbers is often considered an "art" because of the skill required to prevent wafer damage and optimize cleaning efficiency. A poorly adjusted scrubber can do a significant amount of damage. A major disadvantage of these scrubbers is that they do not work well with patterned wafers. The brushes that are several millimeters in diameter ride over a micrometer feature on a wafer, missing debris lodged between features. As brushes wear with use, they tend to shower the wafer with brush fragments and particles collected from prior cleaning operations.

5.2 Hydrodynamic Cleaning

Low (<100 psi) and high (>1000 psi) pressure liquid jets have been used for cleaning surfaces for a very long time. While they are still used in the aerospace and precision machining industry for cleaning large surfaces and metallic parts, they are seldom found in a wafer fabrication facility. The liquid jet, moving over a particle on a flat surface, imparts a drag force F on the particle which is given by:

$$F = C\rho AV^2/2$$

where C is the drag coefficient, A is the particle frontal area, ρ is the particle density, and V the relative fluid-particle velocity. This force must overcome the forces of adhesion. The force of particle removal is proportional to fluid density, fluid velocity, and particle size. Liquid jets are preferred to gas jets owing to the higher density of liquids. Under comparable conditions, Grant et al. (60) report that the drag force on a particle in water is approximately 55 times that in air.

As a fluid moves over a surface, a boundary layer is created wherein the velocity at the surface is zero, whereas it is equal to the bulk velocity at a distance sufficiently far from the surface. Figure 24 illustrates the flow profile of a laminar liquid jet (61). As particle size increases, the velocity, and hence the drag force, increases. Thus, hydrodynamic cleaning works better for large particles. Pressures as high as 3,000 psi may be required for removing submicrometer particles from a wafer. At such high pressures, the wafer could suffer substantial damage. The boundary layer can also be decreased by using low viscosity fluids.

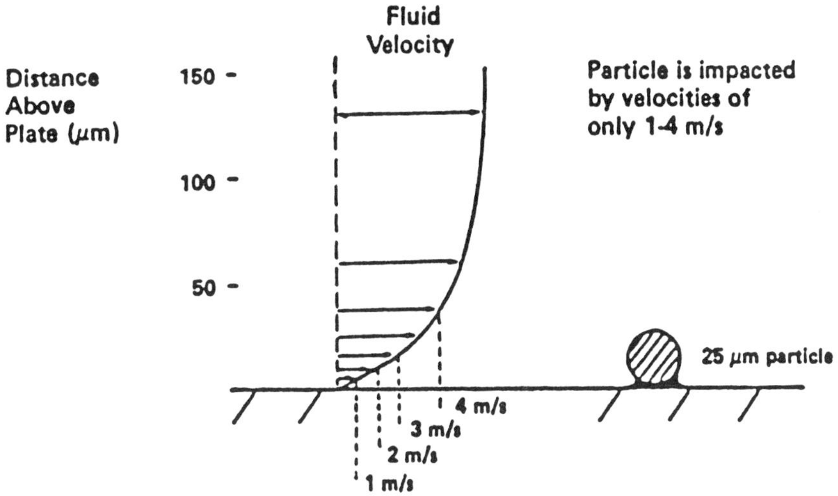

Figure 24. Effect of boundary layer on flow profile (61). *(Reprinted with permission from the* Journal of the Institute of Environmental Sciences, *p. 53, Jan/Feb. 1987)*

Wafer cleaning applications with low pressure water jets are brush scrubbing and spray rinsing. During brush scrubbing, water jets with pressures less than 35 psi are employed to assist in lubricating the region between the brush and the wafer, and to remove dislodged debris from the wafer surface. In spin rinsers, nozzles direct jets of water onto the spinning surface, thus combining hydrodynamic drag with centrifugal force to remove particles. For more details on this subject, see Menon (58) and Nagarajan and Welker (62).

5.3 Ultrasonic Cleaning

Ultrasonics and megasonics (discussed in the next section) are the most commonly used particle removal techniques for silicon wafer cleaning. In ultrasonic cleaning, sonic energy in the range of 20 - 40 kHz is applied to a liquid within which the wafers are immersed. The force required to remove a particle in a sonic field, in the absence of any cavitation is:

$$F_r = ma$$

where m is the mass of the particle and $a = 4\pi^2 f^2 A_m$. A_m is the amplitude and f the frequency of the sonic vibration.

In most commercial ultrasonic cleaners, it is this force, combined with the force created by cavitation that removes particles. Cavitation is the rapid formation and explosion of tiny gas bubbles in the liquid due to the pressure waves generated by the vibrating transducer. During this process a great deal of energy is generated. Particle removal can be accomplished without surface damage when cavitation occurs uniformly across the wafer surface and its density is controlled. However, in many commercial systems cavitation occurs non-uniformly, often at certain specific sites on the wafer, leading to pitting and damage.

An ultrasonic cleaner made by the J. M. Ney Company (63) reportedly provides the ability to control cavitation by modulating both amplitude and frequency while sweeping through a programmed bandwidth of frequencies. Niemczewski (64) compared the cavitation intensity in various liquids to that in water. Table 9 shows that water is more active than most of the Freon®-based cleaning solvents. Busnaina et al. (65) have studied the effect of various parameters such as temperature, frequency, sweep time, etc. on cleaning efficiency. Menon et al. (66) reported on removal efficiencies for various types of particles on bare and oxidized wafers. Their studies correlated with the cavitation intensities, the greatest removal efficiencies being observed for water, and the least for Freon-TF®.

Table 9. Comparison of cavitation intensity in various liquids (64). *(Reprinted from* Particle Control for Semiconductor Mfg., *p. 396, by courtesy of Marcel Dekker, Inc., 1990)*

Liquid	Maximum cavitation intensity (%)
Water	100
Toluene	71
Methanol	52
Ethanol-acetone	45
Freon-TMS	17
Freon-TF	15

5.4 Megasonic Cleaning

The use of megasonic energy in a tank containing the SC-1 solution greatly enhances the particle removal capability of this solution. Commercial megasonic wafer cleaning systems typically operate at a frequency of 700 - 1500 kHz. The megasonic transducers are piezoelectric crystals, which are usually mounted at the bottom of the chemical tank. The sonic pressure waves travel through the liquid in a direction parallel to the wafer surface. Wafer cleaning is accomplished by the impact of these pressure waves on the particle surface.

The following paragraphs are excerpted with permission of the Electrochemical Society, Inc. from a previous paper by Menon and Donovan (67):

> The force required to remove a particle from the wafer surface must equal, or exceed, the force of adhesion. This force is a function of particle size, particle and wafer surface composition, and the nature of the liquid medium (50). For a silica particle of 1 micron diameter (and mass = 5 x 10^{-13} g), adhered to a bare silicon surface, the force of adhesion in water is approximately 4 x 10^{-4}dynes (68). This is the van der Waals force of adhesion. The applied megasonic force acting on this particle is
>
> $$F_{meg} = \text{Mass x Acceleration.}$$
>
> The acceleration produced by a megasonic transducer vibrating at a total power of 300 watts is approximately 2.5 x 10^{8} cm/s^{2} (69). Thus, the megasonic force = 1.25 x 10^{-4} dynes. The megasonic force is approximately equal to the force of adhesion, hence, 1 micron diameter particles should be removed from silicon wafers in a megasonic tank containing water.

When SC-1 chemistry is used in lieu of water, the particle removal efficiency is further increased. Figure 26 shows a comparison of the megasonic cleaning efficiencies of DI water, SC-1 solution, and a choline-hydrogen peroxide solution (50), for particles of size greater than 0.5 microns. The use of SC-1 was found to consistently produce high cleaning efficiencies and relatively small variability between runs. The lowest cleaning efficiencies, and the largest variabilities were seen with DI water. The results for the choline-peroxide system (without surfactant) were intermediate to those of DI water and SC-1.

Megasonic cleaning is a very effective method for removing submicron particles from silicon wafers. However, its effectiveness at particle sizes below 0.2 microns has not been demonstrated. One possible approach to removing submicron particles from wafers is to combine the inherent advantages of chemical dissolution with megasonic energy. Chemical dissolution as a means to particle removal becomes more effective as the particle size decreases. Megasonic cleaning, on the other hand, generally becomes more effective as particle size increases. For example, wafers contaminated with metal particles are subjected to megasonic cleaning in a solution containing hydrochloric acid. The acid will dissolve the small particles, while megasonic energy will dislodge the large particles. Or, the acid treatment can precede or follow megasonic cleaning using a conventional SC-1 solution. For particles that are predominantly made up of SiO_2, a mild HF treatment before megasonic cleaning may be more effective than megasonic cleaning alone. In a recent study using polystyrene particles on silicon wafers, Rooney et al., (70) show that megasonic cleaning followed by a modified RCA clean produces higher cleaning efficiencies than megasonic cleaning alone.

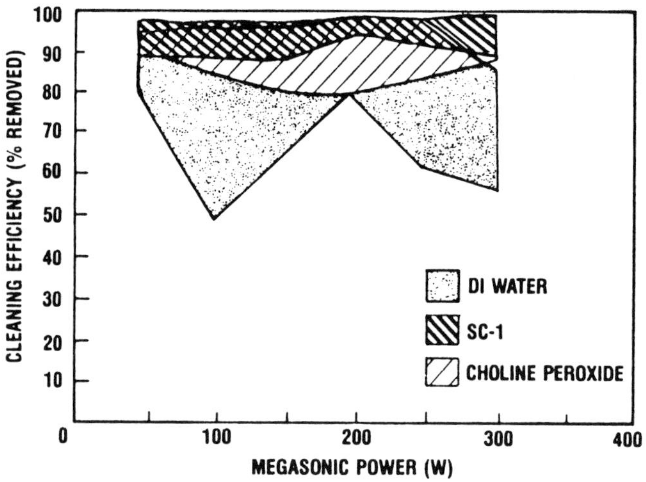

Figure 25. Variation of particle removal efficiency with megasonic power for different cleaning media (50). *(Reprinted from* Microcontamination, *©1989, Canon Communications Inc., Santa Monica, CA)*

Syverson et al. (71) used SC-1 *and* SC-2 in megasonic tanks to remove particles from various types of wafers. They concluded that the application of megasonic energy in both solutions provided better results than spray processing. They found better cleaning efficiencies when the wafers were moved slowly over the transducers at 150 watts per transducer with a bath temperature of 70°C. In a related study, Gow et al. (72) used hot SC-1 solutions with and without megasonic energy to evaluate the cleanability of wafers exiting from different processes. Table 10 compares the pre- and post-counts on wafers for wafers cleaned after they were removed from an oxide etch/strip, or nitride strip (phosphoric acid) operation. In all instances, the addition of megasonic energy improved the particle removal efficiency.

Table 10. Megasonic cleaning for various applications (72). *(Reprinted with permission from* Extended Abstracts, Fall Meeting, The Electrochemical Society, Inc., *Vol. 91-2, p. 821, Table 1, Pennington, NJ, 1991)*

	With Megasonic Energy		Without Megasonic Energy	
	Precount	Postcount	Precount	Postcount
BHF Strip	79	11	100	64
	117	11	87	46
	65	15	65	117
BHF Etch	27	7	87	79
	29	30	27	21
			49	43
Oxide	334	50	704	453
(Hot H_3PO_4)	250	39	500	202
	322	48		
Oxide/Nitride	1289	125	388	185
(Hot H_3PO_4)	818	97	495	129
	873	73	502	300

The megasonic wafer cleaner is an effective tool for removing particles, but it is often not realized that this system can also generate particles. Deteriorating seals and gaskets, and bad transducer bonding materials can shed particles when the transducers are operating. A worthwhile experiment would be to monitor, as a function of operating duration, the particles in a megasonic tank using DI water instead of SC-1.

If the particle count in the bath increases with time, the megasonic unit is obviously generating particles.

5.5 Other Techniques

Wafer scrubbing with CO_2 snow, ice pellets, laser energy, etc., are new approaches that are being explored for removing particles that are adhered to the wafer surface. In all instances, the objective is to provide the mechanical force required to overcome the forces of particle adhesion. The CO_2 snow (73) and ice scrubber (74) techniques work on the same principle. Small pellets of dry ice (CO_2) or water ice (H_2O) are generated and impinged on the contaminated surface using nozzles. Ohmori et al. (74) generated the ice particles (typically between 30 and 300 microns in diameter) by freezing water droplets in liquid nitrogen (see Fig. 26). These particles are then directed at the wafer surface at application pressures of around 3 kg/cm^2 for 30 seconds to 1 minute. Removal efficiencies for polystyrene particles have been reported to be comparable to those of megasonic systems (see Table 11). Both dry ice and ice scrubbing have been shown to be quite effective in removing organic films, such as oils and grease, from surfaces. They are used for cleaning parts in aerospace applications but have not been used in semiconductor processing because of the success of megasonic cleaners, and because of the concern that impinging particles could cause surface and subsurface wafer damage.

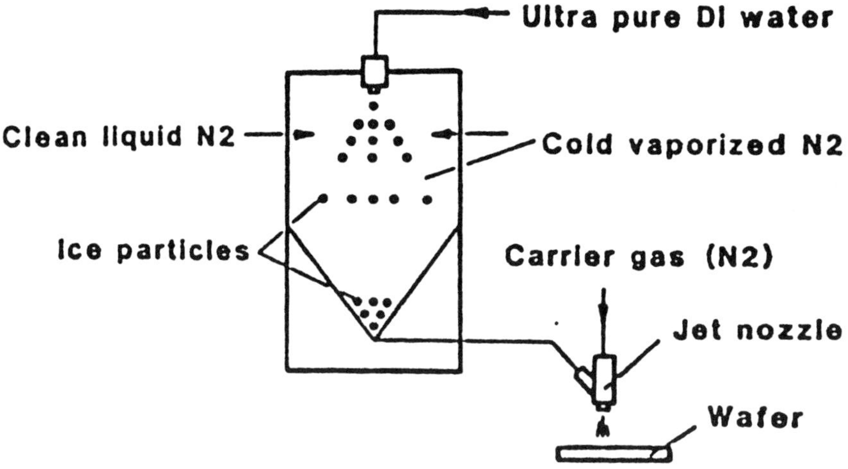

Figure 26. Principle of producing fine ice particles (74).

Table 11. Comparison of ice scrubber performance with other cleaning techniques (74). *(Reprinted with permission from* Technical Proceedings Semicon/Kansai Kyoto, *pp. 142-149, Kyoto, Japan, 1990)*

Cleaning Methods	Removal Rate (%)	Condition	Time
Ice Scrubber	97.6	Pressure: 3kg/cm^2 Angle: 80°	30 sec
	94.0	Pressure: 2kg/cm^2 Angle: 80°	30 sec
Megasonic	95.0	Frequency: 950 kHz	10 min
Brush Scrubber	87.4	Brush pressure: 0.8 kg/cm^2	40 sec
High Pressure Water	84.4	Water pressure: 100 kg/cm^2	40 sec
Ultrasonic	83.9	Frequency: 27 kHz	15 min
Dry-Ice Scrubber	68.9		30 sec

Recently, McDermott et al. (75) reported on the use of an argon aerosol jet for wafer surface cleaning. Gaseous argon and nitrogen were purified, cryogenically cooled, and expanded through a nozzle to create an aerosol. These aerosol particles (around 1 micron in diameter) were then directed at a contaminated wafer surface. They were able to remove submicrometer latex spheres, silicon debris and grease films from bare and patterned wafers. Except for photoresist, most other surfaces showed no obvious signs of damage on the basis of an SEM evaluation.

Laser-based particle removal is a novel approach proposed by Allen (74). A thin film of moisture is condensed on the wafer surface, where it penetrates the region between an adhered particle and the surface. Subsequently, the surface is "zapped" by a laser whose wavelength is tuned to the wavelength for explosive evaporation of water. The instantaneous evaporation of moisture dislodges the particles, which are then carried away in an air stream. Allen (76) reports removal efficiencies approaching 100% for 1 micron polystyrene latex particles on silicon wafers.

6.0 PARTICLE MONITORING PRACTICES

Successful particle control during wafer cleaning requires periodic monitoring of the performance of the cleaning equipment and processes. It is common practice to conduct particle-per-wafer-pass measurements of cleaning systems once a shift, or once a day. The measured particle counts are then used as part of a statistical process control (SPC) chart to make a decision whether the process is behaving "normally." If the particle counts are outside the bounds of SPC for a certain number of measurements, corrective action has to be taken.

For day-to-day operation, the particle per wafer pass measurement combined with routine data on chemical, water, and nitrogen quality is adequate to alert an operator when the process begins to go out of control. When the particle-per-wafer-pass measurement is not in SPC, the following is a list of process parameters that could be monitored to determine the cause of the problem. This list is for an immersion wet station (the parameters are not very different for spray processors or other designs of equipment):

1. Particle-per-wafer-pass for entire process, cleaning, rinsing, and drying.
2. Particle-per-wafer-pass for rinser and dryer. By comparing this result to #1, one can determine if the process bath is the source of the contamination. If the process bath is indeed the source of contamination, obtaining particle count data for the liquid chemicals is recommended. This can be done at the central distribution facility, or at the point of use with a particle counter for liquids.
3. Particle-per-wafer-pass for dryer. This identifies particle sources in the dryer, including possibly the nitrogen used for drying.
4. Particle-per-wafer-pass with and without recirculation of process bath. This determines if the filter is shedding particles.
5. TOC and particle count for water in the rinse tank. For hydrophobic wafers, the particle count increases with increasing TOC.
6. Discoloration of tubing, bath materials etc. This identifies if materials of construction are shedding particles, or reacting adversely with the chemicals. Metallic contamination in acid lines will gradually produce a greenish-brown color in PFA pipes.
7. Continuous drip of rinse tank nozzles for low flow. This prevents bacteria buildup in the nozzles.

8. Laminarity of airflow in and around wet bench, and adequacy of exhaust flow. Inadequate exhaust could cause cross contamination of vapors from adjacent baths, e.g., HCl and NH_4OH vapors could react to form NH_4Cl particles.
9. Integrity of O-rings, seals, etc. This is especially true for ozonized water systems where the ozone has been known to attack polymeric materials.

7.0 CONCLUSIONS

This chapter provides the fundamental information required for the selection and optimization of a contamination-free wafer cleaning system. Common particle measurement techniques for liquids, such as optical light scattering and wafer particle counting are reviewed, and a new approach using a nonvolatile residue monitor is discussed. Methods for controlling particles in chemicals through improved chemical production methods, filtration, chemical recirculation, and DI water quality control are presented. Particle levels in incoming liquids and at the point of use show that a 500-fold decrease in particle levels at the point of use can be achieved through optimization of chemical delivery systems.

This chapter also presents the factors affecting particle control during wafer cleaning. The effects of process chemistry, hardware configuration, and rinse and dry methodologies are reviewed with special emphasis on the hydrophobic/hydrophilic nature of the wafer surface. Finally, post-process particle removal methodologies such as ultrasonic and megasonic cleaning, brush scrubbing, and hydrodynamic cleaning are compared. Megasonic cleaning using an SC-1 solution appears to be the most popular particle removal approach for silicon wafer processing.

ACKNOWLEDGMENT

The assistance of Ms. Genie Kane in the preparation of this chapter is gratefully acknowledged.

REFERENCES

1. Knollenberg, R. G., and Veal, D. L., *Proceedings of the Institute of Environmental Sciences, Annual Technical Meeting,* pp. 751-771, Mount Prospect, Illinois (1991)
2. Grant, D. C., *J. Institute of Environmental Sciences,* pp. 32-37 (July/August, 1990)
3. Dillenbeck, K., *Microcontamination,* 5 (2):31-65 (1987)
4. Batchelder, J. S., and Taubenblatt, M. A., *Applied Phys. Lett.,* 55:215 (1989)
5. Taubenblatt, M. A. and Batchelder, J. S., *Applied Optics,* 30(33):4972-4979 (1991)
6. Blackford, D. B., Belling, K. J. and Sem, G. J., *J. Institute of Environmental Sciences,* pp. 43-47 (July/August, 1987)
7. Hill, E. A., Ensor, D. S., Lawless, P. A., and Donovan, R.P., *Proceedings of the Institute of Environmental Sciences, Annual Technical Meeting,* pp. 489-491, Nashville, Tennessee (1992)
8. Blackford, D. B., Sem, G. J., and Kerrick, T., *Proceedings of the International Conference of Particle Detection, Metrology and Control,* pp. 546-560, Arlington, Virginia (1990)
9. Blackford, D. B., Zarrin, F., Sem, G. J., and Kerrick, T., *Proceedings of the Microcontamination 91 Conference and Exposition,* pp. 39-51, San Jose, California (1991)
10. Tullis, B., in: *Particle Control for Semiconductor Manufacturing,* (R. P. Donovan, ed.), pp. 359-382, Marcel Dekker, New York (1990)
11. Johnson, R., "Surface Contamination Detection: An Introduction," Tencor Instruments, Inc., Company Brochure, p. 7, Mountain View, California (1990)
12. Turner, F. J., personal communication (1991)
13. Rosenfeld, E., DeSelms, B., and Menon, V. B., *Proceedings of the Microcontamination 90 Conference and Exposition,* Tutorial No. 107, pp. 225-258, San Jose, California (1990)
14. Menon, V., "Microcontamination Trends in the 1990's," Keynote Address at the Microcontamination '90 Conference and Exposition, San Jose, California (1990)
15. Matthews, R., *Proceedings of Semiconductor Pure Water and Chemicals Conference,* pp. 3-15, Santa Clara, California (1992)

16. Drab, G., Tichich, J., and Nichols, L., in: *Particle Control for Semiconductor Manufacturing,* (R. P. Donovan, ed.), pp. 183-202, Marcel Dekker, New York (1990)
17. Miki, N., and Ohmi, T., *Proceedings of the Semicon/East Technical Symposium,* pp. 1-7, SEMI, California (1989)
18. Riley, D., and Carbonell, R., *Proceedings of The Institute of Environmental Sciences, Annual Technical Meeting,* pp. 224-228, New Orleans, Louisiana (1990)
19. Jan, D., Ali, I., and Raghavan, S., *Proceedings of the Institute of Environmental Sciences, Annual Technical Meeting,* pp. 849-855, San Diego, California (1991)
20. Shadman, F., Governal, R., and Bonner, A., *Proceedings of Institute of Environmental Sciences, Annual Technical Meeting,* pp. 221-223, New Orleans, Louisiana (1990)
21. Governal, R., Bonner, A., and Shadman, F., *Proceedings of the Institute of Environmental Sciences, Annual Technical Meeting,* pp. 791-795, Mount Prospect, Illinois (1991)
22. Yabe, K., Motomura, Y., Ishikawa, H., Mizuniwa, T., and Ohmi, T., *Microcontamination,* 7(2):37-46 (1989)
23. Krussell, W. C., and Golland, D. I., in: *Semiconductor Cleaning Technology/1989,* (J. Ruzyllo and R. E. Novak, eds.), 90-9:23-32, The Electrochemical Society, Inc., Pennington, New Jersey (1989)
24. Hashimoto, S., Kaya, M., and Ohmi, T., *Microcontamination,* 7(6):25-106 (1989)
25. Menon, V. B. and Donovan, R. P., *Microcontamination,* 8(11):29-66 (1990)
26. Krygier, V., *Microcontamination,* 4(12):20-26 (1986)
27. Skidmore, K., *Semiconductor International,* 11(11):66-71 (1988)
28. Gotlinsky, B., *Microelectronics Manufacturing and Testing,* 10(13):1-5 (1987)
29. Grant, D. C., and Schmidt, W. R., *Proceedings of the Seventh Annual Millipore Microelectronics Technical Symposium,* pp. 1-27, San Jose, California (1989)
30. Gruver, R., Silverman, R., and Kehley, J., *Proceedings of the Institute of Environmental Sciences, Annual Technical Meeting,* pp. 312-315, New Orleans, Louisiana (1990)
31. Menon, V. B., Michaels, L. D., Clayton, A. C., and Donovan, R. P., *Solid State Technology,* 32(10):S9-S11 (1989)

32. Courson, D., Alameda Instruments, private communication (1991)

33. Davison, J., Hsu, C., Trautman, E., and Lee, H., in: *Semiconductor Cleaning Technology/1989* (J. Ruzyllo and R. E. Novak, eds.), Vol. 90-9, pp. 83-92, The Electrochemical Society, Inc., Pennington, New Jersey (1990). (This paper was originally presented at the Fall 1989 Meeting of the Electrochemical Society, Inc., held in Hollywood, FL)

34. Kern, W., and Puotinen, D. A., *RCA Review,* 31(6):187-205 (1970)

35. Menon, V. B., Michaels, L. D., Donovan, R. P., and Ensor, D. S., *Solid State Technology,* 32(10):S9-S11 (1989)

36. Milner, T. A., and Brown, T. M., *Proceedings of Microcontamination 86 Conference and Exposition,* pp. 146-156, San Jose, California (1986)

37. Dillenbeck, K., in: *Particle Control for Semiconductor Manufacturing,* (R. P. Donovan, ed.), pp. 405- 411, Marcel Dekker, New York (1990)

38. Kikuyama, H., Miki, N., Takano, J., and Ohmi, T., *Microcontamination,* 7(4):25-51 (1989)

39. Menon, V. B., Michaels, L. D., Clayton, A. C., and Donovan, R. P., *Proceedings of the Institute of Environmental Sciences, Annual Technical Meeting,* pp. 320-324, Chicago, Illinois (1989)

40. *Handbook of Chemistry and Physics,* 63rd edition, pp. F23-F29, CRC Press, Boca Raton, Florida (1984)

41. Kern, W., *Semiconductor International,* 7(4):94-99 (1984)

42. Niida, T., *Chemical Engineering Institute of Japan,* pp. 14-25, Kansai Branch, Osaka, Japan (1989)

43. Mishima, H., Yasui, T., Mizuniwa, T., Abe, M., and Ohmi, T., *IEEE Transactions on Semiconductor Manufacturing,* 2(3):69-75 (1989)

44. Ohmi, T., Ultra Clean Wafer Processing Presentation to SEMATECH, October 11 (1991)

45. Meuris, M., Heyns, M., and Philipossian, A., *Extended Abstracts, The Electrochemical Society Fall Meeting,* 91-2:775-776, Pennington, New Jersey (1991)

46. Hariri, A., and Hockett, R. S., *Semiconductor International,* 12(9):74-77 (1989)

47. Dillenbeck, K., in: *Particle Control for Semiconductor Manufacturing,* (R. P. Donovan, ed.), pp. 203-210, Marcel Dekker, Inc., New York (1990)

48. McConnell, C. F., *Microcontamination,* 9(2):35-40 (1991)

49. McConnell, C. F., *Microcontamination,* 8(10):45-50 (1990)

50. Menon, V. B., Clayton, A. C., and Donovan, R. P., *Microcontamination,* 7(6):31-108 (1989)
51. Tonti, A., *Extended Abstracts, The Electrochemical Society Fall Meeting,* 91-2;758, Pennington, New Jersey (1991)
52. Ohmi, T., Mishima, H., Mizuniwa, T., and Abe, M., *Microcontamination,* 7(5):25-108 (1989)
53. Olesen, M. B., in: *Proceedings of the Institute of Environmental Sciences, Annual Technical Meeting,* pp. 229-241, New Orleans, Louisana (1990)
54. Marra, J., *Extended Abstracts, Third Symposium on Particles in Gases and Liquids: Detection, Characterization and Control,* p. 52, San Jose, California (1991)
55. Leenaars, A. F. M., Huethorst, J. A. M., and Van Oekel, J. J., *Langmuir* 6(11):1701-1703 (1990)
56. Singer, P. H., *Semiconductor Inernational,* 15(1):24 (1992)
57. Witowski, R., Gordon, M., Chacon, J., and Menon, V. B., *Extended Abstracts, the Electrochemical Society Fall Meeting,* 91-2:882-823, Pennington, New Jersey (1991)
58. Menon, V. B., in: *Particle Control for Semiconductor Manufacturing,* (R. P. Donovan, ed.), pp. 359-382, Marcel Dekker, New York (1990)
59. Ohmi, T., and Shibata, T., *Extended Abstracts, The Electrochemical Society Fall Meeting,* 91-2:788-789, Pennington, New Jersey (1991)
60. Grant, D. C., Liu, B. Y. H, Fischer, W. G., and Bowling, R. A., *Proceedings of the Institute of Environmental Sciences Annual Meeting,* pp. 882-891, Mount Prospect, Illinois (1988)
61. Musselman, R. P., and Yarbrough, T. W., *J. Environmental Sciences,* 51 (Jan/Feb, 1987)
62. Nagarajan, R., and Welker, R. W., *Proceedings of the Institute of Environmental Sciences, Annual Technical Meeting,* 91-2:868-881, Mount Prospect, Illinois (1991)
63. J. M. Ney Company, Bloomfield, Connecticut, Technical Bulletin (1988)
64. Niemczewski, B., *Ultrasonics,* 8:107-110 (1980)
65. Busnaina, A., Abuzeid, S., *Proceedings of the Institute of Environmental Sciences, Annual Technical Meeting,* pp. 130-139, San Diego, California (1991)

66. Menon, V. B, Michaels, L. D., Donovan, R. P., and Ensor, D. S., in: *Particles on Surfaces 2: Detection, Adhesion and Removal,* (K. L. Mittal, ed.), pp. 297-306, Plenum Press, New York (1989)

67. Menon, V. B. and Donovan, R. P., *Extended Abstracts, The Electrochemical Society Fall Meeting,* 89-2:89-92, Pennington, New Jersey (1989)

68. Ranade, M. B., *Aerosol Science and Technology,* 7:161-180 (1987)

69. Pre-Tech, Inc. Japan, Technical Bulletin (1988)

70. Rooney, J. L., Pui, D. Y. H., and Grant, D. C., *Microcontamination,* 8(5):37-99 (1990)

71. Syverson, W. A., Fleming. M. J., and Schubring. P. J., *Extended Abstracts, The Electrochemical Society Fall Meeting,* 91-2:751-752, Pennington, New Jersey (1991)

72. Gow, C. J., Smith, R. E., Syverson, W. A., Kunesh, R. F., Buker, E. D., Albaugh, K. B., and Whittingham, L. S., *Extended Abstracts, The Electrochemical Society Fall Meeting,* 91-2:820-821, Pennington, New Jersey (1991). (This paper was originally presented at the Fall 1991 Meeting of the Electrochemical Society, Inc. held in Phoenix, AZ.)

73. Layden, L., and Wadlow, D., *J. Vacuum Science and Technology,* A8(5):3881-3883 (1990)

74. Ohmori, T., Kanno, I., Fukumoto, T., Komiya, H., Tada, M., and Kawaguchi, T., *Technical Proceedings Semicon/Kansai Kyoto*, pp. 142-149, Kyoto, Japan (1990)

75. McDermott, W. T., Ockovic, R. C., Wu, J. J., and Miller, R. J., *Microcontamination,* 9(10):33-95 (1991)

76. Allen, S. D., *Scientific American,* 262(6):86-87 (1990)

10

Silicon Surface Chemical Composition and Morphology

Gregg S. Higashi and Yves J. Chabal

1.0 INTRODUCTION

The chemical state in which a surface is left subsequent to a clean is as important as the clean itself. A surface that becomes recontaminated before the next processing step will not be useful. The best cleaning techniques are therefore the ones which chemically *passivate* the semiconductor surface in the act of cleaning it.

There are two predominant ways to clean and passivate silicon surfaces chemically. The first is to grow a thin layer of oxide in the act of cleaning. This is best accomplished using acidic or basic solutions mixed with hydrogen peroxide and is the basis of the RCA Standard Clean developed by Kern in 1965 (1). These cleans leave 10 - 15 Å of hydroxylated oxide on the silicon surface which prevents recontamination of the silicon. Such surfaces are hydrophilic in nature and are easily wetted by aqueous solutions. The second way to clean and passivate the surface is to dissolve the surface oxide completely in hydrofluoric acid. Indeed, the early electronic measurements of Buck and McKim (1958) (2) demonstrated the high degree of passivation of HF-treated Si surfaces. These surfaces are now known to be oxide-free and passivated with hydrogen. The H-terminated surfaces are hydrophobic in nature and are *not* wetted by aqueous solutions. The body of this chapter is thus broken up into two parts: The surfaces produced by cleans that involve chemical oxidation (hydrophilic surfaces) are discussed in Sec. 2 and the surfaces produced by HF acid etching (hydrophobic surfaces) are discussed in Sec. 3.

The chemical composition of the silicon surface subsequent to a clean is fundamental to its passivation. The chemically grown oxides exhibit a more complex chemical composition than the high quality stoichiometric SiO_2 grown thermally. Thermal oxides have been studied extensively because of their application as gate insulators in the MOS technology (Sec. 2.1). However, as gate oxide thicknesses decrease, the importance of the cleaning procedures employed is increasing, motivating in-depth studies of chemically grown oxides. Section 2.2 presents what is currently understood about the chemical composition of these hydrophilic surfaces.

The chemical composition of HF-etched silicon surfaces has been the subject of some confusion. It has long been held that the hydrophobic nature of these surfaces was explained by fluorine termination and that the surfaces were hydrophobic because they resembled Teflon™. In actuality, these surfaces are terminated with hydrogen and are therefore more paraffin-like due to the non-polar nature of the Si-H bond. Because of the confusion, a detailed history of the evidence for H-termination is presented in Sec. 3.1. A chemical mechanism which explains how the silicon surfaces become hydrogen-terminated is then described.

The micro-structural state of the surface has been shown recently to affect subsequent device properties (3)-(5). Surface roughness causes degradation of the breakdown field strengths of thin gate oxides (4)(5) and leads to decreased channel mobilities (3)(5). Both hydrophilic and hydrophobic surface cleans can affect the morphology of the Si surface and will be discussed separately in Sec. 2.3 and Sec. 3.2, respectively. Chemically grown oxides are non-crystalline, limiting the information obtained from most techniques for structural analysis. In contrast, HF etching leads to hydrogen termination of the bulk crystal and can be studied in great detail. In order to understand the variety of structures produced by different HF solution chemistries, the principles and limitations of the most useful structural analysis techniques are first briefly reviewed in Sec. 3.2 . The huge differences observed between the surface structures of Si(100) and Si(111) wafers after HF etching are then discussed. In particular, the structural morphology on both surfaces varies with solution pH, favoring the formation of (111) facets at high pH. The mechanism of this preferential etching is presented in Sec. 3.2.

Contamination is also an important issue for any cleaning or passivation process because trace amounts of impurities can drastically influence subsequent materials properties. Contaminants can be intrinsic (e.g., H, OH, H_2O, F) or extrinsic (e.g., C, Fe, Ni, Cr, Cu) to the solutions used and will be treated separately. Contamination issues for oxide-covered surfaces are discussed in Sec. 2.4. The contamination and the resulting loss of passivation of H-terminated surfaces are discussed in Sec. 3.3.

2.0 OXIDE-TERMINATED SURFACES

2.1 Introduction

As early as the 1950s, Atalla, Tannenbaum, and Scheibner (1959) (6) recognized the significance of oxide passivation of silicon semiconductor devices. Their discovery of the unique passivation properties of thermally grown oxides, which led to remarkable improvements in device performance, is a cornerstone of the modern silicon technology. It is noteworthy that the importance of surface chemical cleaning prior to oxidation was already being stressed. In particular, the distinction between cleans resulting in hydrophilic versus hydrophobic surfaces was noted, but not understood at that time. Although this chapter focuses on oxides grown chemically during cleaning, many analogies to the thermally grown oxides exist. Therefore, a brief discussion of the thermal oxides is useful to an overall understanding of these issues.

Thermally grown oxides have been characterized extremely well because of their technological significance. Entire books have been devoted to this subject (7)-(9), hence only a brief summary follows. Thermally grown oxides are formed by heating silicon wafers in either an O_2 or an H_2O environment. The resulting oxide is a non-crystalline, stoichiometric, low defect density form of SiO_2. The SiO_2 layer is largely impervious to contamination and protects the underlying Si substrate. A high degree of chemical passivation of the Si/SiO_2 interface is observed in the electronic properties of the interface. "Dangling" bond densities below 10^{11} cm^2 are routinely achieved at the Si/SiO_2 interface (10^{10} cm^{-2} subsequent to H_2 annealing) with fixed charged densities on the order of 10^{10} cm^{-2}. The electronic perfection of this interface is the basis of the MOS technology and may be related to the atomic order of the Si/SiO_2 boundary.

Although the chemical perfection of the Si/SiO_2 interface is well established, the molecular structure is less well understood. It is clear that there is a transition region between the bulk stoichiometric amorphous SiO_2 and the crystalline Si substrate of 3 - 7 Å where suboxides are observed by XPS (10). It is less clear, however, what the exact structure and composition are for this transition region. The various models proposed have recently been reviewed by Ourmazd and Bevk (1988) (11) and fall into three classes involving: a) An epitaxial SiO_2 intervening layer; b) a disordered sub-stoichiometric oxide layer; and c) an abrupt transition directly from the crystalline Si to the continuous random network of amorphous SiO_2. Although the exact structure of the Si/SiO_2 interface is still being debated, the transmission electron micrographs (TEM) of Ourmazd and Bevk (1988) (11) demonstrate that atomically flat interfaces can be obtained from atomically flat Si(100) starting surfaces.

Chemically grown oxides are produced by various cleaning techniques that are based on the use of acidic and/or basic hydrogen peroxide solutions. The most widely used system is the RCA Standard Clean developed by Kern and Puotinen (1970) (12). It is a sequential two-step clean where wafers are immersed first in a 5:1:1 solution of H_2O:H_2O_2:NH_4OH at 80°C (SC-1) and then in a 5:1:1 solution of H_2O:H_2O_2:HCl at 80°C (SC-2). The Piranha etch, developed earlier, consists of an immersion in 4:1 H_2SO_4:H_2O_2 at a temperature somewhat in excess of 100°C. Other cleans involve exposure to hot chemicals, such as nitric acid or other acid mixtures. The oxides left behind after these surface treatments are similar to the thermally grown oxides in some respects but are quite different in others. The discussion of these chemically grown oxides is divided up into three parts: chemical composition (Sec. 2.2), structural morphology (Sec. 2.3), and contamination issues (Sec. 2.4).

2.2 Chemical Composition

The properties of chemically grown oxides produced by the various cleaning techniques (Piranha etch, SC-1, or SC-2, etc.) are quite similar and have recently been reviewed by Deal (1987) (13). The oxides tend to be ~10 - 15 Å thick, depending on the process temperature as well as the solution chemistry used (14)(15). These films are largely stoichiometric but, because they are so thin, exhibit properties with many of the characteristics of the interfacial transition regions of thicker, thermally grown oxides. The large suboxide content characteristic of these chemically grown oxides is shown in the Si 2p core level spectra of Sugiyama et al. (1990) (16) in Fig. 1. For two different chemical preparations, the spectra are dominated by the Si crystal substrate (Si^0) and by stoichiometric SiO_2 (Si^{4+}). There is, however, a relatively strong and unambiguous Si^{2+} contribution corresponding to an interfacial transition region similar to that observed for the thermal oxides. In this case, the Si^{2+} contribution is 10 - 20% that of the Si^{4+}, indicating that the transition region is a large fraction of the surface layer. Besides the Si^{2+}, one observes varying degrees of other suboxides which depend on the exact surface treatment used (15)(16), but will not be discussed here. Dangling bond defects at the Si/SiO_2 interface have been quantified using minority carrier lifetime measurements to extract the surface recombination velocity and surface defect density (17). Defect densities in the range of 10^{12} cm^{-2} are typical for these clean "native" oxides, placing these surfaces at a level ~2 orders of magnitude higher than for the thermal oxides.

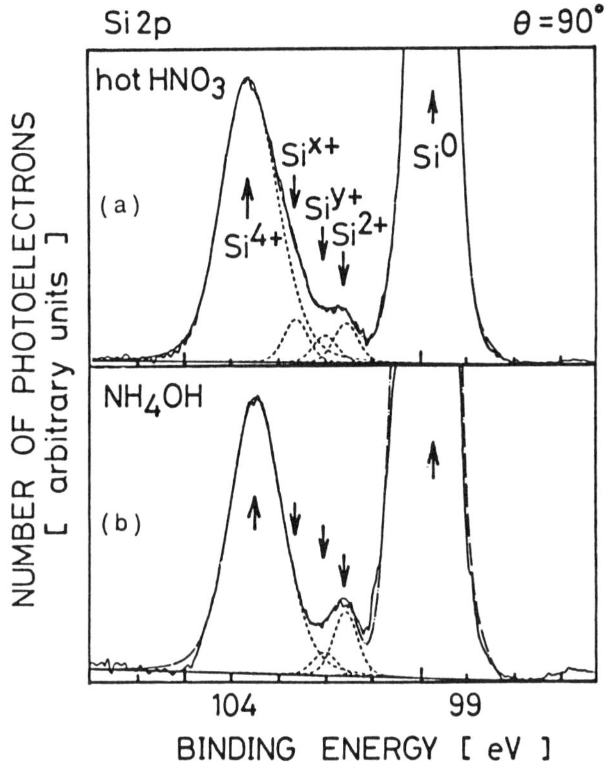

Figure 1. X-ray photoelectron spectra of the Si $2p_{3/2}$ core level associated with native oxides on silicon formed by immersion in (a) HNO_3 at 45 - 60°C for 5 min, and (b) 4:1:1 H_2O:H_2O_2:NH_4OH at 63 - 80°C for 10 min. A thorough rinse in DI water for 10 min was performed in both cases before introduction into the UHV chamber. The spin orbit splitting of the Si 2p level is removed, the data smoothed and the background subtracted. The dashed lines are the result of a spectral deconvolution performed by assuming that the chemical shifts and the values of FWHM for the various components are the same as for the Si^{4+}, which is associated with stoichiometric SiO_2. (From Ref. 16.)

Chemical oxidation involves species other than Si and O. All aqueous solutions, of course, are predominantly composed of H_2O and are, therefore, sources of H_2O, OH and H. Thus, these chemical species must be incorporated in the oxide layer to some extent. There is an extensive literature on the infrared spectroscopy of OH and H_2O in and on the surfaces of silica glasses (18). These hydroxyl units are found in the form of Si-O-H or H-O-H. When the silica glasses are treated at elevated temperatures (≥550°C) under vacuum, a narrow (<50 cm^{-1}) infrared absorption band is observed at ~3750 cm^{-1}. This absorption is assigned to the O-H stretching vibration and is characteristic of isolated Si-O-H units. Without such a thermal treatment, the infrared absorption of the O-H stretch is extremely broad (~400 cm^{-1}) and peaks near 3400 cm^{-1}, indicating a strong interaction between neighboring hydroxyl groups. This interaction, commonly referred to as hydrogen bonding, is simply a weak bonding between a hydrogen atom and an oxygen atom of neighboring molecules or molecular complexes. The exact nature of this bond is not completely understood. It is partially ionic and partially covalent and weakens the O-H bond, causing inhomogeneous broadening on the low frequency side of the isolated O-H stretching vibration. Hydrogen bonding is common to all aqueous solutions and is intimately related to the heat of solvation, as well as to the hydrophilic nature of oxide-covered Si surfaces.

Hydrogen-bonded OH is observed on all chemically grown oxides. An example of an IR absorption spectrum of such a surface is shown in Fig. 2, where a hydrogen terminated silicon surface was chemically oxidized using the SC-2 step of the RCA clean. The spectra presented in Fig. 2 are ratios of oxidized and H-terminated silicon surfaces, and therefore display a negative absorption for the Si-H stretch bands (~2100 cm^{-1}) associated with the reference surface. The hydrogen bonded OH stretching vibration peaks at ~3300 cm^{-1} and is ~400 cm^{-1} wide (FWHM) with an asymmetric lineshape. It is quite difficult, however, to distinguish between Si-O-H and H_2O in these spectra without further spectral information from the scissor mode of the H_2O molecule in the 1600 - 1700 cm^{-1} region. The intensity difference observed between the spectra taken in s- and p-polarizations indicates that the OH groups must reside in or on the oxide layer. Although it is unclear from these spectra, one knows in very general terms that most of the IR signal comes from H_2O on the surface of the oxide, since gentle heating (100°C) decreases the OH absorption substantially. Quantifying the amount of Si-O-H in and on the surface of these oxides is a direction for future research.

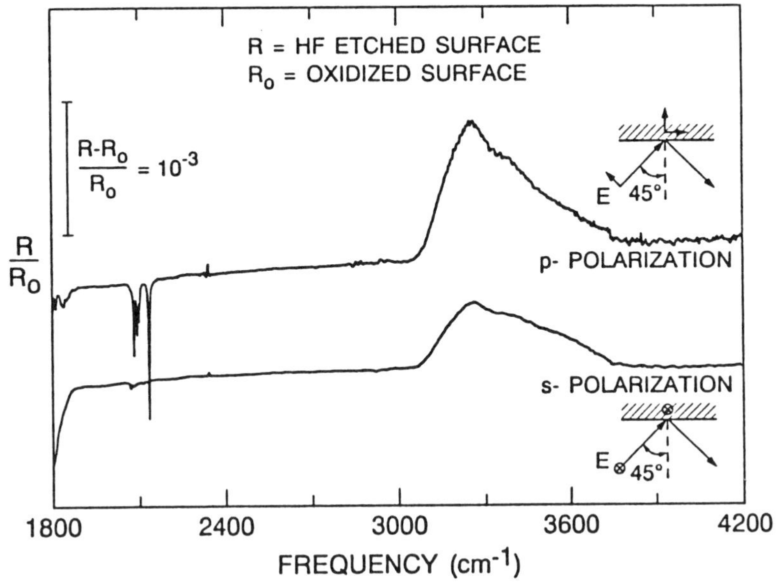

Figure 2. IR absorption spectra of silicon wafer chemically oxidized in a 4:1:1 solution of H_2O:H_2O_2:HCl at 80°C for 10 min. The spectra were ratioed to the corresponding spectra of the H-terminated silicon by etching in a buffered HF solution (see Ref. 89). A multiple internal reflection geometry was used with 75 reflections at a 45° internal angle of incidence, as shown in the inset.

It has now become apparent that Si-H units reside also on Si surfaces upon which chemical oxides have been grown (15)(16). The most convincing evidence comes from the IR spectra of Ogawa et al. (1992), (15) shown in Fig. 3, where Si-H stretching vibrations were identified at ~2260 cm^{-1}. Si-H stretches in that region of the spectrum originate from Si-H where the Si atom is back-bonded to O atoms (19). This evidence clearly indicates that the Si-H resides within the oxide matrix. The areal density is estimated to be 2 - 3 x 10^{13} cm^{-2}. XPS data from the same group suggests that these Si-H units are actually localized near the surface of the oxide. If this is indeed the case, these units may be residual Si-H bonds from the original H-terminated hydrophobic surface before the chemical oxidation. This kind of picture agrees well with the idea that oxidation proceeds via O atom insertion between the Si-Si back-bonds of the surface and is consistent with the observations of Nagasawa et al. (1990) (20) on the initial stages of oxidation of hydrophobic surfaces. Also observed in the spectra of Fig. 3 are Si-H stretches in the range of 2080 cm^{-1} which are best explained by H-atoms bonded to substrate Si (i.e., back-bonded to Si atoms rather than to O atoms). The high frequency shoulder of this band (~2140 cm^{-1}) is most likely associated to Si-H stretches where some of the Si back-bonds are attached to an oxygen atom. If the mode at 2080 cm^{-1} does arise from Si-H at the Si/SiO_2 interface, an interesting direction for future studies will be to determine its formation mechanism.

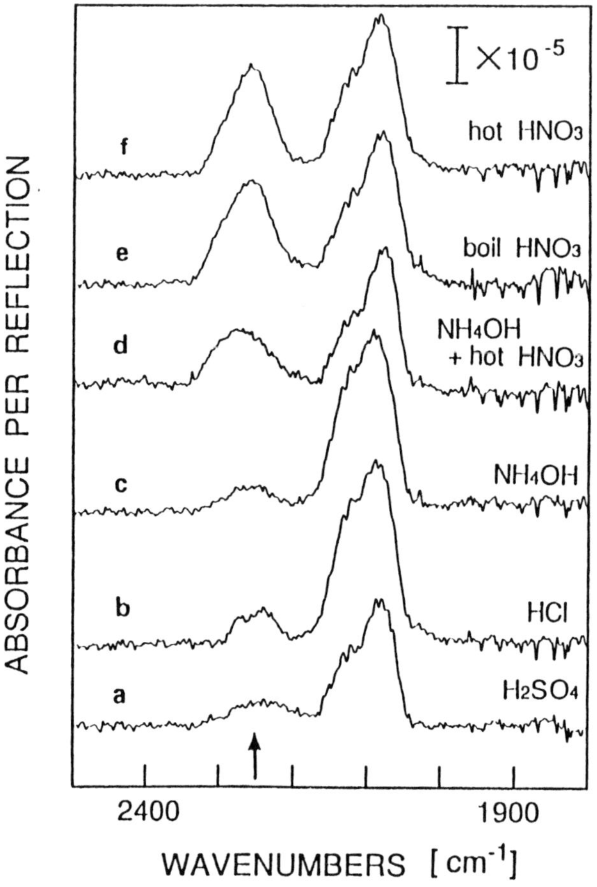

Figure 3. Infrared absorption spectra of six different native oxides on silicon wafers: (a) "H_2 SO_4" corresponds to a 10 min treatment in 4:1 H_2 SO_4:H_2O_2 at 85 - 90°C, (b) "HCl", 10 min in 4:1:1 H_2O:H_2O_2:HCl at 37 - 65°C, (c) "NH_4OH", 10 min in 4:1:1 H_2O:H_2O_2:NH_4OH at 63 - 80°C, (d) "NH_4OH + hot HNO_3", in "NH_4OH" followed by "hot HNO_3". (e) "boil HNO_3", in HNO_3 at 115 - 125°C, (f) "hot HNO_3", 5 min in HNO_3 at 45 - 60°C, The absorption (indicated by the arrow) which peaks at ~2260 cm^{-1} arises from Si-H stretches where the Si is back-bonded to O (i.e. Si-H in or on the SiO_2). (From Ref. 15.)

2.3 Structure and Morphology

The work of Hahn and Henzler (1984) (3) and more recently of Heyns et al. (1989) (4) and Ohmi et al. (1991) (5) have correlated electronic device properties with surface structural properties. While it is intuitively obvious that surface roughness must be detrimental to semiconductor devices at some scale, the main contribution of these workers has been to define at what scale surface roughness is important to device yield and reliability. The evidence for degradation of thin gate oxide (<100 Å) breakdown field strengths and channel mobilities with surface roughness on a microscopic scale is quite convincing and has captured the attention of the industry.

The morphology that a surface exhibits tends to be a function of the complete processing history experienced by the wafer; it is therefore quite complicated. The initial surface polish, any chemical cleaning, and processing steps such as thermal oxidation or any kind of etching all influence what the surface looks like. This section begins with a discussion of substrate wafers, including chemo-mechanical polishing and epitaxy. It then considers the effects of chemical cleans on surface morphology. Finally, future trends in controlling oxidation and interfacial structure are briefly discussed.

Near atomic perfection is achieved in surface chemo-mechanical polishing. Commercial wafers exhibit a typical surface roughness on the order of 2 Å rms and surface finishes produced in the laboratory have approached 1 Å rms (21). A scanning tunneling microscope (STM) image of such a surface is shown in Fig. 4. Although STM images can characterize surface roughness on length scales covering the range from one to a thousand angstroms, these surfaces were also characterized with a variety of other techniques spanning length scales up to 1 mm with good correlations observed on all length scales (21). The STM image shown in Fig. 4 was taken subsequent to removal of any surface oxide in HF, although chemo-mechanically polished wafers are hydrophobic and are already believed to be hydrogen-terminated (22). This process is not very well understood but will be discussed briefly in Sec. 3.1.

Hydrogen-terminated surfaces are not as stable as oxide-terminated surfaces, and thus it is not surprising that wafer vendors ship wafers in the hydrophilic (oxide covered) state. Vendor polishes and cleans are proprietary, but presumably the wafers receive something akin to an RCA clean before they are shipped. Another technique which provides atomically perfect surfaces uses Si molecular beam epitaxy, (23) although this technique has not yet been commercialized. Surfaces formed during commercial Si epitaxial growth by chemical vapor deposition (CVD), however, might also be made atomically perfect under the proper conditions. Presently, as-received CZ-Si and epi-Si substrates seem to exhibit a surface roughness of ~2 Å rms (24).

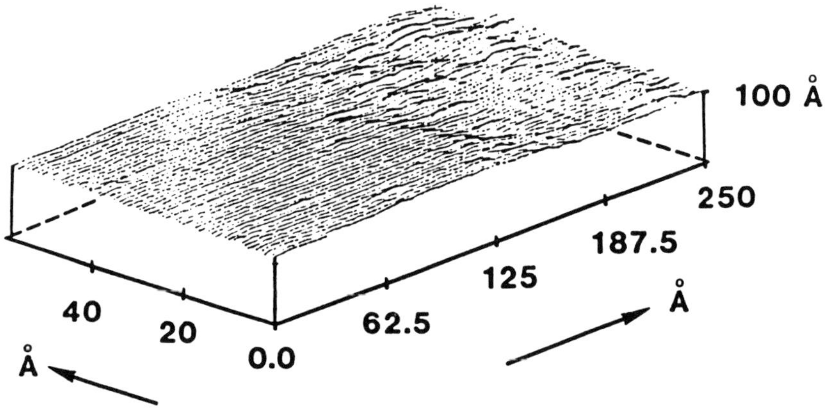

Figure 4. Scanning tunneling microscope image of a polished Si(100) surface exhibiting 1.2 Å rms roughness. The image was taken immediately following an HF dip. *(Courtesy of P. O. Hahn, Wacker-Chemitronic GmbH, Germany, Ref. 21.)*

The consensus right now is that the acidic peroxide cleans (Piranha etch or SC-2) do not cause a substantial increase in the microscopic roughness of as-received wafers. The basic peroxide clean (typically 5:1:1 H_2O:H_2O_2:NH_4OH at 80°C), on the other hand, has been found to substantially increase the surface roughness (25). A comparison of the surface roughness of a control wafer and a wafer cleaned in a standard SC-1 process is shown in the STM images of Fig. 5 (5). Control wafers exhibit an rms roughness of 2 Å. The SC-1 treatment more than doubles the observed roughness and repetitive SC-1 cycles can increase it by as much as a factor of five, approaching 10 Å rms. This kind of roughness has been shown to decrease breakdown field strengths by as much as 30% (26) and to degrade channel mobilities by factors of two to three (27).

(a)

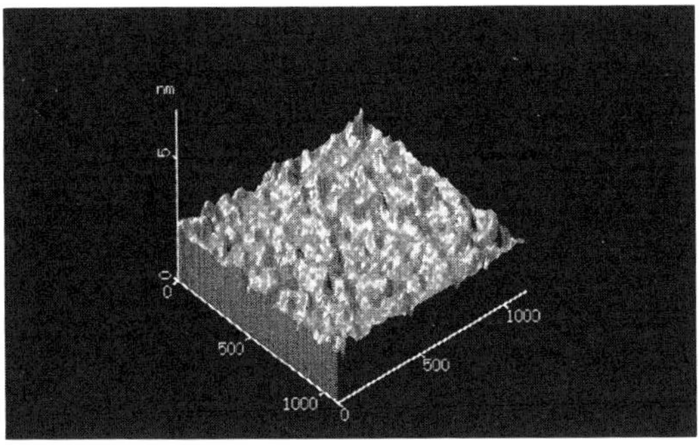

BEFORE CLEANING

(b)

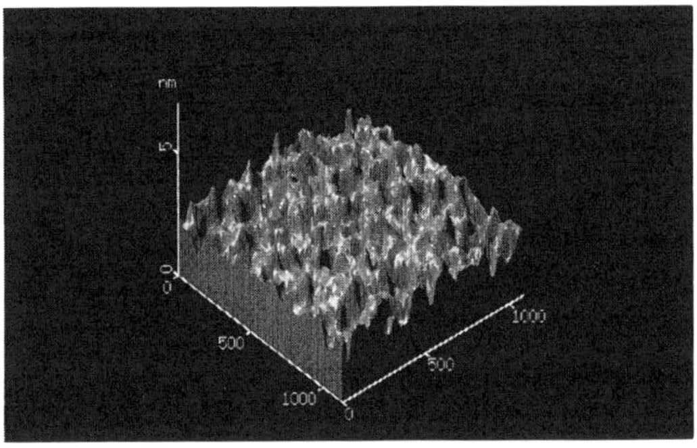

AFTER RCA CLEANING

Figure 5. Typical scanning tunneling microscope images of Si(100) surfaces taken before (a) and after (b) an RCA Standard Clean. Images taken after a newly developed buffered HF treatment where the authors observed minimal increases in surface roughness due to the BHF. (From Ref. 5.)

The mechanism leading to the roughening in the basic peroxide solution is not completely understood but is related to the slow but finite silicon etch rate in the SC-1 solution (~8 Å/min at full strength at 80°C) (5). The acidic peroxides, on the other hand, do not etch and hence do not roughen. A proposed solution to the basic peroxide roughening problem is to reduce the etch rate by reducing the concentration of NH_4OH in the SC-1 solution (25). The etch rate drops to 1 Å/min if the concentration of NH_4OH is decreased by a factor of a hundred. Figure 6 shows a plot of the measured rms roughness as a function of NH_4OH concentration (28). One might wonder why the industry is working so hard to keep the standard SC-1 solution when it is clearly detrimental to the surfaces. The reason is simple:

SC-1 is one of the most efficient particle removal agents known. Further, the fact that the basic peroxide solution slightly etches both SiO_2 and Si may be precisely why it is such an efficient particle remover. This phenomenon is being investigated in the hopes that an optimum concentration can be found to minimize damage and retain particle removal efficiency (5). The microscopic mechanism by which the etching roughens the surface is also being investigated. It should be mentioned that etching alone does not necessarily mean that the surface roughness will be increased. It is non-uniform etching which is really the culprit. For example, Verhaverbeke et al. (1991) (29) has found that the Ca concentration in the SC-1 dramatically changes the degree to which the surface roughens. Studies like these should be the direction of future work.

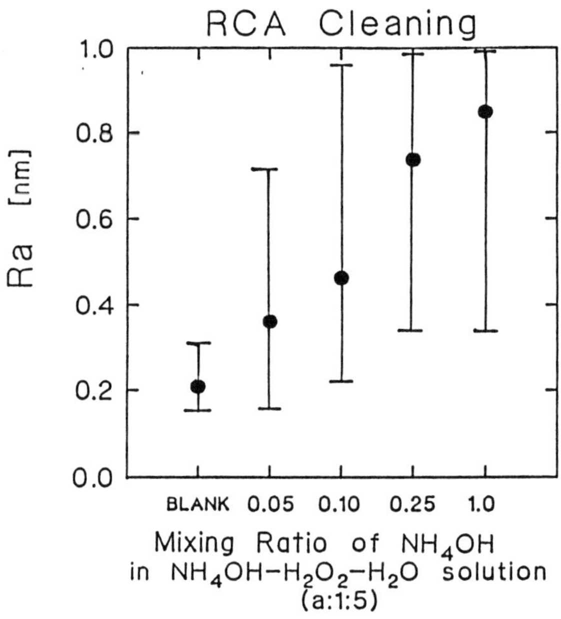

Figure 6. Surface roughness plotted as a function of NH_4OH concentration in a 10 min. ammonia-peroxide solution treatment at 85°C. (From Ref. 28.)

The exact molecular structure of these chemically grown Si/SiO_2 interfaces is very difficult to deduce. In the past few years, however, "native" oxide growth has been shown to occur in an extremely controlled manner (23)(30), leading to atomically ordered interfaces under the right conditions. This phenomenon appears in fact to be more general. In very careful XPS studies of surface oxygen concentration as a function of time, layer by layer initial oxidation of Si has been observed (31)(32). Layer by layer oxidation, of course, requires that one layer finishes before the next layer begins oxidation and leads by necessity to the conclusion that some form of order must exist at the Si/SiO_2 interface.

2.4 Contamination Issues

One of the main objectives of the development of the RCA clean was to remove organic and metal contaminants from the surface of silicon wafers (1). Although the RCA clean was developed over twenty-five years ago, it has functioned extremely well and is still the dominant clean used prior to gate oxidation. As gate oxides move into the sub-100 Å regime, there has been renewed interest in how trace metal contamination in the pre-gate clean affects gate oxide properties (33). Residual trace metal contamination at the 10^{10} cm^{-2} level is observed after RCA cleaning, depending on the quality of the chemicals used and is discussed in detail in Ch. 12. It seems as though metals can get trapped in the oxide formed during the SC-1/SC-2 cleaning process. These metals can subsequently lead to leaky junctions and to yield and reliability problems in gate oxides (33)(34). It is beyond the scope of this discussion to address this issue further, but the reader should understand that the industry needs to be concerned with how cleaning is to be handled in the future.

Another common contaminant on these oxide covered surfaces is carbon. It is most likely incorporated in or on these surfaces in the form of hydrocarbons and can come from the chemicals, the water used to rinse the wafers, or from the air in the laboratory environment. Trace hydrocarbons have not proven to be detrimental to gate oxides. A predominant sentiment in the industry is that the hydrocarbons get "burnt" off in the oxygen-rich high-temperature environment of the oxidation furnace. If handled improperly, however, SiC precipitates can cause weak spots in the oxides grown (35).

Hydrocarbon contamination is much more of a concern for surface preparation prior to epitaxial growth of silicon. In this case, surfaces that are completely free of contamination are needed to grow defect free Si and carbon contamination is of critical concern. The technique of desorbing the oxide at elevated temperature prior to epitaxy was first discussed by Henderson (1972) (36) where he showed that atomically clean surfaces with only a small carbon residue could be obtained after the RCA Standard Clean. Ishizaka, Nakagawa and Shiraki (1982) (37) reduced the level of carbon entrained in the oxide by repetitively immersing the wafers in boiling HNO_3 acid followed by HF, ending with a concentrated SC-2 type of clean (3:1:1 $HCl:H_2O_2:H_2O$ at 90 - 100°C). Another technique which is extremely efficient at removing hydrocarbons is UV-ozone (38) and is discussed in detail in Ch. 6.

Some contaminants, such as S and Cl, can originate directly from the solution used during the chemical oxidation. S and Cl have been observed from the Piranha and SC-2 cleans, respectively (see Ch. 12). Fluorine, on the other hand, has been observed when the chemical oxidation is preceded by an HF treatment. In this case, F is found to segregate at the Si/SiO_2 interface (39).

3.0 HYDROGEN-TERMINATED SURFACES

3.1 Chemical Composition of HF Treated Surfaces (Wet)

Historical Overview. The unique properties of Si surfaces treated in HF solutions were recognized over thirty years ago, (2)(6) but only recently have these hydrophobic surfaces begun to be understood. It is now quite clear that *hydrogen* and *not fluorine* termination of the dangling bonds on the Si surface explains the hydrophobicity, the high resistance to chemical attack, and the low surface recombination velocity. Fluorine is also found on these surfaces, but only in small quantities and should be thought of as a minor contaminant rather than a major constituent of the surface. At first glance it is difficult to understand why the confusion and controversy about H vs. F termination took so long to sort out. The reader should understand, however, that most conventional surface spectroscopic methods (AES, XPS, etc.) rely on core electrons and cannot measure hydrogen, since it has no core. Thus, researchers tended to concentrate on those things they could detect rather than those they could not. Vibrational spectroscopies (EELS, IR, etc.), on the other hand, are extremely sensitive to hydrogen-containing surface species but have only begun to see widespread use in surface science during the last ten years. Because there has been so much confusion regarding this issue, a detailed experimental chronology of HF treated Si surfaces is presented in the following.

It is remarkable that, as early as 1965, Beckmann (40) had investigated the chemical properties of "stain films" with infrared absorption spectroscopy. Stain films were prepared electrochemically by anodic polarization in a 10 N aqueous solution of HF acid. They could be grown as thick as 20 - 50 μm and thus could be studied using *transmission* infrared absorption spectroscopy. A thorough investigation with deuterated solutions clearly identified hydrogen as the main chemical species in the films, mostly as pure hydrides Si-H_x and also as hydrides with silicon back-bonded to oxygen (of the form O-Si-H), characterized by silicon-hydrogen stretching vibrations in the 2100 cm^{-1} and the 2275 cm^{-1} regions, respectively. Beckmann also assigned all the absorption bands measured in the lower frequency region (400 - 1100 cm^{-1}). In particular, the presence of an absorption band at 910 cm^{-1} made him consider fluorine, since it could be assigned to the Si-F stretch mode. This band, however, disappeared upon deuteration, casting serious doubts on the Si-F stretch assignment; the Si-F stretch has more recently observed at ~800 cm^{-1} instead of 910 cm^{-1} (41). However, Beckmann cautiously concluded that fluorine contamination could not be ruled out altogether because a strong interaction could result from the existence of several normal modes in the 910 cm^{-1} range: the deformation vibrations of SiH_2 and SiH_3, as well as the SiF_x stretch modes. He made a more positive identification of OH and a variety of silicon oxides in his spectra. Although the hydrocarbon region could not be studied because of the spectrometer, it is likely that the stain films had a substantial carbon concentration.

Almost ten years later, Harrick and Beckmann (1974) (42) reported on an infrared absorption study of silicon *dipped* in an HF solution (not *electrochemically* prepared as the stain films). The spectra, recorded with an internal reflection geometry (180 reflections), are summarized in Fig. 7. They are characterized by a spectral signature that is very similar to that observed in stain films, namely a strong Si-H_x stretch band with evidence for some oxygen present in the back-bond of the surface silicon atoms. Based on a comparison with the data on stain films, Harrick and Beckmann concluded that the layers formed by HF solution dipping were 20 Å thick on average. With today's knowledge, as will be shown in the next section, their spectra actually correspond to a monolayer coverage of an *atomically rough* surface, possibly with a 20 Å scale roughness. Although only the 2000 - 2200 cm^{-1} spectral region was reported, it is likely that the surface contained carbon in addition to hydrogen and oxygen, due to relatively dirty chemicals used at the time. Regardless of the contamination issue, however, this was a clear indication that HF etched surfaces were terminated with hydrogen. This work was unfortunately either *buried* in the wrong literature or *ignored* because Harrick and Beckmann mistakenly assigned the Si-H stretching vibration they observed to an Si-H polymer deposited on the surface of the wafer.

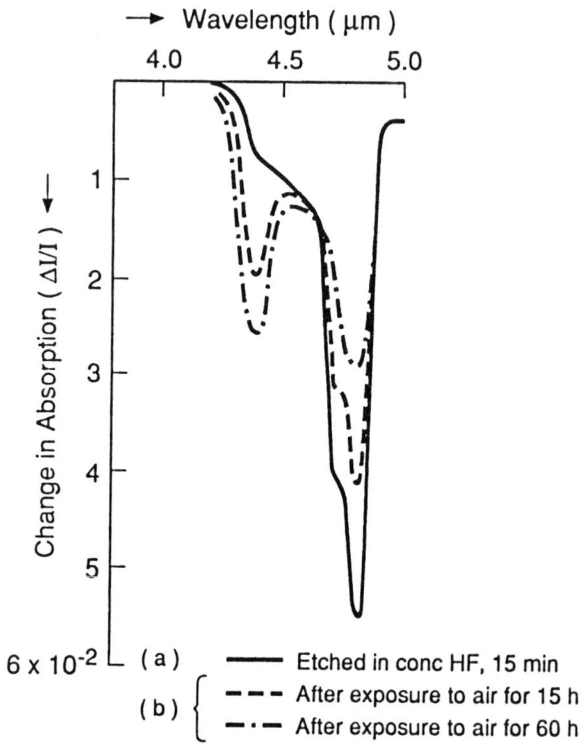

Figure 7. Multiple internal reflection (180 reflections, $\theta_{internal}$ = 45° IR absorption spectra of Si surfaces after (a) HF etching in 10 N solution, and (b) subsequent exposure to air for 15 h (dashed line) and for 60 h (dash-dotted line). The main absorption peak for freshly etched Si (solid spectrum) at 4.75 μm (~2100cm^{-1}) is characteristic of H-termination of Si. The absorption band developing around 4.4 μm (~2250 - 2300 cm^{-1}) upon oxidation in air is characteristic of Si-H with oxygens in the Si-Si back-bonds. (From Ref. 42.)

In 1984, Ubara, Imura and Hiraki (1984) (43) *also* clearly showed that the removal of silicon oxide in HF solutions results in the formation of silicon hydrides. Their results obtained on hydrogenated microcrystalline silicon (μc-Si:H) (44)(45), are summarized in Fig. 8. The as-grown samples (a) are thermally oxidized at low temperature (b), then HF solution etched (c), reoxidized at much higher temperature (d), and finally HF solution etched (e). The main result is best summarized in curves (d) and (e). After the high temperature thermal oxidation in air at 600°C, all traces of hydrogen have disappeared (see curve d), as evidenced by the absence of the SiH_x bands centered at 2100 cm^{-1}. This treatment clearly produced strong Si-O and Si-O_2 bands centered at 1100 cm^{-1}, confirming the formation of an oxide layer. Subsequent dipping in HF solution eliminated all the bands associated with silicon oxide and gave a spectrum with prominent silicon-hydrogen bands at 2100 cm^{-1} (SiH_x stretch), 900 cm^{-1} (SiH_2 scissor), and 650 cm^{-1} (SiH_x bends) (see curve e). These data clearly show the removal of silicon oxide in HF with subsequent termination of the surface by hydrogen. Ubara et al. (1984) (43) postulated a mechanism to account for the hydrogen termination, which was later confirmed by ab-initio calculations (46) and will be discussed in detail later in this section.

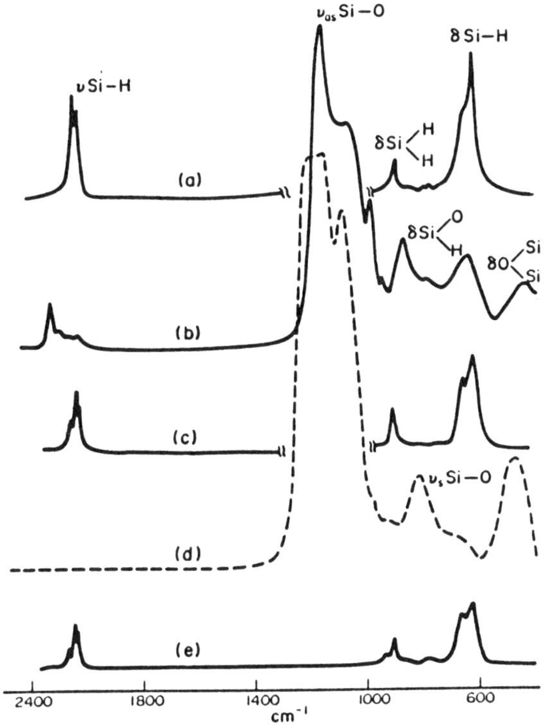

Figure 8. Transmission infrared absorption spectra of hydrogenated microcrystalline silicon (μc-Si:H): (a) as grown using RF-reactive sputtering in H_2 at 250°C, (b) after thermal oxidation in air at 200°C for 5 h, and (c) after subsequent HF etching of the previously oxidized surface; (d) after a thermal oxidation at 600°C for 1 h, and (e) subsequent HF etching of this previously thermally oxidized surface. (From Ref. 43.)

Despite these early pieces of work clearly pointing to hydride formation upon HF solution etching, it was still widely believed that fluorine was in fact the passivating agent. Raider et al. (1975) (47) attributed the hydrophobic nature of HF-etched surfaces to the presence of Si-F bonds or adsorbed HF. Licciardello et al. (1986) (48), indicated that the presence of an organic (hydrophobic) overlayer and not the Si-F surface per se was responsible for the hydrophobicity, although they still assumed that this overlayer was deposited on top of the *F-terminated* silicon surfaces. In addition, Weinberger et al. (1985) (49) explained the electronic passivation of HF-treated Si wafers as being due to fluorine termination of the Si dangling bonds. They further argued that the strength of the Si-F bond would make it an extremely stable surface, a point which will be discussed further in this section. Indeed, indirect support for F-passivation was drawn from the observation that F is a stable adsorbate on the surfaces of silicon prepared in UHV (41). Furthermore, fluorine was directly identified by XPS on HF-treated silicon surfaces, (50) suggesting that a monolayer of fluorine is present at the surface.

The year of 1986 was perhaps the turning point in the H- vs. F-termination debate. Yablonovitch et al. (1986) (51) observed that the F concentration, as measured by XPS, was highly variable and could not be the correct explanation for the remarkable surface passivation achieved subsequent to HF etching. Using infrared spectroscopy, they showed that a monolayer of hydrogen was adsorbed on the surface of HF-etched Si, with a spectrum characteristic of clean Si-H bonds (i.e., without oxygen or fluorine as a nearest neighbor). Another clear demonstration of the variability of the fluorine concentration on the surface came from the XPS data of Grunthaner and Grunthaner (1986) (52) showing no detectable oxygen, fluorine, nitrogen, or sulfur on HF/ethanol spin-etched silicon samples. Figure 9, for instance, shows the Si 2p core level spectra measured after different chemical treatments. The spin etched samples are characterized by an unshifted Si 2p core level, making it possible to set an upper limit of 0.1% of a monolayer of Si directly bonded to electronegative elements, such as fluorine or oxygen, for the spin-etched samples (53).

Independently, Grundner and Jacob (1986) (54) published the results of detailed EELS and XPS studies of oxidized and HF-etched silicon surfaces. The EELS data, shown in Fig. 10, consistently showed strong hydrogen vibrations at 2100 cm^{-1}, 900 cm^{-1} and 650 cm^{-1}, although hydrocarbons (2800 cm^{-1}) and OH (3400 cm^{-1}) vibrations contributed also to the spectra (55). The intensity of the fluorine 1s line measured in XPS data (55)(56) indicated that the fluorine concentration, after a de-ionized water rinse, was always less than 1 - 2% monolayer. This work clearly confirmed the formation of silicon hydrides upon HF etching and suggested that fluorine was a contaminant that could be removed by rinsing in water.

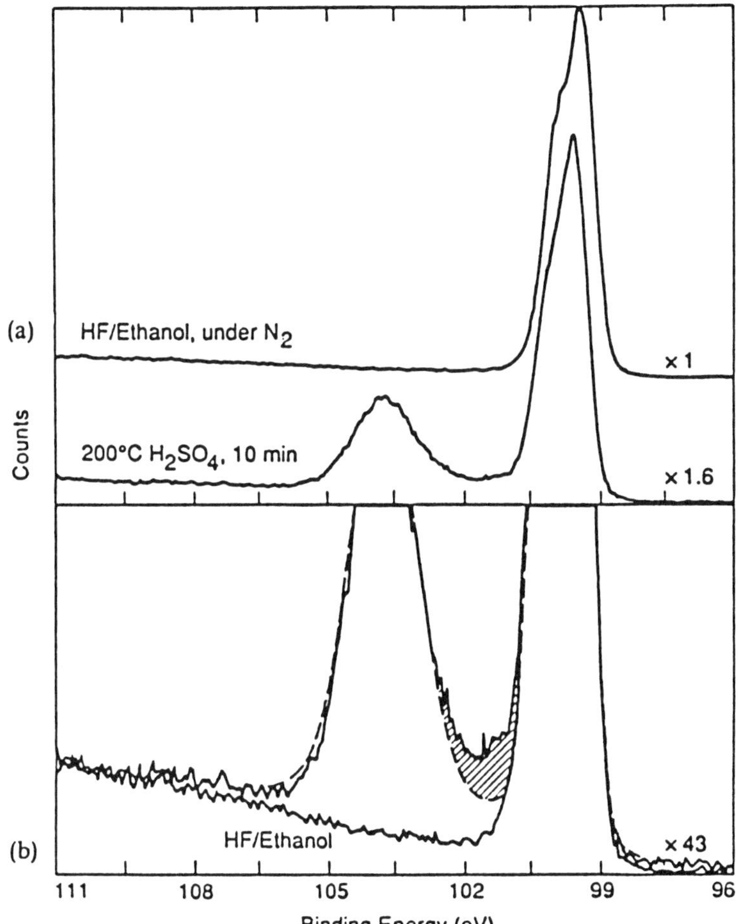

Figure 9. (a) Typical x-ray photoelectron spectra of the Si 2p core level obtained before (lower spectrum) and after (upper spectrum) a spin-etch in N_2 by using HF in ethanol. (b) Expansion and overlay of the data shown in (a) to emphasize the region between the peaks that correspond to the Si substrate and to the SiO_2. The dashed line represents a least-square fit to these two major components. The cross-hatched region corresponds to Si suboxide species. (From Ref. 53.)

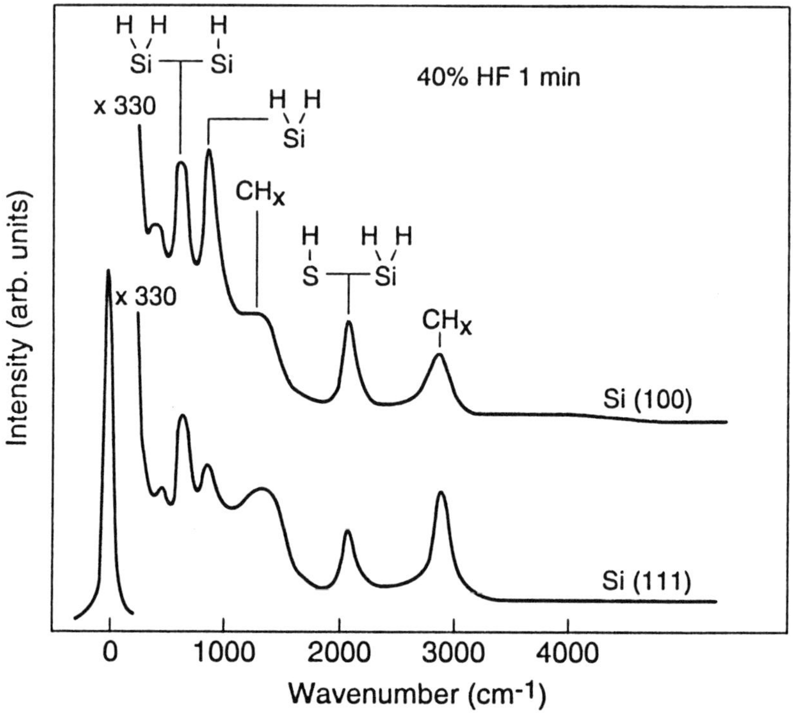

Figure 10. Electron energy loss spectra of Si(100) and Si(111) surfaces after 1 min. immersion in 40% HF solution (no rinsing) and subsequent introduction into UHV. The assignment of the main observed losses is summarized schematically above the spectra. (From Ref. 55.)

Both the surface-recombination velocity measurements of Yablonovitch et al. (1986) (51) and the extensive XPS and EELS studies of Grundner et al. (54)(55) motivated a number of photoelectron emission and infrared absorption studies (57)-(60). Takahagi et al. (1988) (57) used XPS, UPS and IR absorption to quantify the chemical species on the silicon surface after a combination of UV/ozone cleaning and HF dipping. Using dilute HF solutions, they achieved low levels of contamination (totaling less than 5% of a monolayer of O, C and F) and detected both Si-H and Si-H_2 species on the Si(100) surface. A typical XPS survey scan is shown in Fig. 11, which emphasizes the chemical purity of HF cleaned surfaces.

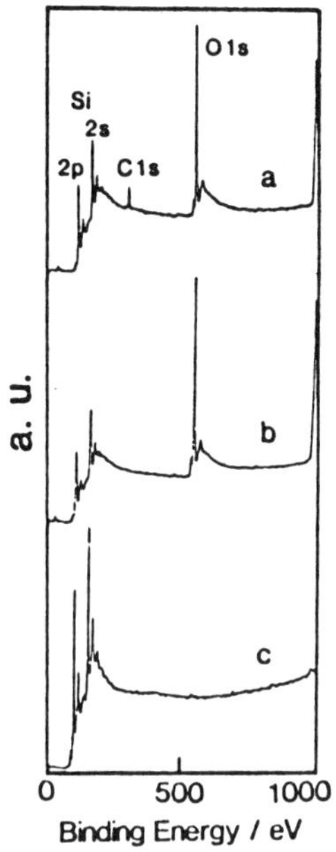

Figure 11. XPS Survey scan spectra of Si wafer surfaces (a) before UV-ozone cleaning (as purchased with 7Å thick native oxide and a 2Å thick organic contamination), (b) after UV-ozone cleaning with a low-pressure mercury lamp (184.9 nm and 253.7 nm emission) in an oxygen atmosphere, and (c) after subsequent HF dipping of the UV-cleaned sample in a 1% HF solution. (From Ref. 57.)

Burrows et al. (1988) (58) and Chabal et al. (1989) (59) investigated the silicon-hydrogen stretch vibrations (2000 - 2200 cm^{-1} region) to quantify the surface morphology of silicon etched in dilute HF (no rinsing) and kept in a purged environment. The spectra for both Si(100) and Si(111) surfaces displayed a variety of hydrides (mono-, coupled mono-, di- and tri-hydrides) consistent with atomically rough surfaces covered with roughly one monolayer of hydrogen, discussed in detail later in this chapter. Although no detectable absorption bands were observed in the O-H stretch (3600 cm^{-1}) and CH_x stretch (2800 cm^{-1}) region, the sensitivity of these IR absorption measurements only placed an upper limit of 10% monolayer (ML) for these species.

Fenner et al. (1989) (60) applied much more sensitive techniques (XPS and AES) for the detection of C, O, F and N to samples prepared by various wet-chemical techniques, cleaving in UHV, and ion sputtering. They found that, among the various wet-chemical techniques, spin-etched samples with

HF-alcohol mixtures (HPLC grade) exhibited the lowest contamination levels (0.03 ML C, and 0.005 ML O and F), close to levels found on cleaved Si in UHV. By comparison, samples dipped in HF solutions or sputtered and annealed showed a tenfold increase in surface residue.

Very recently, Dumas and Chabal (1991,1992) (61)(62) have used Energy Electron Loss Spectroscopy (EELS) to characterize silicon surfaces etched in buffered HF solutions. For samples rinsed in de-ionized water after the etching, they find that the concentrations of impurities, such as Si-F, Si-C, Si-O, Si-OH and Si-CH_x, are less than 1% of a monolayer (see Fig. 12). All losses in the EEL spectra can be assigned to hydrogen or silicon vibrations (61). This is in contrast to the early EELS data (55) (see Fig. 10) where relatively intense losses around 800 - 1100 cm^{-1} (oxide) and around 2900 cm^{-1} (hydrocarbons) were apparent. Possible reasons for the discrepancy are (a) variations in the purity of the chemicals used, (b) the exact handling and rinsing procedures, and (c) the wafer introduction and evacuation procedures used for these UHV studies.

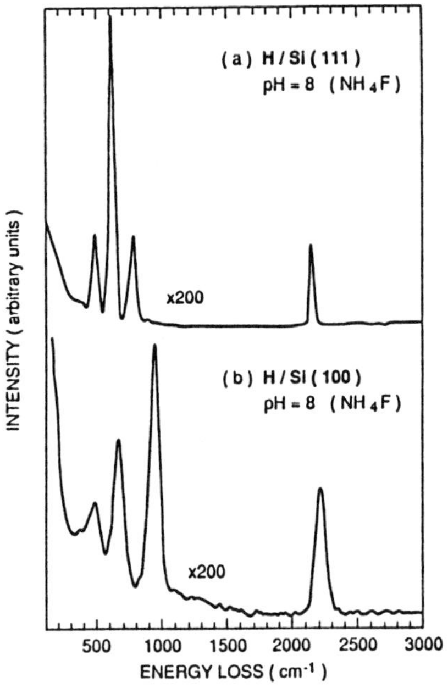

Figure 12. Electron energy loss spectra of (a) Si (111) and (b) Si(100) after etching in an ammonium fluoride solution (NH_4F, pH = 7.8) and a brief (~10 sec) rinsing in DI water. The x200 factor corresponds to the magnification of the spectrum relative to the elastic peak. (From Ref. 61.)

The main conclusion of the above studies is that hydrogen acts as the passivating agent and is the direct result of HF etching. The concentration of contaminants, such as carbon, oxygen and fluorine, depends on the details of processing. In particular, the rinsing procedure after the last etching step directly affects the concentrations of fluorine and oxygen.

Mechanism of Hydrogen Termination. Part of the confusion concerning fluorine termination of the Si following HF etching has arisen not only because of the stability of the Si-F bonds, but because the accepted explanation of the mechanism for SiO_2 dissolution leads automatically to fluorine terminated Si. The dissolution of SiO_2 by HF can be depicted in its simplest form in the following reaction:

$$SiO_2 + 4HF \rightarrow SiF_4 + 2H_2O$$

Notice that the above reaction involves HF molecules and not F^- ions in the solution. HF is a weak acid having an equilibrium constant such that it does not dissociate readily in concentrated solutions (63). In addition, Judge (1971) (63) showed clearly that even if F^- ions are available, they have an etch rate which is negligible compared to HF and HF_2^- species. Thus, only HF in its associated form need be considered in the dissolution mechanism. HF molecules attack Si-O bonds by inserting themselves between the Si and O atoms. This reaction is depicted schematically in Fig. 13a as if it were the last Si-O bond to be broken before reaching the Si substrate. This insertion occurs with a low activation barrier because the reaction is highly exothermic and conserves the number of broken and reformed bonds. The reaction is also greatly facilitated by the highly polar nature of the Si-O bond, which the highly polar HF molecule can use to its advantage during attack. The Coulomb attraction naturally leads to having the positively charged H-atom associated with the negatively charged O-atom, and the negatively charged F-atom associated with the positively charged Si-atom of the Si-O bond. This liberates H_2O into the solution and leaves Si-F in its place on the surface (Fig. 13b). The Si-F bond is the strongest single bond known in chemistry with a bond energy of ~6 eV. The bond strength of the Si-H is only ~3.5 eV and leads one, based on these thermodynamic considerations, to conclude that the F-terminated surface must be more stable than the H-terminated surface.

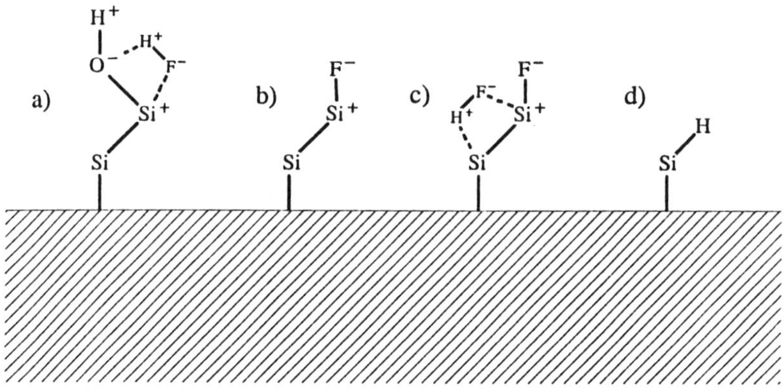

Figure 13. Schematic representation of silicon etching and hydrogen passivation by HF.

Ubara, Imura and Hiraki (1984) (44) were the first to propose a reaction mechanism to get around this dilemma. They recognized that the Si-F bond must be highly polar because of the large electronegativity difference between these atoms. They suggested that the Si-F bond causes bond polarization of the Si-Si back-bond allowing HF attack of the back-bond, as illustrated in Fig. 13c. This kinetically favorable pathway results in the release of stable SiF_x species into the solution leaving Si-H behind on the surface, as shown in Fig. 13d. The validity of this proposed pathway was confirmed using first principles molecular orbital calculations of the activation energies of these types of reactions on model compounds by Trucks et al. (1990) (46). In these calculations, an activation energy of ~1.0 eV was found for reactions of the type shown in Fig. 13c. Low activation energies such as these are due to the charge transfer between the silicon and fluorine atoms, as originally suggested by Ubara et al. (44). In the absence of charge transfer, as is the case for the nonpolar Si-H bonds, the activation energy of the Si-Si back-bond attack is 1.6 eV, which is 0.6 eV higher in energy than for that of fluorinated Si species. The impact of the Coulomb interaction could also be observed by inverting the HF molecule, making the attack occur in opposition to the Coulomb force. In that case, an activation energy of 1.4 eV is obtained. In summary, HF attacks polar species very effectively but is much less effective against nonpolar species. Also, the reactant must attack the bonds in a specific orientation to take advantage of the Coulomb interaction between the positively and negatively charged atoms. These concepts allow us to understand why it is hydrogen and not fluorine that terminates the Si dangling bonds after HF solution etching, and why HF dissolves oxide so readily but leaves the Si relatively untouched.

The preceding arguments give us a basic understanding of HF etching. In reality, however, the situation is much more complex, with HF, HF_2^-, F^-, H_3O^+, OH^-, NH_4F species together in the solution, in chemical equilibrium with one another, not to mention the steric constraints at the surface or the effects of solvation on the reactions. A complete understanding of the detailed chemistry is not available at this time, but certain conjectures can be made with a reasonable degree of confidence. The calculations mentioned above were performed for molecules in free space and thus can accurately describe only gas phase reactions. We know, however, that water vapor is needed to initiate SiO_2 etching reactions with anhydrous HF (64). One can argue that the main effect of placing the polar HF molecule into water is to surround it, on the average, with water molecules in the proper orientation to minimize the Coulomb energy. This, in turn, weakens the H-F bond, facilitating all HF reactions that must break the HF bond. One can, therefore, postulate that solvation simply lowers the activation barriers that exist for the gas phase reactions described above. Reaction rate data are not available for the Si-Si back-bond attack, but HF dissolution of SiO_2 has an activation energy of approximately 0.35 eV (63), to be compared with the 0.55 eV calculated for the gas phase reaction (46). The heat of solvation to place an HF molecule into solution is ~0.4 eV and is consistent with the observed 0.2 eV lowering of the energy barrier. This point of view also allows one to rationalize the HF and HF_2^- etching behavior observed by Judge (1971) (63). One can think of HF_2^- simply as a more highly solvated form of HF with a weaker bond strength that explains the lower activation energy for SiO_2 dissolution (0.31 eV) as well as the increased rate of dissolution (factor of 4 - 5). The question of steric hindrance at the surface will be discussed in a later section. In general terms, confidence should be placed on the chemical trends discussed above while remaining skeptical of the exact activation energies, since modifications due to steric constraints or solvation can be expected.

One last point needs to be clarified regarding HF solution chemistry. It is now clear that the OH^- concentration has a drastic effect on the etching that occurs with HF solutions (65). Experiments show that water rinsing alone can remove dihydride species at steps (65)(66) and leads to monohydride termination on Si(111) surfaces (66). In this regard early workers noticed that samples which were rendered hydrophobic in HF remained so even after boiling in water for extended periods of time (6). It is equally curious that chemo-mechanically polished silicon wafers (polished in slurries with pH ≤ 13) are hydrophobic and are terminated with hydrogen (22). These observations taken together suggest that Si surface reactions with OH^- can also lead to hydrophobic hydrogen terminated Si surfaces once the surface oxide is removed. This leads one to the speculation that HF and OH^- chemistry may be similarly removing Si atoms bonded to electronegative elements by back-bond attack of the polarized Si-Si bond. It is also interesting to note that HF and OH^- in solution may have similarities in their reaction pathways at the surface. Pursuing this idea is clearly a direction for future research.

3.2 Structure and Morphology

Techniques. The atomic structure of silicon surfaces can only be inferred indirectly using diffraction, imaging and vibrational techniques. The information is therefore subject to the limitations of each technique. The aim of this section is to summarize briefly the information obtainable from each technique and their respective limitations.

Diffraction Techniques: LEED, RHEED and X-Rays. Low Energy Electron Diffraction (LEED) and Reflection High Energy Electron Diffraction (RHEED) have been used extensively to study the nature of the long range order of crystalline surfaces. With normal incidence LEED and grazing incidence RHEED, the electrons penetrate 2 - 5 layers into the substrate depending on the electron energy. The diffraction pattern is therefore dominated by the bulk, i.e., most of the diffracted intensity is concentrated into the bulk diffraction beams (integral order beams). When the surface is reconstructed or when regular arrays of steps are present on the surface, additional diffraction spots are present in the pattern, reflecting the extra periodicity. Using electron optics, these additional spots can be Fourier transformed to give real space images of the reconstructed or stepped areas (Low Energy Electron Microscopy) (67). This technique is particularly useful when the surfaces are macroscopically inhomogeneous but locally ordered.

LEED and RHEED give much more limited information when the surfaces are disordered on an atomic scale, particularly if the surface atoms are in a bulk-like position (i.e., are not reconstructed). The diffraction pattern is then due to the bulk order only. The surface atoms modulate the diffracted intensity and contribute to the diffuse background. Quantification of this background is difficult and is meaningful only in a statistical framework. Therefore, the observation of a 1x1 LEED or RHEED pattern does not imply the existence of an atomically flat, unreconstructed surface, but merely the *lack* of an ordered, reconstructed surface.

When surface contamination is eliminated and extraneous scattering is low, quantitative information on the surface morphology can be obtained using LEED spot profile analysis (68). This technique, based on the measurement of the LEED spot profile as a function of electron energy, gives a quantitative measure of the step density, and therefore of the surface roughness.

Grazing Incidence x-ray diffraction is a particularly useful diffraction method because there is no requirement for the experiments to be performed under vacuum conditions. At the same time the penetrating nature of x-rays renders the technique the least surface sensitive but makes it possible to probe buried interfaces. Another advantage of x-rays, particularly those produced by electron storage rings at facilities like the National Synchrotron Light Source (NSLS) at Brookhaven, is their very high resolution. The typical resolution function of a 3-axis diffractometer fills only 5×10^{-10} of the Brillouin zone of Si. Such a resolution is necessary because the bulk diffraction is localized in point-like Bragg peaks that are smaller than this resolution element. Point defects and bulk thermal diffuse scattering (TDS) are diffuse in all reciprocal space directions. In contrast, surfaces and interfaces, by virtue of their 2-dimensional translational symmetry, give rise to rod-like lines of scattering (69). Therefore, 3-D Bragg peaks can be filtered out by avoidance and diffuse scattering by background subtraction. The remaining problem is to distinguish the buried interface from the surface through which it is measured. Typically, what is measured is the intensity profile of one of these surface-symmetry-sensitive rods (70). This can then be fit to specific models of statistical disorder that describe the roughness on an atomic level, involving a distribution of occupancy in each layer Si in the region of the interface. As a result, specific information can be obtained on the structure of chemically prepared surfaces as well as that of the Si/SiO_2 interface.

Imaging Techniques: STM, AFM and TEM. Unlike diffraction methods, Scanning Tunneling Microscopy (STM) provides images of surfaces with atomic resolution in real space. The microscope produces images by bringing a conducting probe (the tip) up to the sample surface until the separation distance is small enough (<10 Å) that the electron tunneling probability is appreciable ($>10^{-4}$). At this point, a measurable current flow may be induced by applying a small bias voltage (1 mV → 4 V) between the tip and sample. A feedback system regulates the tunneling current by varying the tip height to maintain a constant *tunneling probability* as the tip is rastered parallel to the average surface plane. Under these bias conditions, the transiting electrons do not follow classical trajectories, but instead obey paths dictated by the laws of quantum mechanics. The STM owes its vertical resolution (~0.1 Å) to the fact that the tunneling probability increases by an order of magnitude for each angstrom increase in tip-sample separation, and its lateral resolution (~2 Å) to the atomic nature of the probe tip and sample. This allows unusual features in the electronic density of state of the tip to influence the tunneling images under some conditions (71). Furthermore, the appearance and even symmetry of the images are dependent on the polarity and magnitude of the bias voltage (72), especially for highly reconstructed semiconductor surfaces, such as those of silicon and germanium. One must therefore bear in mind that the STM produces images of integrated electronic density of states, *not atom core positions*.

The Atomic Force Microscope (AFM) is a related scanning probe microscope which generates real-space images in a similar fashion to the STM, namely by sensing the attractive/repulsive force between a probe tip and the sample surface as the distance between the two is reduced. For this reason, it holds much promise in atom-by-atom imaging of *insulating* materials. Currently, atomic resolution has only been achieved when the AFM is operated in its repulsive mode (thus with the tip "touching" the sample), raising the question of possible damage. The best resolution obtained in the noncontact attractive mode has been ~100 Å. Stable operation in the attractive mode has been limited by the difficulty in producing a reliable mechanical system with a large enough dynamic range.

Transmission Electron Microscopy (TEM) uses high energy (100 keV) electron scattering to image interfaces and bulk materials. Both plan-view and cross-sectional TEM gives useful structural information about surfaces and interfaces. Like x-ray diffraction, the penetration of the probe is relatively deep (~500 Å), which makes it one of the techniques that can be applied to buried interfaces.

Vibrational Spectroscopies: Inelastic He Scattering, EELS and IRAS. Vibrational spectroscopies are the least obvious structural techniques. Yet, detailed information about local and extended structures is contained in the vibrational spectra of clean surfaces (substrate phonons) and of adsorbate layers (overlayer modes). This is particularly true when the overlayer contains light elements (e.g., H), because the associated vibrational modes are found at a high frequency and are well separated from the substrate phonons.

Inelastic He atom scattering is a high resolution vibrational spectroscopy, best suited for the study of low frequency (<500 cm^{-1}) phonons on *well-ordered* crystalline surfaces (3). A monoenergetic beam of thermal energy (~0.04 eV) He atoms is completely reflected by the surface (no penetration into the bulk). The reflected beam is analyzed by use of time-of-flight techniques yielding high spectral resolution (~0.1 cm^{-1}). The large mass of the He atom, relative to that of an electron, makes it possible to impart a large momentum together with energy to the surface phonons. This technique can thus investigate the *dispersion* of surface phonons over the whole surface Brillouin zone (SBZ) (~$Å^{-1}$). The dispersion results from surface interactions within a well ordered domain and is, therefore, a sensitive measure of surface order. The technique is restricted to low frequencies (<500 cm^{-1}) because the turn around time of He atoms is slow (10 fs) at thermal energies. Increasing the He atom energy shortens the turn around time, but greatly complicates the spectrum with multiphonon effects.

Electron Energy Loss Spectroscopy (EELS) and Infrared Absorption Spectroscopy (IRAS) are very useful techniques to study the microscopic structural arrangement of H-terminated silicon surfaces because they can resolve different hydride phases: monohydrides (relaxed and strained), dihydrides, and trihydrides. In studies of adsorbates with vibrational EELS, 2 - 10 eV electrons are incident on the surface; the reflected electrons (0.1 to 1% of the incident beam) are collected and their energy is analyzed using electrostatic analyzers. A small fraction of the reflected electrons lose a quantum of energy by exciting surface vibrations, through either a long-range interaction (dipole scattering) or a short-range interaction (impact scattering). Dipole scattering is observed only in the specularly reflected direction. For dipole scattering, the spectrum is dominated by the components of vibration perpendicular to the surface (74), just as is the case for optical *external* reflection spectroscopy (75). Indeed, the ratio of the dipole scattering cross-sections of parallel to perpendicular vibrations is proportional to n^{-4}, where n_{Si} = 3.4 is the refractive index of silicon. For impact scattering, the cross-section has a more complex dependence on the mode polarization and has also a more complex angular dependence that can sometimes be used to give symmetry information about the vibrational modes.

An important attribute of EELS is its large spectral range (0.01 - 2 eV) for vibrational studies. This range includes the frequencies of the silicon surface phonons and of all the silicon-hydrogen vibrations. Furthermore, the intensity of the specular beam (and to a lesser extent, the spectral resolution) depend on the surface morphology and can therefore be used to characterize surface roughness. Its main drawback is the limited resolution (~20 cm^{-1}, although recent advances have made it possible to increase it by a factor of three) (76). Both inelastic He atom scattering and EELS require the introduction of the samples into vacuum, which is a disadvantage for wet-chemically prepared silicon samples.

Infrared absorption spectroscopy is especially sensitive when a multiple internal reflection (MIR) geometry can be used (see Fig. 14). For silicon, the sensitivity to parallel and perpendicular components of the adsorbate vibrations (i.e., above the silicon surface plane) is nearly the same (within 20% for $\theta_{internal} = 45°$), so that the orientation of *ordered* surface dipoles can be determined, in principle, by use of polarized radiation (see Fig. 14). The accuracy of such orientation determination depends on the precision available in measuring the electronic screening involved in the dynamic interaction between the adsorbates (77). For disordered surfaces, the IR absorption spectra are useful to quantify the *average* orientation of the surface dipoles, and thereby to learn something about the surface disorder and roughness.

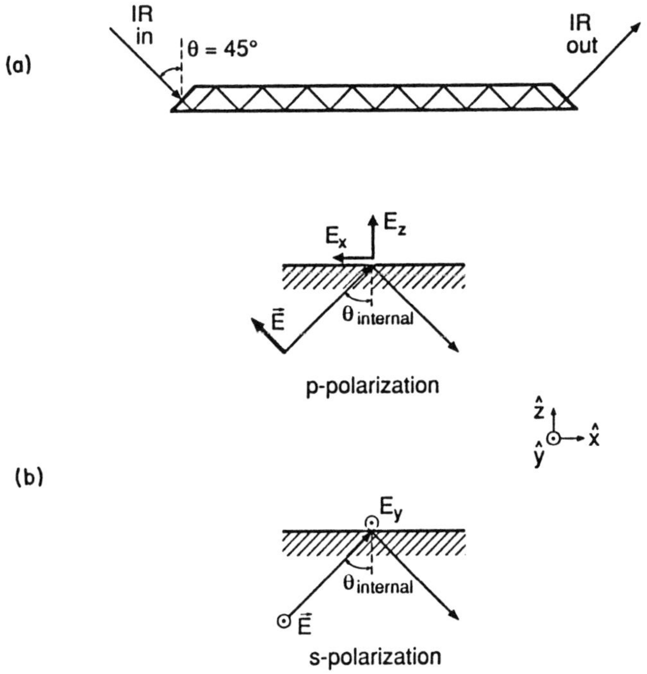

Figure 14. (a) Side view of a silicon crystal used for multiple internal reflection experiments. Typical dimensions are 3.8 cm length, 2 cm width and 0.05 cm thickness. The short sides are beveled at 45° to allow the IR beam to enter and exit as shown. The total number of reflections in this example is 75. (b) Schematic drawing of the electric fields present on the vacuum side of silicon surface for p-polarization and s-polarization, respectively.

When the H-terminated silicon surfaces are homogeneous, the high resolution ($\leq 0.1\ cm^{-1}$) of the IR technique gives a wealth of information. For instance, different hydride species can be identified based on the Si-H stretch spectra alone. Experimentally, the mode assignments are simplified by doing isotopic mixture experiments. In this manner, H-stretches can be studied in the dilute limit, where coupling to near neighbor atoms is suppressed due to the mass difference between H and D. The thus obtained "isolated spectra" are simplified (no dynamical splittings), and are only composed of bands associated with fundamentally different species. As can be seen in Fig. 15, the isolated spectra (a) are simpler than the isotopically pure spectra (b). The three fundamental bands are associated with mono-, di-, and tri-hydride species (59). Unambiguous assignments are based on state-of-the-art first principles cluster calculations of force constants and normal frequencies of vibration. The results of such calculations for the isolated frequencies along with their associated splittings are shown schematically in Fig. 15 as vertical bars. The band positions of the isolated frequencies of the mono-, di-, and tri-hydrides and the mono-, di-, and tri-deuterides are in good agreement with the measured values depicted in Fig. 15. Furthermore, the splittings evaluated from the measured isolated frequencies are in excellent agreement with the measured values for both the hydride and the deuteride stretches (59).

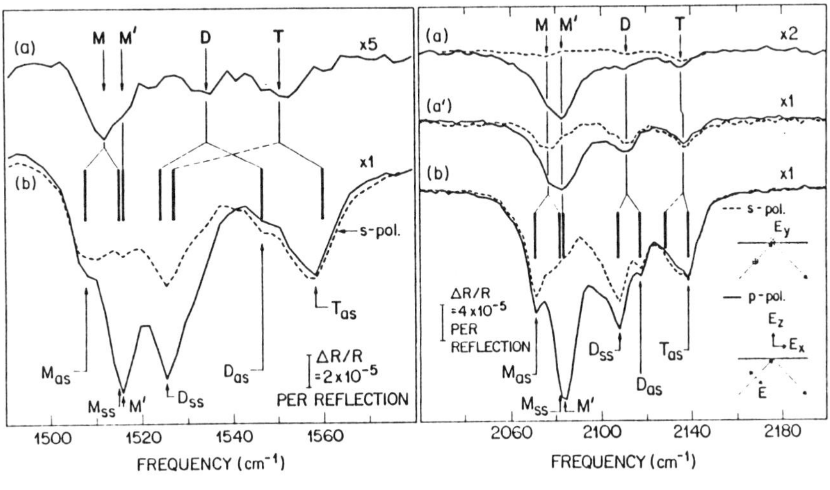

Figure 15. Polarized IR absorption spectra of the Si-D stretch vibrations (left panel) and Si-H stretch vibrations (right panel) for various isotopic concentrations after etching of a Si(111) sample in dilute HF/DF.

In the left panel, (a) corresponds to 7% D and 93% H on the surface and (b) to 95% D and 5% H. The observed isolated frequencies are M = 1512 cm^{-1}, D = 1533.5 cm^{-1}, and T = 1550 cm^{-1}. In the right panel, (a) corresponds to 10% H and 90% D, (a') to 25% H and 75% D, and (b) to 100% H. The observed isolated frequencies are M = 2077 cm^{-1}, D = 2111 cm^{-1}, and T = 2137 cm^{-1}. In both cases, the thick vertical bars represent the calculated coupled mode splittings from the measured isolated frequencies, as shown in the previous figure. (From Ref. 58.)

The transmission cut-off at low frequencies is the main limitation of the multiple internal reflection (MIR) geometry. It is particularly severe for silicon, which is opaque below 1500 cm^{-1} due to multiphonon absorption (78). To study modes below this frequency, different schemes can be used (Fig. 16). If silicon can be grown epitaxially on germanium, then the MIR geometry can still be used with the germanium as the main substrate material (transparent above 650 cm^{-1}), as shown in Fig. 16 (top). Alternatively, the silicon sample can be pressed against a germanium MIR plate (see Fig. 16, middle), giving good sensitivity to modes *normal* to the surface (79)-(81). Otherwise, a transmission or external reflection geometry can be utilized to minimize the substrate absorption. In general, the transmission geometry (e.g., at the Brewster angle shown in Fig. 16, bottom) gives a better sensitivity than the reflection geometry (82).

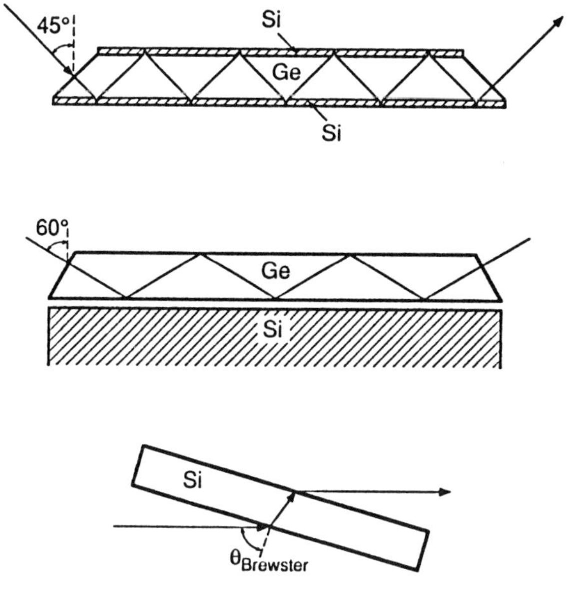

Figure 16. Various configurations for studying surface vibrations in a range of substrate absorption.

Results. This section starts with the effects of aqueous HF (concentrated and dilute HF solutions) on the surface morphology of Si(100) and Si(111). We then examine separately the etching of Si(100) and Si(111) in buffered HF solutions. Throughout these sections, the nature of the starting

Si/SiO_2 interfaces and the role of water (e.g., rinsing) in the etching process are also considered since they are relevant to the final surface structure

Si(100) and Si(111) Etched in Aqueous HF Solutions. The first information available on the morphology of H-terminated Si(100) and Si(111) was the study by Hahn and Henzler (1984) (3). Their goal was to use LEED for investigating the Si/SiO_2 interface morphology as a function of oxidation conditions and polishing methods. As with most surface analysis techniques, LEED requires SiO_2 film removal prior to the measurement. This was achieved by dissolving the oxide layer in concentrated HF followed by rapid load-locking into the analysis chamber. In this manner, the densities of step-atoms could be studied as a function of oxidation conditions. An example of how the analyses were performed is found in Hahn (1986) (83), where LEED spot-profile-analysis was used to study the atomic structure of chemo-mechanically polished Si(111) wafers. The original data, shown in Fig. 17, demonstrated that chemo-mechanically polished Si(111) etched in concentrated HF is step-free over 100 Å distances, on the average. In such an experiment, the electron wavelength is varied by varying the electron energy, leading to in-phase and out-of-phase scattering conditions from one atomic layer to another. The lack of broadening as a function of the electron wavelength shown in Fig. 17 indicates that the width is instrumentally limited and gives a lower limit for the average length of the terrace of 100 Å.

Next, Grundner and Schulz (1988) (55) used the vibrational frequency information from their electron energy loss spectra (EELS) on HF-treated Si(111) and Si(100) to investigate the nature of the H-termination. The main conclusion of this study was that Si(100) was dihydride terminated and Si(111) was monohydride terminated. Conclusive evidence for the dihydride species is the presence of a loss peak at 900 cm^{-1}, corresponding to the SiH_2 scissor mode. The EEL spectrum shown in Fig. 10 (top) for Si(100) is characterized by a strong loss peak at 900 cm^{-1} and compares well to spectra obtained in UHV upon atomic H dosing of clean Si(100) (84). This finding and the observation of a 1x1 LEED pattern (21) led these authors to the conclusion that a uniform dihydride phase had been obtained. For HF-etched Si(111) surfaces, the strong Si-H stretch loss at 2080 cm^{-1} in Fig. 10 (bottom), together with the "good quality" 1x1 LEED pattern (21), led them to conclude that the surface was ideally monohydride-terminated. A weak loss at 900 cm^{-1} was attributed to dihydride at steps.

High resolution infrared reflection absorption spectroscopy (IRRAS) is a powerful technique that provides detailed information not available from EELS. Polarized spectra, taken by multiple internal reflection (MIR), are particularly useful to elucidate the surface structure of H-terminated silicon (see Sec. 3.2, Vibrational Spectroscopies). Contrary to the conclusions drawn from EELS, the infrared studies indicate that both Si(100) and Si(111) surfaces are atomically rough after similar etching treatments.

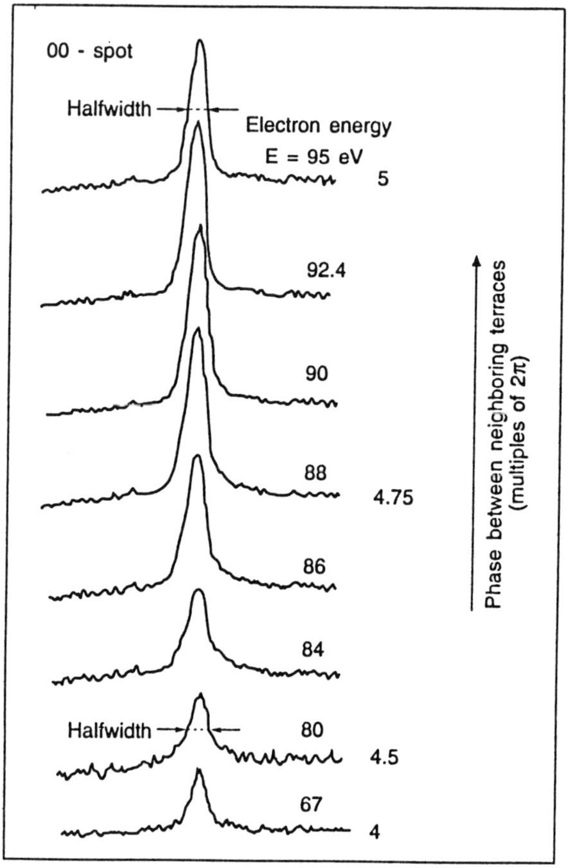

Figure 17. LEED spot profiles of Si(111) polished wafer after HF treatment. (From Ref. 83.)

The roughness is evident from the polarized IR absorption spectra shown in Figs. 18 and 15 for Si(100) and Si(111) respectively. The complexity of the hydrogen stretch spectra is clearly observed for both the Si(100) and Si(111) surfaces. An *ideally terminated* Si(100) surface would be characterized by two modes, the symmetric and anti-symmetric dihydride stretch. Furthermore, the symmetric stretch should be polarized normal to the surface and the anti-symmetric stretch should be parallel to the surface, contrary to what is experimentally observed in Fig. 18. Similarly, an ideally terminated Si(111) would be characterized by a single monohydride stretch mode polarized normal to the surface but is clearly not observed in Fig. 15. Although these surfaces are not ideally terminated, structural information can still be extracted by providing complete assignments of the observed bands. As pointed out in Sec. 3.2 (Vibrational Spectroscopies), the mode assignments of the various hydride species can be performed by using isotopic substitution experiments combined with force constant/normal mode analyses on model compounds (59). The silicon-hydrogen stretch spectra in Fig. 18a and Fig. 15 (a and a′, right frame) are simpler because the major species is deuterium, i.e., the hydrogen atoms are isolated from each other. The three main bands in the isolated spectra are assigned to the monohydrides (coupled M or not coupled M′), to dihydrides (D) and trihydrides (T). The corresponding spectrum for the silicon-deuterium stretch (a) for the isolated deuterium, and the coupled spectra (b), shown in Fig. 15 (left frame), confirm these assignments. The solid vertical bars in Figs. 15 and 18 correspond to the theoretical predictions for the splittings associated with each hydride structure, starting from the measured value of the isolated frequency (58)(59). With these mode assignments the structure of these H-terminated surfaces can be obtained.

The isolated spectra in Fig.18a show that dihydrides are the dominant species on the Si(100) surface, an observation consistent with the EELS studies. It is important, however, to point out that the dihydride observed in Fig. 18 is not the same as the dihydride observed on atomically flat Si(100) prepared in UHV by exposure to atomic H. The latter dihydride has its axis normal to the surface and is surrounded by strained monohydride units (85). The exact morphology of these chemically prepared Si(100) surfaces, however, cannot be inferred accurately from the IR spectra alone. The relative concentrations of mono-, di-, and tri-hydrides give qualitative information of the average surface morphology and are only useful for comparison.

For Si(111), the spectra in Fig. 15a are dominated by monohydride stretches, in agreement with the EELS data. In this case, part of the monohydride spectrum (M′) is found to be polarized normal to the surface and can be associated with ideal termination (monohydride on a Si(111) terrace) (77)(86). Having identified the M′ mode, the relative intensities of M, D, and T modes can be used to deduce the average surface structure, schematically represented in Fig. 19. In this figure, M terminates the side of adstructures and T terminates the (111) terraces. There are two types of uncoupled monohydrides (M″) and the ideal monohydride, (M′) and two types of dihydrides (D and D′). M″ has not been measured separately, either because its frequency is too close to that of the M′ and/or because its concentration is too low to be detectable. These data do show that only a quarter of the surface is covered with M′. Step edge monohydrides (coupled monohydride, M) comprise another quarter of the surface. Together, they make up approximately half of the surface hydrides leading to the strong monohydride EELS signature. Although, there appears to be an apparent contradiction to the EELS results, where no dihydrides were observed, EELS suffers from problems with interpretation depending on whether the scattering mechanism is believed to be dipolar (87)(88) or non-dipolar (62). The IR spectra give quantitative information of the dihydride stretch.

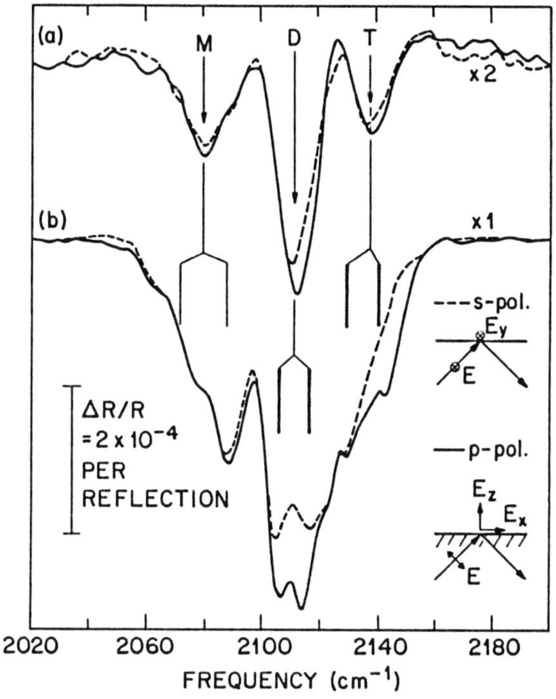

Figure 18. Polarized IR absorption spectra of the Si-H stretch vibrations for two isotopic concentrations after etching of a Si(100) sample in dilute HF/DF. (a) corresponds to 20% H and 80% D, and (b) to 100% H. The thick vertical bars represent the calculated coupled mode splittings from the measured isolated frequencies (labeled M, D and T). (From Ref. 59.)

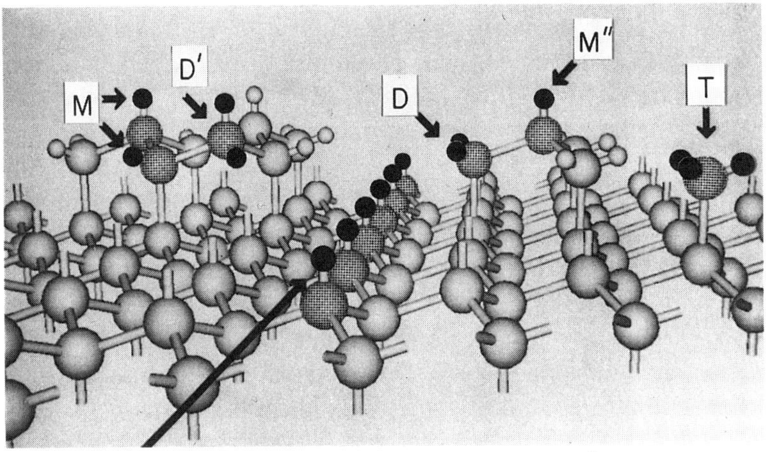

Figure 19. Schematic representation of possible surface structures on the Si(111) surface with their associated hydrogen termination. The ideal monohydride and trihydride termination are possible for an atomically flat (111) plane. The "horizontal" dihydride (D) terminates the corner of a small adstructure where an isolated monohydride (M′′) may exist. Both the "vertical" dihydride (D′) and coupled monohydride (M) can terminate larger adstructures of the type shown here. These are all the possible structures that do not involve surface *reconstruction*. (From Ref. 58.)

The "horizontal" dihydride (D in Fig. 19.) has been observed on the rough Si(111) at a frequency slightly shifted from that of the relaxed dihydride on Si(100) (85). The second type of dihydride (D′), in a plane perpendicular to the surface, has been extensively studied in the context of vicinal surfaces (miscut toward the $<\bar{1}\bar{1}2>$) (89). Notice that in Fig. 19 many dangling bonds have been left unterminated for clarity. HF etched surfaces, however, are completely hydrogen terminated. This implies that there would be a strong interaction between the "vertical" dihydride (D′) and the terrace monohydride directly beneath it. This steric interaction leads to the appearance of three stretch modes, at 2094, 2101, and 2135 cm^{-1}, (89) instead of only two in the 2110 cm^{-1} region for the unconstrained dihydride.

The observation of trihydride on the Si(111) surfaces is both interesting and controversial. It is clear from the isolated spectra of Fig. 15a that trihydride is present. Its concentration, however, is relatively low. It is now understood that a fraction of the strength of the mode labeled T_{as} at 2139

cm^{-1} in Fig.15b arises from the 2135 cm^{-1} mode of the vertical dihydride (D′). Recent STM images of Si(111) etched in 1% HF solutions (90) have shown a pattern more consistent with the presence of large regions terminated by trihydrides than of the small adstructures described above. This conclusion is based on the observation of threefold symmetry and 2.2 Å periodicity in the STM images, which is inconsistent with a Si lattice spacing. The most likely close-packed trihydride arrangement involves a rotation of neighboring trihydrides leading to an average H-H distance of ~2.2 Å (91). If the trihydride concentration is larger than 10% of a monolayer, these STM results are not consistent with the IR absorption spectra (59)(66) which indicate that mono- and di-hydrides are more numerous than trihydrides on Si(111) surfaces etched in dilute HF. The IRRA spectra also fail to show the mode at 2154 cm^{-1} associated with a uniform trihydride phase, observed upon adsorption and decomposition of di-silane (92). In addition, previous STM studies of Si(111) surface etching in dilute HF, while showing the surface roughness predicted by the IR data, did not show any evidence for a uniform trihydride phase (93). In view of these contradictions, the nature of the trihydride must be clarified in future work.

The main conclusion of the results presented above is that etching in dilute HF leads to atomically rough surfaces. Mono-, di-, and tri-hydrides coexist on both Si(100) and Si(111) surfaces. STM images of Si(111) (93) show structures of 10 - 20 Å diameter and 3 Å in height, accounting for about 50% of the surface, consistent with the IR data.

Si(100) Etched in Buffered HF Solutions. Buffered HF is composed of various mixtures of 50 wt% HF in H_2O and 40 wt% NH_4 F in H_2O. A common mixture used in the industry is 7:1 buffered HF, which has a pH of 4.5 and is composed of 7 volumes of NH_4F and 1 volume of HF. The main difference between aqueous HF and buffered HF is the solution pH, which is the object of the following discussion.

Raising the pH of the HF solution increases the etch rate of the H-terminated silicon surfaces. Figs. 20 and 21 clearly show that the morphology of chemically prepared Si(100) surfaces changes as the pH of the etching solution varies from 2 to 8. For a pH of 2 (Fig. 20a), the IR absorption spectra are dominated by dihydrides. In buffered HF (pH ~5), the spectrum sharpens and is dominated by coupled monohydrides (Fig. 20b). For higher pH's, the etching proceeds quickly, as evidenced by the gas bubbles forming at the sample surface. After etching in an ammonium fluoride solution (pH = 7.8), the dihydride contribution is again dominant (Fig.21a).

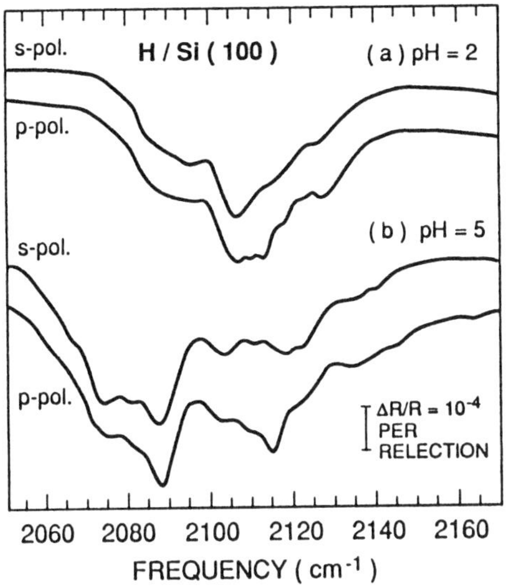

Figure 20. Polarized IR absorption spectra of Si(100) surfaces etched in: (a) dilute HF (1%, pH = 2) and (b) buffered HF (pH = 5). The chemically oxidized surface is used as a reference and the spectral resolution is 1 cm^{-1}. (From Ref. 62.)

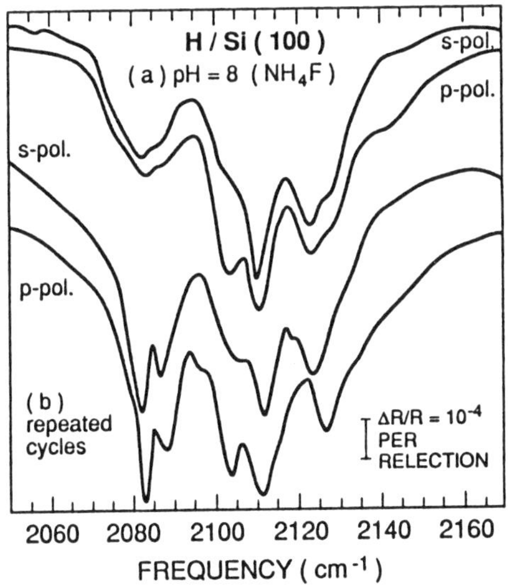

Figure 21. Polarized IR absorption spectra of Si(100) surface etched in a 40% NH_4F solution (pH = 7.8) once (a) and after repeated cycles of etching in a 40% NH_4F solution, chemical reoxidation in H_2O:H_2O_2:HCl (4:1:1) at 80°C, and a final etching in NH_4F (b). (From Ref. 62.)

The dihydride dominated spectra shown in Fig. 20a (pH = 2) are the result of a treatment in dilute HF, and represent a surface which is rough on an atomic scale, as previously discussed. The dominance of the monohydride modes in Fig. 20b suggests the formation of microfacets on the Si(100) surface (62)(65). At high pH (Fig. 21a), although the spectra revert back to being dominated by the dihydride stretch, the polarization of this mode is quite different from the pH = 2 spectra. In this case, the symmetric stretch (2105 cm^{-1}) is polarized normal to the surface and the anti-symmetric stretch (2112 cm^{-1}) is polarized parallel to the surface. Although, these polarizations would be correct for terrace dihydride, for the reasons given above, these surfaces are not believed to be atomically flat because of the existence of spectral contributions from mono- and tri-hydride that are still very strong. Furthermore, the monohydride spectrum is now centered around 2085 cm^{-1}, indicating the growth of (111) facets.

After several etching cycles, Fig. 21b shows that two sharper modes are resolved at 2084 and 2088 cm^{-1}. The first is assigned to the Si-H stretch of the ideal Si(111) monohydride, confirming that (111) facets develop in solutions of high pH. The second is probably associated with the symmetric stretch of coupled monohydrides; the asymmetric stretch mode is more highly screened and therefore not observed. Both types of monohydrides have symmetric stretches pointing away from the normal of the macroscopic surface plane and are therefore unscreened. In contrast, the dihydride modes in Fig. 21 are characteristic of dihydrides with their axis pointing along the surface normal. The simplest atomic arrangement consistent with these observations is a distribution of tent-like structures with a row of dihydrides at the roof top, (111) facets terminated with ideal monohydride on the sides and coupled monohydride at the periphery of the facets. Since the facets are small, the concentration of coupled monohydrides is as high as that of ideal monohydrides.

The use of buffered HF may be ill-advised in attempting to prepare atomically flat (100) surfaces, since (111) facets develop upon etching. Increased surface roughness has been directly observed after buffered HF etching (24)(28). As shown in Fig. 22, a control wafer is relatively smooth with ~2 Å rms roughness, whereas, a wafer treated in buffered HF is characterized by ~5 Å rms roughness. In attempting to smooth Si(100) surfaces, one might use thermal oxidation that is known to result in high-quality Si/SiO_2 interfaces. In such an experiment, a 1000 Å thick dry O_2 oxide was grown at 1000°C and post-annealed in Ar at this temperature for 30 min. The oxide was then removed with concentrated HF. The complex spectra obtained give conclusive evidence that atomically inhomogeneous surfaces again result.

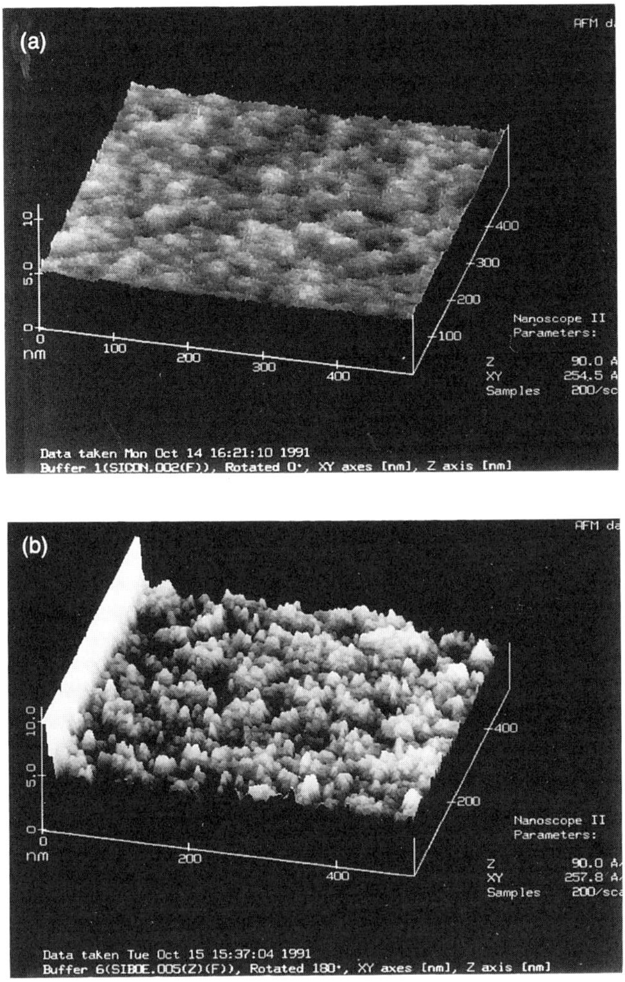

Figure 22. Atomic force microscope images of (a) a chemo-mechanically polished Si(100) control wafer (~2 Å rms) and (b) a Si(100) wafer etched in 7:1 buffered HF solution for 10 min (~5 Å rms).

In summary, Si(100) surfaces are microscopically rough when treated in either dilute or concentrated HF. These surfaces are macroscopically roughened by buffered HF solutions due to (111) facet formation. To date, little is known about the nature of such surfaces. The potential impact on the quality of subsequent interfaces formed after further processing will motivate future work in this area.

Si(111) Etched in Buffered HF Solutions. For the Si(111) surfaces, increasing the pH of the solution leads to a *preferential* etching of the H-terminated surfaces, making it possible to flatten the surface on an atomic scale (65). For instance, Fig. 23 shows the difference between a Si(111) surface etched in dilute HF and buffered (pH ~8) HF solutions. Whereas the dilute HF etched surface is atomically rough (Fig. 23a) with all forms of hydrides, the surface etched in a 40 wt% NH_4F solution is characterized by a single sharp absorption line at 2083.7 cm^{-1}, polarized perpendicular to the surface (Fig. 23b). The obvious implication is that atomically flat surfaces have been obtained with ideal monohydride termination. In the first report, (65) it was noted that the measured linewidth, $\Delta\nu$ ~0.9 cm^{-1}, was the narrowest line ever measured for a chemisorbed atom or molecule on a surface *at room temperature*. Part of the width, however, could still be due to thermal broadening. A method was therefore devised to introduce the wafers samples into UHV, making it possible to cool them so that the inhomogeneous linewidth could be measured. The data confirmed that most of the linewidth measured at room temperature was thermally induced, due to an harmonic coupling of the Si-H stretch mode to surface silicon phonons (86). At present, the best samples are characterized by an extremely small (0.05 cm^{-1}) inhomogeneous broadening (77)(94). This result indicates a high degree of homogeneity and has motivated thorough characterization by most of the techniques described in Sec. 3.2.

The LEED patterns obtained after careful introduction into UHV show a 1x1 pattern with resolution limited integral order spots and a background below the detection limit of conventional LEED systems (Fig. 24). This unreconstructed and ideally H-terminated surface is often referred to as the H/Si(111)-(1x1). Quantitative diffraction studies with SPA-LEED and x-ray diffraction are underway that should be particularly useful to characterize the roughness of Si(111) etched at lower pH, using H/Si(111)-(1x1) as a reference.

STM images (93)(95) such as the one shown in Fig. 25, have confirmed that the surface is nearly contamination free (<1 ML), atomically flat, and well ordered with 1x1 (3.84 Å) periodicity. The electronic structure obtained from (dI/dV) measurements displays no states in the gap, which is expected for a hydrogen-covered surface. Further support for the bulk-like character and the ideal monohydride termination comes from electron-stimulated desorption experiments (96) showing the formation of the π-bonded chains (2x1) reconstruction after the H is desorbed. This surface has also been imaged with an atomic force microscope, confirming the 1x1 periodicity (97).

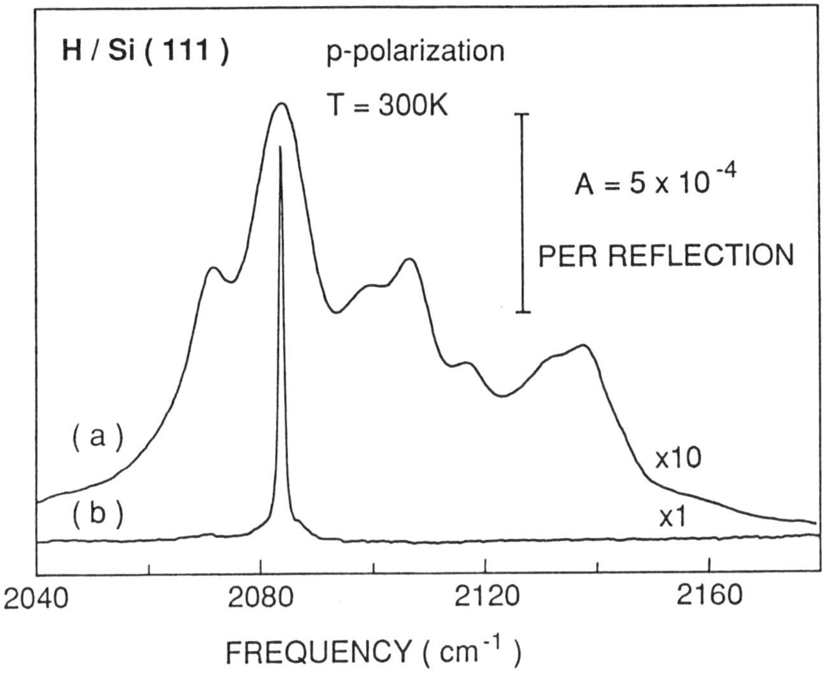

Figure 23. P-polarized IR absorption spectra of Si(111) after (a) etching in dilute HF (pH = 2), and (b) a 40% NH_4F solution (pH = 7.8). (From Ref. 62.)

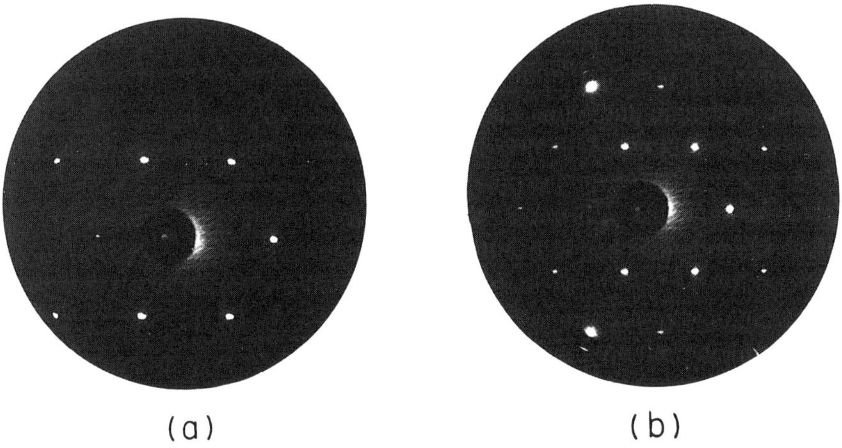

Figure 24. Photograph of LEED pattern of Si(111) surface after NH_4F treatment: (a) 82 eV and (b) 125 eV. (From Ref. 93.)

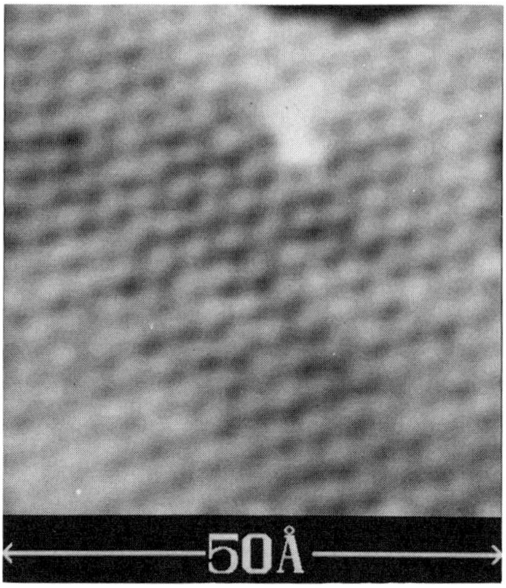

Figure 25. STM picture of H/Si(111)-(1x1). The vertical grey scale is 0.5 Å. (From Ref. 93.)

An alternate means of characterization is to measure the low frequency surface vibrations (Si phonons) of this surface. Theoretical calculation of the vibrational manifold (98) predict strong dispersions for the modes of this ordered surface. Inelastic helium atom scattering measurements, performed by Doak et al. (1990) (99), show phonon losses up to 30 meV (250 cm^{-1}). Two phonon branches (S_8 and S'_8) at 27.4 and 23.7 meV, respectively, are observed to disperse in addition to the Raleigh wave between 0 and 16.3 meV, consistent with a well-ordered surface. The incoherent elastic scattering and broad inelastic backgrounds are almost two orders of magnitude lower than that measured for H-terminated Si(111) prepared using standard UHV techniques (100). This confirms the perfection of the surface compared to surfaces prepared in UHV.

The EELS studies of the H/Si(111)-(1x1) are similarly characterized by a very high specular beam intensity and a low background (61). Recently, the dispersion of both the silicon phonons and the H vibrations has been measured (62)(101) in good agreement with the calculated curves (98).

High resolution IR absorption spectra recorded at low temperatures (<50 K) have also been used to characterize the extent of the perfect 1x1 domains (77)(94). Below 50 K, the lifetime (~0.005 cm^{-1}) (102) and thermal (<0.001 cm^{-1}) (86) broadenings are negligible compared to the measured linewidth (0.07 cm^{-1}) (Fig. 26a). After deconvolution of the resolution function (0.04 cm^{-1}), the natural linewidth (0.05 cm^{-1}) and the line shape are obtained and can be related directly to surface inhomogeneities.

In considering the line shapes, Jakob et al. (1991) (77) pointed out that a distribution of point defects leads to a symmetrical broadening (such as Lorentzian or Gaussian), whereas the presence of finite domains leads to an asymmetric broadening. This asymmetry is dominated by effects associated with dipole coupling between the hydrogen atoms. For a finite domain containing N atoms, there are N normal modes. The strongest IR active mode is the "in-phase" normal mode. This normal mode is at the highest frequency, and this frequency increases with domain size because Si-H is oriented perpendicular to the surface. A few other normal modes, however, also have a strong enough IR cross section (~3 - 5% of the in-phase mode) to be detectable and give a low-frequency tail to the absorption band. Furthermore, the measured absorption associated with a distribution of domain sizes is proportional to P(N)xN, where P(N) is the distribution function (the bigger domains contribute more to the absorption). Therefore, a symmetric distribution of domain sizes leads to an asymmetric absorption line shape, characterized by a low frequency tail for the Si-H system. When both effects are taken into account, the spectra of Fig.26a can be well fit with $N = 2 \times 10^4$ Si-H units with a 30% distribution in domain size (77). Recent STM images show that the average linear terrace size is 500 Å on flat samples in excellent agreement with the IR absorption line shape analysis (~600 Å) (103).

The nature and origin of point defects is not completely clear at present; triangular pits, with 5 - 20 Å sides and a double layer deep, and "white balls" have been observed with STM (95)(103). These defects are not observed on all samples and could depend on process and materials parameters, such as doping, solution pH, etching rate, sample vicinality, etc.

The sensitivity of the IR absorption technique makes it possible to investigate various preparations of the surface in detail. In particular, the nature of the oxide prior to the HF etching has been found to be important. Fig. 26 shows high-resolution spectra associated with H/Si(111)-(1x1) surfaces prepared in two different ways: (a) by stripping the thick thermal oxide in buffered HF (pH = 5) and directly dipping the H-passivated sample into a 40 wt% NH_4F solution for 4 min, and (b) by reoxidizing chemically (SC-2) and then stripping and re-etching in a 40 wt% NH_4F solution for 6 min. The first absorption line was discussed in detail above. The second spectrum peaks at lower frequency, indicating a smaller average domain size. It is also broader, indicating a larger distribution of domain sizes. Therefore, thermally grown oxides result in a smoother interface than chemically grown oxides.

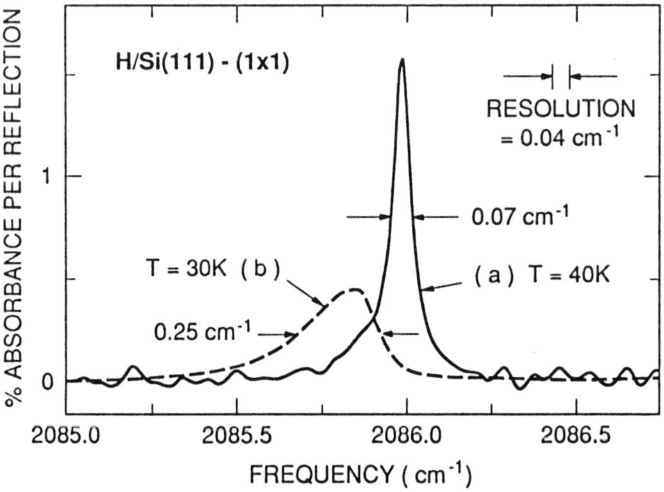

Figure 26. P-polarized IR absorption spectra of H/Si(111)-(1x1) prepared by: (a) thermal oxidation followed by etching in buffered HF (pH = 5) for 2 min and subsequent etching in 40% NH_4F for 4 min, and (b) chemical oxidation followed by etching in a 40% NH_4 F solution for 6½ min. Both samples are thoroughly rinsed in DI water after the last etching step. (From Ref. 94.)

This phenomenon is well known in the literature and is investigated further by treatment in concentrated HF where etching of the H-terminated silicon is minimized. Previous researchers have found that, when a thick thermal oxide (~1000 Å) is grown, with post-annealing in an inert gas at the growth temperature (~1050°C), a very smooth Si/SiO_2 interface is formed (3)(30)(104)(105). Dissolution of this oxide in concentrated HF produces a H-terminated Si(111) surface, characterized by a multimode spectrum which indicates atomic roughness (see Fig. 27, top, a). This roughness disappears upon rinsing (see Fig. 27, top, b), as evidenced by the dominance of the monohydride peak afterwards. In contrast, the rough Si(111) surface produced by HF etching of a *chemical* oxide is not removed upon simple rinsing. These observations confirm that the thermally grown oxide has a smoother interface than the chemically grown oxide and suggests that water rinsing alone can remove small surface defects preferentially. Identical experiments were also performed on Si(100) surfaces and are shown in Fig. 27, bottom. As discussed earlier, the multi-mode spectra imply that the surfaces are atomically rough. Certain spectral changes do occur, however, after rinsing. The monohydride modes decrease in strength while the dihydride modes increase. Note that this is opposite to the behavior observed on Si(111).

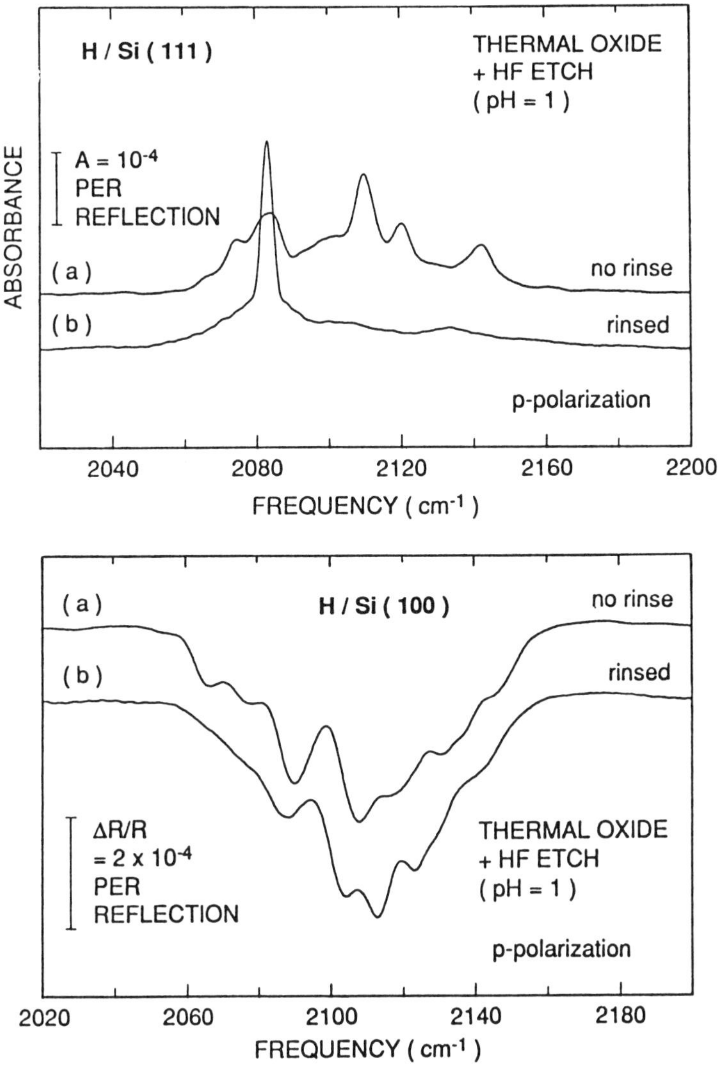

Figure 27. Top: P-polarized IR absorption spectra of thermally oxidized Si(111) after (a) etching in concentrated HF, and (b) subsequent rinsing in de-ionized water. Bottom: P-polarized IR absorption spectra of thermally oxidized Si(100) after (a) etching in a concentrated HF solution, and (b) subsequent rinsing in de-ionized water.

Preferential etching by water, resulting in flat H-terminated Si(111) surfaces, has recently been demonstrated by Watanabe et al. (1991) (66). These authors studied the effects of water rinsing (65) as a function of water temperature, finding that hot water (100°C) increases the rate of removal of (111) surface defects while maintaining the H-termination. The spectrum obtained is shown in Fig. 28. A single mode, polarized perpendicular to the surface, dominates the spectrum. However, using the analysis developed by Jakob et al. (1991) (77), the spectra indicate that the average domain size is 20 Å, a factor of fifteen smaller than for the sample presented in Fig. 26a. It is surprising that boiling water rinsing did not lead to the growth of a surface oxide. A possible mechanism is discussed in the following section.

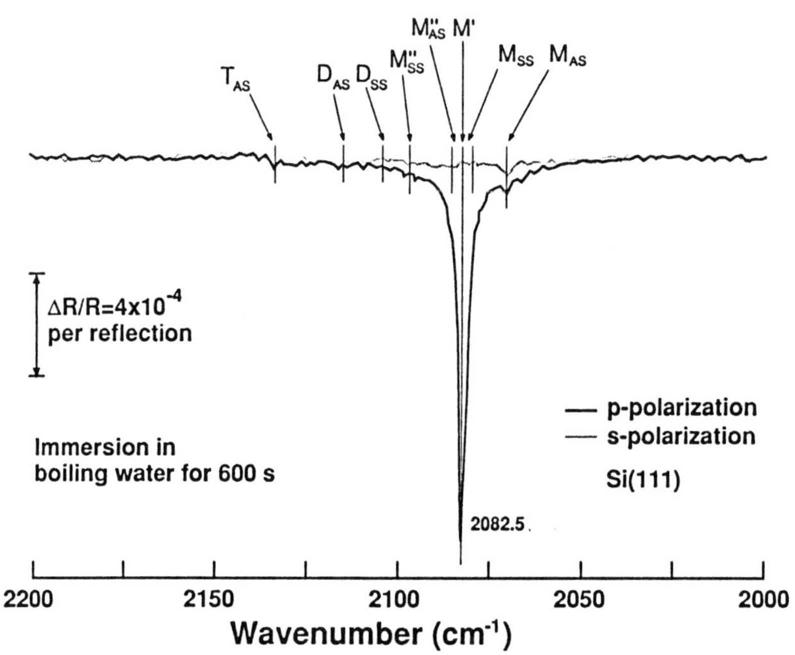

Figure 28. Polarized IR absorption spectra of a chemically oxidized Si(111) sample after etching in 1.5% HF solution and subsequent boiling in DI water at 100°C for 10 min. The resolution is 0.5 cm^{-1}. (From Ref. 66.)

Mechanism of Preferential Etching. A solution of concentrated HF dissolves silicon oxide efficiently and passivates the silicon surface with hydrogen. Once hydrogen passivation is achieved, etching stops. As a result, the morphology of the original Si/SiO_2 interface is preserved.

Clearly, the data presented in the previous section indicate that dilution of concentrated HF with water or buffering with ammonium fluoride induces a *slow* etching reaction of the H-passivated silicon surfaces. The overall etch rate increases with the pH of the solution, as evidenced by the increasing formation of small H_2 bubbles as the pH is raised. The bubbles are probably formed during oxidation of the surface by OH^-, according to the following reaction:

$$\equiv Si - H + OH^- \rightarrow \equiv Si - O^- + H_2$$

Once oxidized, the surface is subject to HF attack through HF insertion into the Si-O bond, according to the schematic reaction

$$\equiv Si - O^- + HF \rightarrow \equiv Si - F + HO^-$$

with subsequent removal of the surface Si atom (now labeled Si* to distinguish it from the underlying bulk Si atoms), and passivation of the second layer silicon atoms by hydrogen, according to the mechanism proposed in Sec. 3.1:

$$(3Si) \equiv Si^* - F + 3HF \rightarrow 3\ (Si - H) + Si^*\ F_4$$

In the above processes, the last two steps are fast compared to the initial oxidation of the H-passivated surface. As a result, the surface is always H-terminated.

The role of OH^- is clearly a key ingredient in the attack and etching of H-terminated silicon surfaces. It is also important to note that silicon can be etched without HF. Silicon can also be etched with alkaline solutions, such as KOH or NaOH, (106) and even with water (66). These observations indicate that, once oxidized, the silicon surface can be attacked by OH^-. A plausible reaction pathway involves the silicon back-bond attack by OH^-:

$$(3Si) \equiv Si^* - O^- + 3OH^- \rightarrow 3(Si - H) + Si^*\ O_4^-$$

where the last species is actually unstable and decomposes into other soluble products. Confirmation and quantification of the above reaction steps should be possible using first principles cluster calculations, as was done to understand H-passivation of silicon (46). In addition, the influence of surface charges on the anisotropic etching of silicon needs to be understood (107).

The key point in considering *preferential* etching is to realize that oxidation of the H-terminated silicon surface is extremely slow and is the rate-determining step. It takes many collision between OH^- ions and the surface Si-H to effect a reaction because the reaction barrier is large. When this is the case, relatively minor factors may affect the reaction probabilities greatly. For instance, if some surface structures are strained, they may be more easily attacked because the reaction barrier is lowered only a small degree. Alternatively, if a surface structure is more accessible for the OH^- ion in solution, it may be attacked faster because of an increased reaction probability (larger prefactor).

To understand and quantify the etch rates of various surface structures, the chemical etching of stepped silicon (111) surfaces was studied by Jakob and Chabal (1991) (89) by utilizing IR absorption spectroscopy and STM images to characterize the surface structures after each chemical treatment. The results are summarized in the schematic drawing of Fig. 29 (108). At low pH (pH = 1 - 3), the HF solutions do not modify substantially the original Si/SiO_2 interface which usually displays a fair degree of atomic roughness. The (111) terraces have many small adstructures, and the more extended steps are wandering with a high density of kinks. The IR absorption spectra are very similar to those in Fig. 15 with a relatively low concentration of ideal monohydride termination of the (111) planes. As the pH is increased (pH = 5 - 6), the small adstructures and defects present on the (111) terraces are etched away, leaving atomically flat, ideally monohydride terminated (111) terraces. The step edges, however, remain rough on an atomic scale with a high concentration of kinks. Solutions of higher pH (pH ≈6.6) are needed to remove kinks and generate atomically straight steps. After three minutes of etching at room temperature, for instance, the steps are straight with a very small (1%) concentration of kinks which is probably accounted for by the imperfection in the azimuth of the miscut.

The above observations indicate a step flow etching mechanism (108)(95). As the pH is increased beyond 7.0, the etching of stepped surfaces increases drastically as evidenced by a large formation of bubbles. The surface then roughens, partly because of increasing fluctuations in the terrace length leading to step bunching, and partly because of the more inhomogeneous conditions at the surface (bubbles, fluctuation in the concentrations of various chemical species, etc.). The result is the formation of large, three-dimensional roughness as evident in STM images (103).

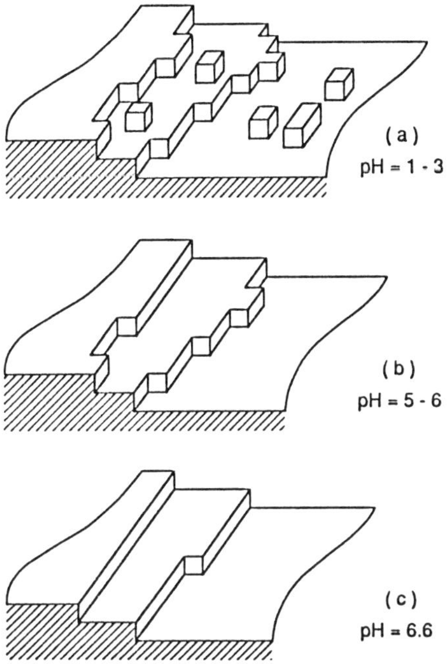

Figure 29. Schematic picture of the changes of the Si surface morphology as the etch rate is increased by increasing the pH of the etching solution: (a) pH <3, (b) pH = 5 - 6, and (c) pH = 6.6. A total etch time of 3 min, including the removal of approx. 10 Å SiO_2, is assumed. A pH higher than 6.6 leads to step bunching, and therefore to the formation of multiple steps and facets (not shown here).

At a microscopic level, many possible etching mechanisms must be considered. Steric constraints, strain, and bonding configuration can all play a role in determining etch rates. The IR studies of intentionally miscut wafers (89) show that the stability of two types of steps with different bonding configurations and different levels of strain is similar. The unstrained monohydride step, with three Si back-bonds, has an etch rate similar to that of the *strained* dihydride step, with only two back-bonds; and both types of steps remain stable in water. Therefore, the etch rate does not appear to depend strongly on surface strain or on the number of Si back-bonds. On the other hand, the progression summarized in Fig. 29 suggests that the accessibility of surface Si-H entities to OH^- ions in solution influences the etch rate directly. To discuss the observations, we use the nomenclature defined in Fig. 19. Isolated trihydrides (labeled T) and "horizontal" dihydrides (labeled D) are removed very efficiently, even by simple rinsing with water (65). Note that such dihydrides are always at the tip of small adstructures and are therefore almost as exposed to OH^- ions as isolated trihydrides. The coupled monohydrides (labeled M) and the "vertical" dihydrides (labeled D′) are much more resistant to etching. They are part of bigger adstructures and are therefore less accessible to OH^- ions. When coupled monohydrides or dihydrides are part of an extended step, they become even more resistant to etching. Finally, the flat and monohydride terminated (111) planes are the most stable.

The chemistry on these surfaces is obviously complex. The simple mechanisms described above are meant only to give a framework in which to address the problem. They are based solely on HF and OH^- chemistry. Surface interactions can also be influenced by species such as HF_2^-, H^+ and NH_4^+. In addition, the differences between OH^- and HF are not clearly understood but are being addressed theoretically and experimentally by different approaches.

3.3 Contamination Issues

An important contaminant after HF etching is fluorine. Fluorine contamination is the most misunderstood of all the contaminants found on HF etched surfaces. When wafers are treated with concentrated HF without water rinsing, XPS reveals that monolayer quantities of fluorine are present on the surface (50). This could lead to the conclusion that fluorine is the fundamental surface termination responsible for the amazing surface passivation (49). Indeed, fluorine from the HF was identified as playing an important role in reducing the number of defects in gate oxides formed on these surfaces (109). Although there is no doubt that fluorine is present, the form in which it is found and how it is bound to the surface are uncertain. An important experimental result by Gräf et al. (1989) (56) states that rinsing in water strongly reduces surface fluorine concentrations for immersions as short as a few minutes. Surface fluorine concentrations on silicon wafers that were immersed in dilute HF are in the range of a few percent of a monolayer, (55) showing that the surface fluorine concentration can vary vastly. The hydrogen content on the surface, on the other hand, does not vary substantially with rinsing or with HF concentration in the solution. Thus, it is clear that fluorine must be thought of as a surface contaminant rather than the fundamental surface termination.

While it is now generally accepted that the hydrogen termination resulting from HF etching explains the hydrophobicity and the passivation of silicon surfaces, there is still disagreement as to the nature of the fluorine on the surface. It is clear that some of the fluorine must be physisorbed to the hydride covered Si surface in chemical forms, such as HF, H_2SiF_6, $(NH_4)_2SiF_6$, or as ions such as F^-, HF_2^-, SiF_6^{2-} (110). The reason for postulating physisorbed species is that the concentration of these contaminants changes greatly with rinsing without changing the surface hydrogen concentration. Thus, physisorbed fluorides can explain many of the observations of fluorine on silicon. All observations of fluorine may not, however, be attributed to physisorbed fluorine, especially after extended rinsing. Many authors believe that some fluorine is directly bonded to substrate Si after HF etching (57)(56)(32)(109). The water rinsing experiments of Gräf et al. (1989) (56) show that fluorine disappears from the surface at the same time that oxygen is taken up by the surface, indicating that an exchange reaction is occurring. Sunada et al. (1990) (32), on the other hand, argue that fluorine termination prevents oxidation until the fluorine concentration falls below a critical value, after which oxidation can proceed normally. Interesting new data suggests that the fluorine is in fact subsurface and only comes to the surface upon oxidation of that surface (39). Clearly, the chemistry of fluorine on the Si surface after HF etching is an important issue still being debated and needs to be studied further.

HF-etched Si surfaces can be prepared to be oxygen free. As mentioned above, however, oxygen concentrations are found to grow as a result of prolonged water rinsing (56)(32). The atomically flat Si(111) wafers produced by etching in NH_4F, on the other hand, are amazingly resistant to oxidation and show negligible amounts of oxygen after water rinses as long as 10 min (61). The conclusion from this result is that the oxidation rate must somehow be related to the step density on the surface. This might lead to the conclusion that H_2O or OH^- attack the step edges more readily and lead to oxidation, but the boiling in water experiments of Watanabe et al. (1991) (66) suggest otherwise. Their surfaces are found to be terminated completely with hydrogen, with little if any residual oxygen, demonstrating that step edge attack in water does not necessarily lead to oxide formation. Oxidation appears to be more complex than simple attack by water. Oxidation in liquid solution has been found to be related to the oxygen concentration in the water used during the rinsing process as well as the doping type of the wafers used (31). It is interesting to note that lightly doped wafers can be rinsed for 10^4 min without appreciable oxide growth when water with a low O_2 concentration is used. Further research regarding this issue is certainly warranted. Oxidation due to storage in air is also of current interest. Gräf et al. (1990) (111) have reported that wafers stored in standard moist air (relative humidity 35 - 40%), oxidize extremely slowly, growing less than 1 Å of oxide in seven days (10^4 min) (Fig. 30). After seven days of storage the oxidation rate increases abruptly, with the oxide thickness approaching 10 Å after forty-five days. This bimodal oxidation rate distribution is not understood at this time but has been observed in other work (31)(32) and has been postulated to be related to the fluorine coverage on the surface (32). It should be noted that hydrogen-terminated surfaces have been prepared that show less than a monolayer of oxygen after seven days of storage both in air (111) and in water (31), given the fact that many researchers still believe that the growth of native oxide begins immediately. The cleanliness of the chemicals available at the time may have influenced the observed oxidation rates, leading to the erroneous conclusion that the oxide growth was an intrinsic rather than extrinsic effect.

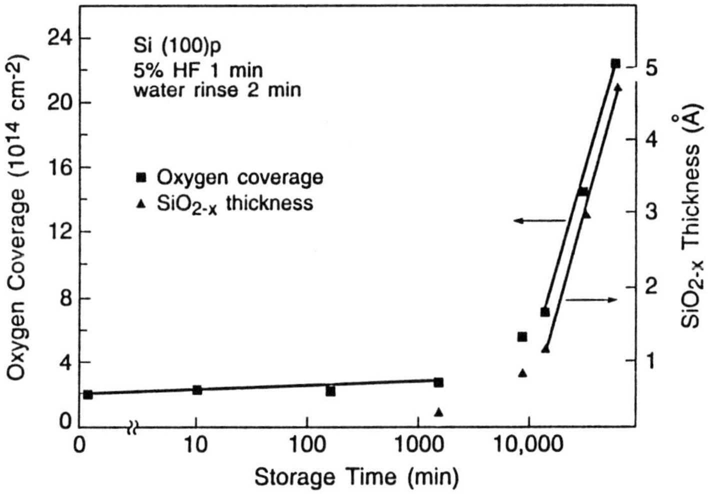

Figure 30. Oxygen coverage (left scale) and oxide thickness (right scale) as a function of storage time in moist air. Oxygen coverage was determined by x-ray photoelectron spectroscopy. (From Ref. 114.)

Hydrocarbon contamination is a particularly insidious problem for hydrogen-terminated HF treated wafers. Whereas hydrophilic oxide-passivated surfaces are relatively resistant to hydrocarbon contamination, H-terminated Si surfaces can be covered with a monolayer of hydrocarbons extremely quickly if exposed to a contaminated environment. Hydrocarbon contamination can come from the water rinse used (57), from contaminants in laboratory air, or from the environments of processing equipment. Infrared spectra have shown that H-terminated Si samples, prepared in dilute HF and rinsed in high-quality water, transported through air and loaded into a nitrogen purged vacuum chamber remain hydrocarbon free, only to be contaminated the second one begins evacuating the chamber (112). In fact, clean surfaces have been preserved only in the case where the load locks of the analysis chambers were pumped down to a few times 10^{-10} torr prior to venting for sample introduction (90)(61). It is obvious that these surfaces are highly sensitive to hydrocarbons and become contaminated unless extreme precautions are taken (55)(57). Part of the problem is that many common contaminants desorb at temperatures above the hydrogen desorption temperature (113). When this happens silicon carbides are formed and are impossible to remove at reasonable temperatures. In contrast, hydrocarbons on top of oxide coated surfaces can be made to desorb before the oxide desorbs. Oxide passivation is obviously preferred if the 850 - 950°C desorption temperatures can be tolerated (37).

Metal contamination is also of crucial importance and is discussed in other chapters of this book. In brief, wafers etched in HF following the RCA standard clean can be free of metal contamination ($<10^{10}$ cm^{-2}) (4). In other instances, however, Cu was found to be a problem and was eliminated by adding some H_2O_2 to the HF (5). Cu has been found to be extremely detrimental to the stability of the H-termination by causing increased oxidation rates during water rinses, resulting in 3 Å of oxide growth in just 2 min (114). This reaction has been attributed to Cu-catalyzed oxidation (114) of the Si, which causes bulk etching and surface roughness (5) upon exposure to contaminated HF solutions. Cu coverages of up to a half a monolayer are obtained on exposure to HF solutions contaminated with only ppm levels of Cu (114) . Copper in HF is known to plate out on semiconductor surfaces; the use of ultrapure HF (and other reagents) in semiconductor processing is, therefore, imperative (1)(12).

4.0 SUMMARY AND FUTURE DIRECTIONS

Two distinct types of surface cleans for silicon wafers have been discussed. The first leaves a thin chemical oxide behind and the second results in a hydrogen-terminated silicon surface. Both techniques are extremely effective at cleaning and passivating the Si surface. Three major issues were discussed for each of these classes of cleans: chemical composition, structure and morphology, and contamination issues.

The composition of chemically grown oxides on silicon surfaces is similar to that of the interfacial transition regions of thermally grown oxides. The oxides are largely composed of SiO_2 containing a fraction of sub-oxide species, dominated by Si^{2+}. The largest intrinsic contaminant found on such surfaces is hydrogen-bonded OH and H_2O. Minor levels of hydrogen in the form of Si-H are also observed. The structure and morphology of chemo-mechanically polished silicon wafers approaches atomic perfection with a surface roughness of 2 Å rms. The SC-1 clean is found to increase levels of surface roughness, whereas the Piranha and SC-2 cleans tend to leave surface topographies unchanged. The oxide interfacial structure is found to vary with preparation technique with the indication that the chemically grown oxide interfaces are substantially rougher than their thermal oxide counterparts. Extrinsic contaminants, such as metals and hydrocarbons, are efficiently removed using RCA cleaning, but there is an increasing concern that the 10^{10} cm^{-2} level achieved for most metals today will have to be improved in the future. Other areas where future work is necessary include: gaining an understanding of how contaminants in the solutions used influence etching and surface topography; elucidating the mechanism of chemical oxidation in the various solutions employed; and learning, finally, how to control the surface structure and composition of oxide-terminated silicon surfaces.

Hydrogen-terminated silicon surfaces are composed of hydrogen atoms covalently bonded to the substrate wafer. The electronic perfection of this interface is unsurpassed, indicating an extremely low dangling bond density. Fluorine is now understood to be a minor constituent of the surface and is one of the major intrinsic contaminants observed. The mechanism of H-termination involves back-bond attack by HF molecules that is facilitated by the charge transfer caused by Si bonding to highly electro-negative elements, such as fluorine or oxygen. Si(100) and Si(111) surfaces are found to be atomically rough when etched in concentrated or dilute HF. This result was originally inferred by the coexistence of many types of silicon-hydrides on the surface and was later confirmed using a variety of techniques. High-pH HF solutions lead to increased roughness on Si(100) with the formation of microfacets, but can produce atomically smooth Si(111). These ideally monohydride-terminated Si(111) surfaces have been characterized by use of a variety of surface science techniques, demonstrating the surface quality which can be achieved using solution chemistry alone. Such surfaces allow highly detailed studies which can be used to elucidate the fundamental mechanisms of etching. In this regard, OH^- has been identified as being responsible for the anisotropic etching observed. Surprisingly, water rinsing alone can also be used to achieve a certain level of atomic perfection on Si(111) surfaces and leads to the conclusion that anisotropic etching can occur in rinsing operations. Hydrogen-terminated silicon surfaces are particularly susceptible to hydrocarbon contamination and extreme precautions must be taken in order for this contaminant to be avoided. H-terminated surfaces have been prepared which are essentially metal free ($<10^{10}$ cm^{-2}), but solutions contaminated with Cu must be avoided when using HF. Recent findings indicate that the growth of native oxide occurs much more slowly than previously believed. Researchers find that silicon wafers remain essentially oxide free for up to seven days in both in air and water. Future work should be focused on understanding the role of fluorine on these surfaces and the chemistry of HF and OH^- in these solutions. Understanding the role of metal contaminants in HF solutions is also of critical importance. Learning to control Si(100) surface chemistries and morphologies at a level similar to that achieved for Si(111) will be of greatest importance.

The foundation of the advances made in our understanding of these surfaces comes not only from old techniques, such as infrared absorption spectroscopy, but also from several newer experimental tools, such as the scanning tunneling microscope, the atomic force microscope, or high-resolution electron energy loss spectroscopy and high-resolution x-ray photoelectron spectroscopy. Furthermore, new theoretical techniques, such as first principles molecular orbital calculations of chemical activation barriers have provided a quantitative basis for understanding important surface chemical reactions. In this chapter, we have stressed not only the technology of wet chemical cleaning, but have tried to emphasize the science behind the wet chemistry. This area of research is growing rapidly and we hope that better understanding will be central to successful semiconductor processing in the future.

ACKNOWLEDGMENTS

It is a pleasure to acknowledge our long time collaborators in this work, R. S. Becker, P. Dumas, and K. Raghavachari. We would also like to thank V. A. Burrows, P. Jakob, and G. W. Trucks, for stimulating discussions and their contributions to our present understanding of this problem. P. Jakob deserves special thanks for providing many of the IR spectra used in this manuscript. Thanks also go to M. A. Hines, I. K. Robinson, and J. E. Rowe for critical reading of this manuscript. S. B. Christman, E. E. Chaban, A. J. Becker, and R. D. Yadvish are also gratefully acknowledged for their technical support.

REFERENCES

1. Kern, W., *J. Electrochem. Soc.* 137:1887 (1990)
2. Buck, T. M., and McKim, F. S., *J. Electrochem. Soc.,* 105:709 (1958)
3. Hahn, P. O. and Henzler, M., *J. Vac. Sci. Technol.* A2:574 (1984)
4. Heyns, M., Hasenack, C., De Keersmaecker, R., and Falster, R., *Proc. of the 1st Int. Symp. on Cleaning Technology in Semiconductor Device Manufacturing,* (J. Ruzyllo and R. E. Novak, eds.), PV 90-9:293 Electrochemical Society, Pennington, NJ, (1990)
5. Ohmi, T., Miyashita, M. and Imaoka, T., *Proc. of the Microcontamination Meeting,* San Jose, CA, p. 491, Canon Communications, (October 16-18,1991)
6. Atalla, M. M., Tannenbaum, E. and Scheibner, E. J., *Bell System Tech. Journal,* 38:749 (1959)
7. *The Physics of SiO_2 and Its Interfaces,* (S. T. Pantelides, ed.) Pergamon Press, NY (1978)
8. *The Physics and Chemistry of SiO_2 and the Si-SiO_2 Interface,* (C. R. Helms and B. E. Deal, eds.) Plenum Press, NY (1988)
9. Nicollian, E. H. and Brews, J. R., *MOS (Metal Oxide Semiconductor) Physics and Technology,* Wiley-Interscience, NY (1982)
10. Grunthaner, F. J. and Maserjian, J., *IEEE Trans. on Nuclear Science,* NS 24:2108 (1977)
11. Ourmazd, A. and Bevk, J., in *The Physics and Chemistry of SiO_2 and the Si-SiO_2 Interface,* (C. R. Helms and B. E. Deal, eds.), p. 189, Plenum Press, NY (1988)
12. Kern, W. and Puotinen, D., *RCA Rev.* 31:187 (1970)
13. Deal, B. E. and Kao, D.-B., *Proc. of the 1986 Tungsten and Other Refractory Metals for VLSI Applications II,* (E. K. Broadbent, ed.), p. 27, Materials Research Society, Pittsburgh, PA (1987)
14. Mikata, Y., Inoue, T., Takasu, S., Usami, T., Ohta, T., Hirano, H., *Proc. of the 1st Int. Symp. on Si Molecular Beam Epitaxy,* (J. C. Bean, ed.), p. 45, Electrochemical Society, Pennington, NJ (1990)
15. Ogawa, H., Terada, N., Sugiyama, K., Moriki, K., Miyata, N., Aoyama, T., Sugino, R., Ito, T. and T. Hattori, *Appl. Surf. Sci.,* 56-58:836 (1992)
16. Sugiyama, K., Igarashi, T., Moriki, K., Nagasawa, Y., Aoyama, T., Sugino, R., Ito, T. and Hattori, T., *Jpn. J. Appl. Phys.,* 29:L2401 (1990)
17. Yablonovitch, E. and Gmitter, T. J., "Diagnostic Techniques for Semiconductor Materials and Devices," *Fall ECS* (1988)

18. Kiselev, A. V. and Lygin, V. I., *Infrared Spectra of Adsorbed Species* (L. H. Little, ed.), pp. 213, 228, Academic Press, NY (1966)

19. Schaefer, J. A., Frankel, D. J., Stucki, F., Göpel, W. and Lapeyre, G. J., *Surf. Sci.* 139:L209 (1984)

20. Nagasawa, Y., Ishida, H., Takayagi, T., Ishitani, A. and Kuroda, H., *Solid-State Electronics,* 33:129 (1990)

21. Hahn, P. O., Grundner, M., Schnegg, A. and Jacob, H., *The Physics and Chemistry of SiO_2 and the Si-SiO_2 Interface,* (C. R. Helms and B. E. Deal, eds.), p. 401, Plenum Press, NY (1988)

22. Schnegg, A., Lampert, I. and Jacob, H., *Electrochemical Society (ECS) Extended Abstracts,* 85-1:394, Toronto (1985)

23. Ourmazd, A., Taylor, D. W., Rentschler, J. A., and Bevk, J., *Phys. Rev. Lett.,* 59:213 (1987)

24. Green, M. P., Hanson, K. and Higashi, G. S., to be published.

25. Mishima, H., Yasui, T., Mizuniwa, T., Abe, M. and Ohmi, T., *IEEE Trans. Of Semiconductor Manufacturing,* 2:69 (1989)

26. Miyashita, M., Itano, M., Imaoka, T., Kawanabe, I. and Ohmi, T., *Technical Digest of the 1991 Symp. on VLSI Technology,* Oiso, Japan, p. 45., (May 28-30, 1991)

27. Ohmi, T., Kotani, K., Teramoto, A., and Miyashita, M., *IEEE Electron Dev. Lett.,* 12:652 (1991)

28. Ohmi, T., Miyashita, M., Itano, M., Imaoka, T. and Kawanabe, I., *IEEE Trans. on Electron Dev.,* 39:537 (1992)

29. Verhaverbeke, S., Meuris, M., Mertens, P. W., Heyns, M. M., Philipossian, A., Gräf, D. and Schnegg, A., *Proc. Int. Electron Devices Meeting,* p. 71 (1991)

30. Gibson, J. M., Lanzerotti, M. Y. and Elser, V., *Appl. Phys. Lett.,* 55:1394 (1989)

31. Morita, M., Ohmi, T., Hasegawa, E., Kawakami, M., and Suma, K., *Appl. Phys. Lett.,* 55:562 (1989)

32. Sunada, T., Yasaka, T., Takakura, M., Sugiyama, T., Miyazaki, S. and Hirose, M., *Ext. Abstracts of the Conf. On Solid State Devices and Materials,* Sendai, p. 1071 (1990)

33. Heyns, M., Hasenack, C., De Keersmaeker, R. and Falster, R., *Microelectronic Engineering,* 10:235 (1991)

34. Ohsawa, A., Honda, K., Takizawa, R., Nakanishi, T., Aoki, M. and Toyokura, N., *Semiconductor Silicon 1990,* (H. R. Huff and K. G. Barraclough, eds.), p. 601, Electrochemical Society, Penningtion, NJ (1990)

35. Murrell, M., Sofield, C., Sugden, S., Verhaverbeke, S., Heyns, M. M., Welland, M. and Golen, B., *Proc. Silicon Ultra-Clean Processing Workshop,* Oxford (Sept. 1991)

36. Henderson, R. C., *J. Electrochem. Soc.,* 119:772 (1972)

37. Ishizaka, A., Nakagawa, K. and Shiraki, Y., *Second Int. Symp. on MBE and Clean Surface Related Techniques,* (R. Ueda, ed.), Jpn. Soc. of Applied Physics, Tokyo, p. 183 (1982)

38. Vig, J. R., *J. Vac. Sci. and Technol.,* A3:1027 (1985)

39. Kasi, S. R., Liehr, M. and Cohen, S., *Appl. Phys. Lett.,* 58:2975 (1991)

40. Beckmann, K. H., *Surf. Sci.,* 3:314 (1965)

41. Shinn, N. D., Morar, J. F. and McFeely, F. R., *J. Vac. Sci. and Technol.,* A2:1593 (1984)

42. Harrick, N. J. and Beckmann, K. H., *Characterization of Solid Surfaces,* (P. F. Kane and G. B. Larrabee, eds.), p. 243, Plenum Press, NY (1974)

43. Ubara, H., Imura, T. and Hiraki, A., *Solid. State Comm.,* 50:673 (1984)

44. Imura, T., Mogi, K., Hiraki, A., Nakashima, S. and Mitsuishi, A., *Solid State Comm.,* 40:161 (1981)

45. Miyasato, T., Abe, Y., Tokumura, M., Imura, T. and Hiraki, A., *Jpn. J. Appl. Phys.,* 22:L580 (1983)

46. Trucks, G. W., Raghavachari, K., Higashi, G. S. and Chabal, Y. J., *Phys. Rev. Lett.* 65:504 (1990)

47. Raider, S. I., Flitsch, R. and Palmer, M. J., *J. Electrochem. Soc.* 122:413 (1975)

48. Licciardello, A., Puglisi, O. and Pignataro, S., *Appl. Phys. Lett.* 48:41 (1988)

49. Weinberger, B. R., Deckman, H. W., Yablonovitch, E. Gmitter, T., Kobasz, W. and Garoff, S., *J. Vac. Sci. Technol.* A3:887 (1985)

50. Weinberger, B. R., Peterson, G. G., Eschrich, T. C. and Krasinski, H. A., *J. Appl. Phys.* 60:3232 (1986)

51. Yablonovitch, E., Allara, D. L., Chang, C. C., Gmitter, T. and Bright, T. B., *Phys. Rev. Lett.* 57:249 (1986)

52. Grunthaner, F. J. and Grunthaner, P. J., *Mat. Sci. Reports,* 1:65 (1986)

53. Grunthaner, P. J., Grunthaner, F. J., Fathauer, R. W., Lin, T. L., Hecht, M. H., Bell, L. D., Kaiser, W. J., Schowengerdt, F. D. and Mazur, J. H., *Thin Solid Films* 183:197 (1989)

54. Grundner, M. and Jacob, H., *Appl. Phys.* A39:73 (1986)

55. Grundner, M. and Schulz, R., *AIP Conf. Proc. No. 167,* (G. W. Rubloff and G. Lucovsky, eds.), American Institute of Physics, pp. 329-337, NY (1988)

56. Gräf, D., Grundner, M. and Schulz, R., *J. Vac. Sci. Technol.* A7:808 (1989)

57. Takahagi, T., Nagai, I., Ishitani, A., Kuroda, H. and Nagasawa, Y. , *J. Appl. Phys.* 64:3516 (1988)

58. Burrows, V. A., Chabal, Y. J., Higashi, G. S., Raghavachari, K. and Christman, S. B., *Appl. Phys. Lett.* 53:998 (1988)

59. Chabal, Y. J., Higashi, G. S., Raghavachari, K. and Burrows, V. A.. *J. Vac. Sci. Technol.* A7:2104 (1989)

60. Fenner, D. B., Biegelsen, D. K. and Bringans, R. D., *J. Appl. Phys.* 66:419 (1989)

61. Dumas, P. and Chabal, Y. J., *Chem. Phys. Lett.,* 181:537 (1991)

62. Dumas, P., Chabal, Y. J. and Jakob, P., *Surf. Sci.,* 269/270:867 (1992); Dumas, P. and Chabal, Y. J., *J. Vac. Sci. Technol.* A10:2160 (1992)

63. Judge, J. S., *J. Electrochem. Soc.* 118:1772 (1971)

64. Novak, R. E., *Solid State Technol.,* p 39., (March 1988)

65. Higashi, G. S., Chabal, Y. J., Trucks, G. W. and Raghavachari, K., *Appl. Phys. Lett.* 56:656 (1990)

66. Watanabe, S., Nakayama, N. and Ito, T., *Appl. Phys. Lett.* 59:1458 (1991)

67. Bauer, E., *Ultramicroscopy,* 17:51 (1985); Telieps, W. and Bauer, E., *Ultramicroscopy,* 17:57 (1985)

68. Henzler, M., *Topics in Current Physics: Electron Spectroscopy for Surface Analysis,* (H. Ibach, ed.) 4:117, Springer, Berlin (1977); Henzler, M., *Advances in Solid State Physics,* (J. Treusch, ed.) (Festkörperprobleme X, Vieweg, 1979) p. 193; Henzler, M., *Surf. Sci.* 36:109 (1973)

69. Robinson, I. K., *Phys. Rev.* B33:3830 (1986)

70. Robinson, I. K., Waskiewicz, W. K., Tung, R. and Bohr, J., *Phys. Rev. Lett.* 57:2714 (1986)

71. Klitsner, T., Becker, R. S. and Vickers, J. S., *Phys. Rev.* B41:3837 (1990)

72. Becker, R. S., Swartzentruber, B. S., Vickers, J. S. and Klitsner, T., *Phys. Rev.* B39:1633 (1989)

73. Doak, R. B., Single Phonon Inelastic Scattering, in *Atomic and Molecular Beam Methods,* Vol.II, (G. Scoles, ed.), Ch. 14, p. 384,Oxford Univ. Press, NY (1991)

74. Ibach, H. and Mills, D. L., *Electron Energy Loss Spectroscopy and Surface Vibrations,* p. 94, Academic Press, London, (1982)

75. Chabal, Y. J., *Surf. Sci. Reports* 8:211 (1988)

76. Ibach, H., *Electron Energy Loss Spectrometers, The Technology Of High Performance,* Springer Verlag, Berlin (1991)

77. Jakob, P., Chabal, Y. J., and Raghavachari, K., *Chem. Phys. Lett.* 187:325 (1991). Note that the analyses were carried out for isolated domains. For adjacent domains, as is the case for real surfaces, the fits must be re-normalized leading to N = 2 x 10^4 Si-H units (i.e., 600 Å domains).

78. Collins, R. J. and Fan, H. Y., *Phys. Rev.* 93:674 (1954)

79. Harrick, N. J., *Internal Reflection Spectroscopy* (Wiley, NY (1967); Second printing by Harrick Scientific Corporation, Ossining, NY (1979)

80. Olsen, J. E. and Shimura, F., *Appl. Phys. Lett.* 53:1934 (1988); *J. Appl. Phys.* 66:1353 (1989)

81. Sawara, K., Yasaka, T., Miyazaki, S. and Hirose, M., *Proc. Int. Workshop on Science and Technol. for Surface Reaction Process,* p. 93, Tokyo (Jan. 22-24, 1992)

82. Richardson, H. H., Chang, H-C., Noda, C. and Ewing, G. E., *Surf. Sci.* 216:93 (1989)

83. Hahn, P. O., *Mat. Res. Soc. Symp. Proc.* 54:645 (1986)

84. Schaefer, J. A., Stucki, F., Frankel, D. J., Göpel, W. and Lapeyre, G. J., *J. Vac. Sci. Technol.* B2:359 (1984)

85. Chabal, Y. J. and Raghavachari, K., *Phys. Rev. Lett.* 54:1055 (1985). The frequency associated with the isolated vibrations of the dihydrides, 2111 cm^{-1} at room temperature on HF etched Si(100), is substantially different from that of the corresponding dihydride mode on the UHV prepared H/Si(100), 2096 cm^{-1} at room temperature. In addition, the similarity between the dihydride spectra of Fig. 18 in p- and s-polarization clearly indicates that the dihydrides are inclined with respect to the surface normal. It is not surprising, therefore, that the modes associated with the strained monohydride (anti-symmetric = 2087 cm^{-1} and symmetric = 2099 cm^{-1}) are absent in the spectra of the chemically prepared Si(100). These two observations clearly show that the chemically prepared Si(100) surfaces are not atomically flat. Also, no spectral feature can be ascribed to an ideal dihydride termination of the surface as suggested by Grundner and Schulz from EELS data, in Ref. 55. In fact, calculations have shown that the strain associated with an "ideally" dihydride terminated surface would render it unstable.

86. Dumas, P., Chabal, Y. J., and Higashi, G. S., *Phys. Rev. Lett.* 65:1124 (1990)

87. Kobayashi, H., Edamoto, K., Onchi, M. and Nishijima, M., *J. Chem. Phys.* 78:7429 (1983)

88. Froitzheim, H., Kolher, U. and Lammering, H., *Surf. Sci.* 149:537 (1985)

89. Jakob, P. and Chabal, Y. J., *J. Chem. Phys.,* 95:2897 (1991)

90. Morita, Y., Miki, K. and Tokumoto, H., *Appl. Phys. Lett.* 59:1347 (1991)

91. Morita, Y., Miki, K. and Tokumoto, H., *Ultramicroscopy* (May 1992); Morita, Y., Miki, K. and Tokumoto, H., *Jpn.. J. Appl. Phys.,* 30:3570 (1991)

92. Uram, K. J. and Jansson, U. *Surf. Sci.* 249:105 (1991)

93. Higashi, G. S., Becker, R. S., Chabal, Y. J. and Becker, A. J., *Appl. Phys. Lett.* 58:1656 (1991)

94. Jakob, P., Dumas, P. and Chabal, Y. J., *Appl. Phys. Lett.* 59:2968 (1991)

95. Hessel, H. E., Feltz, A., Reiter, M., Memmert, U. and Behm, R. J., *Chem. Phys. Lett.* 186:275 (1991)

96. Becker, R. S., Higashi, G. S., Chabal, Y. J. and Becker, A. J., *Phys. Rev. Lett.* 65:1917 (1990)

97. Kim, Y. and Lieber, C. M., *J. Am. Chem. Soc.* 113:2333 (1991)

98. Miglio, L., Ruggerone, P., Benedek, G. and Colombo, L., *Physica Scripta* 37:768 (1988)

99. Doak, R. B., Chabal, Y. J., Higashi, G. S. and Dumas, P., *J. Electron Spectr. Related Phenomena* 54/55:291 (1990)

100. Harten, U., Toennies, J. P., Wöll, C., Miglio, L., Ruggerone, P., Columbo, L. and Benedek, G., *Phys. Rev.* B38:3305 (1988)

101. Stuhlmann, C., Bogdanyi, G and Ibach, H., *Phys. Rev.,* B45:6786 (1992)

102. Guyot-Sionnest, P., Dumas, P., Chabal, Y. J. and Higashi, G. S., *Phys. Rev. Lett.* 64:2156 (1990)

103. Becker et al., to be published.

104. Hahn, P. O. and Henzler, M., *J. Appl. Phys.* 52:4122 (1981)

105. Ogura, A., *J. Electrochem. Soc.* 138:807 (1991)

106. Holmes, P. J., *The Electrochemistry of Semiconductors,* (P. J. Holmes, ed.), p. 329, Academic Press, London (1962)

107. Seidel, H., Csepregi, L., Heuberger, A. and Baumgärtel, H., *J. Electrochem. Soc.,* 137:3626 (1990)

108. Jakob, P., Chabal, Y. J., Raghavachari, K., Becker, R. S. and Becker, A. J., and *Surf. Sci.,* 275:407 (1992)

109. Yu, B.-G., Arai, E., Nishioka, Y., Ohji, Y., Iwata, S. and Ma, T. P., *Appl. Phys. Lett.,* 56:1430 (1990)

110. Yota, J. and Burrows, V. A., *Mat. Res. Soc. Symp. Proc.,* 204:345 (1991); Yota, J. and Burrows, V. A., *J. Appl. Phys.,* 69:7369 (1991)

111. Gräf, D., Grundner, M., Schulz, R. and Mühlhoff, L., *J. Appl. Phys.,* 68:5155 (1990)

112. Hydrocarbon content was measured by infrared absorption before and after evacuation.

113. Kasi, S. R., Liehr, M., Thiry, P. A.. Dallaporta, H. and Offenberg, M., *Appl. Phys. Lett.,* 59:108 (1991)

114. Gräf, D., Grundner, M., Mühlhoff, L. and Dellith, M., *J. Appl. Phys.,* 69:7620 (1991)

11

Analysis and Control of Electrically Active Contaminants by Surface Charge Analysis

Emil Kamieniecki and G. (John) Foggiato

1.0 INTRODUCTION TO SURFACE CHARGE ANALYSIS TECHNIQUE

Chemical contamination and process induced damage of the wafer surface are major factors in limiting yield in device manufacturing. Effective control of these phenomena requires techniques which non-destructively, yet in real-time, can monitor contamination and process induced damage on a production line. In this chapter the technique of Surface Charge Analysis (SCA) (1) for monitoring the effectiveness of cleaning processes is discussed.

The SCA is an electro-optical method which allows for rapid and non-destructive characterization of the electronic properties of a bare semiconductor surface or one covered with an insulator (2). Of particular importance for the monitoring of cleaning processes is the ability to measure the insulator charge and the interface state density during various cleaning steps and the properties of films grown on cleaned surfaces. In addition, the method provides a new means for simultaneous determination of other critical parameters such as the surface doping concentration and conductivity type.

The SCA technique in the most basic implementation may be treated as an electro-optical equivalent of the metal-oxide-semiconductor capacitance-voltage (MOS C-V) technique. However, unlike the MOS C-V technique, which requires presence of an oxide and the preparation of special test structures, SCA uses neither a gate electrode nor direct electrical contacts, thus eliminating any additional processing steps. It can be used with a bare wafer surface or with dielectric coatings of several microns thickness. This feature offers the potential for utilization of the technique as an in-line monitoring tool for a number of common cleaning processes in current manufacturing environments.

The SCA is a relatively new technique intended as a production monitoring tool. More effective utilization of the SCA as a diagnostic tool requires better understanding of the surface characteristics and their effects on electrical parameters, as well as the correlation of the electrical characteristics with contamination and process induced structural damage of the wafer surface. The SCA feature, critical in both diagnostic and monitoring, is its ability to separate different electrical parameters that characterize the wafer surface.

New wafer characterization methods, such as the SCA, need correlation with currently used methods to gain acceptance by users. In conjunction with outlining the principles of operation of the SCA technique, a summary of recent experimental data is given to show the capability for monitoring and diagnostics.

2.0 PRINCIPLES OF OPERATION

The dependence of surface photovoltage (SPV) on the photon energy of the irradiating light has been used for more than thirty years in determining the minority carrier diffusion length (3)-(5). This method was found to be very effective in detection of heavy metal contamination in silicon (6). The SPV induced by sufficiently high intensity illumination can be used as a direct measure of the initial surface barrier height (7). The surface potential barrier is formed at the surface due to presence of interface (surface) state charge and charge in the dielectric film (oxide charge) on the wafer. The height of this barrier depends not only on those charges but also on the doping concentration of the wafer. If this doping concentration is independently established, the knowledge of the initial barrier height allows determination of the initial surface charge, which is the sum of the interface state charge and the oxide charge that may include slow state charge. A change in the oxide charge, in the density or energy distribution of the interface states, and in the doping concentration of the wafer will affect initial barrier height and hence initial surface charge. Whereas the measurements of the initial barrier height may allow monitoring reproducibility of the wafer surface conditions, they do not permit separation of the specific electronic parameters characterizing the surface region of the wafer. This limits usefulness of the initial barrier height measurements, as well as of other nonselective SPV methods, in the detection and identification of contaminants and/or process induced damage due to wafer cleaning and other IC production processes. The approach taken in the SCA technique removes this limitation by determining separately different parameters characterizing the surface of the wafer. Specifically, it determines doping concentration of the wafer, conductivity type, density and energy distribution of the interface states, oxide charge, and the change in oxide charge due to slow states.

In the SCA method, the electronic properties of the surface are determined from measurements of the alternating current surface photovoltage (ac-SPV) as a function of an electrical field capacitively applied to the semiconductor-insulator structure (1)(2). The schematic of SCA measurements technique and the various charges at the semiconductor surface are depicted in Fig. 1. The ac-SPV signal is generated with a beam of chopped light, incident on the wafer surface, with a photon energy greater than the semiconductor bandgap. The semiconductor surface is kept in depletion or inversion with the superimposed electric field. The illumination is adjusted to a level at which the measured ac-SPV signal is proportional to the incident light intensity. Under such conditions, the induced ac-SPV signal is proportional to the depletion layer width (8)(9). By measuring the depletion layer width dependence on induced charge, while the surface is externally biased, the surface doping concentration, the oxide charge, and the density and energy distribution of the interface states can be determined.

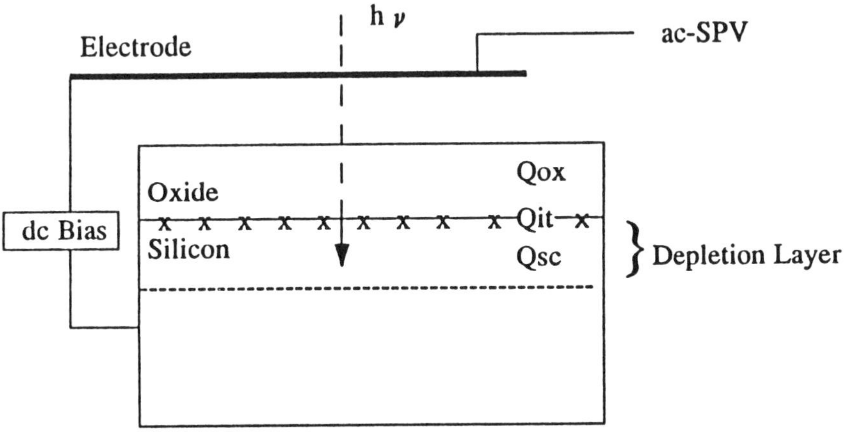

Figure 1. Schematic of SCA Measurement Technique. Q_{ox} = Oxide Charge, Q_{it} = Interface Trapped Charge, Q_{sc} = Semiconductor Space Charge.

2.1 Basic Relationships

The basic theory of ac-SPV generation under low intensity, short wavelength illumination is discussed in detail in Refs. 8 and 9. Here we will concentrate on outlining theoretical aspects essential for understanding of operation as well as current limitations of the surface charge analysis technique.

Consider a n-type semiconductor, as illustrated in Fig. 2, illuminated with light of photon energy higher than the band-gap of the semiconductor. Photogeneration of electron-hole pairs in the semiconductor leads to an increase in the density of free minority carriers (holes) at the surface. The recombination of carriers at the surface proceeds mainly through the surface/interface states. The rates of change of the total density (per cm^2) of free holes at the surface, P_s, and of the density of electrons in the surface/interface states, n_t, are given by the continuity equations:

Eq. (1) $$dP_s/dt = (J_h/q) - R_h$$

and

Eq. (2) $$dn_t/dt = R_e - R_h$$

where J_h is the light-induced current density of holes through the depletion layer, and R_e and R_h are the net capture rates at the surface of electrons and holes respectively. The rate of change of the surface charge density (neglecting, under depletion and inversion conditions, the charge due to free electrons at the surface) becomes

Eq. (3) $$dQ_s/dt = qd(P_s - n_t)/dt$$

In order that charge neutrality be maintained, the change of the surface charge density dQ_s must be compensated by an equivalent change in the net charge density in the space-charge region dQ_{sc}, thus $dQ_s = -dQ_{sc}$. Combining this expression with Eqs. (1) through (3):

Eq. (4) $$-dQ_{sc}/dt = J_h - qR_e$$

Note that in this approach, the charge of the photogenerated minority carriers collected at the surface, qP_s, is included in the surface charge, Q_s, and not in the space-charge, Q_{sc}. Therefore, in parallel with the notation used in the analysis of the capacitance-voltage characteristic of ideal metal-insulator-semiconductor structure (10), the differential capacitance of the semiconductor depletion layer is given by

Eq. (5) $$C_{sc} = -dQ_{sc}/d\psi_s$$

and

Eq. (6) $$dQ_{sc}/dt = (dQ_{sc}/d\psi_s)(d\psi_s/dt) = -C_{sc}(d\psi_s/dt)$$

where ψ_s is the height of the surface potential barrier or the surface potential.

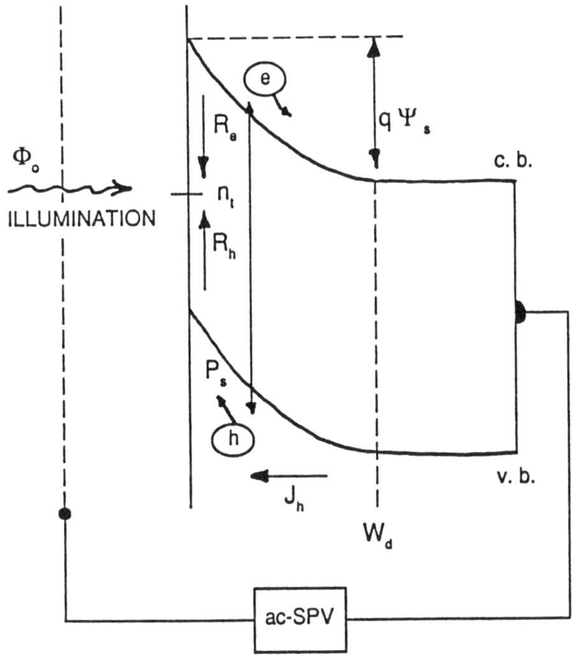

Figure 2. Energy band diagram at the surface of a n-type semiconductor illustrating processes that control generation of an ac-SPV.

Solving the Continuity Equation. The expression for the ac surface photovoltage is derived here for conditions specific to the SCA measurements. It is further assumed that the wafer is uniformly doped single crystal material and that the semiconductor surface is under depletion or inversion conditions. The surface is uniformly illuminated with a low intensity light modulated sinusoidally at angular frequency ω. The ac component of the photon flux inside the semiconductor at the distance x from the surface, $\delta\Phi$, can be written as

Eq. (7) $$\delta\Phi(x,t) = \Phi_o(1 - R)\exp(j\omega t - \alpha x)$$

where Φ_o is the incident photon flux, R the reflection coefficient of the wafer surface and α the absorption coefficient of the semiconductor.

The total photocurrent of holes in the depletion region, J_h, consists of the component due to holes generated within the depletion region and the component due to holes generated in the adjacent bulk semiconductor that diffuse into the depletion region (11). Solving the diffusion equation for minority carriers and assuming that the electric field in the depletion layer is large enough to prevent carrier recombination in this region, the ac component of the total hole current in the depletion region is determined as (12)(13):

Eq. (8) $$\delta J_h = q\Phi_o(1 - R)[1 - \exp(-\alpha W_d)/(1 + \alpha L)]\exp(j\omega t)$$

where the modified diffusion length is

Eq. (9) $$L = L_p/(1 + j\omega\tau_p)^{1/2}$$

and $L_p = (D_p\tau_p)^{1/2}$, D_p and τ_p are the minority-carrier diffusion length, diffusion coefficient and lifetime, respectively. In the abrupt junction approximation, the width of the depletion layer, W_d, is related to the space-charge capacitance as

Eq. (10) $$C_{sc} = \varepsilon_s/W_d$$

where ε_s is the permittivity of the semiconductor.

The net rate of electron capture (i.e., total capture minus thermal emission) at the surface, R_e, can be derived using the recombination statistics of Shockley and Read (14). In thermal equilibrium the concentration of free carriers at the surface is in equilibrium with the bulk semiconductor. Free carriers at the surface are also in equilibrium with the surface/interface states, with the capture rate balanced by an equal thermal emission rate. Illumination of the wafer disturbs this situation. Collection of the holes at the surface lowers the surface potential barrier and increases the concentration of electrons at the surface, in turn increasing the capture rate of electrons by the surface states.

The SCA measurements are performed at sufficiently low illumination to assure that the induced change of surface potential barrier, $\delta\psi_s$, is much smaller than the initial barrier height, ψ_s, and much smaller than kT/q (0.0259 V at room temperature). Under such conditions, as a first order approximation, the net rate of electron capture at the surface is proportional to the deviation of the electron concentration at the semiconductor surface, δn_s, from its equilibrium value, $n_s = n_b \exp(q\psi_s/kT)$ and

Eq. (11) $$R_e = s\delta n_s = s(q\delta\psi_s/kT)n_b \exp(q\psi_s/kT)$$

where s is the surface capture velocity in cm/sec and n_b the concentration of electrons in the bulk semiconductor (8)(9). In general, s depends on the equilibrium occupation of surface/interface states and, therefore, on the height of the surface potential barrier in thermal equilibrium. This neglects change by illumination of the barrier under steady-state conditions by assuming low enough illumination and sufficiently high surface recombination.

The equation describing the ac component of surface photovoltage under steady-state conditions, measured by the SCA method, can be written by combining Eqs. (4) and (6) in the form

Eq. (12) $$C_{sc}(d\psi_s/dt) = dJ_h - qR_e$$

and

Eq. (13) $$C_{sc}(d\psi_s/dt) = J_1 \exp(j\omega t) - qs^*\delta\psi_s$$

where following Eqs. (8) and (11)

Eq. (14) $$J_1 = q\Phi_o(1 - R)[1 - \exp(-\alpha W_d)/(1 + \alpha L)]$$

and

Eq. (15) $$s^* = s(q/kT)n_b \exp(q\psi_s/kT)$$

The solution of Eq. (13) has the form

Eq. (16) $$\delta\psi s = \psi_{s1} \exp(j\omega\tau)$$

where

Eq. (17) $$\psi_{s1} = C_{sc}^{-1}[\tau/(1 + j\omega\tau)]J_1$$

and

Eq. (18) $$1/\tau = qs^*C_{sc}^{-1}$$

Equation (17) represents the relatively complex dependence of the small signal alternating current surface photovoltage (ac-SPV) on lifetime and diffusion length of the minority carriers. However, focusing our attention on the materials typically used for semiconductor processes, we can substantially simplify this expression. Let us consider three cases: *(i)* Si of short diffusion length of approx. 1 µm, *(ii)* medium, where L_d = 35 µm, and *(iii)* long, with L_d = 200 µm. With the diffusion coefficient for holes D_p at room temperature close to 12 cm^2/sec, the lifetime of holes, τ_p, is about 1 nsec in case *(i)*, 1 µsec in case *(ii)*, and 35 µsec in *(iii)*. At a modulation frequency of illumination at 50 kHz, the term $\omega\tau_p$ in Eq. (9) is approximately 3×10^{-4} in case *(i)*, about 0.3 in case *(ii)*, and about 10 in the case of long diffusion length *(iii)*. To further analyze the modified diffusion length, it is convenient to express it the form

Eq. (19) $$L^2/L_p^2 = 1/(1 + \omega^2\tau_p^2) - j\omega\tau_p/(1 + \omega^2\tau_p^2)$$

In the case of a short lifetime *(i)*, $\omega\tau_p$ in the first term, as well as the entire imaginary term of Eq. (19), can be neglected and the modified diffusion length becomes equal to a hole diffusion length, $L = L_p$. Similarly, in the intermediate case of medium diffusion length *(ii)*, whereas there is an increase in the imaginary term to about 0.3, the absolute value of L^2/L_p^2 decreases only a few percent below unity and again a good approximation for our purpose can assume that $L = L_p$. The situation drastically changes for long diffusion lengths *(iii)* when the first term in Eq. (19) can be neglected and the second term becomes $1/\omega\tau_p$, so that the modified diffusion length becomes a complex number but is independent of the hole lifetime, $L^2 = -jD_p/\omega$, as illustrated in Fig. 3. In this case, the absolute value of L reaches its saturation at a value of about 60 µm.

The impact of carrier diffusion on the magnitude of surface photovoltage depends on the absorption coefficient of the incident illumination, given by Eq. (14). For the wavelength of 560 nm, the absorption coefficient of silicon a is 7×10^3 cm^{-1}(15). The maximum value of the term related to the diffusion length and absorption coefficient in Eq. (14), independent of doping concentration, will not exceed 3 and 5 percent for the long *(iii)* and medium *(ii)* diffusion length cases, respectively. For a depletion layer width W_d = 1 µm (corresponding to the maximum depletion layer width in silicon with a

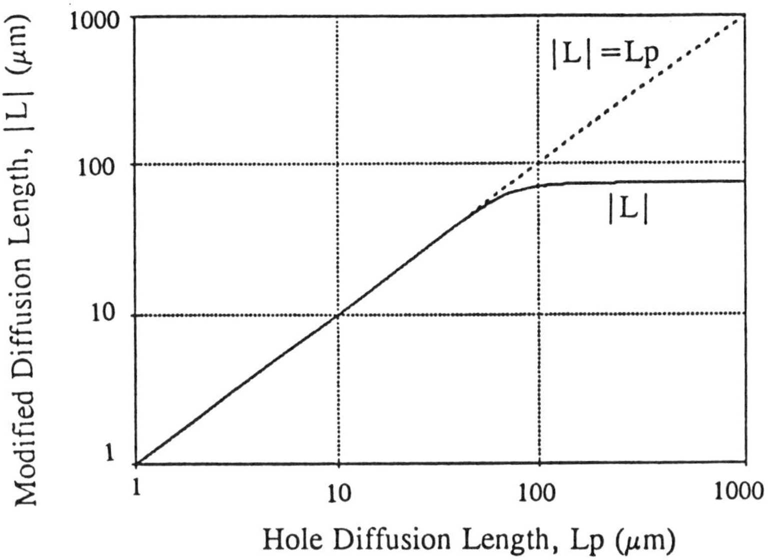

Figure 3. Dependence of the absolute value of the modified diffusion length, |L|, on the diffusion length of minority carriers (holes), L_p, at a light modulation frequency of 50 kHz.

doping concentration close to 10^{15} cm^{-3}) this value will be further reduced by a factor of two. Therefore, for the Si wafers with long enough diffusion length, at least corresponding to the medium case *(ii)*, the term related to the carrier diffusion and the absorption coefficient in Eqs. (14) can be neglected resulting in

Eq. (20) $$J_1 = q\Phi_o(1 - R)$$

In case *(i)*, the short diffusion length becomes comparable to the penetration depth of the 560 nm light in Si, $1/\alpha = 1.5$ μm, and the diffusion-absorption term in Eq. (14) cannot be neglected. In Si with doping concentrations about 10^{15} cm^{-3} at the maximum depletion layer width, $W_{max} = 1$ μm, the reduction of the ac-SPV signal due to the effect of absorption and carrier diffusion will be close to 30%. For smaller depletion layer widths (e.g., at higher doping concentration or at lower surface potentials), this difference will increase further. In the case of the short diffusion length Eq. (14) can be written in the form

Eq. (21) $J_1 = q\Phi_o(1 - R)[1 - \exp(-\alpha W_d)/(1 + \alpha L_p)]$

Note that in Eq. (21) the modified diffusion length, L, was replaced by a minority carrier (hole) diffusion length, L_p. Equation (21) is a general expression for J_1, which can be used for both short and long diffusion lengths, whereas Eq. (20) is useful for cases of longer diffusion lengths that exceed the penetration depth of light in Si.

The real and imaginary parts of ψ_{s1}, described by Eqs. (17) and (21) or (20), are

Eq. (22) $Re(\psi_{s1}) = J_1 C_{sc}^{-1}[\tau/(1 + \omega^2\tau^2)]$

and

Eq. (23) $Im(\psi_{s1}) = -J_1 C_{sc}^{-1}[\omega\tau^2/(1 + \omega^2\tau^2)]$

and the ratio

Eq. (24) $Re(\psi_{s1})/Im(\psi_{s1}) = -1/\omega\tau$

The dependence of $Im(\psi_{s1})$ and $Re(\psi_{s1})$ on applied voltage is illustrated in Fig. 4.

According to Eqs. (18) and (15), τ is proportional to $\exp(-q\psi_s/kT)$. Therefore, τ will reach its highest value at the onset of inversion conditions (in n-type material ψ_s is negative under depletion conditions). In the SCA method the light modulation frequency, ω, is selected high enough to assure that $\omega\tau \ll 1$ at the onset of inversion, and the real part of ψ_{s1} under such conditions is negligible compared to the imaginary part.

The width of the depletion layer, W_d, in the surface charge analysis technique is determined from the expression obtained by substituting J_1 from Eq. (21) and C_{sc} from Eq. (10) in Eq. (23), resulting in

Eq. (25) $Im(\psi_{s1}) = -(q\Phi_o/\varepsilon_s\omega)(1-R)[1-\exp(-\alpha W_d)/(1+\alpha L_p)][\omega^2\tau^2/(1+\omega^2\tau^2)]W_d$

where $\omega\tau$, given by Eq. (24), is determined from the ratio of the real and imaginary parts of ψ_{s1}.

The linear dependence of the low-illumination ac-SPV on the photon flux (Eq. 25) is a consequence of the assumed small disturbance approximation. However, the dependence on the illumination intensity is also hidden in the depletion layer width, W_d. Incomplete recombination of the minority carriers during the "dark" portion of the modulation cycles will lead to accumulation of minority carriers at the surface over a period of many cycles. The accumulation of minority carriers at the surface will reduce the surface barrier and hence W_d, since W_d is proportional to the square root of the surface potential barrier (10). This in turn will lead to a non-linear dependence on illumination intensity of the ac-SPV signal. Therefore, linearity of the ac-SPV signal measured under depletion conditions indicates that the difference between a steady-state value of W_d under illumination and a dark, thermal equilibrium value is negligibly small, i.e., much smaller than its equilibrium value with the corresponding change in the surface potential barrier and much smaller than kT/q (8)(9).

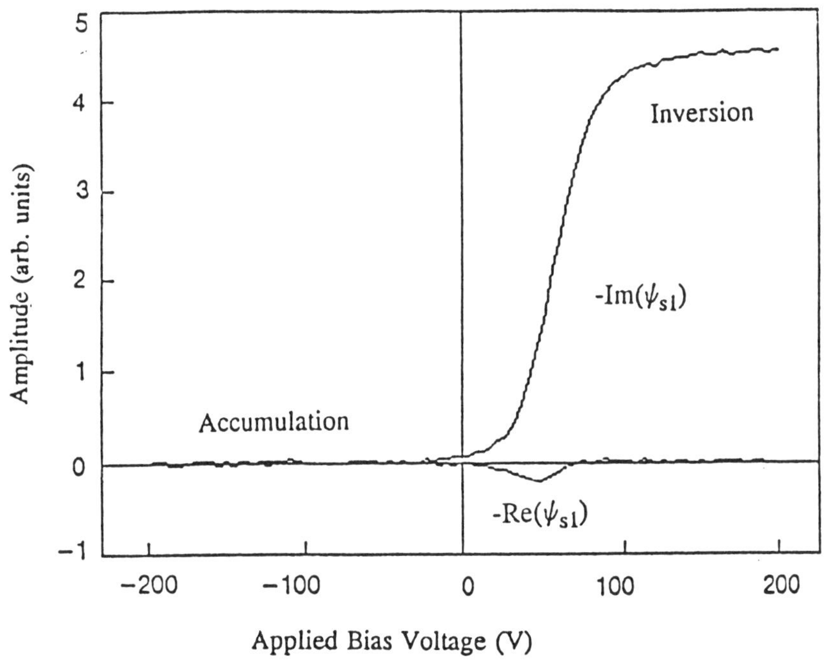

Figure 4. Dependence of the real and imaginary parts of the ac-SPV on bias voltage for a thermally oxidized wafer.

SCA Measurements. In typical SCA measurements, the low-illumination ac-SPV signal is detected using a pick-up electrode placed parallel to the wafer surface and separated from it by an insulating spacer. The illumination of the wafer area under the electrode is uniform. The equivalent electrical circuit, as shown in Fig. 5 for the ac-SPV, can be deduced considering that Eqs. (16) and (17) describe the ac voltage $\delta\psi_s$ generated by a current $\delta J_h = J_1 \exp(j\omega\tau)$, across a parallel RC network. In this network capacitance C_{sc} corresponds to the space-charge capacitance (per unit area) and resistance $R = \tau/C_{sc}$ represents the surface recombination. The time constant, τ, is described by Eq. (18). The coupling between the semiconductor surface and the pick-up electrode is purely capacitive, as represented in Fig. 5 by C_p. The current, δJ_m , through the coupling capacitance C_p is measured by using a current amplifier and corrected for the coupling capacitance C_p:

Eq. (26) $$\delta\psi_s \propto \delta J_m / C_p$$

Correction for the coupling capacitance compensates for eventual changes of geometry and distance of the pick-up electrode from the semiconductor surface.

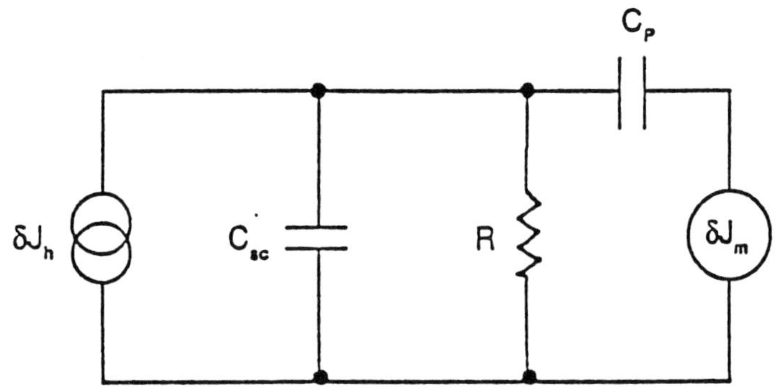

Figure 5. Equivalent circuit for the ac-SPV.

In the SCA technique the real and imaginary components of the ac-SPV signal are measured as a function of the total charge induced in the semiconductor, Q_{ind}. The charge is induced by applying a slowly varying (relative to the frequency of light modulation) bias voltage to the pick-up electrode. The value of charge induced in the semiconductor is computed as a product of applied bias voltage and the coupling capacitance C_p:

Eq. (27) $$Q_{ind} = -C_p V_{bias}$$

Whereas the coupling capacitance, C_p, represents the total capacitance between the pick-up electrode and the semiconductor, in most applications it is dominated by the capacitance of the spacer, which is usually much thicker than the oxide or other dielectric coating on the semiconductor. Assuming that the oxide charge is independent of bias, the charge induced in the semiconductor represents the sum of the changes in the semiconductor space-charge and in the surface (interface) trapped charge.

A schematic representation of the SCA system is shown in Fig. 6 (16). The electrical coupling to both the front and back surface of the wafer is purely capacitive. The chopped illumination from a green (560 nm) light emitting diode (LED) is brought to the surface of the wafer via a Mylar spacer (about 12 μm thick) coated on the side opposite the wafer with an optically transparent, conductive film. The conductive film on the Mylar spacer forms a pick-up electrode and is used to capacitively measure the ac-SPV signal. It is also used to apply the variable voltage bias (up to 1000 V) that induces the charge in the semiconductor. The bias voltage ramp has a triangular form starting always at zero, next going to maximum either positive or negative (maximum negative or positive Q_{ind}, respectively), and then going to the opposite extreme and back to zero.

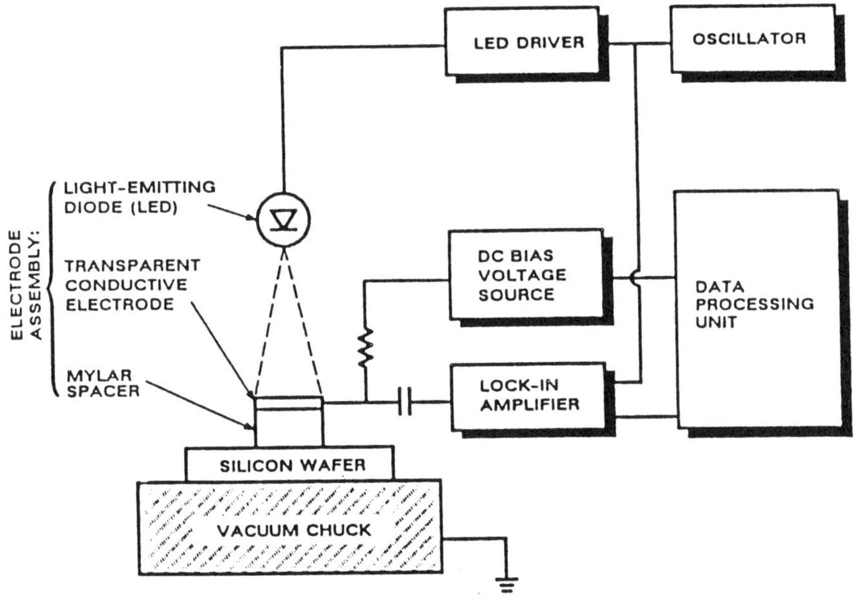

Figure 6. Schematic diagram of the Surface Charge Analyzer.

2.2 Overview of Measured Parameters

The typical SCA plots, illustrated in Fig. 7, represent dependence of the depletion layer width, W_d, (determined from the low light intensity ac-SPV) on charge induced in the semiconductor, Q_{ind}. The ac-SPV signal vanishes in accumulation (positive Q_{ind} for p- and negative for n-type semiconductors), which corresponds to the zero value of the depletion width. On the other hand the ac-SPV signal and depletion layer width maximize at the onset of strong inversion. Increase of the induced charge in the silicon beyond the onset leads only to small increases of the depletion layer width due to the shielding by the inversion layer that prevents further expansion of the depletion region.

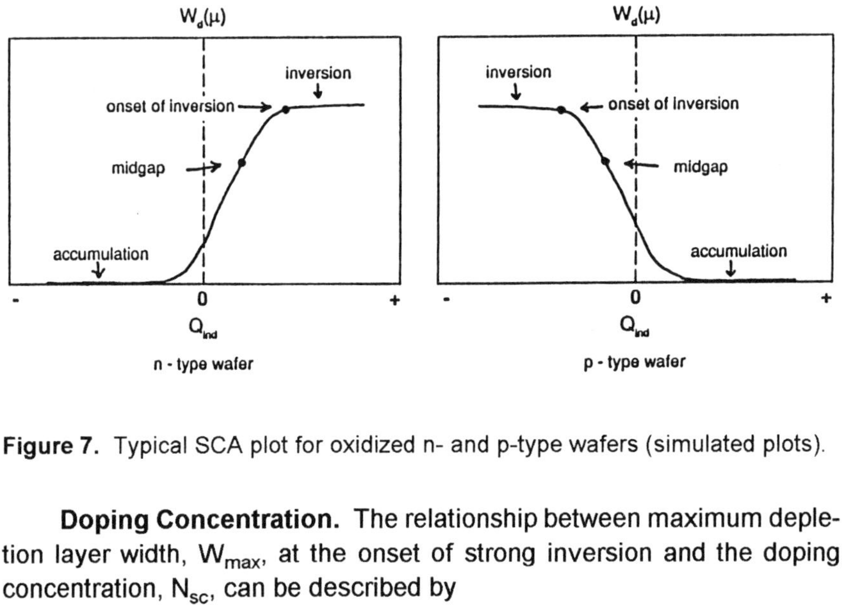

Figure 7. Typical SCA plot for oxidized n- and p-type wafers (simulated plots).

Doping Concentration. The relationship between maximum depletion layer width, W_{max}, at the onset of strong inversion and the doping concentration, N_{sc}, can be described by

Eq. (28) $$W_{max} = |2\varepsilon_s\psi_L/qN_{sc}|^{1/2}$$

where

Eq. (29) $$\psi_L = \pm\, kT/q\,(2.10\,\ln|N_{sc}/n_i| + 2.08)$$

which is the Lindner surface potential barrier, positive for a p-type and negative for a n-type semiconductor, at which minority carriers pin the surface in inversion (17). N_{sc} is the net doping concentration in the space-charge region, $N_{sc} = N_D - N_A$; N_D and N_A are donor and acceptor concentrations, respectively. Since doping concentration can be determined relatively easily, it can be used to calibrate the SCA measurement system by using a reference wafer of a known doping concentration. It should be noted that the doping value determined by SCA represents the concentration in the surface region of the wafer that corresponds to the depletion layer width under inversion conditions.

Conductivity Type. As it is shown in Fig. 7, the variation of W_d vs. Q_{ind} depends on the conductivity type of the wafer. Inversion in a p-type semiconductor occurs at more negative Q_{ind} than accumulation, whereas in a n-type inversion occurs at more positive Q_{ind}. This difference can be used to establish the wafer conductivity type. However, in our case it was found that the change in polarity of the ac-SPV signal offers a more reliable approach. Since the surface potential barriers under depletion conditions for n- and p-type semiconductors have the opposite sign and illumination causes a decrease of the magnitude of the barrier, the phase of the ac-SPV signal will differ by 180°C, depending on the conductivity type of the wafer. Use of the phase for establishing conductivity type of the wafer surface was found to be very effective even for very thin layers at the silicon surface (18).

Oxide Charge. The induced charge, Q_{ind}, represents the sum of the charges from zero bias conditions in the semiconductor space-charge, Q_{sc}, and in the interface trapped charge, Q_{it},

Eq. (30) $$Q_{ind} = \Delta Q_{sc} + \Delta Q_{it}$$

At zero bias $Q_{ind}(0) = 0$ and the oxide charge, Q_{ox}, is compensated by the total of the space-charge and the surface/interface trapped charge,

Eq. (31) $$Q_{ox} = - [Q_{sc}(0) + Q_{it}(0)]$$

where $Q_{sc}(0)$ and $Q_{it}(0)$ are values at zero bias. Since the oxide charge is independent of bias, and at any bias

Eq. (32) $$\Delta Q_{sc} = Q_{sc} - Q_{sc}(0)$$

and

Eq. (33) $$\Delta Q_{it} = Q_{it} - Q_{it}(0)$$

the induced charge represents the sum of the oxide charge and the charges in the space charge region and the interface states,

Eq. (34) $$Q_{ind} = Q_{ox} + Q_{sc} + Q_{it}$$

To determine the oxide charge using Eq. (34), we need to eliminate the component due to the interface trapped charge, Q_{it}. This could be done by biasing the semiconductor surface to a condition, referred to as the midgap (MG), when the Fermi level at the Si surface (Si/SiO_2 interface) is at the silicon midgap position (2). The use of the midgap bias condition to characterize the Si/SiO_2 system has been reported previously in analysis of metal-oxide-semiconductor capacitance-voltage (MOS C-V) characteristics (19). This approach is based on consideration that the Si/SiO_2 interface states located above the midgap are acceptor-like while those below midgap are donor-like with an equal number of each type. Therefore, when the Fermi level is at midgap, the interface is electrically neutral and the interface trapped charge can be neglected in Eq. (34) resulting in

Eq. (35) $$Q_{ox} = Q_{ind}(MG) - Q_{sc}(MG)$$

In deriving Eq. (35), it was assumed that the density of interface states is symmetrical about the midgap. However, this assumption is not valid at very high densities of interface states ($>10^{12}$ eV^{-1} cm^{-2}) (19) as well as when contamination or process-induced damage produces local peaks in the energy distribution of the interface state density (20). In such cases there can be a contribution from the interface trapped charge, $Q_{it}(MG)$, into the SCA measured Q_{ox}. A relatively simple criteria for testing if the distribution of interface state density is symmetrical about midgap is to check the location of the minimum in the distribution of states as measured by SCA. However, even if there is some asymmetry in the distribution of states, the error in Q_{ox} measurements can be usually neglected as the density of states typically reaches its minimum at the midgap.

To determine the oxide charge by using Eq. (35) we have to first find the value of the depletion layer width at midgap conditions. In the abrupt junction approximation, the relationship between the depletion layer width and the surface potential barrier is given by (10)

Eq. (36) $$\psi_s = - qN_{sc}W_d^2/2\varepsilon_s$$

with ψ_s being negative in n-type and positive in p-type material ($N_{sc} = N_D - N_A < 0$). At midgap conditions the Fermi level, E_F, and the intrinsic Fermi level, E_i, (lying very close to the middle of the bandgap) are aligned at the surface and the surface potential barrier is given by

Eq. (37) $$\psi_s(MG) = \pm(kT/q)\ \ln|N_{sc}/n_i|$$

where n_i is the intrinsic carrier concentration in silicon, and $kT\ \ln|N_{sc}/n_i|$ is the absolute value of the difference between the Fermi level, E_F, and the intrinsic level, E_i. Combining Eqs. (36) and (37), the midgap depletion layer width is

Eq. (38) $$W_d(MG) = [2\varepsilon_s kT \ln|N_{sc}/n_i|/(q^2|N_{sc}|)]^{1/2}$$

Knowledge of the doping concentration from the onset of strong inversion allows determination of $W_d(MG)$ and the depletion layer charge from

Eq. (39) $$Q_{sc}(MG) = qN_{sc}W_d(MG)$$

which is positive in n-type and negative in p-type semiconductors. Further, using the SCA measured W_d vs. Q_{ind} dependence, we can determine the total charge induced in the semiconductor under midgap conditions, $Q_{ind}(MG)$, and hence, using Eq. (35) and (39), Q_{ox}.

The oxide charge measured by the SCA technique differs from that determined in the MOS capacitance-voltage (C-V) method. Because the conductive SCA pick-up electrode is separated from the semiconductor by a Mylar spacer, which is much thicker than the insulating film, the SCA measures total charge in the film. On the other hand, since in the MOS C-V configuration the gate electrode is deposited directly on the insulating film, the results of this technique are dominated by a charge located in the proximity of the semiconductor-oxide interface (10).

Location of typical charges at the thermally oxidized silicon surface is pictorially depicted in Fig. 8. The effective net oxide charge at the silicon-oxide interface, usually dominated by the fixed oxide charge, Q_f, can be determined with C-V measurements using the expression $Q_f = -C_{ox}(V_{FB} - \Phi_{MS})$, where V_{FB} is the flat band voltage; Φ_{MS}, metal-semiconductor work function difference; C_{ox}, oxide capacitance (10). To obtain a measure of Q_m, the mobile charge, the wafers are subjected to a heat cycle and bias voltage. With the mobile charges moving to the silicon/oxide interface, an indication of the density of Q_m can be obtained. Any oxide surface charge is masked by the capacitor electrode required for a C-V measurement.

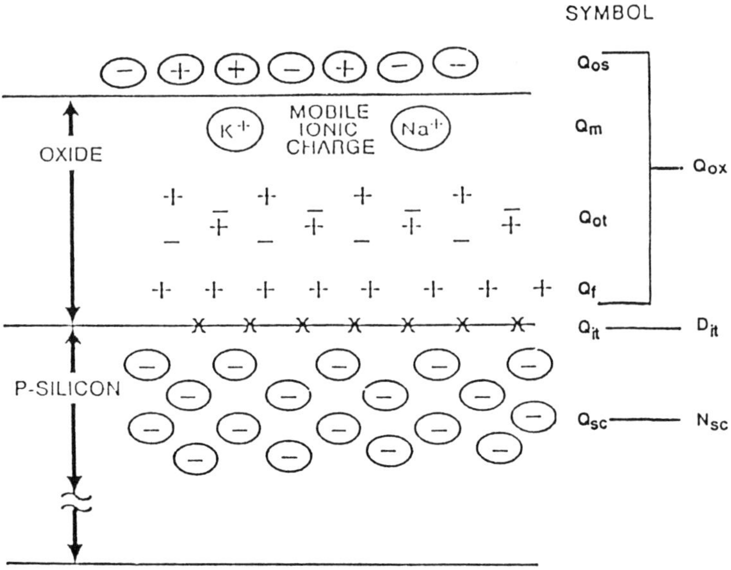

Q_{os} = OXIDE SURFACE CHARGE
Q_m = MOBILE IONIC CHARGE
Q_{ot} = OXIDE TRAP CHARGE
Q_f = OXIDE FIXED CHARGE
Q_{it} = INTERFACE TRAP CHARGE
Q_{sc} = SILICON SPACE CHARGE

Figure 8. Review of charges in Si-SiO_2 structure.

With the SCA technique, the sum of these charges is measured, including any fixed and trapped charges, Q_{ot}, typically found after high energy bombardment of the wafer surface. The SCA measurement gives the complete "status" of resident charges on the wafer surface.

Interface (Surface) State Density. Since the oxide charge is independent of bias, a small change of the external bias under depletion conditions will induce change in the depletion charge, dQ_{sc}, and in the surface (interface) trapped charge, dQ_{it}, resulting, according to Eq. (34), in

Eq. (40) $$dQ_{it} = dQ_{ind} - dQ_{sc}$$

where

Eq. (41) $dQ_{sc} = qN_{sc}dW_d$

The distribution of interface state density, $D_{it}(E)$, as a function of energy is defined as (10)

Eq. (42) $D_{it}(E) = -dQ_{it}/qdE$

with the energy

Eq. (43) $E = E_{Fs} - E_{is} = E_F - (E_i - q\psi_s)$

where, E_{Fs} is the Fermi level at the surface, E_{is} intrinsic Fermi level at the surface, E_F and E_i are the Fermi level and the intrinsic level in the bulk, respectively. From Eq. (43)

Eq. (44) $dE = qd\psi_s$

and using Eqs. (36) and (41),

Eq. (45) $dE = -(q^2N_{sc}W_d/\varepsilon_s)dW_d = -(qW_d/\varepsilon_s)dQ_{sc}$

Hence,

Eq. (46) $D_{it} = (\varepsilon_s/q^2W_d)[(dQ_{ind} - dQ_{sc})/dQ_{sc}]$

Expression (46) was derived starting with the linear approximation for ac-SPV. When the interface state density is very low, close or below about 10^{10} eV^{-1} cm^{-2}, a small difference exists between Q_{ind} and Q_{sc}, which may be outside the accuracy of this ac-SCA approximation. In such cases monitoring of interfaces can be realized by using an interface quality factor, IQF, given by

Eq. (47) $IQF = dQ_{ind}/dQ_{sc}$

which is related to the interface state density, Eq. (46), by

Eq. (48) $D_{it} = (\varepsilon_s/q^2W_d)(IQF - 1)$

IQF may be particularly useful in monitoring high quality Si/SiO_2 interface produced by thermal oxidation.

An example of the energy distribution of the interface state density measured by SCA is shown in Fig. 9. It shows the characteristic for the thermally oxidized silicon U-shaped background interface state density (10).

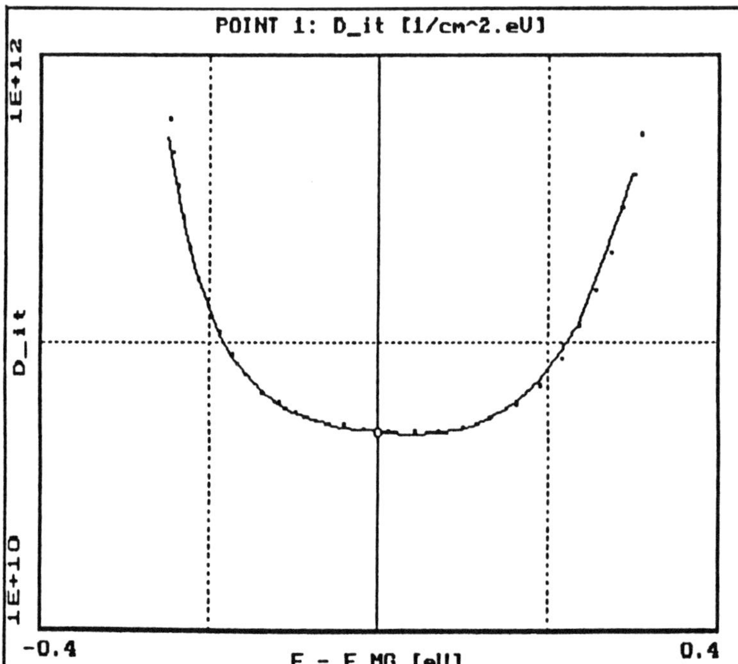

Figure 9. Energy distribution of interface state density for thermally oxidized silicon

3.0 SURFACE CHARGE CHARACTERIZATION OF CLEANING PROCESSES

Of particular importance for the monitoring of cleaning processes is the ability of the SCA technique to measure the oxide charge and the interface state density. These SCA measurements can be performed both on oxidized and bare silicon surface. Since there is no processing involved, the characterization can be performed practically in "real-time", a feature important for understanding the evolution of the wafer surface after cleaning.

The wafer cleaning process typically consists of several steps with the end result depending on the contribution of each individual step. Chemical contamination and surface damage induced by each step depend on the chemical interaction with the silicon surface and the reagents used. This interaction is also sensitive to the processing procedure and construction of the processing equipment that is used. The "RCA Standard Clean" (21) and its modifications are still the most frequently used cleaning processes. The RCA Standard Clean 1 (SC-1) utilizes a hot alkaline and hydrogen peroxide solution. This is followed by the RCA Standard Clean 2 (SC-2) consisting of hot hydrochloric acid and hydrogen peroxide. A frequently used modification of the cleaning procedure adds a preliminary clean-up with a hot sulfuric acid and hydrogen peroxide mixture to the original RCA cleans, followed, for bare silicon wafers, by an etch in very diluted aqueous HF solution (22). Typical formulations and cleaning sequences are given in Tables 1 and 2.

The use of Surface Charge Analysis for characterization and monitoring of the cleaning processes will be described for the typical RCA cleaning sequence. While each step in the sequence is optimized to remove specific contamination, it will also modify the structure of the silicon surface and may result in additional contamination or a complexing of the contamination source (22). This leads to changes of oxide charge and interface state density, as well as in some cases, surface doping concentration.

Table 1. Typical Silicon Wafer Cleans

Name	Chemical Composition	Typical Formulations	Purpose of Clean
SC-1, RCA-1, "Huang 1"	NH_4OH: H_2O_2: H_2O	1:1:5, 75°C	Organics Removal, Metal Ion Complexing
SC-2, RCA-2, "Huang 2"	HCl:H_2O_2: H_2O	1:1:6, 80°C	Alkali Ion Removal, Metal Hydroxides Dissolution, Residual Trace Metal Removal
Mixture of Sulfuric Peroxide, SPM, "Piranha"	H_2SO_4: H_2O_2	2:1, 90°C	Organics Removal
Diluted HF	HF: H_2O	1:10 to 100, 25°C	Native Oxide Removal

Table 2. Typical Silicon Wafer Cleaning Sequences

Wafer with Oxide	Bare Wafers
SPM	SPM
	Diluted HF
RCA SC-1	RCA SC-1
	Diluted HF
RCA SC-2	RCA SC-2
Diluted HF	Diluted HF

Since the chemical oxide formed in peroxide-based cleans may trap some impurities, it was found beneficial in some applications to remove it prior to subsequent treatments using an HF-H_2O solution, as indicated in Table 2 (22).

The hydrophilic silicon surface, passivated with chemical oxide several angstroms thick, is found beneficial in numerous fabrication processes by providing, for instance, protection of the wafer surface during transportation and storage. However, the more refined growth of thin thermal oxides required for ULSI CMOS technologies necessitates removal of this passivating chemical oxide. It was established that etching off the oxide by using HF leaves hydrophobic silicon surfaces which for a limited time are chemically stable against oxidation (23). Stability of the HF treated surface has been mainly attributed to passivation due to hydrogen. Fluorine remains only a minor species (23)-(25). The importance of the HF treated surfaces necessitates characterization and monitoring of the effectiveness of both oxide removal and residual surface contamination.

As noted, there are numerous combinations of cleans which may be employed, thus characterization becomes critical in assessing their effectiveness. Further, the reagents used in these cleans are not stable, leading to degradation with usage and time. To provide sufficient effectiveness in cleaning, solutions are heated, which increases the rate of degradation. Not only are solutions changing with time, the wafer surface also changes as exposure to air provides oxygen for oxidation of the highly reactive, just cleaned, wafer surface. In addition, chemicals in air introduce carbon, sulfur and minute amounts of metal ions. To determine to some degree what may be occurring, a rapid measurement technique requiring no other processing must be employed, leading to the use of the previously described SCA technique. Early results obtained from the evaluation of various cleans will be presented.

3.1 HF Etch

Cleaning processes that end by using diluted HF as the last step display SCA characteristics, shown in Fig. 10, that are easily distinguishable from those observed for hydrophilic silicon surfaces. These characteristics do not depend on the processes proceeding an HF oxide strip (26). Soon after etching in HF the wafer surfaces show a very strong hysteresis frequently exceeding 10^{12} q/cm^2. A typical SCA plot (depletion layer width, W_d, vs. charge induced in the semiconductor, Q_{ind}) for wafer after a clean ending with an HF dip and DI water rinse is shown in Fig. 10. The characteristic shows a hysteresis, which is much more pronounced when the bias sweep corresponds to positive Q_{ind} than a negative one. The equilibrium oxide (surface) charge, Q_{ox}, exhibited soon after etching is positive and typically below mid 10^{11} q/cm^2. The hysteresis and the positive oxide charge are due to a high density of slow states. The centers responsible for the slow states are believed to be associated with the defects identified with ionized silicon atoms which, as in the case of thermal oxide, are located in the proximity of the interface (27). Both the magnitude of the oxide charge and hysteresis decreases with elapsed time after the HF treatment. This latter effect was attributed to the disappearance of surface defects due to the growth of native oxide (28).

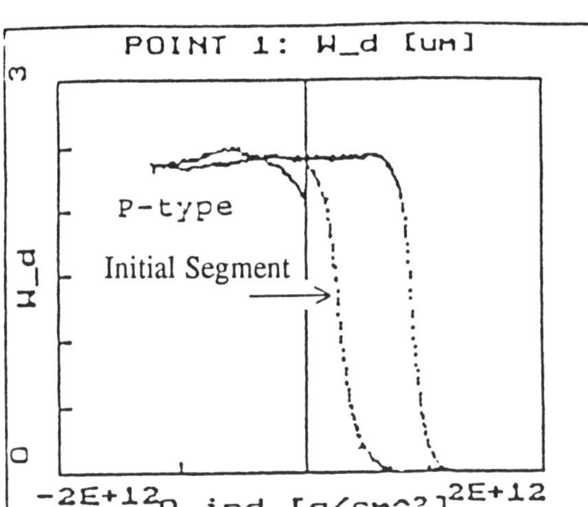

Figure 10. Typical SCA plot for HF-etched silicon wafer.

Detailed studies of the variation of oxide charge with ambient air exposure time after HF etching were reported recently for both liquid and vapor treated surfaces (29). The paper reports the time dependence of the equilibrium oxide charge, determined from the initial segment of the characteristic acquired upon the first bias ramp, from zero to the first extreme, as shown in Fig. 10. The monitor wafers were prepared by growing approximately 150 Å of sacrificial thermal oxide on 1 to 10 ohm-cm, p-type, <100> oriented, CZ silicon substrates followed by liquid-HF/H_2O etching in a conventional wet bench. In contrast, vapor-HF/H_2O etching was performed in an Advantage Production Technology model EDGE-2000 system. In both cases the procedure insured complete removal of the sacrificial oxide.

The dependence of the equilibrium oxide (surface) charge on the time of exposure to the ambient air is shown in Fig. 11 for both liquid-HF/H_2O and vapor-HF/H_2O etched silicon wafer surfaces (29). Three evolutionary periods exist, with characteristics which are nearly the same for both treatments.

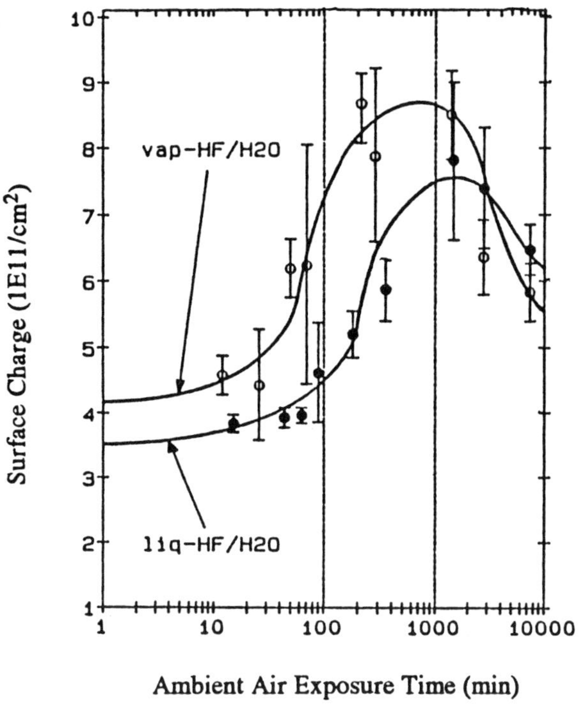

Figure 11. Surface charge vs. ambient air exposure time for liquid HF/H_2O (filled circles) and vapor HF/H_2O (circles) etched silicon wafer surfaces. (From Ref. 29.)

During an initial period of about one hour, the oxide (surface) charge is very stable, indicating a low level of interaction with ambient air. This effect is believed to be mainly due to hydrogen passivation of the silicon surface (23)(25)(29). During the next hour the oxide (surface) charge rapidly increases; this is believed to be due to the rapid annihilation of the hydrogen passivation layer and subsequent formation of trivalent silicon centers. At the end of this intermediate period, the rate of change of oxide (surface) charge decreases. This can be attributed to the reaction of the trivalent silicon centers with hydroxyl groups present in air (29). Formation of Si-OH bonding is believed to annihilate positive charge of the trivalent silicon centers, and therefore reduce effect of hydrogen desorption on the oxide (surface) charge. The rate of formation of trivalent silicon centers decreases with decreasing density of hydrogen atoms on the surface, reaching equilibrium with the rate of charge neutralization due to adsorption of the hydroxyl groups (end of the intermediate period in Fig. 11). A further increase in the adsorption of the hydroxyl groups and oxidation of the silicon surface (30) leads to decreased density of trivalent silicon and consequently in the SCA measured oxide (surface) charge. In Fig. 11 this corresponds to period 3, which extends over a long period of time characterized by a gradual decreasing of the oxide (surface) charge.

The variation of the oxide (surface) charge with exposure time of wafers to ambient air were found to be in good agreement with the time dependence of the oxide thickness (29). However some differences may be observed in the transition between initial (first) and intermediate (second) period by comparing results of Ref. 29 and previously reported studies in Ref. 23. Similarly, it seems that a transition between the initial and intermediate periods in oxide (surface) charge variation indicates that this transition happens earlier for vapor-HF/H_2O etched surfaces than for liquid-HF/H_2O etched surfaces. These effects may be due to differences in composition of the passivating hydrogen layer, such as differences in the concentration of fluorine (Si-F groups), oxygen, organic residues, or metallic contamination. Factors such as ambient air humidity and temperature may also influence the effectiveness of hydrogen passivation.

As shown in Fig. 11, the silicon wafer surfaces treated with liquid-HF/H_2O versus vapor-HF/H_2O exhibit a lower net charge in the initial and intermediate evolutionary periods. This was attributed to a higher content of oxygen on the liquid-treated surfaces (29). The XPS (X-Ray Photoelectron Spectroscopy) analysis of wafers immediately following HF/H_2O treatment has shown that liquid treatment leaves about 2.2 times more oxygen on the silicon surface compared to the vapor treatment. The source of oxygen has been traced to the DI water rinsing following the liquid-HF/H_2O step. The vapor treated surfaces were not subjected to DI water rinsing. The oxygen dissolved in the water may react with a small fraction of the Si-H groups to form negatively charged Si-O centers, thus resulting in a net loss of positive oxide charge during the first two periods (29). Nearly complete oxidation of the surface in the final (third) period of the wafer evolution leads to the disappearance of oxide charge differences between liquid and vapor treated surfaces (29).

To verify that there are three characteristic periods of surface evolution, well synchronized with the periods characterizing variation of the oxide charge, measurements of contact angle were performed. Initial contact angle measurements indicated a hydrophobic wafer surface, with subsequent oxidation causing the contact angle to decrease. Reference 29 shows the distinct periods with an initial period of nearly 100 minutes having a high contact angle and a subsequent rapid change in angle asymptotically approaching the equivalence of a thermally oxidized surface. These three periods represent stages of wafer oxidation in ambient air. The values of the contact angle observed in the final (third) period, for vapor treated surfaces with lower content of organic residue than liquid treated surfaces, approximate those measured for thermally oxidized silicon (29).

The stability of the surface for an appreciable period, referred to as period 1, allows monitoring of the chemical cleaning procedures. The rapid variation observed in phase 2 would not make the phase 2 period suitable for consistent process monitoring. In phase 3, the electrical parameters vary slowly as the surface conditions are becoming well stabilized. The phase 3 period could be used for monitoring, however, wafer storage conditions may dominate resultant surface characteristics, thereby reducing sensitivity to the cleaning method. It is important to use a consistent time period for measurements of the surface stability regardless of which phase is used for monitoring.

3.2 RCA Standard Clean 1 (SC-1)

Unlike the case for HF etch, treatment of the surface with the RCA standard clean 1 (SC-1) results in the formation of several angstroms of chemical oxide. As noted in Table 1, the treatment in the typical RCA SC-1 solution is performed at elevated temperatures, typically about 75° - 80°C for 10 min (22). The process results are very sensitive to the composition of solution, temperature, and time of exposure to the solution. Modifications in the DI water rinsing procedure following the treatment may also affect the electrical characteristics of cleaned wafers.

The components of the RCA SC-1 solution have opposite effects on the silicon surface: the peroxide oxidizes the silicon surface, while the ammonium hydroxide is dissolving this chemical oxide. This results in oxidation of new atomic layers of silicon and an advancement of the silicon/oxide interface deeper into the silicon substrate, resulting in formation of trivalent silicon centers in proximity with the interface. The progressing chemical oxidation, however, annihilates these centers by forming bonds with oxygen or hydroxyl groups.

Typical SCA data for RCA SC-1 treated wafers are shown in Fig. 12. Similar to the case of HF treated wafers, the data show a characteristic hysteresis that is much more pronounced for positive than for negative induced charge, Q_{ind}. The positive hysteresis is associated with the slow states located in the oxide in proximity of the interface. These states are believed to be due to trivalent silicon centers. It should be noted that the hysteresis effect observed after RCA SC-1 treatment is much smaller than for the case of HF treated surfaces This indicates a lower density of the slow states that can be attributed to surface passivation by the chemical oxide. Excess oxygen incorporated into chemical oxide may contribute the negative charge (30), thus compensating positive charges due to trivalent silicon centers. While there is a high density of surface coverage with the hydroxyl groups (31), their contribution to the net oxide charge measured by SCA has not been determined. On the other hand, it could be speculated that excess oxygen and free hydroxyl groups may each contribute negative charge to the oxide charge.

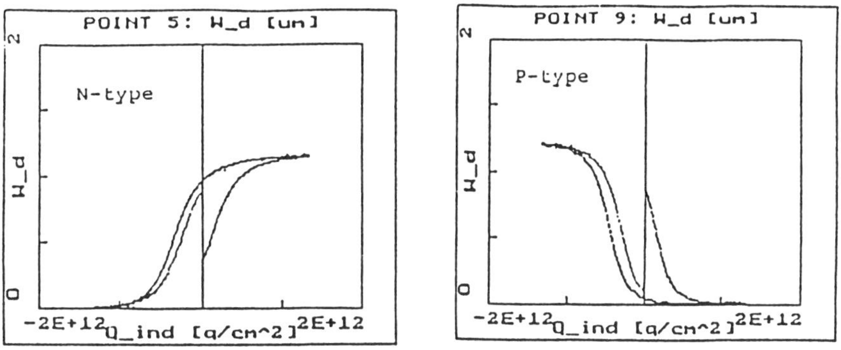

Figure 12. Typical SCA traces showing hysteresis effects after RCA SC-1 treatment of bare silicon wafers.

For the data shown in Fig. 12, the equilibrium (initial) oxide charge, Q_{ox}, is negative (about -3 x 10^{11} q/cm^2). The negative oxide charge is typical for the RCA SC-1 treatment. It is not unusual for this charge to exceed a negative 10^{12} q/cm^2. An example of RCA SC-1 process details reported in Ref. 26 shows a high negative charge, as seen in Fig. 13. It should be noted that growth of the thermal oxide (130 Å) and annealing in nitrogen for an hour at 1050°C still left significant negative charge, although reduced by a factor of three from its initial value. It was determined that increasing the water content of the RCA SC-1 pre-oxidation clean results in a decrease of negative charge, as measured before and after oxidation. The dependence on dilution and presence of the negative charge after high temperature annealing suggested that the origin of the negative charge was contamination from impure RCA SC-1 solution that was being incorporated into the surface chemical oxide.

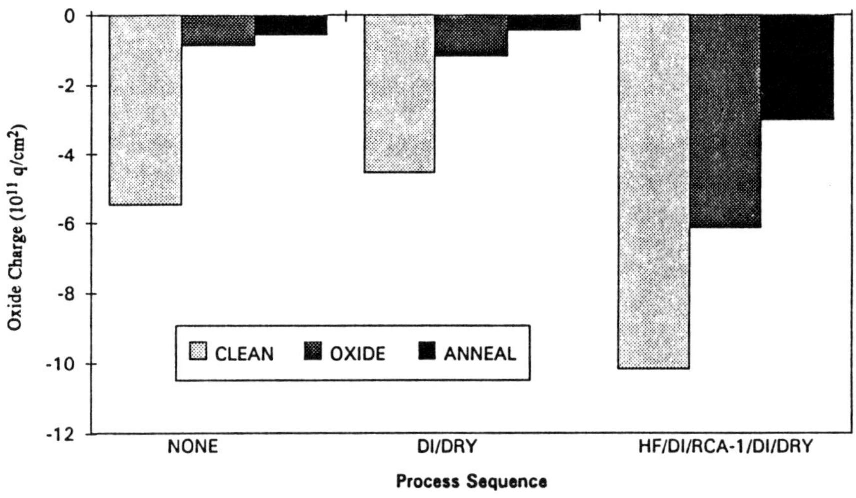

Figure 13. Oxide charge after RCA SC-1 clean of silicon wafers

In summary, a better understanding of the resulting wafer surface is being gained through measurement of changes in its electronic state. These are important factors in establishing process sequences and periods of time during which wafers may be stored after cleaning and prior to subsequent processing.

3.3 RCA Standard Clean 2 (SC-2)

In the sequence of RCA SC-1/SC-2 treatments, RCA SC-1 clean was designed to remove primarily organic surface films by oxidative breakdown and dissolution. Also, several metal ion species are complexed and removed during the procedure. The RCA SC-2 treatment was designed as the second step of the RCA clean to remove alkali ions, metal hydroxides, and metal contaminants remaining after the first cleaning step (22). The removal of aluminum and heavy metals, including iron, is of particular importance. The composition and conditions for RCA SC-2 solution have been noted in Table 1.

The removal of metallic contaminants left in the chemical oxide after RCA SC-1 is associated with regrowth of this oxide and formation of the silicon/chemical oxide interface with trivalent silicon centers. Most of these centers will be passivated due to oxidation, however, some of them will contribute to the net oxide charge. Presence of residual Si-H groups after RCA SC-2 treatment (31) suggests stronger participation of hydrogen in surface passivation as can be expected from the acidic character of the solution used. Similarly, as in the cases of chemical oxides formed during RCA SC-1 treatments, we can expect the presence of negative charge due to excess oxygen and other sources, in addition to the positive charge due to trivalent silicon.

Figure 14 (32) shows an example of SCA results that demonstrate effects of an RCA SC-1/SC-2 sequential clean. The evaluation was done on p- and n-type wafers with <100> and <111> orientation. All incoming wafers had negative Q_{ox}; this is commonly observed on hydrophilic wafers supplied by wafer manufacturers. The wafers were subjected to the RCA SC-1/SC-2 clean in a spray tool and measured within five minutes after completion of DI wafer rinsing and drying. All wafers showed negative charge, however, it should be noted that this charge was relatively low (less negative) compared to that before cleaning. Wet oxidation at 950°C rendered the charge more positive in all wafers; this change was strongly dependent on the crystallographic orientation of the wafers. The last group of bars in Fig. 14 is for the wafers on which aluminum capacitors were formed and annealed at 450°C. As shown in Fig. 15, oxide charge, Q_{ox}, measured by SCA and conventional C-V show very good agreement (32).

Preliminary evaluation of the effect of the air exposure on the oxide charge, Q_{ox}, after RCA SC-1/SC-2 clean indicated that this charge is usually stable for several hours after cleaning. The oxide charge is also relatively low, depending on the modification of the RCA solutions, between -2×10^{11}

and +2 x 10^{11} q/cm^2. However it should be noted that further experiments are needed to gain better understanding of all the variables which affect the final charge.

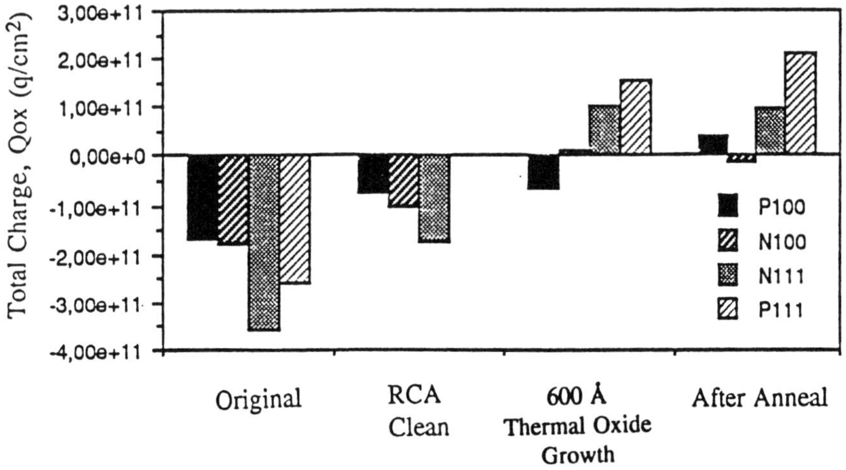

Figure 14. "Oxide charge" on silicon wafers after complete RCA cleaning, oxidation, and anneal. *(Reprinted with permission: CNET, Meylan Cedex, France.)*

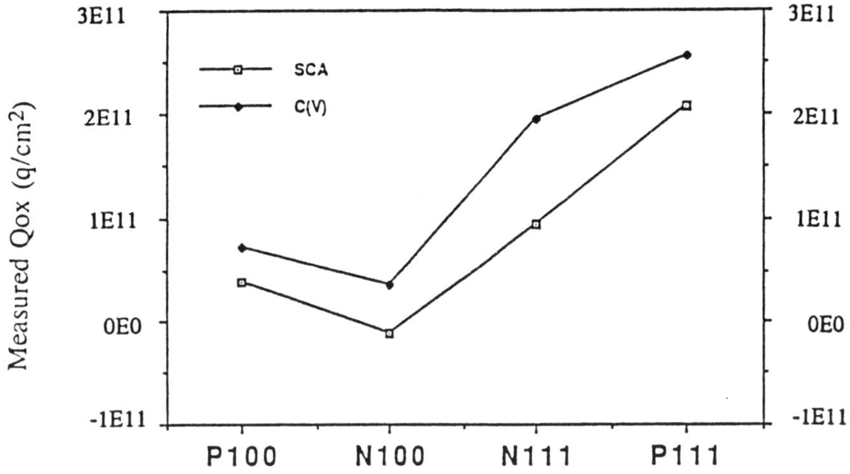

Figure 15. Comparison of the oxide charge measured by SCA and by conventional C-V methods. *(Reprinted with permission: CNET, Meylan Cedex, France.)*

In a manner similar to that for the RCA SC-1 clean, the RCA SC-2 clean generates a chemical oxide that has an associated electrical charge. It now remains to investigate, in a precise experimental manner, what the differences are between wafers with only the RCA SC-1 clean, those that received the RCA SC-2 clean, and those that were subjected to the sequence of RCA SC-1 followed by RCA SC-2.

3.4 Effects of Metallic Contaminants

It has been known for a long time that metallic and organic contaminants have a deleterious effect on oxides grown on silicon wafers. Whereas the effects of organics, as they decompose during thermal oxidation, are still being investigated, their contribution of carbon to the oxides is known to introduce microvoids (33). Metal ions have been shown to affect the quality of oxides in a number of ways, including the formation of voids and the reactivity with silicon to enhance or slow oxide formation (34). Of particular interest is the way such contaminants affect the electrical characteristics of the oxides and the Si/SiO_2 interface.

Studies concentrating on effects of aluminum have found its presence on the native oxide surface (35). SIMS analysis indicated that Al is incorporated onto the wafer surface only in the presence of an oxidizing medium (36). SCA results confirmed this conclusion. With wafers subjected to aluminum contaminated sulfuric-peroxide mixtures, the SCA measured an increase of the oxide charge with higher levels of contamination (37). Shown in Fig. 16a are the results for bare wafers. Similar charge changes are seen on wafers with 500 Å of oxide; these are illustrated in Fig. 16b. However, with wafers exposed to contaminated buffered hydrofluoric acid (BHF), no change in Q_{ox} was observed (Fig. 16a). Subjecting wafers to other cleaning processes have also indicated sensitivity to aluminum contamination. In an oxidizing environment (RCA SC-1, or thermal oxidation) Al contamination results in negative oxide charge, as illustrated in Fig. 16c (38). The associated charge is believed to come from oxidized aluminum (28). Through repeated partial etchback of the oxide film, as depicted in Fig. 16d, it was shown that Al accumulates at the oxide surface opposite to Si (38). Aluminum cannot be detected by the C-V measurements because it is masked by the gate electrode formed on the oxide on which the Al is located.

Negative oxide charge is also produced by RCA SC-1 solution contaminated with Fe. Dependence of this charge on the Fe concentration is similar to that observed for Al. Whereas Fe can be detected effectively when diffused into bulk Si (6), SCA may allow detection of Fe contamination at the silicon surface prior to high temperature processing.

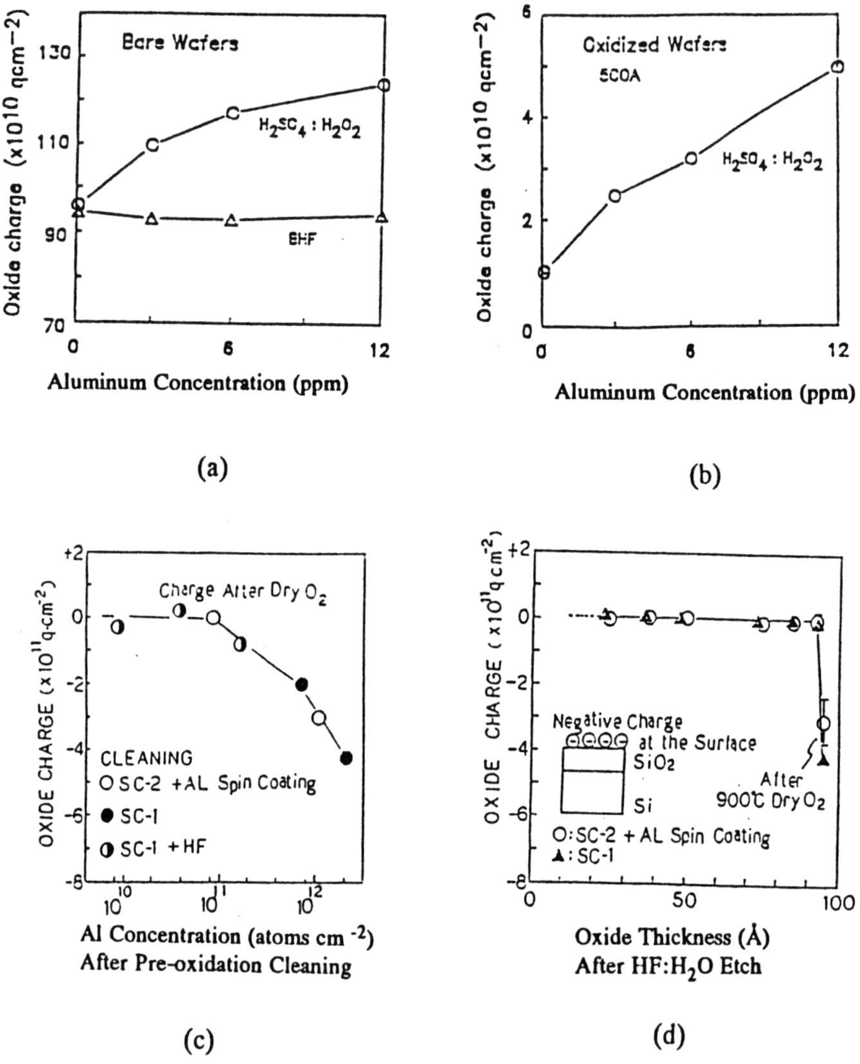

Figure 16. Effects of aluminum contamination on silicon wafer surface characteristics after various chemical processing conditions: (a) Bare wafers subjected to Al-contaminated sulfuric-acid/hydrogen-peroxide (H_2SO_4-H_2O_2) and buffered hydrofluoric acid (BHF); (b) Oxidized wafers (500 Å oxide), after exposure to Al-contaminated H_2SO_4-H_2O_2; (c) Repeated etchback of contaminated oxide showing accumulation of the negative charge at the oxide surface opposite Si; (d) Wafers subjected to indicated cleans and Al contamination, then oxidized. *(Reprinted with permission. Figs a, b from Ref. 37, ©1990 IEEE. Figs c, d from Ref. 38, ©1991.)*

Although only oxide changes have been discussed so far, the interface state density, D_{it}, also needs to be characterized. When measuring bare wafers, slow states are present that introduce a hysteresis effect. Resultant interface state densities introduced by metallic ion contamination have to be investigated.

3.5 Rinsing and Drying

DI water rinsing, because of the frequency of use, is one of the most critical steps in wafer cleaning. Following the rinse, a drying cycle is typically performed by spinning the wafers. Wafers are left in cassettes which are mounted off- or on-axis within the spin dryer. The water residing on the wafer is removed, the effectiveness being related to the radial velocity of water droplets. In a process under good control, rinsing and drying are expected to have a neutral effect on a cleaned-control wafer, leaving the oxide charge unchanged. However, water rinsing very frequently adds negative charge to the processed wafers. As is illustrated in Fig. 17, wafers processed through an on-axis spin rinser/dryer show an increase of negative charge in the center of the wafers while the charge along the edges remains unchanged* (37). This additional charge can be attributed to residues left by the drying DI water. It is believed that colloidal silica is the most probable source of this charge. This conclusion is supported by a strong correlation between the number of particles added to the wafers during cleaning and the oxide charge (39). Positively charged wafers, cleaned in an HF solution, picked up more particles than negatively charged wafers cleaned with an RCA sequence. Since most particles in deionized water are negatively charged, they would tend to be attracted to a positively charged, HF-etched, surface. The inert colloidal silica may become negatively charged as it passes through insulated de-ionized water piping or comes in contact with the spray processor itself (40). Colloidal silica is one of most troublesome particles in the wafer cleaning process. Fortunately, SCA technology offers a new tool that allows for early warning of buildup of colloidal silica, which is particularly important for the production of high density silicon integrated circuits.

An important application of the SCA technology is monitoring the cleanliness of wafers exposed to contaminating solutions, such as photore-

* Preliminary comparison of IPA and spin/rinse drys shows that both processes result in negative Q_{ox}, with IPA dry characterized by a lower charge magnitude and a more uniform distribution.

sist developers. Since SCA does not require any additional processing it can be used as an effective tool for monitoring surface contamination with alkali metals before wafers are subjected to a high temperature process (37), therefore allowing the reduction of manufacturing losses. Fig. 18 illustrates dependence of the oxide charge on the number of rinse cycles after wafers were exposed to AZ 353 photoresist developer for one minute. These results demonstrate effective monitoring of the rinsing process that leads to limiting the number of rinse cycles needed to remove aqueous photoresist stripping solutions, resulting in a reduction of DI water consumption.

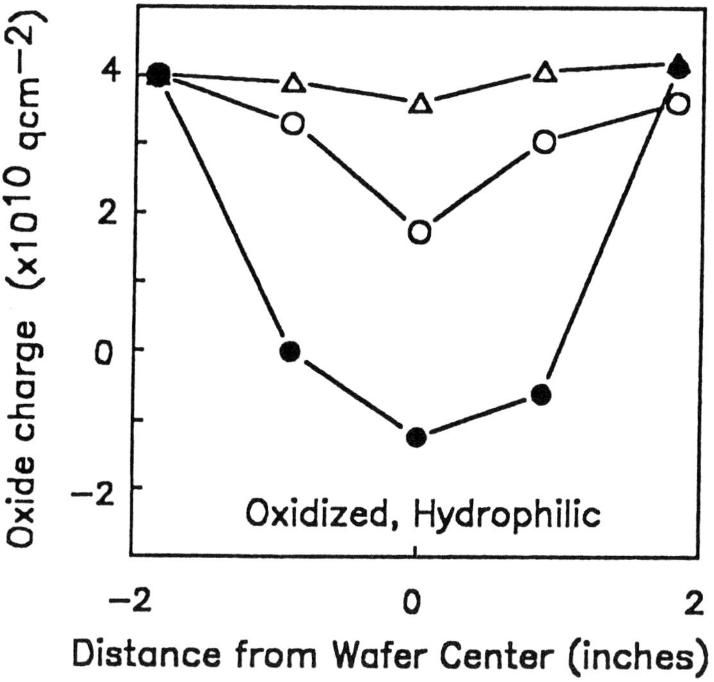

Figure 17. Oxide charge distribution for wafers subjected to a dump rinse, followed by rinse/spin dry vs. spin dry only. Δ – prior to dump rinse, O – spin dry, ● – rinse/ spin dry. (From Ref. 37.) *(Reprinted with permission. ©1990 IEEE.)*

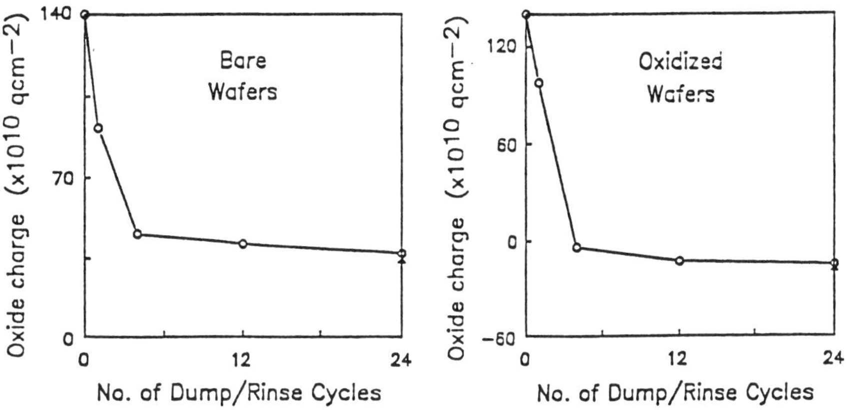

Figure 18. Oxide charge of silicon wafers as a function of number of dump/rinse cycles after exposure to AZ 353 developer for one minute. Triangles indicate wafers that received 24 dump/rinse cycles without any exposure to the developer. (From Ref. 37.) *(Reprinted with permission. ©1990 IEEE.)*

4.0 INCOMING WAFER SURFACE QUALITY MONITORING

The need to monitor wafer contamination starts when wafers are initially received. Here we give examples of contamination observed by SCA on a large number of bare wafers having both hydrophobic and hydrophilic surfaces before being subjected to high-temperature processing.

For several months, SCA-measured surface doping concentrations, N_{sc}, were monitored for wafers from various manufacturers. The SCA measurements were compared with the manufacturers' specifications that are typically based on the four-point probe method. Results for n-type wafers were typically in good agreement, but SCA results for p-type wafers were very often much lower than those given by a supplier. RCA cleaning did not affect doping concentration, however thermal oxidation caused return of N_{sc} to the supplier specified value. Similarly, baking of the bare p-type wafers in the room air at temperatures close to 200°C restored doping concentrations to the supplier specified value. The discussed effect is illustrated in Fig. 19 (41). The bare p-type wafer (hydrophilic surface) was baked at 150°C for 15, 35, 80, 180, and 420 minutes. Each time after cooling to room temperature the doping concentration of the wafer was measured with SCA. The results for bare wafer is compared in Fig. 19 with the SCA-measured doping concentration of the oxidized wafer from the same lot.

This observation has been initially attributed to the well-known deactivation of acceptors by atomic hydrogen. The hydrogen atom (proton) is known (42)(43) to diffuse up to several microns into the silicon bulk and neutralize the dopant in this region. Spreading resistance profiles show this effect very clearly; however, the four-point probe which is measuring bulk resistivity and, unlike the SCA method, is not sensitive to the surface doping profile and does not show this effect. It was shown that deactivation of acceptor centers can be reversed by annealing at temperatures exceeding about 150°C to 200°C (44). Recently, researchers who previously tried to explain acceptor deactivation by the introduction of atomic hydrogen during the polishing process have shown that copper in the polishing slurry is actually responsible for this effect (45). They have shown that, in wafers stored for several weeks, a considerable increase of the Cu concentration near the surface occurred.

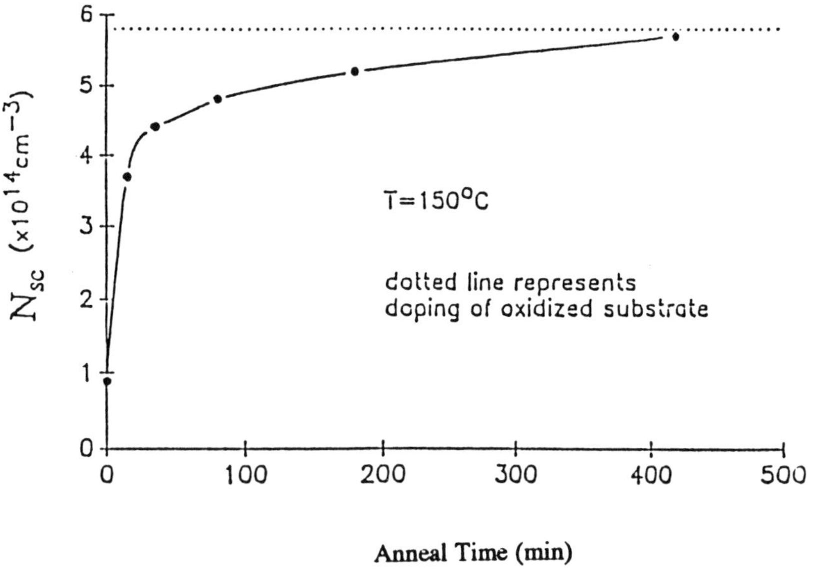

Figure 19. Dependence of the SCA-measured surface doping concentration on anneal time. (From Ref. 41.) *(Reprinted with permission. ©1991 SEMI.)*

5.0 SUMMARY

The Surface Charge Analysis (SCA) technique has been described as a new way to quantitatively characterize semiconductor surfaces and certain effects introduced through various processing steps. The method allows for immediate and nondestructive characterization of the electronic properties of bare silicon as well as of semiconductor/insulator structures. Unlike conventional capacitance techniques, it does not require formation of a gate electrode and direct electrical contacts. The SCA method is based on determination of the depletion layer width from ac-SPV (alternating current surface photovoltage) measurements and its dependence on an external electrical field. The method allows for determination of the doping type, doping concentration at the semiconductor surface, oxide charge, and energy distribution of the interface state density. Applications to characterize and monitor HF etchants and RCA cleans, including rinsing and drying, have been discussed to illustrate the method.

The dependence of the HF-etched wafers on the time of exposure to ambient air indicates that the optimum time for monitoring this process is within a half-hour after treatment. Formation of native oxide on the silicon wafer modifies its surface properties, making it more important to control the time of measurements and condition of wafer storage. However, with native oxide growth, the silicon wafer surface becomes more stable and its properties change more slowly, allowing greater latitude to perform repeatable measurements.

In the case of RCA SC-1 treated wafers, our measurements have indicated that approximately one day after treatment the wafers are electrically stable; however, the stability immediately after treatment has not yet been established and may depend on specific contamination and processing conditions. The situation seems to be clearer for RCA SC-1/SC-2 treated wafers that were found to be stable starting within a few minutes after treatment. In both cases long-term stability exceeding several days will need to be further investigated.

The results obtained to date have demonstrated the usefulness of SCA in monitoring different cleaning procedures as well as in comparing equipment used in wafer processing. Effects of metallic contamination of cleaning solutions or polishing slurry, as well as nonmetallics such as colloidal silica, have been discussed to illustrate the diagnostic capabilities of the SCA.

REFERENCES

1. Kamieniecki, E., U.S. Patent 4,827,212 (May 1989)
2. Kamieniecki, E., *Semiconductor Cleaning Technology/1989,* (J. Ruzyllo and R. E. Novak, eds.), 90-9:273, Electrochemical Soc., Pennington, NJ (1990)
3. Moss, T. S., *J. Electrochem. Control,* 1:126 (1955)
4. Quilliet, A. and Gosar, M. P., *J. Phys. Radium,* 21:575 (1960)
5. Goodman, A. M., *J. Appl. Phys.* 32:2550 (1961)
6. Ryoo, K. and Socha, W. E., *J. Electrochem. Soc.* 138:1424 (1991)
7. Many, A., Goldstein, Y. and Grover, N. B., *Semiconductor Surfaces,* North-Holland, Amsterdam (1965)
8. Kamieniecki, E., *J. Vac. Sci. Technol.* 20:811 (1982)
9. Kamieniecki, E., *J. Appl. Phys.* 54:6481 (1983)
10. Sze, S. M., *Physics of Semiconductor Devices*, Ch. 7, Wiley, NY (1981)
11. Gartner, W. W., *Phys. Rev.* 116:84 (1959)
12. Blood, P. and Orton, J. W., *Rep. Prog. Phys.* 41:157 (1978)
13. Ukah, C. I., Perz, J. M. and Zukotynski, S., *J. Appl. Phys.* 65:3617 (1989)
14. Shockley, W. and Read, W. T., Jr., *Phys. Rev.* 87:835 (1952)
15. Jellison, G. E., Jr. and Modine, F. A., *J. Appl. Phys.* 53:3745 (1982)
16. Surface Charge Analyzer SCA Model 2000 Operation Manual, SemiTest Incorporated (1991)
17. Nicollian, E. H. and Brews, J. R., *MOS Physics and Technology*, Ch. 2, Wiley, NY (1982)
18. Wu, A. T., Chan, T. Y., Murali, V., Lee, S. W., Nulman, J. and Garner, M., *IEEE Conf. on Electron Devices,* 1(1):271 (Dec. 1989)
19. Buchanan, D. A. and DiMaria, D. J., *J. Appl. Phys.* 67:7439 (1990)
20. Murali, V., Wu, A. T., Fraser, D. B., Kamieniecki, E. and Nulman, J. , *Symp. on Silicon Nitride and Silicon Dioxide Thin Insulating Films,* (S. Bibyk and V. Kapoor, Eds.) 89-7:252, Electrochemical Soc., Pennington, NJ (1989)
21. Kern, W. and Poutinen, D. A., *RCA Rev.* 31:187 (1970)
22. Kern , W., *J. Electrochem. Soc.* 137:1887 (1990)

23. Morita, M., Ohmi, T., Hasegawa, E., Kawakami, M. and Ohwada, M., *J. Appl. Phys.* 68:1272 (1990)

24. Grundner, M. and Jacob, H., *Appl. Phys.,* A 39:73 (1986)

25. Yablonovitch, E., Allara, D. L., Chang, C. C., Gmitter, T. and Bright, T. B., *Phys. Rev. Lett.* 57:249 (1986)

26. Resnick, A., Kamieniecki, E., Philipossian, A. and Jackson, D., *Semiconductor Cleaning Technology/1989,* (J. Ruzyllo and R. E. Novak, eds.), 90-9:335, Electrochemical Soc., Pennington, NJ (1990)

27. Deal, B. E., *J. Electrochem. Soc.* 121:198C (1974)

28. Shimizu, H. and Munakata, C., *Semicond. Sci. Technol.* 5:842 (1990); *Jpn. J. Appl. Phys.* 30:2466 (1991)

29. Philipossian, A., *Cleaning Technology in Semiconductor Device Manufacturing,* (J. Ruzyllo and R. E. Novak, eds.), 92-12:234, Electrochemical Soc., Pennington, NJ (1992). This paper was originally presented at the Fall 1991 meeting of the Electrochemical Society in Phoenix, AZ.

30. Iwamatsu, S., *Appl. Phy. Lett.* 48:1542 (1986)

31. Grundner, M., Hahn, P. O., Lempart, I., Schnegg, A. and Jacob, H., *Semiconductor Cleaning Technology/1989,* (J. Ruzyllo and R. E. Novak, eds.), 90-9:215, Electrochemical Soc., Pennington, NJ (1990)

32. Gimine, G. and Lutti, C., *CNET Internal Report,* (February 1991)

33. Liehr, M., Lewis, J. E. and Rubloff, G. W., *J. Vac. Sci. Technol.* A5:1559 (1987)

34. Dallaporta, H., Liehr, M. and Lewis, J. E., *Phys. Rev. B,* 41:5075, (1990)

35. Slusser, G. J. and MacDowell, L., *J. Vac. Sci. Technol.* A5:1649 (1987)

36. Morita, E., Yoshimi, T. and Shimanuki, Y., *ECS Ext. Abstr.,* Vol. 89-1, Abstracts No. 237 and 238 (1989)

37. Murali, V., Wu, A. T., Chatterjee, A. and Fraser, D. B., *Digest of Technical Papers, 1990 Symp. VLSI Technology,* Honolulu, Hawaii, p. 103, (June 1990)

38. Kato, J. and Maruo, Y., *The 38th Spring Mtg., Jpn. Soc. of Appl. Phys. and Related Societies,* No. 2, 30p-V-7 (1991)

39. Menon, V., DeSelms, B., Chacon, J. and Kamieniecki, E., *Microcontamination,* 23:104 (Oct. 1990)

40. Burggraaf, P., *Semicond. International,* 52 (Oct. 1990)

41. Kamieniecki, E., *Semicon/Europa 91 Technical Proceedings,* 94 (1991)

42. Vieweg-Gutberlet, F. G. and Siegesleitner, P. F., *J. Electrochem. Soc.* 126:1792 (1979)
43. Seager, C. H., Anderson, R. A. and Panitz, J. K. G., *J. Mater. Res.* 2:96 (1987)
44. Heddleson, J. M., Horn, M. W., Fonash, S. J. and Nguyen, D. C., *J. Vac. Sci. Technol.* B6:280 (1988)
45. Prigge, H., Gerlach, P., Hahn, P. O., Schnegg, A., and Jacob, H., *J. Electrochem. Soc.* 138:1385 (1991)

12

Ultratrace Impurity Analysis of Silicon Surfaces by SIMS and TXRF Methods

Richard S. Hockett

1.0 INTRODUCTION

The analytical measurement of ultratrace impurities on silicon surfaces is described in this section, with emphasis on SIMS and TXRF techniques. The chapter is divided into five parts. First the analytical problem is defined. Second, currently available surface analytical techniques are considered. Third, SIMS measurement is explained and three approaches are described: static SIMS, dynamic SIMS, and polyencapsulation/SIMS. Examples are presented for characterizing cleaning effectiveness by these techniques. Fourth, TXRF is described. Examples are given for characterizing cleaning efficacy by these methods. Finally in the fifth section, four analytical techniques which may become important in the future are described: VPD/TXRF, VPD/SIMS, VPD-ICP/MS, and TOF-SIMS.

2.0 ANALYTICAL PROBLEM

2.1 Relevant Contamination Levels

A description of the analytical problem for the detection of trace metals on the surface of silicon substrates first requires an estimate of both the lowest and the highest levels of contamination that affect product quality.

The most abundant component of the surface is oxygen, with an areal density of approximately 10^{15} atoms/cm^2 in the native oxide of the silicon substrate; this sometimes is seen as a contaminant, particularly for low temperature epi growth. The most abundant metallic contaminant is typically Al, which is deposited from SC-1 cleaning solutions that contain a high level of Al originally from impure H_2O_2. The surface Al level can be as high as 10^{14} atoms/cm^2. Estimation of the lowest level of contamination that adversely affects the product depends upon many parameters: IC device type (e.g., DRAM versus EPROM) and design (e.g., DRAM trench type, or epi versus non-epi substrate), relevant IC performance parameter (e.g., oxide breakdown versus generation or recombination lifetime). Furthermore, it is difficult to perform meaningful experiments to prove what contamination level is critical.

In spite of these difficulties, there have been a number of reports on the relevant lower level of contamination. For example, in the case of iron contamination, Shimono et al. (1) showed that the minority carrier recombination lifetime of p-type substrates is inversely proportional to the surface Fe areal density, down to less than 10^{10} atoms/cm^2. Lifetime can be limited to less than 100 microseconds in the presence of surface Fe at 1×10^{10} atoms/cm^2.

Matushita and Tsuchiya (2) demonstrated that there is an interaction between surface Fe (or Ni) with implant-induced damage that causes oxidation-induced stacking faults (OISF). Metal contamination introduced by a spin coating method (3) was quantified by VPD/AAS (4). They used an oxidation anneal after the implant. For Fe without implant damage, there is a threshold of about 10^{12} atoms/cm^2 before OISF density begins to increase. If there is damage (after p-well implant), however, this threshold level is reduced to about 10^{11} atoms/cm^2; above the threshold the OSIF density is two orders of magnitude higher than without the damage. The effect also occurs for surface Ni, but the OISF density is not as high. Because of the interaction between surface metals and surface damage, these authors recommend controlling surface Fe down to at least 10^{10} atoms/cm^2. This level is considered to be adequate for present IC manufacturing, but will be inadequate for future IC designs. The perceived need for the immediate future is to control surface metals at the 10^9 atoms/cm^2 level or lower (1).

In addition to Fe or other transition metals on the surface of silicon substrates, there is metallic contamination in the crystal bulk. Levels of Cr, Fe, Ni, and Cu concentrations in bulk silicon crystal have been estimated by Abe et al. (5). Detection limits of analytical techniques for the transition metals are not adequate to directly measure bulk contamination. To circumvent this problem, they measured transition metal concentrations in the polycrystalline residue ("pot bottom") that remains at the bottom of the crucible after crystal growth by dissolving 100 g. of the material in HNO_3/HF and analyzing the liquid by ICP. Metal concentration in the bulk crystal can then be calculated using known segregation coefficients for the metals of interest. Table 1 shows the analytical (pot bottom) and calculated (bulk crystal) results for Cr, Fe, Ni and Cu for intentionally contaminated and uncontaminated crystal growth runs. Bulk crystal levels for intentionally contaminated melts (Case 1) are on the order of 10^{10} atoms/cm^3, and approximately 10^9 atoms/cm^3 for uncontaminated melts (Case 2).

Table 1. Metal Concentrations in CZ Pot Bottom and Bulk Crystal (5)

	Concentrations in Residual Melts of 0.2% ppmw (ICP)		Conversion Values at Fraction Frozen of 80% x 10^9 atoms/cm^3		
Element	Case 1	Case 2	Case 1	Case 2	NAA D.L
Cr	12.9	0.7	49	2.5	3 x 10^{12}
Fe	48.5	3.1	49	3.0	3 x 10^{13}
Ni	5.3	0.4	19	1.5	4 x 10^{13}
Cu	0.6	0.05	31	2.5	1 x 10^{13}

To compare surface areal density with bulk concentration, one can multiply the bulk concentration by the substrate thickness. For example, a 625 μm (6.25 x 10^{-2} cm) thick substrate with a bulk Fe concentration of 3 x 10^9 atoms/cm^3 (Case 2) is equivalent to a surface areal density of 2 x 10^8 atoms/cm^2 if the bulk is the only source of Fe. In other words, if the Fe atoms on the surface are much greater than 2 x 10^8 atoms/cm^2, then surface Fe is the dominant source of Fe for the substrate. The conclusion is that metal atoms introduced to the wafer surface need to be measured down to 10^8 atoms/cm^2 if we want to reduce the surface metals down to the level of bulk areal densities. However, the level of Fe on the polished surface of prime silicon wafers is reported to be on the order of 10^{10} to 10^{11} atoms/cm^2 (6)(7), suggesting that there are other, much more important sources of Fe besides the bulk contribution.

2.2 Analysis Depth and Number of Atoms

The lower level of detection, 10^8 atoms/cm^2, is equivalent to only 1 atom/μm^2. Metals are located in the native oxide region or just below it in the silicon. If we assume the maximum depth of contamination to be about 5 nm, which includes the native oxide and some shallow subsurface layer, then detection of 10^8 atoms/cm^2 in a depth of 5 nm requires either extremely high sensitivity or a very large area for analysis, or both. If the analysis area for a technique is 100 μm^2 (e.g., some SIMS), then the total number of metal atoms in the top 5 nm available for detection is 100. If the analysis area is 1 cm^2 (e.g., TXRF), then 10^8 metal atoms are available for detection. If the entire wafer surface of a 200 mm diameter wafer is analyzed, the total number of atoms available for detection is 3×10^{10} (VPD/AAS). If the analysis depth is greater than 5 nm, the effective concentration detection limit must be even better due to the volume dilution.

2.3 Quantification

Most analytical techniques require some kind of reference sample for quantification. Standardized reference materials are needed if analytical results from different laboratories are to be compared. The creation of stable reference materials and the assignment of their areal densities at contamination levels as low as 10^8 atoms/cm^2 is non-trivial, and so is creation of blanks (controls) down to this level to demonstrate the absence of instrumental backgrounds.

2.4 Composition of Clean Native Oxide

The major (10^{13} to 10^{15} atoms/cm^2) constituents of a clean native oxide are Si, O, H, C, N, and F, and minor (10^{11} to 10^{12} atoms/cm^2) constituents are B, Mg, Al, P, S, Cl, Ca, and Br. Fe, Ni, Cu, and Zn are in the 10^{10} atoms/cm^2 range. Not all native oxides are this clean. Common fluctuations are in surface Al and Zn, which can reach the 10^{13} atoms/cm^2 level.

3.0 AVAILABLE ANALYTICAL TECHNIQUES

In this section the strengths and limitations of currently available analytical techniques are described in order to put SIMS and TXRF into perspective. These techniques are summarized in Table 2. SIMS and TXRF techniques, the major subjects of this chapter, will be discussed in more detail later. In this comparison we have excluded electrical methods and techniques which require a thermal treatment before analysis, such as surface photovoltage, microwave photoconductivity, and deep level transient spectroscopy. We have excluded surface charge analysis, because it is not element-specific; its signal represents the net charge (combination of negative and positive charges) from chemical bonds, not merely trace metals, in the native oxide. We do not mean that these excluded techniques have no value; on the contrary, they can be very useful in a different context.

Table 2. Comparison of Analytical Techniques

Analytical Technique	Probe In	Detection Out	Detection Depth	Analysis Area (cm^2)	D.L. ($atoms/cm^2$)
ESCA	x-ray	electron	5nm	1	10^{13}
AES	electron	electron	5nm	10^{-2}	10^{13}
He-RBS	ion	ion	1 μm	10^{-2}	10^{13}
N-RBS	ion	ion	1 μm	10^{-2}	10^{10}
LIMS	laser	ion	1 μm	10^{-8}	10^{13}
XRF	x-ray	x-ray	1 μm	1	10^{13}
HREELS	electron	electron	5nm	10^{-2}	10^{12}
IR	photon	photon	5nm	10^{+1}	10^{12}
VPD/AAS	photon	photon	5nm	10^{+2}	10^{8} - 10^{9}
SIMS	ion	ion	5 - 50nm	10^{-4}	10^{9} - 10^{11}
TXRF	x-ray	x-ray	5nm	1	10^{10} - 10^{12}

3.1 Electron Spectroscopy for Chemical Analysis

Electron Spectroscopy for Chemical Analysis (ESCA) uses an x-ray beam to excite core electrons in the elements of the sample surface (x-ray in, electron out) (8). The emitted electrons are analyzed according to their energies and counted. Elemental identification is made on the basis of electron energy, and chemical information can be obtained by the energy shift of the core electrons caused by bonding. Although the x-rays penetrate deeply (to μm depths), the analytical depth is determined by the escape depth of the emitted electrons and is approximately 5 nm. The analysis area varies from a few hundred μm^2 to a few cm^2 depending upon the beam size. Detection limits are on the order of 10^{13} atoms/cm^2.

ESCA is useful for surface analysis of O, C, N, and under some conditions, F. It has been used to study chemical bonds in native oxides after cleaning (9) and for determining the O and F levels after HF etch of native oxides (10). In principle, ESCA can be used to estimate the native oxide thickness (11). The ESCA spectrum of a clean silicon wafer is shown in Figs. 1 and 2; the latter spectrum shows the chemical shift between Si-O bonds and Si-Si bonds. ESCA is not useful for measuring surface trace metals after commercial cleaning of silicon, because the detection limits are insufficient.

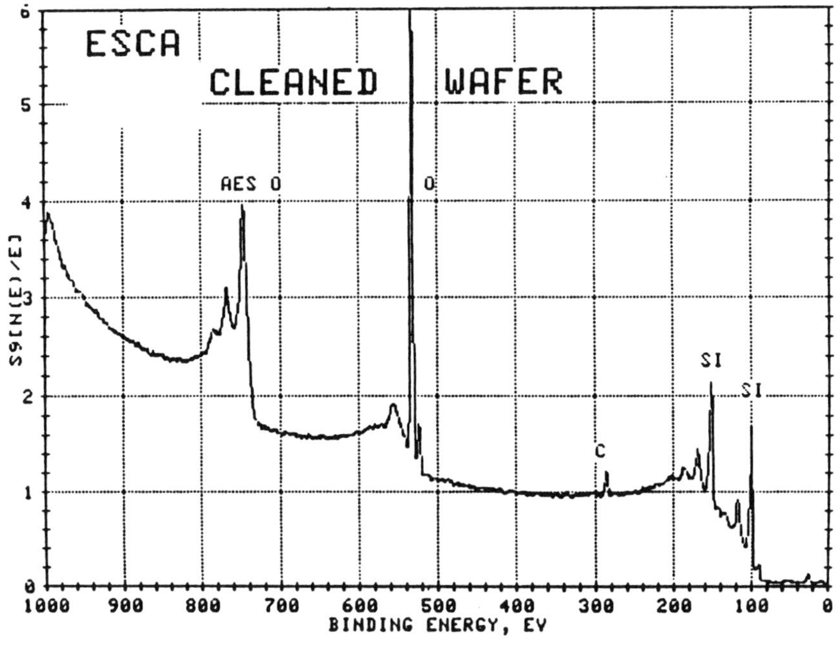

Figure 1. Full ESCA spectrum of a clean silicon substrate.

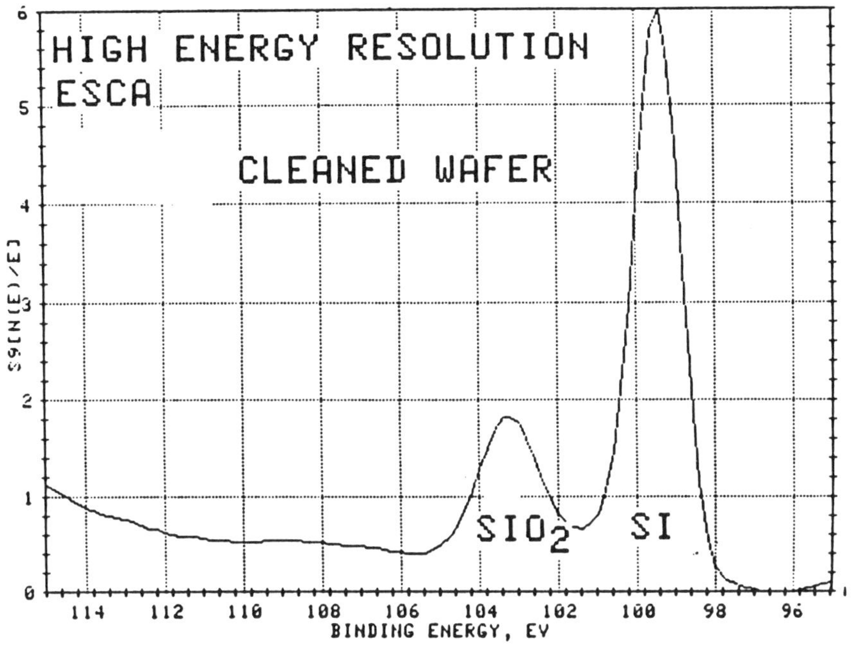

Figure 2. Expanded ESCA spectrum showing the Si and SiO_2 bonding for a clean silicon substrate.

3.2 Auger Electron Spectroscopy

Auger Electron Spectroscopy (AES) uses an electron beam to excite Auger electrons in the elements of the sample surface (electron in, electron out) (12). Elemental identification is made on the basis of energies of the emitted Auger electrons. Limited chemical information can be obtained by the energy shifts caused by bonding. As with ESCA, analytical depth is determined by the escape depth of the emitted electrons and is approximately 5 nm. Analysis area for bare silicon wafers is on the order of 1 cm^2. Detection limits are on the order of 10^{13} atoms/cm^2. A linear correlation between AES and TXRF is shown in Fig. 3 for high levels of surface Fe and Cu (13).

In principle, AES can be used to estimate native oxide thickness (14) by evaluating the ratio of the Si to SiO_2 peaks. There is good correlation between native oxide thickness estimated from an AES spectrum and that measured by ellipsometry; the AES appears to be more sensitive. This particular application requires special care during analysis to avoid decomposing the native oxide with the electron beam.

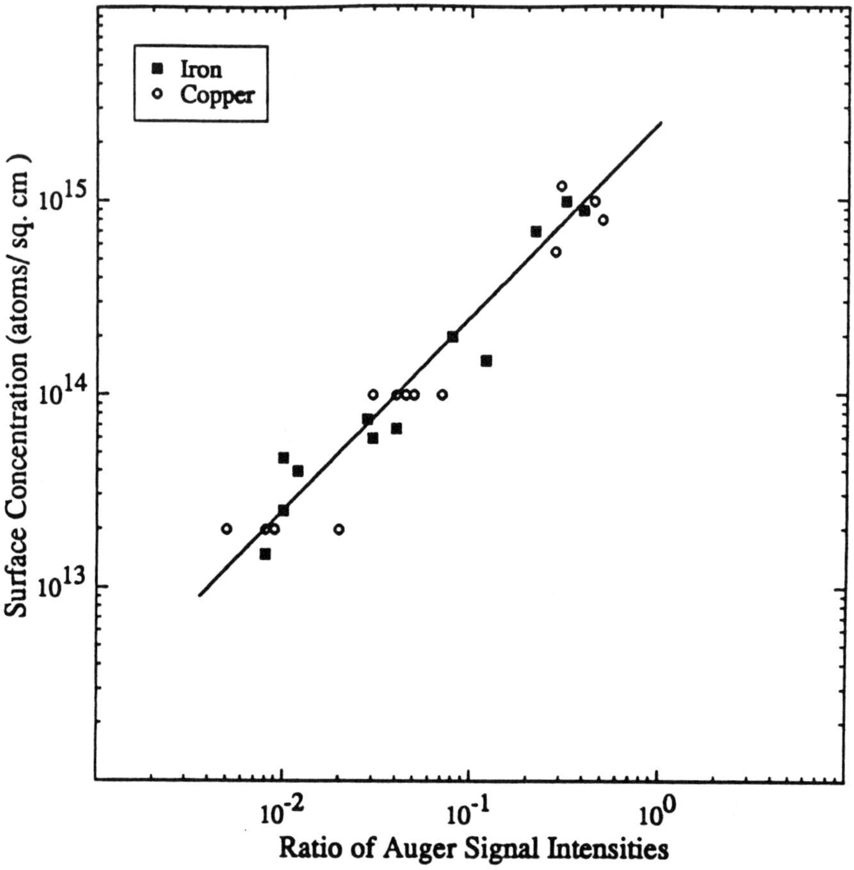

Figure 3. Correlation of Fe and Cu surface concentration between AES and TXRF for intentionally contaminated silicon substrates (13).

AES can be used for the surface analysis of O, C, N, and under some conditions, F, although the electron beam may desorb F. An AES spectrum measured on a clean silicon wafer is shown in Fig. 4. AES is not useful for measuring surface trace metals after commercial cleaning of silicon, because the detection limits for the surface metals are insufficient.

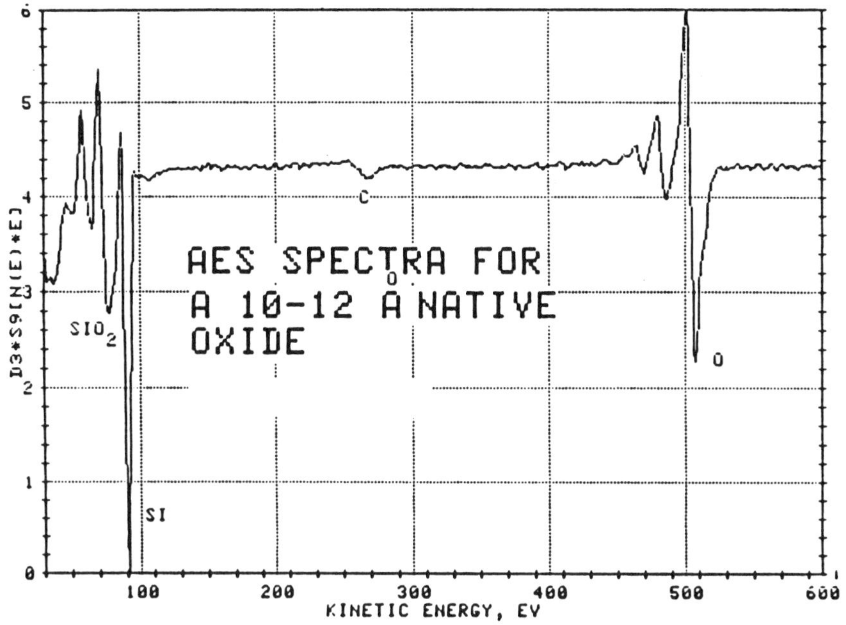

Figure 4. Typical AES spectrum of a clean silicon wafer with 10 - 12 Å native oxide.

3.3 Rutherford Backscattering Spectrometry

Rutherford Backscattering Spectrometry (RBS) uses an MeV ion beam (typically He ions) to bombard the sample. He ions are scattered backwards away from the elements on the sample surface (ion in, ion out) (15). Elemental identification is made on the basis of energy of the backscattered ions. Analytical depth of RBS is determined by the penetration depth and escape depth of backscattered He ions and is on the order of 1 µm. Analysis area is several mm^2. Detection limits by using a He beam are on the order of 10^{13} atoms/cm^2 for the heavy elements above Ar in atomic number.

He-beam RBS uses a tandem accelerator to create an MeV He ion beam. A rarely applied alternative is to use a Van de Graaff accelerator to create an MeV nitrogen ion beam (16). In this case the cross section for scattering is increased, and an interference from backscattering off the Si matrix is greatly reduced so that detection limits can be as low 10^{10} atoms/cm^2.

Quantification does not require reference standards, which is an important advantage of RBS. The major drawback is poor mass resolution; transition metals that are close in atomic number cannot be resolved, for example, Fe and Ni. A He-beam RBS spectrum for a sample with surface Cu on silicon at 10^{14} atoms/cm^2 is shown in Fig. 5.

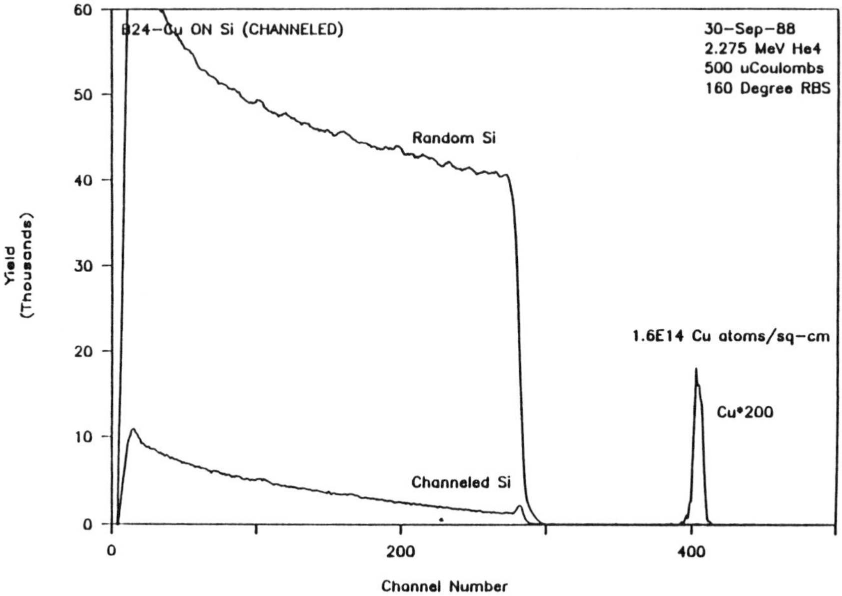

Figure 5. He-beam RBS spectrum of a silicon wafer that was intentionally contaminated with Cu.

This technique is not useful for the analytical measurement of surface trace metals after commercial cleaning of silicon, because He-beam RBS is not sensitive enough and both He-beam and N-beam RBS do not have sufficient mass resolution. N-beam RBS is useful for assigning absolute values to surface areal densities at low levels, if the surface metal identification is already known by another technique (i.e., TXRF) and if mass interferences are not important.

3.4 Laser Ionization Mass Spectrometry

Laser Ionization Mass Spectrometry (LIMS) uses a laser beam to volatilize the elements of the sample surface (laser in, ions out) (17). Emitted ions are analyzed for mass by a time-of-flight drift tube and the elemental identification is made by the mass. The analytical depth of LIMS is determined by the volatilized volume which is on the order of 1 µm deep. Detection limits depend on mass interferences, which are severe for transition metals. Detection limits are estimated to be on the order of 10^{13} atoms/cm^2 for transition metals. LIMS analysis may be improved in the future by selective laser post-ionization of the volatilized neutral atoms. However, the analysis depth is not appropriate for a 5-nm contamination depth. This technique is not useful for measuring surface trace metals after commercial cleaning of silicon because of insufficient detection limits.

3.5 X-Ray Fluorescence Spectroscopy

X-Ray Fluorescence Spectroscopy (XRF) uses an x-ray beam to excite the fluorescence x-rays from the surface elements (x-ray in, x-ray out) (18). Elemental identification is made on the basis of energy of the emitted fluorescence x-rays. Analytical depth is determined by the penetration depth of the x-rays and the escape depth of the fluorescence x-rays, and is several µm. Detection limits are on the order of 10^{13} atoms/cm^2 for transition metals on the surface. These detection limits are seriously affected by high backgrounds resulting from the deep penetration of the incident x-rays into the body of the sample. This technique is not useful for measuring surface trace metals after commercial cleaning of silicon, because the detection limits are insufficient.

3.6 High-Resolution Electron Energy Loss Spectroscopy

High-Resolution Electron Energy Loss Spectroscopy (HREELS) uses an electron beam to scatter electrons from the surface (electron in, electron out) (19). The scattered electrons are analyzed according to their energies (particularly for their energy loss). Chemical information, particularly of hydride bonds, can be obtained by the energy loss of the scattered electrons (20). The analytical depth of HREELS is determined by the escape depth of the scattered electrons and is approximately 5 nm. This technique is not useful for the analysis of surface trace metals on silicon, because it does not detect metals. However, it is indirectly sensitive to metals via the hydride bonding when the bond density is above 10^{12} to 10^{13}/cm^2.

3.7 Infrared Spectroscopy

Infrared (IR) spectroscopy utilizes a polarized infrared beam in an internal reflection mode at the surface to absorb infrared light according to chemical bonding energies (photon in, photon out) (21). Absorption energies provide chemical information about the native oxide, and the internal reflection increases sensitivity to surface contamination. This technique is not useful for measuring surface trace metals after commercial cleaning of silicon, because of insufficient detection limits.

3.8 VPD/AAS

Vapor Phase Decomposition followed by Atomic Absorption Spectroscopy (VPD/AAS) involves wet-HF pre-concentration of the surface metals, followed by graphite furnace AAS measurement of the metals in the pre-concentrated liquid residue (3)(22)-(24). The technique uses a vapor phase HF to decompose the surface native oxide (hence the term Vapor Phase Decomposition). Next, a drop of high-purity water is rolled over the wafer surface to dissolve the metals which were in the native oxide. Finally the water droplet is removed from the wafer and analyzed in an AAS instrument. The collection efficiency of the VPD and water droplet steps has been demonstrated by intentionally contaminating substrate surfaces at very low areal concentrations of metals (3). A schematic diagram of the process is shown in Fig. 6 under the category WSA. Published detection limits vary by several orders of magnitude, suggesting that there are critical steps required to reach the lowest published detection limits of 10^8 to 10^9 atoms/cm^2. In addition, there has been some difficulty collecting surface Cu by this technique, so that alternative vapor phase chemistries are being studied to collect surface Cu (24).

An attempt to estimate the within-laboratory precision was done using two sets (10 and 11 wafers) of silicon substrates intentionally contaminated with Fe, Cu, Zn, and Ni by spin coating process (3) and then performing the VPD/AAS on all the wafers. The results were excellent and were within 10% CV (25).

One drawback of the VPD/AAS technique is that it provides no spatial information across the substrate surface, and if the surface has particulate contamination of heavy metals the VPD/AAS result will not represent a homogeneous concentration. Additionally, the VPD process is closely aligned to some of the new vapor phase cleaning chemistries under development for high-density ICs; it seems questionable that the chemistry used to clean the wafer and that used to collect the impurities for analysis can be essentially the same and still be analytically valid.

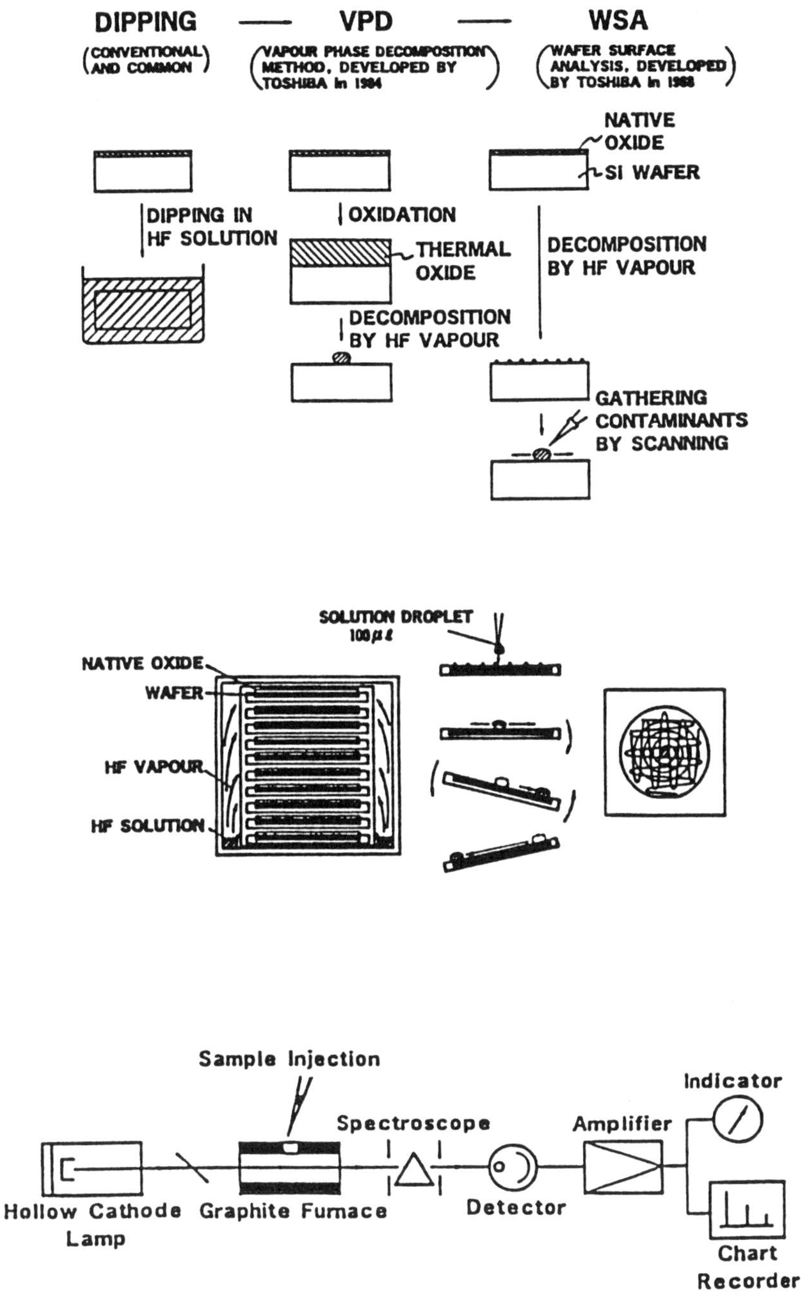

Figure 6. Schematic of the VPD analysis process (24).

The following illustrate some applications of VPD/AAS. Surface Al on silicon from spin-coating was calibrated by VPD/AAS for a series of Al surface concentrations. The wafers were then thermally oxidized, and the final oxide thicknesses were measured. The results showed the final oxide thickness was a function of the surface Al concentration for Al levels above 10^{11} atoms/cm^2, and that the surface Al somehow retarded the oxidation process. This relationship between surface Al and oxide thickness (or oxidation growth rate) has crucial implications for thin gate oxide thickness control, because the surface Al after cleaning can vary depending upon the cleaning chemistry and the Al level in the cleaning solutions. An example of the use of VPD/AAS is presented in Fig. 7 which shows the results of a cleaning development for Al, Na, and Fe down to the 10^9 atoms/cm^2 range. Results of a cleaning process, which includes a slight wet-chemical etch (SE) of the silicon surface to remove surface Fe, are shown in Fig. 8. (26).

The VPD/AAS technique is very useful for the measurement of surface metals, but results can be affected by their chemical bonding, the measurement is destructive, and there is no spatial information available.

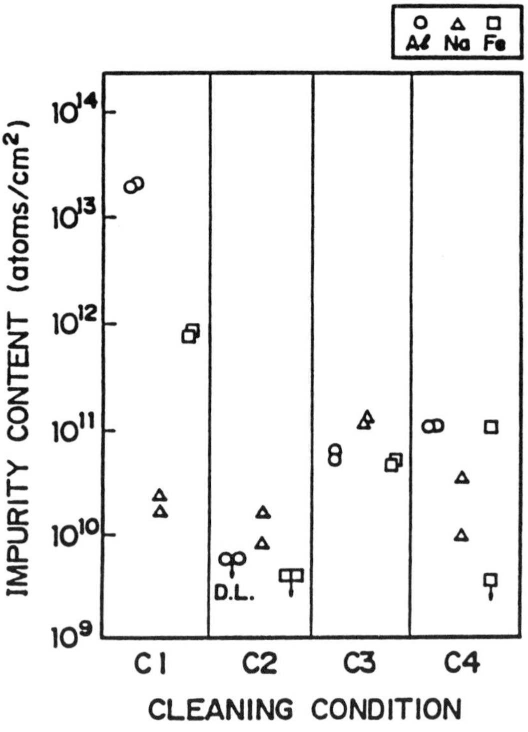

Figure 7. VPD/AAS characterization of different cleaning processes (23).

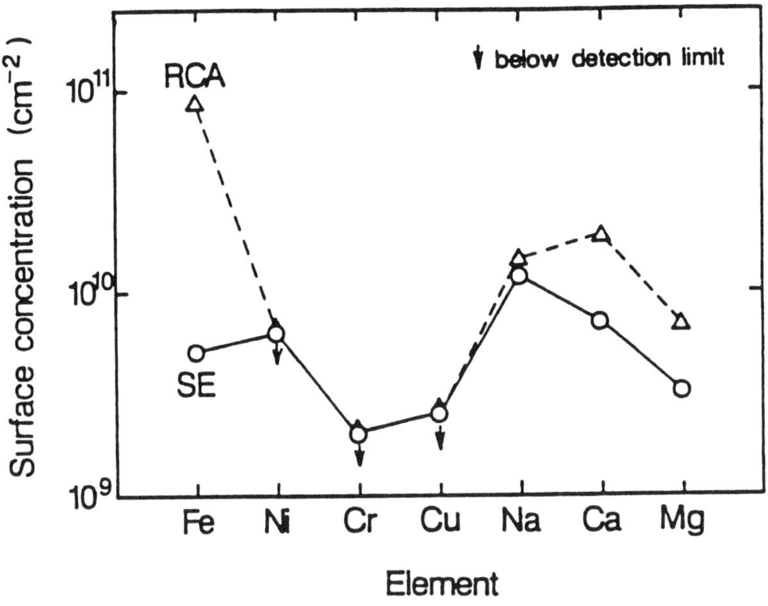

Figure 8. VPD/AAS characterization of the slight etch and RCA cleaning processes (26).

3.9 Secondary Ion Mass Spectrometry

Secondary Ion Mass Spectrometry (SIMS) uses an ion beam to sputter ionize material from the sample surface (ion in, ion out) (27). Elemental identification is made by analysis of the mass-to-charge ratio of emitted ions by quadrupole or magnetic sector mass spectrometers. The analytical depth of SIMS is determined by the sputter rate and the analysis time, and can range from 5 to 500 nm, depending upon the instrument and approach to the analysis. SIMS is a very sensitive technique for many elements. Problems with SIMS analysis have been matrix mass interferences, the quantification for surface analysis, and sensitivity variation of many orders of magnitude among elements. This technique is useful for the measurement of surface metals, but it is difficult to use for transition metals and it is destructive. More is said about SIMS in Sec. 4.0.

3.10 Total Reflection X-Ray Fluorescence Spectroscopy

Total Reflection X-Ray Fluorescence Spectroscopy (TXRF) uses an x-ray beam at very low (glancing) angle to excite the fluorescence x-rays from the surface elements (x-ray in, x-ray out) (28). Elemental identification is made by the energy of emitted fluorescence x-rays. Analytical depth is determined by penetration depth of the evanescent wave x-rays, which is 3 to 5 nm under total reflection conditions. Detection limits are on the order of 10^{10} atoms/cm^2 for transition metals. Analysis area is approximately 1 cm^2. Detection limits are much lower than for conventional XRF because the high backgrounds from XRF are eliminated by the change in x-ray penetration depth. This technique is useful for the measurement of surface metals, but it does not detect the low atomic number metals. This measurement is non-destructive. More is said about this technique in Sec 5.0.

4.0 PRINCIPLES AND METHODOLOGY OF *SIMS* ANALYSIS

Three approaches of SIMS that are used for measuring trace elemental contaminants on the surface of semiconductor wafers after cleaning are described in some detail:

1. Static SIMS of the surface (a qualitative analysis using a primary ion beam with an oxygen jet)
2. Dynamic SIMS of the surface (a semi-quantitative analysis using a primary reactive ion beam with an oxygen jet and relative sensitivity factors for quantification)
3. Polyencapsulation/SIMS (PC/SIMS) (a quantitative analysis useful for low atomic number elements)

4.1 Principles of SIMS

SIMS analysis is performed by means of an ion beam to sputter-ionize material from the sample surface (ion in, ion out) (27). The emitted ions are analyzed for mass-to-charge ratio by a mass spectrometer and elemental identification is based on this ratio. Analytical depth is determined by sputter rate and analysis time. Typical SIMS instruments and approaches to this surface measurement can be divided into two general categories: static SIMS, for which the sputter rate is very low (e.g., 1 nm per 5 min) and the mass spectrometer is a quadrupole; and dynamic SIMS, for which the sputter rate is much faster (e.g., 1 nm per 10 sec) and a magnetic sector mass spectrometer is used.

4.2 Static SIMS

Static SIMS analysis of silicon wafer surfaces after cleaning was developed in the early to mid 1980s (11)(29)(30). A primary beam of noble gas ions (He^+, Ar^+, Xe^+) or reactive ions (O_2^+), is used to sputter-ionize the sample surface. The ejected ions are detected using a quadrupole mass spectrometer. This analysis is often performed with an oxygen jet focused on the analysis area. The purpose of the oxygen jet is to stabilize ion yields while sputtering through the native oxide region. For example, a Xe^+ primary ion beam was used for sputter ionization and a quadrupole mass spectrometer was used for detection of secondary positive ions. Without the oxygen jet, the $^{28}Si^+$ signal changes several orders of magnitude during the analysis due to changes in the $^{28}Si^+$ ion yield as a function of changing oxygen content in the sample as the sputtering moves through the native oxide into the silicon substrate. Use of the oxygen jet largely eliminates this artifact.

An important point in SIMS is that the surface metal contaminants in the native oxide are sputter- or cascade-mixed during the analysis. For illustrative purposes, if the native oxide has 10^{13} Al atoms/cm^2, during the first minute of analysis 90% of the Al is sputtered off the surface and 10% of the Al is driven deeper into the sample by the primary ion beam cascade process. In the second minute of analysis when the sputtering penetrates deeper where the residual 10% Al was driven, the sputtering removes 90% of that 10% residual and drives 10% of the 10% residual Al even deeper. This process keeps repeating itself until the last residual Al is undetectable. Now the 90%/10% is hypothetical in this example, but the concept leads to a SIMS plot where the log of the SIMS Al signal versus sputter time (i.e., depth) is linearly dropping with time. The slope of this line has received much study in the SIMS research community, and appears to be affected by many measurement parameters (e.g., the primary ion species and impact energy, the angle of incidence of the primary ion beam, the surface contaminating element species, and whether an oxygen jet is used). The important point is that either the SIMS analysis must be continued to a depth where the vast majority of the impurity has been sputtered out of the sample, or the slope of an impurity element versus the analysis depth must be the same for all comparative samples.

In a static SIMS measurement the quadrupole mass spectrometer sweeps the mass-to-charge ratio quickly during the sputtering process. In a typical five minute analysis where the sputter rate is about 3 nm in 5 min, the quadrupole sweeps from 1 to 100 amu/q three times. This means, for example, the detector counts the Na (amu/q of 23) three times during the measurement, and presumably for 1 second each time. The electronics stores the total Na counts for a total of 3 seconds taken over three different times. Thus, static SIMS analyzes each element (or amu/q) briefly at three different depths. This also means that the depths at which the Na (amu/q = 23) and the Al (amu/q = 27) are first detected are different. Since for some elements the slope is sharp and most of the impurity signal occurs in the first few seconds, this highest signal may be missed by the analysis, while the quadrupole mass spectrometer is tuned to other elements. Nevertheless, if the slopes are constant in time, and if the behavior is the same for comparison unknowns, the comparative analysis by static SIMS can be useful. However, this effect introduces an inaccuracy to any quantification attempt.

Figures 9 to 12 are expanded scales from 1 to 100 atomic mass units divided by charge Q for a typical static SIMS measurement of a bare silicon wafer after SC-2 cleaning. The measurement conditions were a Perkin Elmer 560 with a SIMS II probe using a Xe^+ primary beam, an oxygen jet, a quadrupole mass spectrometer, and an analysis period of 5 minutes.

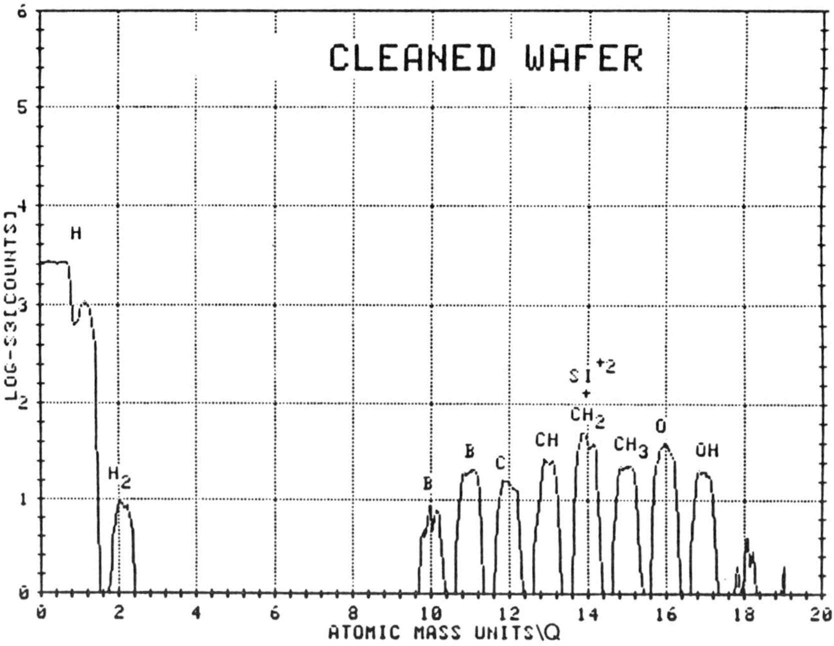

Figure 9. SIMS: expansion of region 1 to 20 amu.

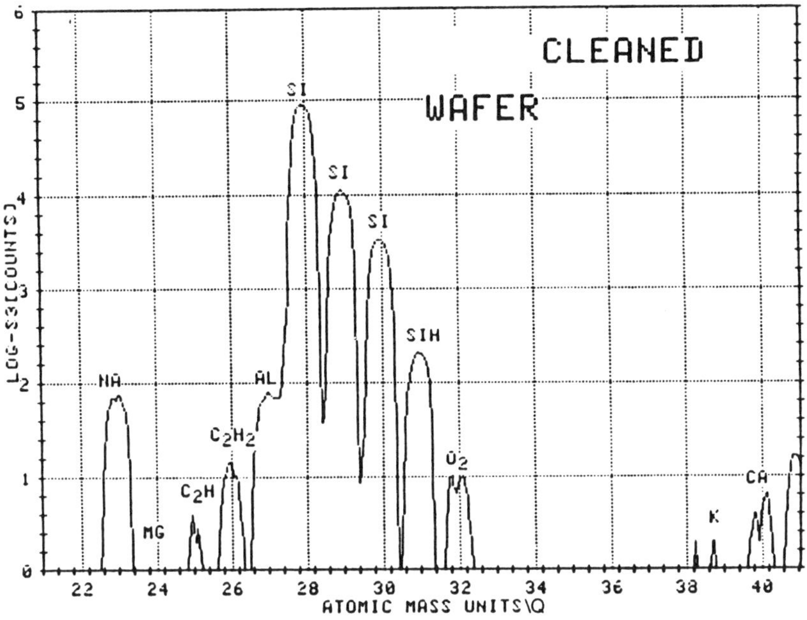

Figure 10. SIMS: expansion of region 20 to 40 amu.

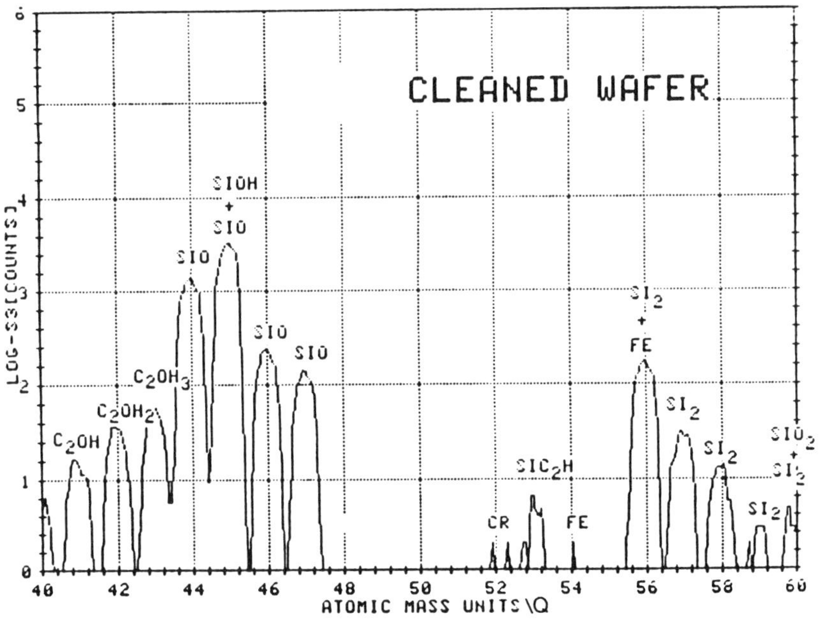

Figure 11. SIMS: expansion of region 40 to 60 amu.

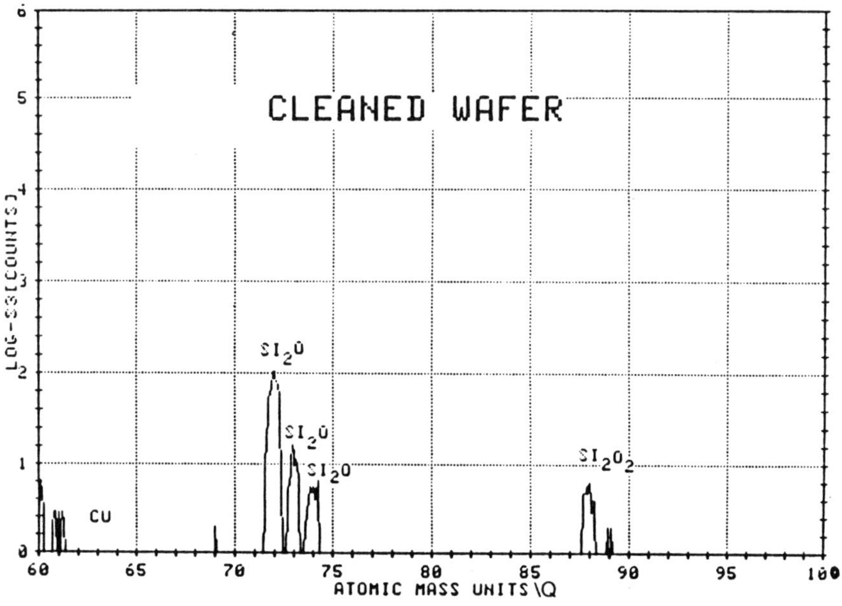

Figure 12. SIMS: expansion of region 60 to 100 amu.

Static SIMS Comparison of Cleaning Processes. As an example of the use of static SIMS to develop commercial scale cleaning processes for silicon wafers, Table 3 lists the relative SIMS signals (relative to a 10^5 ^{28}Si matrix measurement) for Na, K, Al, Ca and Mg+C_2, where C_2 is the carbon dimer, for three cleaning processes: modified RCA, choline/H_2O_2, and a "Monsanto clean". The standard deviation of the measurements is shown in each parenthesis. This work used a Xe^+ primary ion beam, an oxygen jet, and a quadrupole mass spectrometer on a Perkin Elmer Phi 560 instrument with a SIMS II probe.

The role of surface Al after cleaning and its effect upon thermal oxidation rate has received attention both in the U.S. (31) and in Japan (23). The effect of Al in different RCA cleaning solutions on the thickness of subsequently grown thermal oxide is shown in Fig. 13 (14). The data indicate that a small amount of Al added to an NH_4OH/H_2O_2 solution can inhibit thermal oxide growth as much as a much greater amount of Al in an HCl/H_2O_2 bath. A correlation between surface Al in the bath and on the wafer was done by qualitative static SIMS.

Table 3. Qualitative SIMS Comparison of Cleaning Processes (30)

Cleaning Process	(Units are integrated counts ratioed to 10^5 ^{28}Si counts) Na	K	Al	Ca	Mg
Modified RCA	4	5	38	19	10
Choline/H_2O_2	4	13	3530	28	7
Monsanto clean	4	4	10	8	4

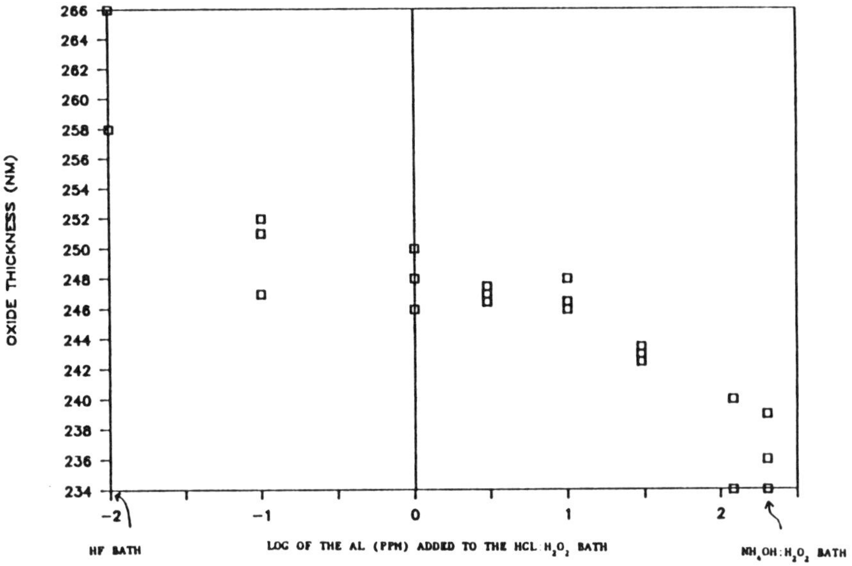

Figure 13. The effect of varying Al in different cleaning baths upon thermal oxide thickness (14).

Static SIMS Deficiencies. In spite of the oxygen jet, the technique is only qualitative, because standards have not been developed that can produce representative ion yields similar to the real analysis. Attempts to rectify this have used "relative sensitivity factors" (RSFs) taken from implants into SiO_2 (32) and have assumed the ion yields are the same during the static SIMS analysis. The reference implanted samples result in only an approximate quantification, because the ion yields in the static

SIMS analysis and in the reference samples are expected to be different to some extent. Furthermore, for elements with low slopes (e.g., Al) the quantification will be in error because most of the Al is not sampled within 5 min; for elements with high slopes (e.g., Na) the mass spectrometer does not analyze the element during the first few seconds even though most of the Na signal occurs at this point.

Probably an even greater deficiency is the mass resolution problem of quadrupoles for some key impurities (e.g., Si_2 for Fe; Si_2 and SiO_2 for the isotopes of Ni; SiCl for the two isotopes of Cu; SiH for P; and SiC_2 for Cr). In addition even for Al at low levels the $C_2H_3^+$ interference becomes a problem, so that in Table 3 for the Monsanto clean, the Al entry of 10 becomes suspect; the actual Al may be much lower. Mass resolution requirements for some metals on the surface of silicon have been published (32), but reports in the literature of surface Cr, Cu, Fe and Ni at low levels have been misleading when this technique was used. The technique has been applied primarily to the qualitative measurement of Na, Mg, Al, K and Ca. However, even these data can at times can be misleading, depending upon mass interferences.

4.3 Dynamic SIMS

An alternative to static SIMS is to sputter at a higher rate and to a greater depth, monitoring a few impurity elements only, and integrating the total counts for each impurity. Normally this is done with a magnetic sector mass spectrometer which has higher mass resolution capabilities than the quadrupole mass spectrometers. This approach is called dynamic SIMS.

Stingeder et al. (33) have performed a systematic study of dynamic SIMS for surface analysis of silicon wafers to demonstrate the importance of the higher mass resolution. For example, a silicon wafer contaminated with surface Al at 10^{13} atoms/cm^2 was profiled for both Al and C_2H_3 interference. The mass resolved C_2H_3 interference signal was about two orders of magnitude lower than the real Al signal, thus giving evidence that the Al measurement by quadrupoles is limited to a detection capability of about 10^{11} atoms/cm^2 when organics are present on the wafer surface.

If an oxygen jet is used to stabilize the ion yields, the dynamic SIMS technique can be made approximately quantitative using RSFs taken from implants into SiO_2 (32) and assuming the ion yields to be the same during the dynamic SIMS analysis and in the reference samples. However, as we indicated for static SIMS, the ion yields in the dynamic SIMS analysis and in the reference samples are expected to be different to some extent. In contrast to static SIMS, however, quantification will be improved because the SIMS signal is integrated over both the near surface and the greater depth until cascade mixing is no longer important.

Normally, dynamic SIMS measurement is done using an O_2^+ primary beam. In contrast, Sumita et al. (34)(35) used a Cs^+ primary beam, but with a quadrupole mass spectrometer. They showed that the quantification (down to where the quadrupole mass resolution became a limitation) could be developed in a linear manner against spin-coated contaminated samples. Figure 14 shows a depth profile of Al, Cu and Cr using a Cs^+ beam, and Fig. 15 shows the integrated intensity normalized to a matrix integral for each impurity versus the spin coated contamination level. The results indicate linearity and show that the working curve slopes for each element are independent from the impurity species. For an O_2^+ beam they have also demonstrated linearity, but the slopes varied with impurity species. Using the quadrupole detector, their detection limits for Fe and Al were limited to the mid to high 10^{11} atoms/cm^2 level. Thus, combining RSFs and dynamic SIMS may work; and the accuracy of the quantification may be checked with spin-coated samples calibrated by VPD/AAS.

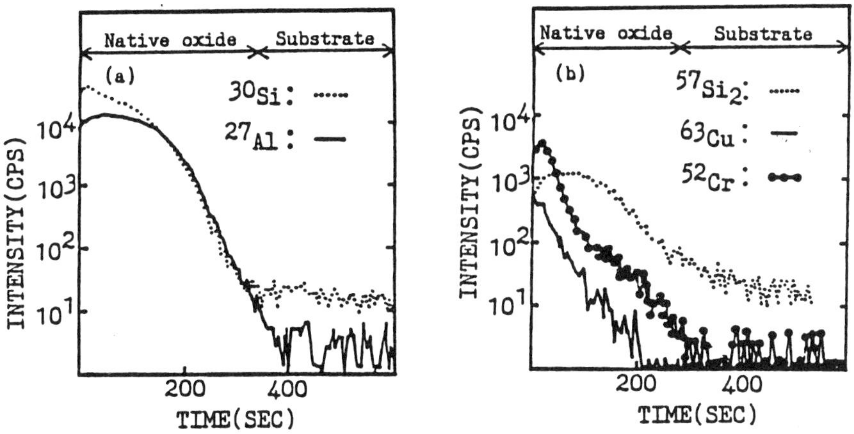

Figure 14. Dynamic SIMS profile of Al, Cu and Cr using a Cs^+ primary ion beam (33).

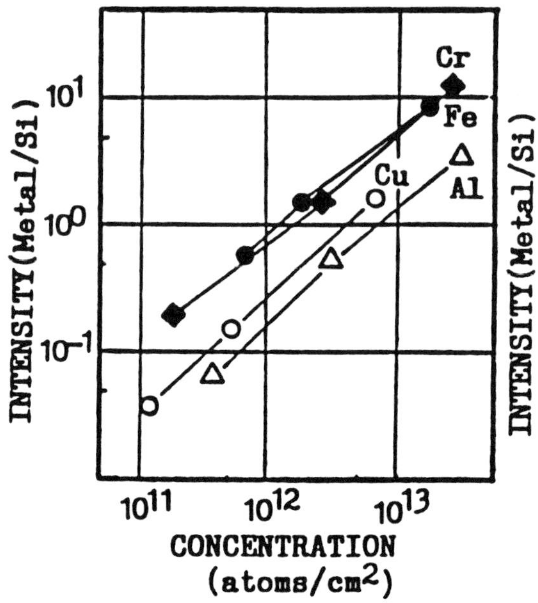

Figure 15. Integrated intensity of Al, Cu and Cr normalized to silicon matrix versus surface areal density as measured by VPD/AAS (33).

The main difficulty using spectrometers with high mass resolution to eliminate the mass interferences is that the ion beam erosion rates must be very small to obtain a significant sample of the native oxide region during the analysis; but under high mass resolution there is often not enough signal to accurately tune the magnet. Furthermore, if electronegative elements are of interest, the oxygen jet reduces ion yields, and therefore, detection limits. Attempts to use a Cs^+ primary beam to detect electronegative surface elements (e.g., F, Cl) have not been successful because varying oxygen content in the native oxide causes varying ion yield damping during the analysis. (The addition of an oxygen jet would stabilize the effect but makes the ion yields not useful for detection.) These difficulties have led to the development of polyencapsulation/SIMS.

4.4 Polyencapsulation/SIMS

Description and Advantages. The polyencapsulation/SIMS (PC/SIMS) technique (36) was developed in 1980 but was not reported in use until much later (37)-(40). In PC/SIMS the silicon wafer is covered with a thin layer of polysilicon or amorphous silicon which transfers the difficulty to a surface region away from the silicon wafer surface of interest by "burying" the original surface deeper than the non-linear region. The encapsulation material is usually polysilicon or amorphous silicon, although CVD oxide can be used if an O_2^+ beam or an oxygen jet is also used for the analysis of electropositive (but not electronegative) elements. The chosen thickness of the encapsulant is a balance between being greater than the thickness of the non-linear ion yield region, but thin enough to avoid loss of the interface signal due to dilution mixing. Typical thicknesses have been 50 to 120 nm. Since the surface impurities are now fully encapsulated in a silicon matrix: (a) the original surface is protected from subsequent contamination, (b) SIMS non-linear ion yield effects are minimized, (c) ion implants can be used to quantify the SIMS ion signals, (d) high mass resolution can be used, and (e) a Cs^+ primary beam can be used to optimize ion yields of the electronegative elements. PC/SIMS is, in principle, capable of detecting all trace elements quantitatively with a wide range of detection limits due to the widely ranging ion yields for elements across the periodic table. However, its limitation is often due to artifacts from the encapsulation itself.

Encapsulation Artifacts. Care must be taken to determine if the encapsulation process itself introduces, or removes, impurities in the analytical region of interest. A recommended procedure to assess the first issue is to perform the encapsulation twice, so that the interface between the first and second polysilicon layers provides an analytical measure of the impurities introduced by the encapsulant process, and the interface between the polysilicon and crystal provides an analytical measure of the original surface impurities, plus those introduced by the encapsulation. Because the reactivity of fresh polysilicon is much greater than that of a native oxide, a subtraction between the two interface measurements is not recommended. Also, because impurities from air can contaminate the fresh polysilicon surface and interfere with the identification of impurities from the encapsulation process itself, wafers should not be exposed to air after the first poly layer. A study of several methods for depositing polysilicon or amorphous silicon has shown that the encapsulation process itself can easily introduce impurities (40).

One must also assess whether the encapsulation process removes impurities of interest. Removal might occur by two different mechanisms: *(i)* desorption to the chamber ambient; or *(ii)* diffusion into the encapsulant or into the crystal. An assessment of this issue was made for heavy metals by measuring the heavy metal areal densities first by total reflection x-ray fluorescence (TXRF), performing a double polyencapsulation via CVD polysilicon, and then measuring the interfacial metals using SIMS. The results revealed that surface Zn was removed by the encapsulation process (probably desorption), that surface Cu diffuses into the polysilicon, and that surface Br was desorbed down to a level of 10^{11} atoms/cm^2. PC/SIMS analysis of surface Al does not lose the Al, as was shown by performing the analysis on intentionally Al contaminated wafers in the 10^{12} to 10^{14} atoms/cm^2 range, as calibrated by VPD/AAS.

PC/SIMS Quantification. The intent of polyencapsulation is to provide a constant ion yield in the analytical region of interest, i.e., the poly-Si/single crystal silicon interface. This region has a native oxide that is cascade mixed (36) into both the polysilicon and the single-crystal by the primary ion beam, so that the amount of oxygen remaining at the interface from the native oxide is about 4 atomic percent, as measured by Cs^+ beam SIMS. This amount may introduce minor ion yield artifacts, which can be experimentally assessed by implanting silicon with oxygen (7×10^{16} atoms/cm^2 at 100 keV) and followed by co-implanting with the element of interest so that the co-implant peak coincides with the oxygen peak. A comparison of SIMS profiles and calculated relative sensitivity factors (RSFs) with and without oxygen implant provides a measure of the ion yield effect of the cascade-mixed native oxide. Studies of this type for a variety of elements have shown that there is no ion yield enhancement for most elements and only up to 20 to 50% enhancement for others. The same kind of oxygen implant can be used to measure the amount of voltage offset required to remove oxygen-containing molecular interferences at the polysilicon/single-crystal interface. Quantification is achieved using ion implants of a known dose of metal into silicon to calculate an RSF which is then used for the polyencapsulation/SIMS measurement. The areal density of the impurity is desired, which is easily calculated from the SIMS calibrated depth profile and the RSF.

PC/SIMS Detection Limits. As with any analytical technique, detection limits are influenced by numerous factors. In our case these include: instrumental background, molecular interferences, signal-to-noise, and impurity introduction or loss by the encapsulation process. Some estimated elemental detection limits for polyencapsulation/SIMS are listed in Table 4, where the practical problem of contamination from polyencapsulation is included.

Table 4. Estimated Detection Limits (DL) for PC/SIMS (40)

		(Units of atoms/cm^2)			
Li	2E09	B	5E10	Al	2E10
Na	1E10	P	1E10	Mg	2E09
K	2E09			Ca	2E10
				Fe	2E11

PC/SIMS Comparison of Cleaning Processes. The surface Al, Ca and Mg after an SC-1 + SC-2 clean sequence were measured by this technique to be 1.3, 4, and 0.7 x 10^{12} atoms/cm^2, respectively (38). In 1989 a comparison of commercial silicon wafers from five vendors after different cleaning processes was completed using PC/SIMS (40). All wafers were encapsulated at the same time. The results are shown in Table 5. Both P and B dopants were detectable even though these wafers were both n- and p-type. We also detected the alkalis Na and K, and the metals Mg, Ca and Al. The wafers had similar levels of P, B, Na, Mg and Ca (D-2 is an exception for Ca). The "A" group had much lower K and Al concentrations compared to the others.

Table 5. PC/SIMS Comparison of Cleaning Processes (40)

	(Units of 10^{12} atoms/cm^2)						
Wafer Group	P	B	Na	K	Mg	Ca	Al
A-1	0.2	p+	0.1	0.009	0.06	0.1	<0.003
A-2	0.1	p+	0.1	<0.009	0.07	0.3	0.003
B-1	0.09	1	0.08	0.2	0.1	0.2	0.07
B-2	0.07	1	0.07	0.2	0.04	0.1	0.1
C-1	0.08	1	0.08	0.02	0.06	0.08	0.06
C-2	0.1	1	0.09	0.1	0.07	0.3	0.3
D-1	0.08	1	0.1	0.2	0.07	0.1	0.08
D-2	0.1	1	0.08	0.3	0.1	3	2
E-1	0.1	1	0.1	0.03	0.07	0.07	0.0
E-2	0.06	2	0.1	0.1	0.1	0.1	0.2

As a general conclusion, dynamic SIMS and PC/SIMS can provide useful detection limits and practical analytical capability for low Z elements, such as Li, B, Na, Mg, Al, K, and Ca. However, the detection capability for the transition metals is limited, and the technique is destructive. An alternative analytical technique for the transition metals complements SIMS very well: TXRF.

5.0 *TXRF* ANALYSIS

5.1 Principles of TXRF

The basic history and principles of total reflection x-ray fluorescence (TXRF) have been described by Schwenke and Knoth (28)(41). The original discovery that x-rays can totally reflect from solids was made by Compton (42); the phenomenon occurs because the real part of the index of refraction is less than unity. Basically, in TXRF x-rays from some source impinge on the sample surface at a very low angle, below the angle for total external reflection, and excite only the top few mono-atomic layers by the x-ray evanescent wave. The fluorescence x-rays from these top few monolayers emit in many directions, and a Si(Li) detector perpendicular and close to the sample surface collects emitted fluorescence x-rays. The Si(Li) detector is an energy dispersive spectrometer which detects the fluorescence x-rays and analyzes them according to energy, which yields elemental identification.

The critical angle, as defined by Snell's law, for the total reflection of x-rays from matter depends linearly on the wavelength of the x-ray (or is inversely proportional to the x-ray energy) and depends upon the square root of the total electron density of the material. For example, the critical angle for silicon is 0.22°, 0.195°, and 0.10°, for the respective x-ray sources of Cu K-alpha (8.04 keV), W L-beta (9.67 keV), and Mo K-alpha (17.5 keV); the critical angle for quartz (similar to SiO_2) is only 3% lower than for silicon.

To develop a qualitative idea of the behavior of total reflection, we will discuss some published theoretical penetration depth curves. In Fig. 16 we see the penetration depth versus glancing angle for two materials: quartz and silicon (43). We note that below the critical angle, the penetration depth is nearly constant at a few nanometers; this is the TXRF regime. Above the critical angle the penetration depth can reach several microns; this is the glancing XRF regime. We note the two curves are similar, since the electron densities are similar. Since quartz is slightly less dense compared to silicon, we note the quartz critical angle is slightly less, and that the high-angle penetration is deeper for quartz than for silicon. It is at the higher angles that the difference mainly shows up. Below the critical angles, the penetration depths cannot be distinguished for all practical purposes.

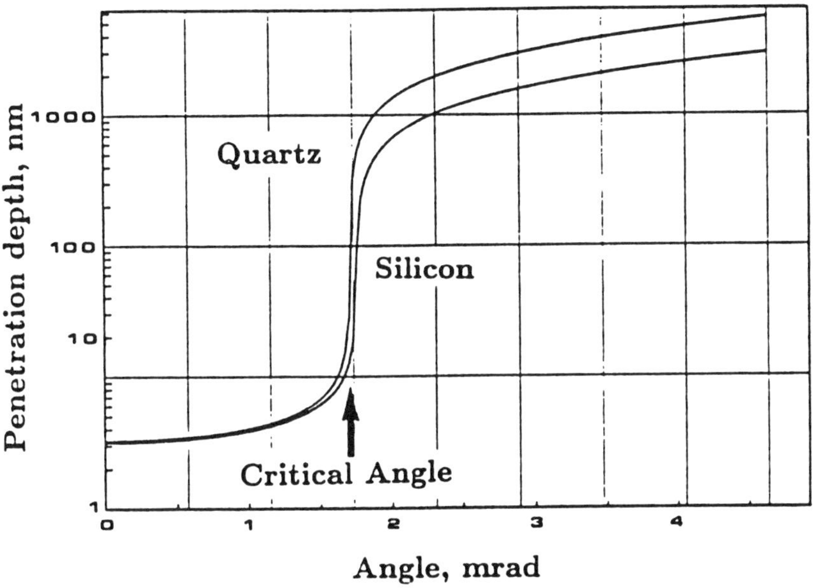

Figure 16. Theoretical TXRF penetration depth versus glancing angle for quartz and silicon using a Mo (17.5 keV) x-ray source (43).

In contrast, Fig. 17 compares the penetration depth for silicon for two different x-ray sources (Mo and Cu) (44). We note the low-angle penetration depth is about the same for both sources even though the critical angle is about two times higher for the Cu than for the Mo source. The shape of the curves is similar, particularly if we scale them with their respective critical angles. The main difference shows up above the critical angles where the penetration of the Cu source x-ray is much less than that for the higher energy Mo source x-ray. The Cu x-rays penetrate only a few hundred nm above the critical angle for the Cu source. The curve for a W source (9.67 keV) would lie somewhere between the Mo and the Cu curves, but closer to the Cu curve.

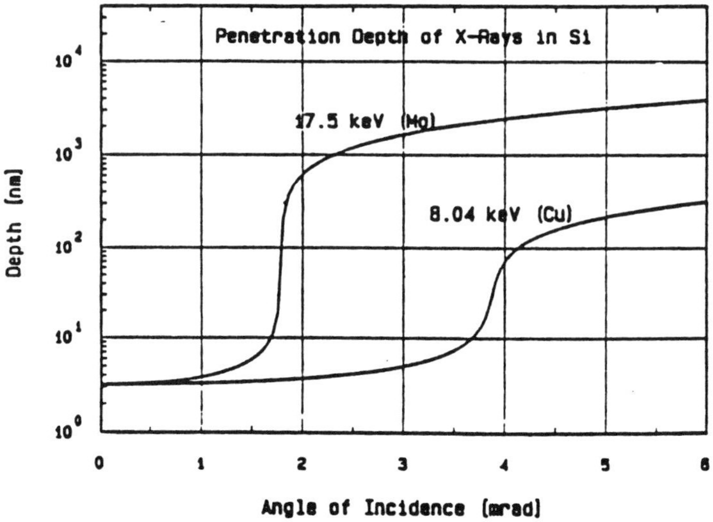

Figure 17. Theoretical TXRF penetration depth versus glancing angle for silicon using two x-ray sources: Mo (17.5 keV) and Cu (8 keV) (44).

Another important aspect of TXRF is the effect of the type of contamination on the surface of the silicon wafer. Bernieke et al. (43) have calculated theoretical curves of the relationship between fluorescence intensity versus glancing angle for different types of contamination, and Eichinger et al. (45) have experimentally verified these curves shown in Fig. 18. For residues on top of the silicon wafer, the fluorescence intensity below the critical angle is constant until the critical angle is approached, whereas for contamination localized in the native oxide region, the fluorescence intensity below the critical angle is rapidly rising as the glancing angle increases up to the critical angle. These differences provide qualitative information on the type of contaminant; they also affect the approaches to quantification and the analysis conditions to attain optimum detection limits.

Because the most common metal contaminants from wafer cleaning are localized in the native oxide region, rather than occurring as a residue, we can see from Fig. 18 that the ability to reproduce a glancing angle will strongly affect the reproducibility of the measurement. The accuracy of the glancing angle, it turns out, will affect the quantification. For this reason we will discuss different current methods for the angle calibration.

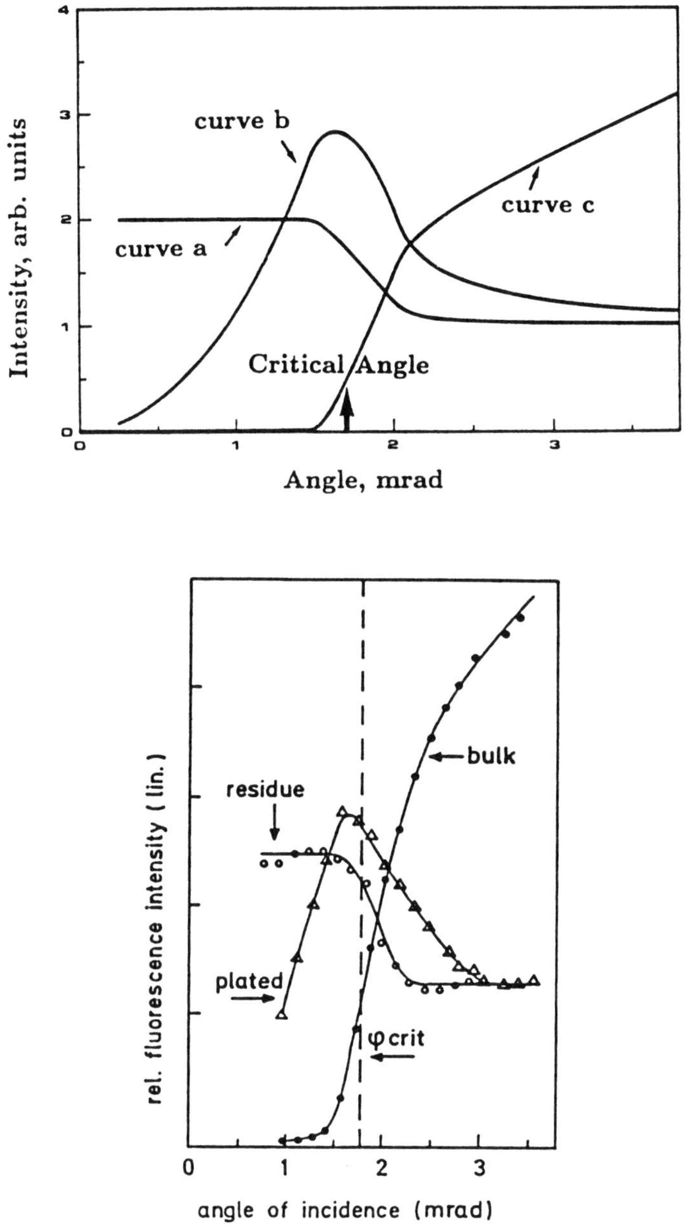

Figure 18. Theoretical (43) and experimental (45) angle-scans for plated and residue contamination on a silicon substrate.

There is no agreement at this time as to which method is best. The first method (43) for angle calibration utilized the fact that the angle dependence of the fluorescence signal from the Si matrix was very steep at the critical angle; this allowed an angle adjustment with an reported accuracy of 0.1 mrad (for a Mo x-ray source). At each analysis location the Si count rate versus some mechanical parameter (e.g., x-ray tube height with respect to an aperture) that controlled the glancing angle was measured. The inflection point was then calculated to assign the critical angle, and by some additional approach the zero angle was determined. This allowed a transformation of mechanical parameter controlling the glancing angle into an angle calibration scale. The first use of this method had the advantage that the frontside reference of the sample was determined by clamping the wafer against a highly polished quartz block, so that the distance from the sample analysis to the detector was always constant and did not have to be determined separately. Unfortunately, this also led to contamination on the silicon sample surface, so that this method has now become more complicated. The wafers are now no longer placed against a polished frontside reference surface and thus have no fixed frontside reference.

The procedures to define a frontside reference without the quartz block involve three laser reflections of the wafer, followed by an iterative procedure to keep the distance from the sample to detector constant and at the same time determine the angle calibration, using the matrix signal inflection point. Research is in progress to understand how the wafer surface roughness, which can be affected by different cleaning processes, may affect the laser reflections and the matrix signal curve, since it is known that for very rough surfaces (such as etched backsides of wafers) the matrix signal versus glancing angle has no inflection point.

Another method for the angle calibration (46) is based on the angle alignment methods used in the x-ray diffraction community. The sample is removed from the path of the x-rays, which are then measured by a scintillation counter. Next the sample is moved into the path of the x-rays, while the x-ray intensity is still being measured with the scintillation counter. An interactive procedure is used to find the sample positioning and alignment conditions so that a small decrease in the sample-to-detector distance results in just beginning to reduce the scintillation counter signal from full count rate, and yet having a tilt axis perpendicular to the x-ray beam. This procedure defines a zero angle plane for the sample surface and allows stepper motors to tilt the sample at a calibrated angular distance. This method is independent of the matrix signal curve and surface roughness, but will be affected by extreme sample warp; the sample is flattened on the backside using an electrostatic chuck.

5.2 Quantification

There are different approaches to quantification, but in all cases, one needs a reference material which must be assigned an areal density of metal contaminants determined by a different analytical technique. The TXRF has been shown to be linear with respect to different levels of surface contamination on a silicon wafer, so the present approaches to quantification reference standards use areal densities on the order of 10^{13} atoms/cm^2. The reason is the difficulty of keeping a sample of lower areal density clean, i.e., no added contamination can be tolerated at levels significant in comparison to the intentional contamination level.

The first method used for TXRF quantification is based on the quantification model described by Schwenke et al. (41). An atomic absorption liquid reference solution, for example Ni of 1000 ppm, is diluted with clean water, and a known amount of this liquid is deposited on a Ni-free silicon wafer. The dilution and liquid volume are chosen so that the dried residue of the droplet remains within the analysis area of the TXRF instrument (6 to 10 mm in diameter). The total number of Ni atoms in the residue should be on the order of 10^{13} atoms/cm^2 where the area term is the analysis area. The intent of the quantification method was to form a residue of nickel ions, so that the TXRF angle-scan would appear like curve (a) in Fig. 18. To achieve this, the residue must have a thickness much greater than a small multiple of the TXRF standing wave pattern which occurs on the surface as a function of glancing angle (47). In contrast, if the droplet residue dries in a manner where the Ni ions are localized in the native oxide, the angle-scan resembles curve (b), which is a result of the standing waves in TXRF. For this reason, some kind of suspension material which does not generate TXRF signals is necessary to suspend the Ni ions above the surface over a height greater than about 0.5 microns. The reference sample now has a known areal density of Ni (from the AAS assignment). The quantification is done at a unique angle (about 70 to 80% of the critical angle) where, for the same areal density, curves (a) and (b) intersect in count rate. This results in a calibration which is valid then for both localized and residue (or particle) contamination.

Another method is based on intentionally contaminating a silicon wafer by spin-coating to attain an areal density of 10^{13} atoms/cm^2, as determined by VPD/AAS. The homogeneity of the contamination is determined by SIMS or TXRF; the local areal density for TXRF calibration is then assigned from the VPD/AAS measurement. The spin-coating can be quite uniform, as shown in Fig. 19. This kind of reference sample gives angle scans like those shown in curve (b) of Fig. 18. For quantification, the reference and unknowns are analyzed at the same angle, typically lower than the 70 - 80% of the critical angle used in the previous method. This will lead to an error in accuracy if the contamination is distributed, which is not common for contamination from cleaning solutions. An alternative to spin coating is the microdroplet method, which is identical to the distributed residue approach, except the contamination is localized near the native oxide as in the spin-coated samples.

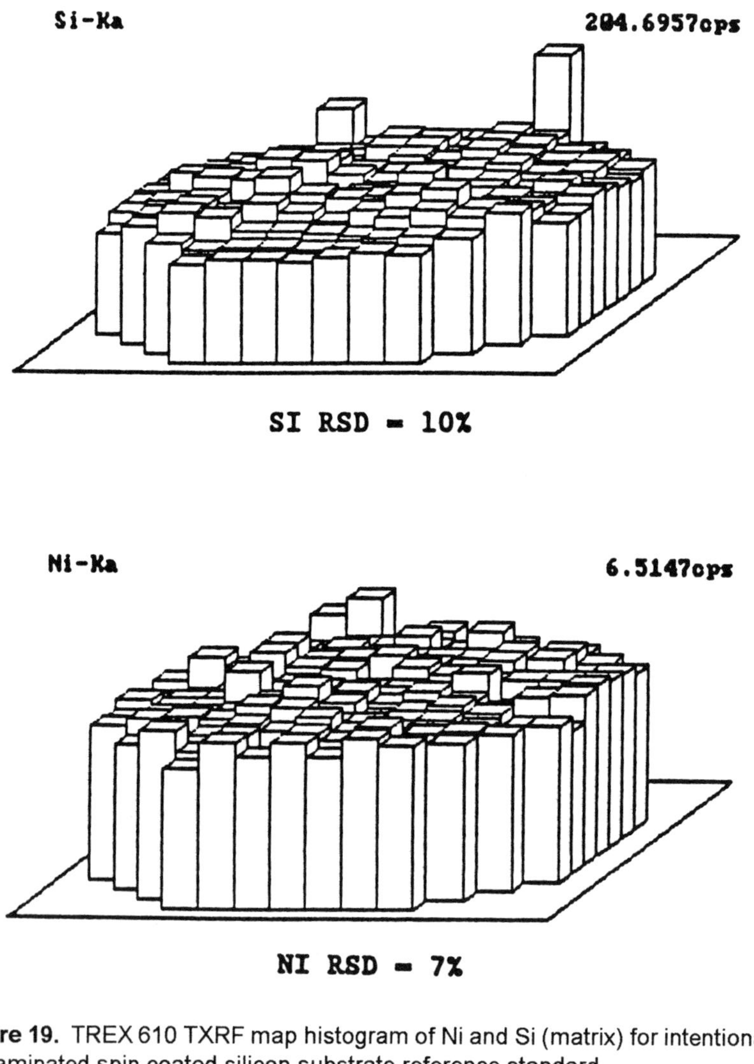

Figure 19. TREX 610 TXRF map histogram of Ni and Si (matrix) for intentionally contaminated spin coated silicon substrate reference standard.

One of the difficulties in quantification has been that not all of the reference materials result in the same quantification. None of the above noted methods use a direct measurement of the surface areal density to assign the areal density. Eichinger (48) has taken a different approach by using nitrogen-beam RBS (16) to determine the metal areal density in the analysis area of the reference material. This is valid as long as the reference metal dominates the RBS spectrum, i.e., there are no significant mass interferences. This, however, is easy to accomplish.

Most of the reference materials in use today use only one of the transition metals (Ni is common, but Fe and Cu have also been used). However, these metals may be added unintentionally at times, thereby destroying the reference standard. An alternative approach is to use vanadium as the reference metal, since it is not found as a common contaminant; and this would extend the lifetime of reference samples.

Each of these approaches to forming reference materials assumes a model for the angle scans of the unknown samples, and these models may be valid for cleaning contamination. However, for some other contamination sources, such as ion implantation or dry etching, the contamination may be subsurface, and here the models fail to give accurate quantification. Weisbrod et al. (44) have tried to extend the quantification by modeling subsurface contamination as well, but a model is still required.

Calibration for quantitative measurements is done using a reference sample as just discussed. Normally this procedure requires only one element for calibration. Calibration for other elements is done using relative sensitivity factors (RSF). TXRF RSFs can be obtained from relative fluorescence yield calculations, but at low Z numbers where the transmission of fluorescence x-rays is less than near 100% an error is introduced by such calculations. Two approaches have been proposed to solve this problem. One is to prepare different residue or spin-coated standards and obtain the RSFs experimentally; this can be done for elements which do not evaporate during the residue drying or which can be spin coated. The other is to use standards with different implanted elements in silicon to obtain the RSFs where the glancing angle is above the critical angle, and is high enough to fully analyze the implant dose; this can be done for any element which can be implanted into silicon. Even with these efforts there will be some bias from one instrument to another, particularly for lower Z elements, because the detectors do not all have the same transmissions. For the transition metals (Ti to Zn) we do not expect much effect from different detectors.

5.3 Quantitative Comparisons

Quantitative correlations (40) have been completed for silicon surface measurement of Fe, Ni, Cu and Zn prepared by spin-coating (3). The surface analytical techniques included: VPD/AAS by Kyushu Electronic Metals Co., TXRF (ATOMIKA XSA-8000) at Charles Evans & Associates and at GeMeTec (Munich, Germany), and RBS (N-beam) at GeMeTec. Groups of four wafers were each intentionally contaminated by Kyushu Electronic Metals Co. with Fe, Ni, Cu and Zn at target levels of 10^{11}, 10^{12}, and 10^{13} atoms/cm^2 per wafer. The contamination process was previously characterized by VPD/AAS. Vapor phase HF decomposition was performed on the native oxide. A blank (control) was also provided with VPD/AAS impurity measurements for Fe, Ni, Cu and Zn of 1.0, <1.0, <0.23, and 0.11×10^{10} atoms/cm^2 respectively, which represented the starting impurity levels for the intentionally contaminated wafers. The correlation between RBS, TXRF and VPD/AAS shows a systematic difference between VPD/AAS and RBS, with RBS reading higher than VPD/AAS. However, RBS reads within a factor of 3x of VPD/AAS.

5.4 Angle Properties

It has been shown both experimentally (45), as indicated in Fig. 18, and theoretically (43) that the change in the TXRF signal as a function of the incident angle is different for plated contaminants (atomic copper) and for residue contaminants. This difference can be used to qualitatively distinguish between particulate and plated contamination in the same series of measurements.

We present here examples of different angle scans (49). TXRF angle variation of the Cu signal of a highly contaminated silicon wafer was measured for several wafers. The results for one wafer are shown in Fig. 20. The graph reveals that the Cu is plated rather than occurring in a particulate or residue form.

The TXRF angle variation shown in Fig. 21 indicates that the Fe and Zn contamination was particulate rather than a heterogeneous plating. In contrast, the TXRF angle variation shown in Fig. 22 of a bare substrate cleaned by an RCA SC-1 type chemistry shows plated Fe contamination, as expected.

The TXRF angle variation for a dry etched and cleaned sample is shown in Fig. 23 and reveals the heavy metal contamination to be neither particulate, nor plated, but imbedded in the oxide surface. This finding is consistent with a model of a dry etcher driving heavy metals from the resist into the oxide surface where the metals are difficult to remove by cleaning.

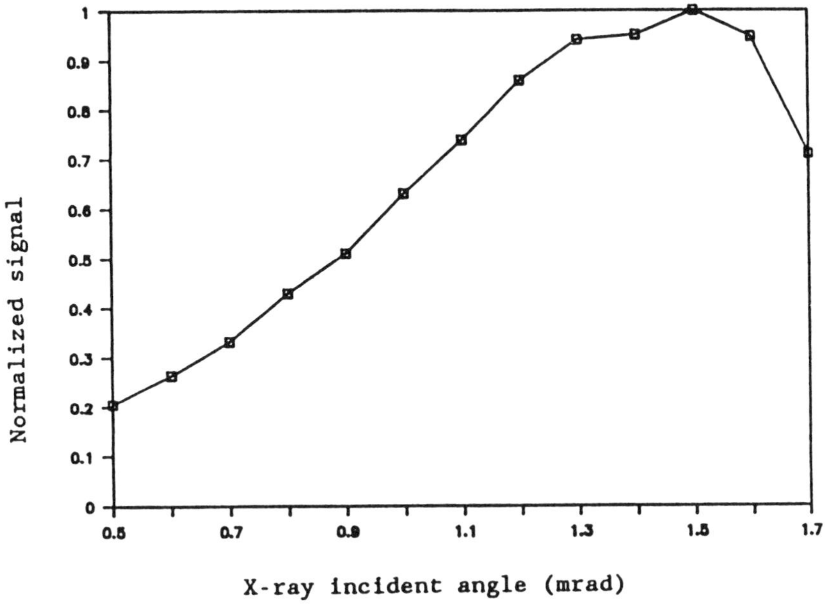

Figure 20. Polychromatic (ATOMIKA XSA-8000 TXRF instrument) TXRF angle-scan of plated Cu on silicon substrate (49).

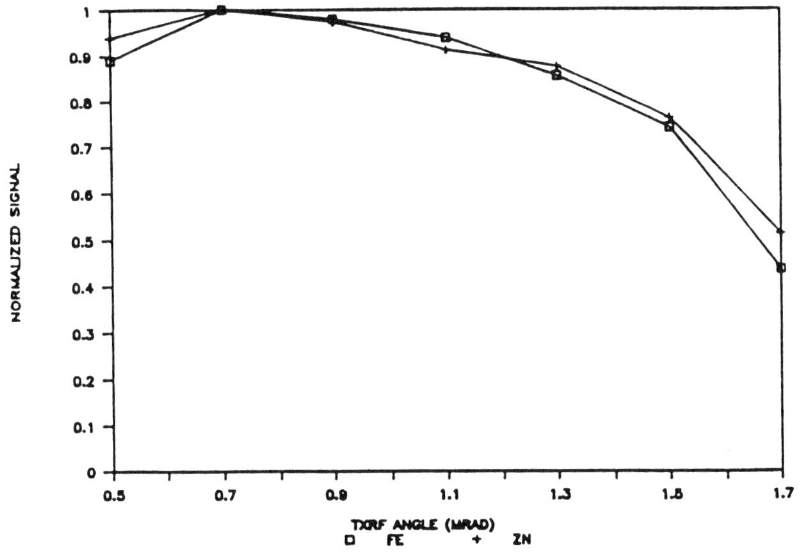

Figure 21. Polychromatic (ATOMIKA XSA-8000 TXRF instrument) TXRF angle-scan of particulate Fe and Zn on silicon substrate (49).

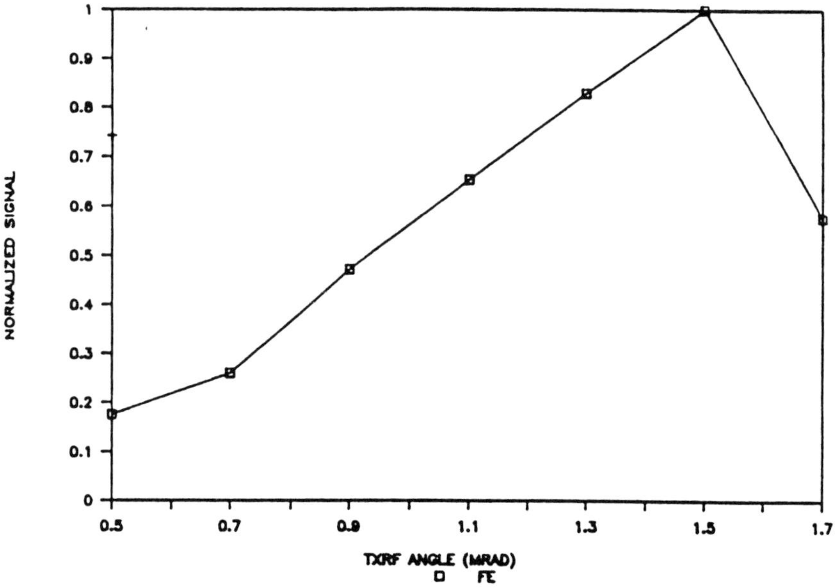

Figure 22. Polychromatic (ATOMIKA XSA-8000 TXRF instrument) TXRF angle-scan of plated Fe on silicon substrate after SC-1 cleaning (49).

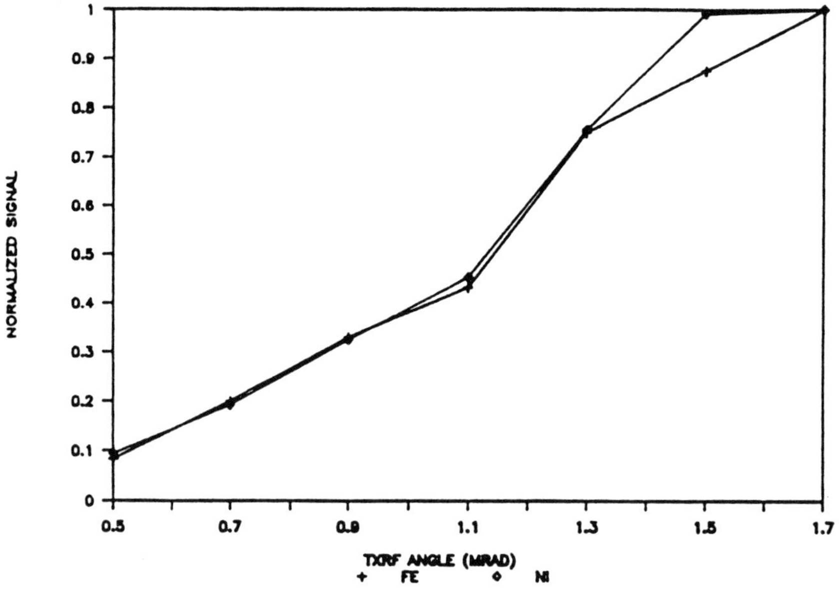

Figure 23. Polychromatic (ATOMIKA XSA-8000 TXRF instrument) TXRF angle-scan of embedded metals on silicon substrate after ashing (49).

Rapid thermal processing equipment has been found to have cross contamination even when the equipment is HF-cleaned quartz. TXRF reveals tungsten cross contamination (from tungsten silicide annealing) onto the surface of silicon wafers. The transfer mechanism uses oxygen and has an activation energy of about 0.8 eV. The TXRF angle variation for a sample processed at 1000°C for 120 seconds in air is shown in Fig. 24. The data demonstrate that the contamination is not particulate, but is not strictly plated either. The increase in signal at the higher angles suggests some of the W diffused into the top few monolayers of either the silicon (unlikely) or an oxide which may have grown during the processing. It may also be that W was incorporated in the growing oxide, rather than diffused.

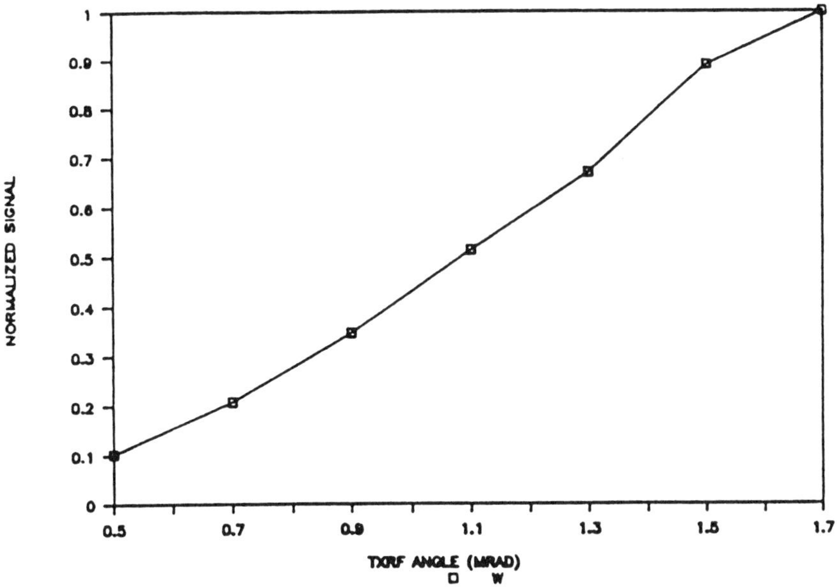

Figure 24. Polychromatic (ATOMIKA XSA-8000 TXRF instrument) TXRF angle-scan of tungsten cross contamination on a silicon substrate in an RTP unit (49).

5.5 Monochromatic TXRF

Schuster (50) described the physics of improving TXRF by using a monochromator between the x-ray source and the sample, which is the key hardware difference between the polychromatic and monochromatic TXRF. Monochromators have included LiF (200), graphite, and multilayer crystals. The net effect is to greatly reduce the background signals and thus improve the detection limits. For example, Fig. 25 shows a comparison between polychromatic and monochromatic TXRF for the detection of Fe. There are presently three monochromatic TXRF instruments manufactured: TECHNOS (51), ATOMIKA (44) and Rigaku (52). The TECHNOS and Rigaku instruments use rotating anodes with tungsten targets, and the monochromator selects the W L-beta line (9.67 keV); the measurements are taken in vacuum. The ATOMIKA instrument uses a Mo or Cu tube, and the monochromator selects the respective k-alpha line; the measurements are made in helium gas. In 1991, all of these instrument had detection limits for surface plated Ni in the low 10^{10} atoms/cm^2, however, this is a rapidly changing technology and hardware improvements are expected to reduce the detection limits even further. Most published data from these second generation instruments are for the TECHNOS TREX 610 TXRF, and it is used here to illustrate monochromatic TXRF.

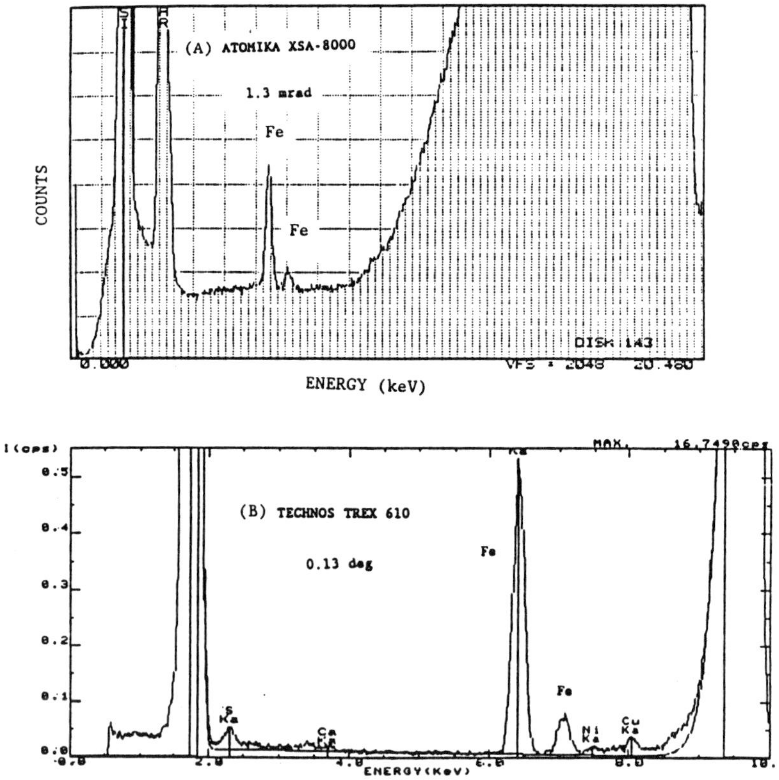

Figure 25. Comparison of polychromatic (ATOMIKA XSA-8000 TXRF instrument) and monochromatic (TECHNOS TREX 610 TXRF instrument) spectra for a silicon substrate cleaned in an SC-1 bath.

Basic Configuration. A schematic of the basic configuration of the TREX 610 is presented in Fig. 26. The rotating anode is a 9-kW x-ray source. The monochromator selects the W L-beta line at 9.67 keV in order to include the detectability of surface Zn. The detector is an 80-mm^2 area Si(Li) energy dispersive spectrometer (EDS). The analysis area is 10 mm in diameter. The instrument uses a scintillation counter to determine the zero angle plane and then tilts the sample to the desired angle. The detection limits are in the low 10^{10} atoms/cm^2 for Fe, Ni, and Cu.

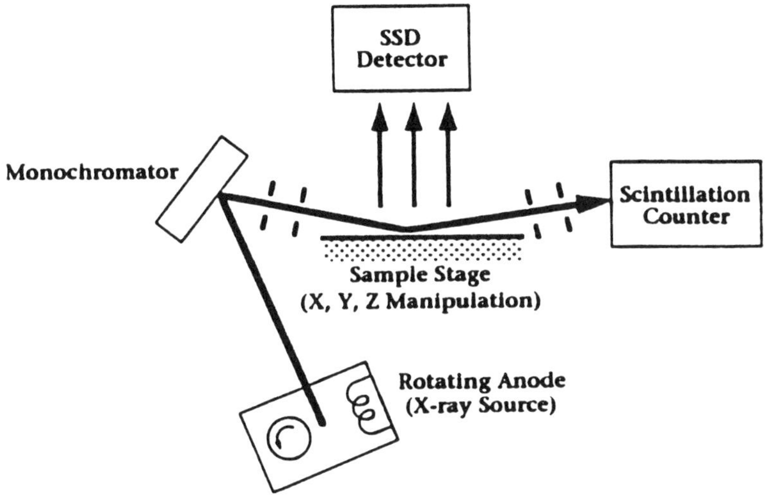

Figure 26. Schematic of the TECHNOS TREX 610 TXRF instrument (51).

5.6 Characterization of Cleaning Processes by TXRF

Using polychromatic TXRF, Hockett and Katz showed (Table 6) that commercial silicon wafers in 1987 - 88 could have high levels of Fe, Cu, and Zn on the order of 10^{12} atoms/cm^2 (53). By 1990 - 1991 the surface of commercial silicon wafers had become much cleaner, to the point where polychromatic TXRF could no longer detect the surface metals. A typical monochromatic TXRF spectrum for a commercial silicon wafer in 1990 is shown in Fig. 27 (7). Here we see a plot of fluorescence intensity (I) in counts per second (cps) versus fluorescence energy (in keV) from 0 to 10 keV. At 1.74 keV there is a large Si peak from the silicon native oxide and substrate, and at 9.7 keV there is another large peak from the diffracted W. In between peaks for S, Fe, Ni, Cu and Zn. The table above the plot lists several elements, the energy of the element peaks, the net integrated intensity in cps, and the areal density of the elements in units of 10^{10} atoms/cm^2. The transition metals are in the mid to high 10^{10} atoms/cm^2. The sulfur is much higher in areal density and is expected to originate from atmospheric $(SO_4)^{-2}$ contamination in clean rooms or from cassettes (54)-(56). This spectrum was taken at 30 kV, 200 mA, in vacuum for 1000 sec. The Fe level for this commercial wafer is consistent with results reported by Zoth and Bergholz who used other techniques (6). Piranha solution cleaning leaves higher levels of sulfur on wafers, depending on the rinse, and is expected to be bound differently than atmospheric sulfur. A TREX 610 spectrum for a silicon wafer cleaned in piranha is shown in Fig. 28, confirming the expected high level of sulfur. A comparison of contamination left on silicon wafers after some commonly used cleans, as analyzed by a TREX 610, is shown in Table 7. The largest differences are in the high sulfur level after H_2SO_4/H_2O_2 and the high Fe and Zn concentrations after SC-1.

Table 6. TXRF Comparison of Cleaning Processes (53)

	(units of 10^{12} atoms/cm^2)				
Cassette	Fe	Cu	Zn	Ca	Br
A	0.5	0.5	2.5	<2	0.3
B	0.6	60	0.5	<2	<0.2
C	0.4	0.8	1.5	3.5	0.2
D	<0.3	0.5	0.5	<2	2

Shimanuki and Ryuta (57) used the TREX 610 TXRF measurement to show the correlation between the concentration of transition metals in the SC-1 solution and the metal areal densities on the silicon wafers cleaned in the SC-1 solution. Fig. 29 shows the log-log correlation plot. In addition they evaluated the time dependence of the Fe and Zn adsorption from the SC-1 solution onto the silicon wafer surface, as shown in Fig. 30.

No.	Element/Line	Energy(keV)	Inten.cps	Conc(xE10atoms/cm2)
1	Si -Kα	1.740	151.7708	82062.10
2	S -Kα	2.308	0.6374	159.06
3	Fe -Kα	6.400	0.1313	6.00
4	Ni -Kα	7.473	0.2348	8.86
5	Cu -Kα	8.042	0.2294	8.00
6	Zn -Kα	8.632	0.3118	10.12

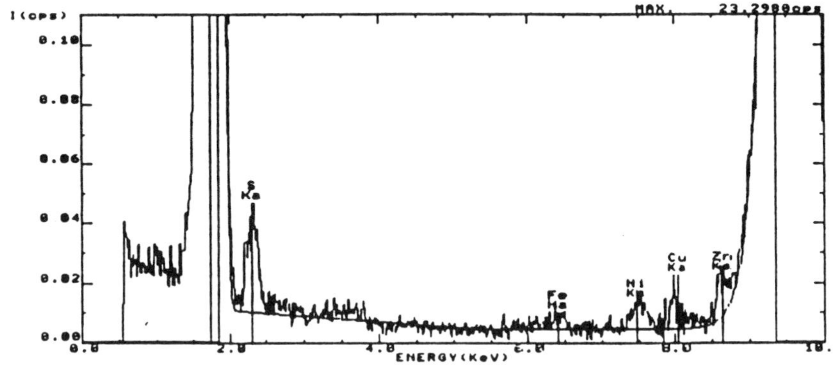

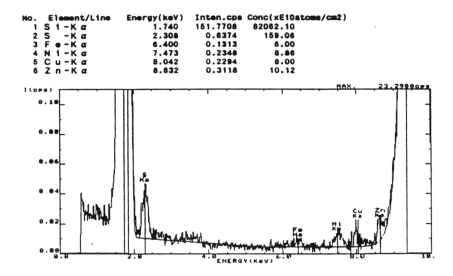

Figure 27. Typical TREX 610 spectrum of a clean commercial silicon substrate in 1991 (7).

No.	Element/Line	Energy(keV)	Inten.cps	Conc(xE10atoms/cm2)
1	Si -Ka	1.740	139.2765	75306.47
2	S -Ka	2.308	1.5036	375.24
3	Fe -Ka	6.400	0.0622	2.84
4	Ni -Ka	7.473	0.1321	4.98
5	Cu -Ka	8.042	0.1230	4.29
6	Zn -Ka	8.632	0.2153	6.98

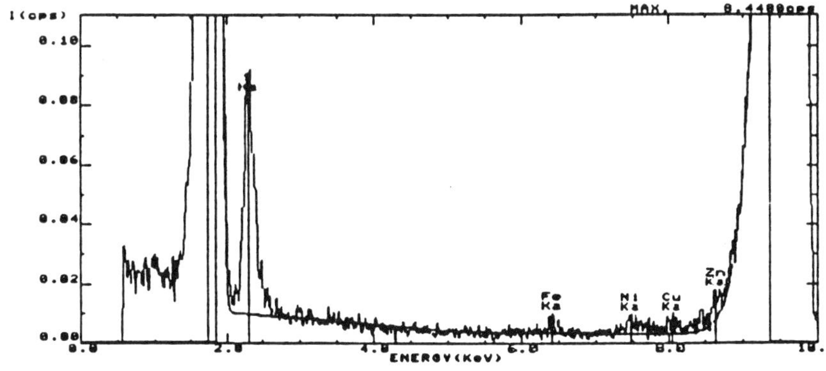

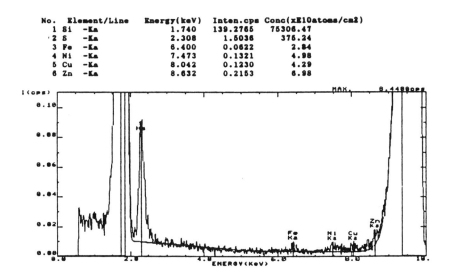

Figure 28. Typical TREX 610 spectrum of a silicon substrate after a piranha bath (H_2SO_4/H_2O_2) clean (7).

Table 7. TREX 610 Comparison of Cleaning Processes

Solution	(units of 10^{10} atoms/cm^2) S	Fe	Ni	Cu	Zn
H_2SO_4/H_2O_2	4000	6	8	10	<5
HF	500	<3	8	10	<5
SC-1	1000	100	5	7	30 - 60
SC-2	1000	7	9	10	5

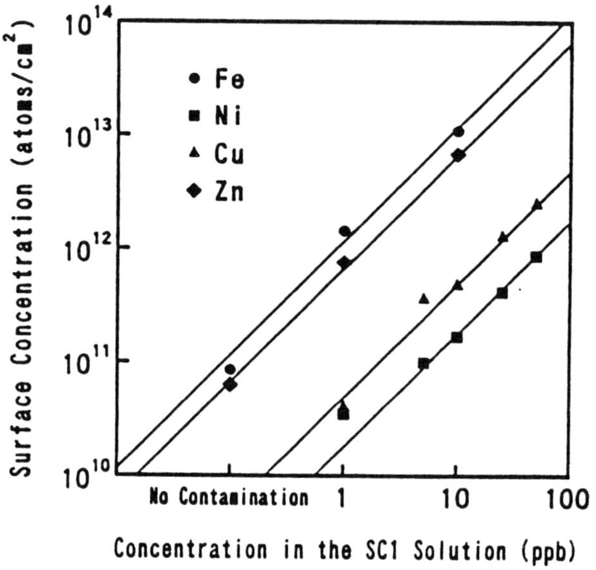

Figure 29. TREX 610 TXRF results showing the effect impurity concentration in the intentionally contaminated SC-1 solution on the surface concentration (57).

Although surface S and Zn are expected to desorb during some thermal processes, ion implantation and other damage-producing processes may result in subsurface S and Zn. By increasing the glancing angle of the x-ray source, the depth of analysis can be increased. Spectra were taken for different glancing angles of an ion implanted silicon wafer. Because the S, Cu and Zn signals increase so much at 0.17 degree as compared to those at 0.13 degree, we conclude the S, Cu, and Zn are subsurface, while the Fe and Ni are on the top surface.

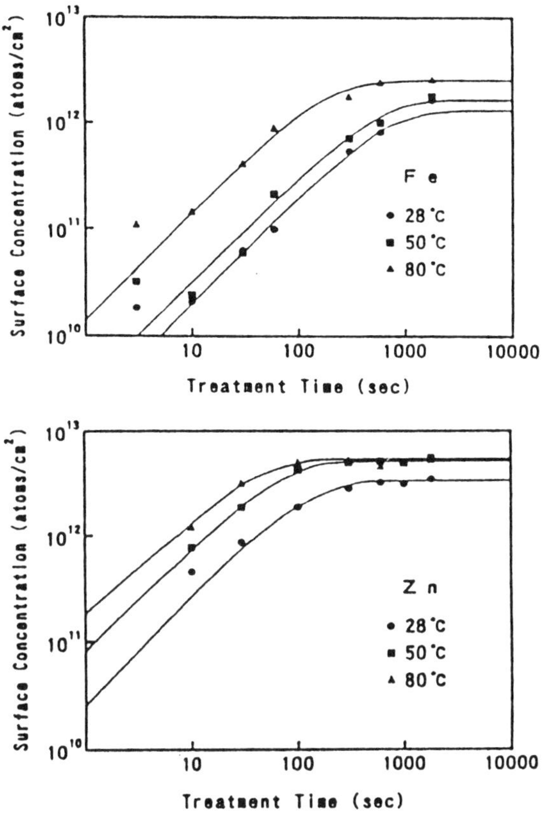

Figure 30. TREX 610 TXRF result showing the variation with treating time of surface concentration of Fe and Zn on initially clean wafers (57).

An example of defect introduction showed up as a radial distribution of surface stacking faults in IC devices, resulting in yield loss. The density of stacking faults was higher in the wafer center, and the cleaning process before the p-well implant was suspected. The TXRF map of the surface contamination detected on a 125-mm diameter silicon wafer after the cleaning process is shown in Fig. 31. The TECHNOS TREX 610 was used in the automatic mode in vacuum under analytical conditions of 30 kV, 200 mA, 400 seconds per analysis point, 97 analysis points, and at a glancing angle of 0.09 degrees. The map shows a radial distribution of Ni, Cu and Zn contamination, whereas the Fe contamination is random and the S and Cl contaminants are distributed more or less uniformly. The Si matrix signal is also constant as it should be. The Ni and Cu levels are in the range of 10^{11}

atoms/cm^2 with the center Ni being near 10^{12} atoms/cm^2; the Zn, S and Cl levels are in the range of 10^{12} atoms/cm^2. Figure 31 indicates that Ni does not correlate with Fe; therefore the Ni source is not expected to be stainless steel. The Cu, Zn, and Ni may have been plated out on the silicon from contaminated cleaning chemicals. An examination of the chemicals as possible sources of contamination and their replacement led indeed to the successful elimination of the defects.

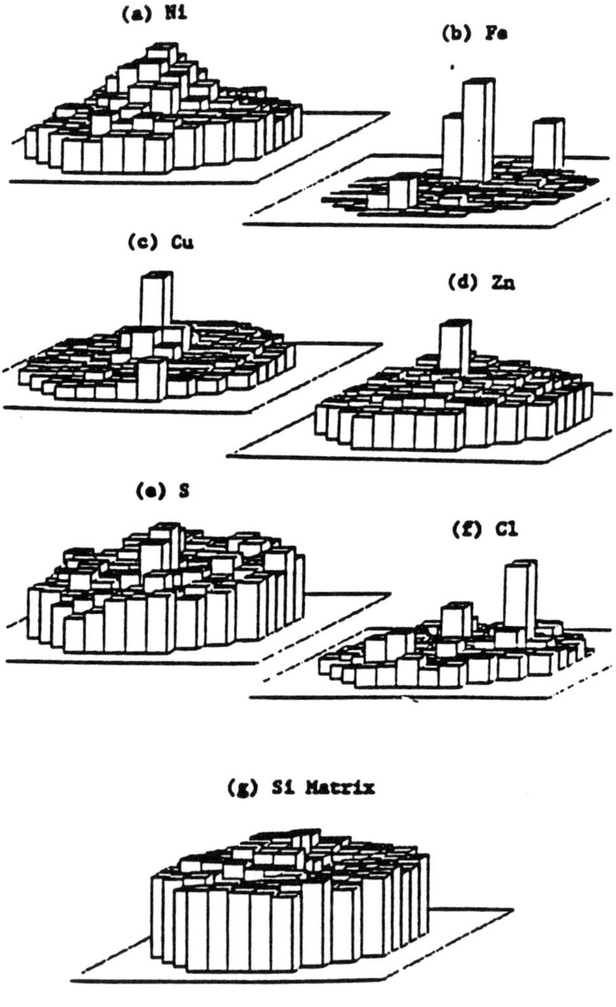

Figure 31. TREX 610 histogram maps of surface contamination on a silicon substrate after cleaning and before implant and anneal. The radial distribution of Ni is suspected of causing a radial distribution of surface stacking faults after anneal (7).

Controlled experiments by Matsushita and Tsuchiya using surface Ni contamination at this level have shown that surface stacking faults can be caused by Ni after p-well implant and subsequent thermal processing (2). They showed, as noted earlier, that surface Fe with implant damage produces a much higher level of OSF after thermal treatment than surface Fe without implant damage. Furthermore, Ni with implant damage introduced OSF in the $100/cm^2$-range at about 10^{12} atoms/cm^2, i.e., at the same Ni level we found at the center of the wafer described in the preceding paragraph where we propose the interaction between the surface Ni and the implant damage to be the cause of the radial pattern of surface stacking faults.

Results on cleaning and dry etching by various IC processing materials are summarized in Table 8 (58). The standard wafer was contaminated with the transition metals Mn, Fe, Ni, Cu and Zn on the order of 10^{11} atoms/cm^2. The Mn on the standard wafer is unusual, but the other levels are typical. The Caros cleaning not only results in a cleaner wafer, but does not deposit Cu and Ti if added to the solution. Cleaning after TEOS etching removes some contaminants from the silicon surface but introduces more contamination. The photo strip etch and clean results in a fairly clean surface. The polysilicon etch from the Model A 9500 system introduces Ni, Fe and Cu, and the clean does not remove this contamination, but introduces Zn and very high levels of Cl. The polysilicon etch from Model B 6000 system introduces more Ni, Cu, Mn and Fe; however, cleaning has some effect in reducing the contamination, but again introduces high Cl concentrations from the cleaning chemicals.

6.0 FUTURE ANALYTICAL TECHNOLOGY

The analytical technology for measuring trace surface metals is rapidly changing. As advanced ICs become more complex and more sensitive, the importance of even lower levels of metal contaminants increases; this contributes to a driving force to achieve lower detection limits. As soon as the contaminants can be detected and measured, the process engineer can now design experiments to find and eliminate the source until the metals can no longer be detected. Then there is a renewed drive to develop better detection limits. In other words, this analytical technology continues to become obsolete at a rapid pace by having nothing left to detect.

TABLE 8. TREX 610 TXRF Results (58).

Wafer Treatment	Cl	Mn	Fe	Ni	Cu	Zn	Ar
	(units of 10^{10} atoms/cm^2)						
1. Standard wafer	150	7	10	8	55		
2. CAROS bath contaminated with AlSiCu,TiW)				30			
3. TEOS etch	400		6	7	10		150
4. TEOS etch + clean	350		7		40	20	150
5. Photo strip			15	6	5	60	
6. Photo strip + clean			7	5	6	15	
7. Poly etch (Model A, Cl_2/He)	9500		9	20	70	40	
8. Poly etch (Model A, Cl_2/He) + clean	8500		6	10	60	40	
9. Poly etch (Model B, HBr/Cl_2/He)	6000	25	55	100	150	50	
10. Poly etch (Model B, HBr/Cl_2/He) + clean	3000	15	15	30	40		

Some of the near-term analytical technology is discussed in this section. One class of improvements relates to vapor phase decomposition (VPD) chemistry and post-VPD methods. Once the VPD pre-concentration has been completed, there are several options besides AAS for analysis of the liquid residue: ICP-MS, TXRF, and SIMS. To make these successful, however, VPD collection efficiencies need to be understood and improved, as discussed below. Apart from VPD related technology, there is the option of a new kind of SIMS, called TOF-SIMS, which we discuss also.

6.1 VPD Chemistries

The original VPD chemistry was based on vapor phase HF that was shown to have high collection efficiencies for Na, Al, Fe, and Ni. However, the collection efficiency of surface Cu was poor. In order to improve it, alternative vapor phase chemistries have been studied (1)(24). In addition, there can be an effect of surface concentration of the metals on the collection efficiency. The scanning speed and temperature for collection can also affect the results.

One of the conceptual difficulties in VPD-related techniques is whether some percentage of a metal is chemically bound at the surface, so that in a cleaning evaluation the VPD collection is 100% for only one form of the metal. One reason for this concern is that collection efficiency studies have been done on spin-coated samples where the chemistry of bonding is expected to be of only one type. For example, in an actual sample Fe can be present partially subsurface and then may not be collected by the HF VPD technique even though Fe collection on spin coated samples by HF VPD has been demonstrated to be near 100%.

In the future, reports of VPD results should indicate the type of chemistry used in the collection. The exact analytical conditions should be reported as well, since they also affect the collection efficiency.

6.2 VPD ICP/MS

Once the VPD collection has been completed, the liquid sample can be analyzed by methods other than AAS, for example, ICP/MS (Inductively Coupled Plasma/Mass Spectrometry). An injection attachment for small volume of sample solution is required, and there are reports (24) of using an ETV (Electro-Thermal Vaporization) attachment for this purpose. The advantage of VPD ICP/MS over VPD/AAS is multi-element capability and improved detection limits (10^7 atoms/cm^2) for 200 and 300 mm diameter wafers.

6.3 VPD/TXRF

Another alternative for a post-VPD analysis of the sample solution is TXRF, where the solution is allowed to dry first. This approach was first introduced by Huber et al. (59), with further details given later by Neumann and Eichinger (60). The advantage over AAS is again its multi-element capability and improved detection limits for some elements. The method is also simpler to quantify than normal TXRF (60). The theoretical improvement in detection limits is a simple scaling due to the preconcentration. For example, the improvement in detection limits using the TREX 610 is shown in Fig. 32. In practice there is some loss in detection capability from the incident x-rays scattering from the dried residue, particularly if the VPD chemistry results in high silica content. Figure 33 shows the effect of an H_2SO_4/H_2O_2 + HF cleaning on as-received silicon substrates (60).

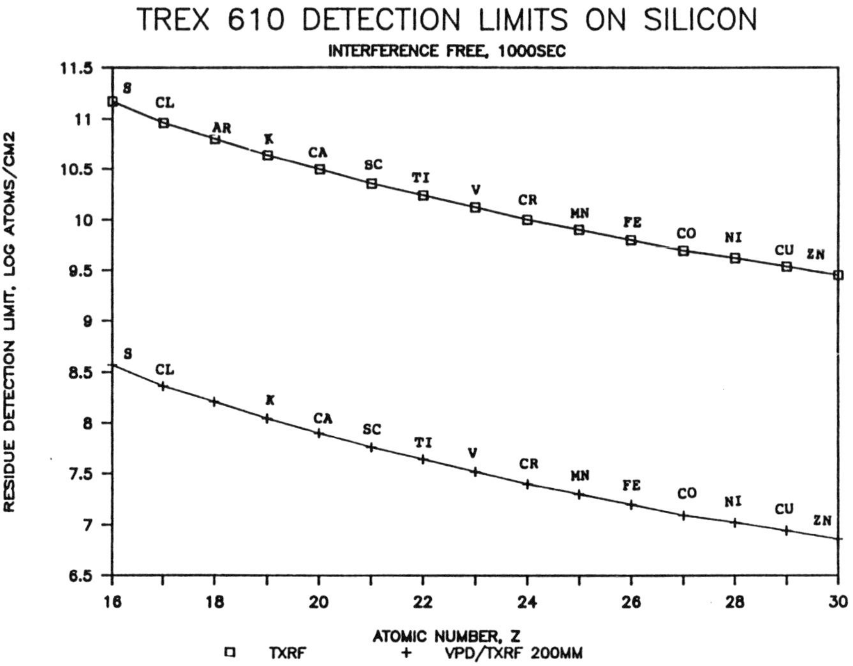

Figure 32. VPD/TXRF detection limits for monochromatic TXRF (TECHNOS TREX 610).

6.4 VPD/SIMS

Another alternative for a post-VPD analysis of the sample solution is SIMS. This approach was first introduced by Chia and Hockett (61) for surface Al. In this method the VPD solution is spiked with a known amount of Y and shrunk by 100x. Then a 20-nl micro-sample of the solution is transferred to a surface and allowed to dry. A dynamic SIMS (CAMECA IMS-3f or -4f) profile is then completed for the Al in the solution residue; quantification of the Al is done using the Y signal. This method appears capable of reaching down to 10^8 atoms/cm^2 for surface Al on silicon wafers. In principle, this method can be used for any other elements which do not desorb during the drying process.

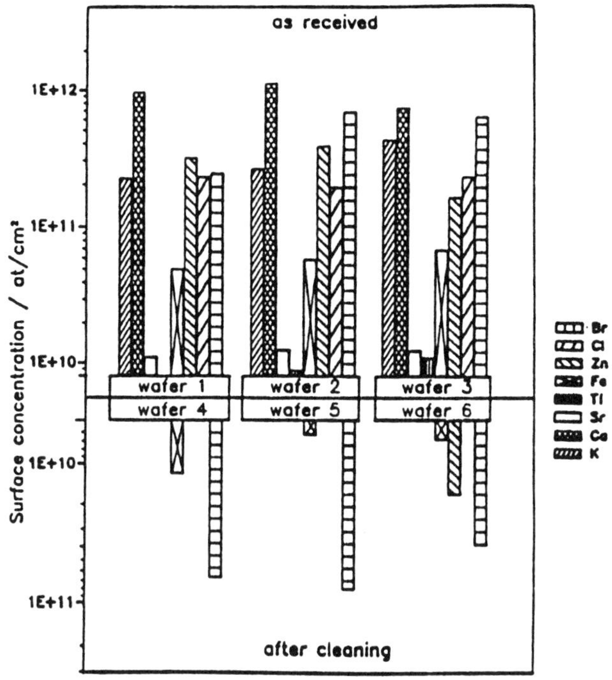

Figure 33. VPD/TXRF (polychromatic TXRF) characterization of H_2SO_4/H_2O_2 + HF cleaning (60).

6.5 TOF-SIMS

A future technique that can perform the analysis directly on the silicon surface without the VPD pre-concentration is TOF-SIMS (Time of Flight SIMS) (62). In TOF-SIMS all ions sputtered from the surface are collected in the time of flight spectrometer, i.e., there is no loss of ions from switching the mass spectrometer from one element to another. The technique also is multi-element and the high mass resolution capability does not result in ion losses, as in conventional SIMS. With a useful yield of 10^{-3} (ratio of the number of ion counts detected to the number of total atoms of the element), it should be possible to reach detection limits of less than 10^8 atoms/cm^2 for an analysis area of 10^{-4} cm^2. Table 9 shows a comparison of TOF-SIMS and TXRF (63). It is also possible to do depth profiling through the native oxide so it may be possible to determine the depth distributions of metals at low depth (64); no other technique has been able to achieve this.

Table 9. Comparison of TXRF and SIMS (64)

	Wafer 1 (As-Received)		Wafer 2 (SIRTL)		Wafer 3 (SC-1)	
Element	TOF-SIMS*	TXRF	TOF-SIMS*	TXRF	TOF-SIMS*	TXRF
Cu	3.3×10^{12}	3.8×10^{12}	--	--	--	--
Cr	--	--	1.5×10^{15}	1.0×10^{15}	--	--
Fe	1.3×10^{12}	0.5×10^{12}	--	--	1.9×10^{12}	1.0×10^{12}
Zn**	--	4.0×10^{11}	--	--	--	4.5×10^{12}
B	2.1×10^{12}	UD	1.8×10^{12}	UD	2.6×10^{12}	UD
Mg	2.2×10^{11}	UD	--	UD	1.3×10^{11}	UD
Al		UD	--	UD	3.8×10^{13}	UD
Ca	6.4×10^{11}	--	2.1×10^{12}	--	9.4×10^{11}	--

UD = Undetectable
-- = Not detected
* TOF-SIMS quantification assumed unit relative ion yields
** Zn has a very poor positive ion yield

7.0 CONCLUSIONS

Present cleaning processes can be characterized for surface metal contamination using SIMS, TXRF and VPD/AAS. The analytical technology for this type of measurement is rapidly improving in detection limits, so that future cleaning processes are expected to be characterized for surface metal contamination by alternate VPD chemistries followed by different analyses, such as, ICP/MS, TXRF, SIMS, and by TOF-SIMS which will require no VPD pre-concentration. The combination of the VPD/ technologies and the TOF-SIMS, particularly for silicon surfaces after VPD removal of some metals, is expected to provide new information about the chemistry of surface metals as well.

REFERENCES

1. Tsuji, T. S., Morita, M. and Muramatu, Y., *Microcontamination '91 Conference Proceedings,* Canon Communications, Inc., Santa Monica, CA, pp. 544-551 (1991)

2. Matsushita, Y. and Tsuchiya, N., *Automated Integrated Circuits Manufacturing VI, ECS Proc.* 91-5:119-131 (V. E. Akins and H. Harada, eds.), Electrochem. Soc., Pennington, NJ (1991)

3. Hourai, M., Naridomi, T., Oka, Y., Murakami, K., Sumita, S., Fujino, N. and Shiraiwa, T., *Jpn. J. Appl. Phys.* 27, L2361-L2363 (1988)
4. Shiraiwa, T., Fujino, N., Sumita, S. and Tanizoe, Y., *Semiconductor Fabrication: Technology and Metrology,* ASTM STP 990, (D. C. Gupta, ed.), pp. 314-323, American Society for Testing and Materials (1989)
5. Abe, T., Itoh, T., Hayamizu, Y., Sunagawa, K., Yokata, S. and Yamashishi, H., *Defect Control in Semiconductors,* (K. Sumino, ed.), pp. 297-303, Elsevier Science Publishers B. V., North Holland (1990)
6. Zoth, G. and Bergholz, W., *J. Appl. Phys.* 67:6764-6771 (1990)
7. Hockett, R. S., *Defects in Silicon II, ECS Proceedings,* (W. M. Bullis, U. Gosele, and F. Shimura, eds.), 91-9:57-64, Electrochem. Soc., Pennington, NJ (1991)
8. Riggs, W. M. and Parker, M. J., *Methods of Surface Analysis,* (A. W. Czanderna, ed.), pp. 103-158, Elsevier Scientific Publishing Company, NY (1975)
9. Wong, C. Y. and Klepner, S. P., *Appl. Phys. Lett.* 48:1129-1230 (1986)
10. Weinberger, B. R., Peterson, G. G., Eschrich, T. C. and Krasinski, H. A., *J. Appl. Phys.* 60:3232-3234 (1986)
11. Zazzera, L. A. and Moulder, J. F., *J. Electrochem. Soc.* 136:484-491 (1989)
12. Joshi, A., Davis, L. E. and Palmberg, P. W., *Methods of Surface Analysis,* (A. W. Czanderna, ed.), pp. 159-223, Elsevier Scientific Publishing Company, NY (1975)
13. Kniffin, M. L. and Helms, C. R., *ECS Ext. Abstr. #526,* 91-2:785 (1991)
14. Falster, R. J., Wingrove, R. D., Hockett, R. S., Craven, R. A. and Golland, D. I., "The Total System Approach to Wafer Ecology," SEMICON/Europe '86, Zurich (March 1986)
15. Chu, W.-K., Mayer, J. W. and Nicolet, M. A., *Backscattering Spectrometry,* Academic Press, NY (1978)
16. Eichinger, P., *Nucl. Instr. and Meth.* A253:313-318 (1987)
17. Odom, R. W., Hitzman, C. J., and Schueler, B. W., *Mat. Res. Soc. Symp. Proc.* 69:265-271 (1986)
18. Gilfrich, J. V., *Characterization of Solids,* (P. F. Kane and G. B. Larrabee, eds.), pp. 275-306, Plenum Press, NY (1974)
19. Kasi, S. R. and Liehr, M., *Appl. Phys. Lett.* 57:2095-2097 (1990)
20. Graf, D., Grundner, M., Muhlhoff, L. and Delith, M., *J. Appl. Phys.* 69:7620-7626 (1991)

21. Burrows, V. A., Chabal, Y. J., Higashi, G. S., Ragahvachari, K. and Christman, S. B., *Appl. Phys. Lett.* 53:998-1000 (1988)
22. Shimazaki, A., Hiratsuka, H., Matsushita, Y. and Yoshii, S., *Ext. Abstr. of the 16th (1984 International) Conf. on Solid State Devices and Materials,* Kobe, pp. 281-284, Jpn. Soc. of Appl. Phys., Tokyo (1984)
23. Shimazaki, A. and Mashimo, N., *SEMICON/Japan Proc.,* Tokyo, pp. 285-293 (Nov. 14, 1989)
24. Shimazaki, A., *Defects in Silicon II, ECS Proc.* (W. M. Bullis, U. Gosele, and F. Shimura, eds.) 91-9:47-56, Electrochem. Soc., Pennington, NJ (1991)
25. Private Communication, S. Sumita, Kyushu Electronic Materials Co., Saga, Japan
26. Takizawa, R., Nakanishi, T., Honda, K., and Oshawa, A., *Jpn. J. Appl. Phys.* 27:L2210-L2212 (1988)
27. Benninghoven, A., Rudenauer, F. G. and Werner, H. W., *Secondary Ion Mass Spectrometry,* John Wiley & Sons, NY (1987)
28. Schwenke, H. and Knoth, J., *Handbook on X-Ray Spectrometry,* (R. Van Grieken and M. Markowitz, eds.), Marcel Dekker, NY (in print 1992)
29. Phillips, B. F., Burkman, D. C., Schmidt, W. R. and Peterson, C. A., *J. Vac. Sci. Technol.* A1:646-649 (1983)
30. Krusell, W. C., Farber, J. J. and Sing, A. L., *ECS Ext. Abstr.* 86-1:331, Electrochem. Soc., Pennington, NJ, (1986)
31. deLarios, J. M., Kao, D. B., Helms, C. R. and Deal, B. E., *Appl. Phys. Lett.* 54:715-717 (1989)
32. Wilson, R. G., Stevie, F. A. and Magee, C. W., *Secondary Ion Mass Spectrometry: A Practical Handbook for Depth Profiling and Bulk Impurity Analysis,* John Wiley & Sons, NY (1989)
33. Stingeder, G., Grundner, M. and Grasserbauer, M., *Surface and Interface Analysis,* 11:407-413 (1988)
34. Sumita, S., Horie, H., Tanizoe, Y., Takeshita, M., Fujino, N. and Shiraiwa, T., *Ext. Abstr., 12th International Congress on X-ray Optics and Microanalysis* (1989)
35. Fujino, N., Horie, H., Hiramoto, K., Tanizoe, Y., Sumita, S., and Shiraiwa, T., *SIMS VII,* (A. Benninghoven, C. A. Evans, K. D. McKeegan, H. A. Storms, and H. W. Werner, eds.), pp. 527-530, John Wiley & Sons, NY (1990)
36. Williams, P. and Baker, J. E., *Appl. Phys. Lett.* 36:842-844 (1980)

37. Slusser, G. J. and MacDowell, L., *J. Vac. Sci. Technol.* A5:1649-1651 (1987)

38. Hockett, R. S., *Mat. Res. Symp. Proc.,* (S. R. Wilson, R. Powell, and D. E. Davies, eds), 92:41-45, Materials Research Society, Pittsburg (1987)

39. Hockett, R. S. and Norberg, J. C., *SIMS VII,* (A. Benninghoven, C. A. Evans, K. D. McKeegan, H. A. Storms, and H. W. Werner, eds.), 491-494, John Wiley & Sons, NY (1990)

40. Hockett, R. S., *Semiconductor Cleaning Technology/1989, ECS Proc.* (J. Ruzyllo and R. E. Novak, eds.), 90-9:227-242, Electrochem. Soc., Pennington, NJ (1990)

41. Schwenke, H., Berneike, W., Knoth, J. and Weisbrod, U., *Advances in X-Ray Analysis,* (C. S. Barrett, J. V. Gilfrich, R. Jenkins, T. C. Huang, and P. K. Predecki, eds.), 32:105-113, Plenum Publishing (1989)

42. Compton, A. H., *Phil. Mag.* 45:1121 (1923)

43. Berneike, W., Knoth, J., Schwenke, H. and Weisbrod, U., *Fresnius Z. Anal. Chem.* 333:524-526 (1989)

44. Weisbrod, U., Gutschke, R., Knoth, J. and Schwenke, H., *Fresnius Z. Anal. Chem.* 341:83-86 (1991)

45. Eichinger, P., Rath, H. J. and Schwenke, H., *Semiconductor Fabrication: Technology and Metrology,* ASTM STP 990, (D. C. Gupta, ed.), pp. 305-313, American Society for Testing and Materials (1989)

46. Private communication, K. Nishihagi, Technos Corporation, Osaka, Japan

47. de Boer, D. K. G., *Spectrochemica Acta,* 46B:1433-1436 (1991)

48. Private communication, P. Eichinger, GeMeTec, Munich, Germany

49. Hockett, R. S., *Proc. of Advanced Semiconductor Processing: Ultra Clean Processing Environments,* pp. 115-124, SEMICON/East, Boston, MA (1989)

50. Schuster, M., *Spectrochimica Acta,* 46B:1341-1349 (1991)

51. Nishihagi, K., Kawabata, A., Taniguchi, T. and Ikeda, S., *Semiconductor Cleaning Technology/1989, ECS Proc.,* 90-9: 243-250, (J. Ruzyllo and R. E. Novak, eds.), Electrochem. Soc., Pennington, NJ (1990)

52. Rigaku Sales Literature.

53. Hockett, R. S. and Katz, W., *J. Electrochem. Soc.* 136:3481-3486 (1989)

54. Shimizu, H., Honma, N., Munakata, C. and Ota, M., *Jpn. J. Appl. Phys.* 27:1454-1457 (1988)

55. Oki, T., Biwa, T., Kudo, J. and Ashida, T., *ECS Ext. Abstr.,* 91-2:790 (1991)
56. Rathmann, D., *ECS Ext. Abstr.,* 91-2:816 (1991)
57. Shimaunki, Y. and Ryuta, J., *Abstr. Symp. on Advanced Science and Technology of Materials,* pp. 293-298, Jpn. Soc. for the Promotion of Science, The 145th Committee, Kona, Hawaii (November 25-29, 1991)
58. Hockett, R. S. and Metz, J., *Ibid.* pp. 287-292.
59. Huber, A., Rath, H. J., Eichinger, P., Bauer, T., Kotz, L. and Staudigl, R., *Diagnostic Techniques for Semiconductor Materials and Devices, ECS Proc.,* (T. J. Shaffner and D. I. Schroder, eds.), 88-20:109 (1988)
60. Neumann, C. and Eichinger, P., *Spectrochemica Acta,* 46B:1369-1377 (1991)
61. Chia, V. K. F. and Hockett, R. S., *RNP Abstract, Recent News Papers,* Electrochem. Soc. Mtg., Phoenix, AZ (October 13-18, 1991)
62. Schueler, B., Sander, P. and Reed, D. A., *Vacuum* 41:1661-1664 (1990)
63. Schueler, B., Sander, P. and Reed, D. A., *SIMS VII,* (A. Benninghoven, C. A. Evans, K. D. McKeegan, H. A. Storms, and H. W. Werner, eds.), pp.851-854, John Wiley & Sons, NY (1990)
64. Schueler, B., Odom, R. W. and Chakel, J. A., *SIMS VIII,* (A. Benninghoven, C. A. Evans, K. D. McKeegan, H. A. Storms, and H. W. Werner, eds.), John Wiley & Sons, NY (in print 1992)

Part V.

Conclusions and Future Directions

13

Future Directions

Werner Kern

1.0 INTRODUCTION

The objective of this final chapter is a projection of the probable direction semiconductor wafer cleaning and its associated technologies are likely to take in the immediate future and up to the year 2000. The opinions expressed are those of the author and are based largely on the expertise and information presented by the writers of this book.*

We consider expected future developments by first defining the purity requirements for ultraclean silicon wafers and then discussing liquid-phase cleaning and gas-phase cleaning processes. We then examine the future prospects of technologies closely associated with wafer cleaning, namely process chemicals, wafer cleaning equipment, and contamination control.

The high degree of interest and activity in this entire field is manifested by the rapidly increasing number of publications and technical meetings. Scientific societies have planned several symposia and conferences that address various aspects of wafer cleaning science and technology, as well as associated surface chemistry, microcontamination control, and methodology for ultratrace impurity analysis. Forthcoming topical conferences have been announced by the European Solid State Device Research Conference (International Symposium of Ultraclean Processing of Silicon Surfaces), The Electrochemical Society, the Material Research Society, the Institute of Environmental Science, the Ultrafine Particle Society, the Microcontamination Conference, and others.

* No specific references are cited in this text; instead, the reader may consult the recent and pertinent ones in Sec. 4.4 of Ch. 1 (pp.47-54) A few recent references have been added in proof at the end of this chapter.

2.0 PURITY REQUIREMENTS FOR ULTRACLEAN SILICON WAFERS

Several key properties of ultraclean silicon wafers that are generally being considered requirements for the manufacture of 64-Mbit DRAMs, (Dynamic Random Access Memories) in the 1994 time frame are listed in Table 1. The numbers show that total surface metallic impurities must not be above 10^{10} surface atoms per cm^2, or above 10^{11} atoms per cm^3 in the bulk of the silicon crystal. Surface ions have to be 10 times lower. These statements are representative of those held by most experts.

Table 1. Future needs of the 1990s: Wafer Cleanliness Requirements

▲ Time frame 1994	▲ Device 64-Mbit DRAM
▲ Minimum feature 0.35 μm	▲ Particle size 0.11 μm
▲ Particle density 0.05-0.10/cm^2	▲ Particles per wafer 15-30/200 mm
▲ Surface metals $\leq 10^{10}/cm^2$	▲ Bulk metals $\leq 10^{11}/cm^3$
▲ Surface ions $\leq 10^{9}/cm^2$	▲ Common metals Cu, Fe, Cr, Ni, Zn, Al

The number of particles of 0.1 micron minimum size per 200 mm diameter wafer is listed generously as 10 to 30, but some experts specify only 10 per process step. The size of particles is critical in assessing their damage potential to IC functionality and impact on device yield. Table 2 relates particle size to minimum device feature size for circuits up to the year 2000, and defines what size particle should be considered a potential circuit "killer" defect.

Future studies are needed to correlate the relationship between chemical trace contaminants, their types and concentration levels, wafer cleaning operations, and the effects on finished devices. More attention needs to be paid to the often neglected wafer backside, which can be a serious source of recontamination.

Table 2. Killer Defect Size vs. Minimum Feature size. *(Source: SEMATECH)*

Year →	*1991*	*1994*	*1997*	*2000*
▲ DRAM (Mbit) generation	16	64	256	1000
▲ Min. feature size (μm)	0.5-0.6	0.35	0.25	0.15
▲ Killer defect size (μm)	0.5-0.2	0.035-0.12	0.025-0.08	0.015-0.05

3.0 FUTURE OF LIQUID-PHASE WAFER CLEANING PROCESSES

It can be safely predicted that liquid-phase wafer cleaning processes will continue to be used for the fabrication of less demanding semiconductor devices. The advantages of liquid-phase (wet-chemical) cleaning often outweigh their generic problems in many production applications. For example, the ability to remove both exposed and oxide-entrapped trace metallic impurities of many types by wet-chemical cleaning is unsurpassed by gas-phase processes. The same is true of the removal efficacy of particles from wafer surfaces by megasonic treatments with an aqueous reagent, such as diluted SC-1. Furthermore, liquid-phase procedures are material selective and simpler to implement in a production environment than dry cleaning methods, and they are more easily adaptable to multi-wafer batch processing. Improvements in the efficiency and performance of liquid-phase cleaning equipment will continue to be made, and refinements in the chemical processes will be introduced for specific applications. In short, the chip manufacturing industry will continue to utilize the well-established wet cleaning technology for many IC production lines for as long as it performs efficiently and more economically than other alternatives.

Specific changes in wet-chemical cleaning are likely to be introduced in the fab. The concentration of NH_4OH in the RCA 5:1:1 SC-1 solution will be reduced by at least 75% to avoid silicon micro-etching effects (surface microroughening); slightly milder processing conditions (10 min. at 70 - 75°C) will be used for the same reason.

The RCA SC-2 H_2O-H_2O_2-HCl solution may be replaced by dilute mineral acid (e.g., 1:10^{-4} H_2O-HCl) to avoid deposition of particles, if ultrahigh-purity chemicals are available for preparing the preceding SC-1 that then do not lead to wafer contamination by traces of certain metals (Al, Fe, Mg, Ca, Zn).

Etching before and after SC-1/SC-2 cleaning with superpure and particle-free aqueous dilute solutions of HF, HF-HCl, or HF-H_2O_2 will be more widely used to strip metallic contaminants trapped in oxide films on silicon wafer surfaces. HF-last immersion without water rinsing for passivation of the silicon surface by hydrogen will be used in some applications.

The benefits of ozonized DI water and various aqueous chemicals will be further explored, and the potentials of high-purity choline-surfactant-H_2O_2-H_2O mixtures are likely to be examined in more detail. Control of the zeta potential in cleaning solutions and on the wafer surface by chemical means may be introduced to minimize deposition of particles on wafers during cleaning treatments.

More extensive use will be made of megasonic cleaning in recirculating aqueous cleaning solutions at elevated temperatures and of overflow rinsing in ultrapure DI water to combine the removal of particles with chemical decontamination.

Semiconductor wafer drying after cleaning and rinsing in pure DI water is in a state of refinement and development. The IPA vapor-drying method is well established in industry, but potentially superior and more economically attractive alternatives are being explored for future applications, including wafer withdrawal from hot ultrapure DI water and the surface-tension-effect technique known as Marangoni drying.

Finally, the exact chemical reaction mechanism underlying the oxidation of silicon in aqueous oxidizing cleaning reagents should be investigated in detail to gain a better understanding of the process.

4.0 FUTURE OF GAS-PHASE WAFER CLEANING PROCESSES

There is no doubt that silicon wafer cleaning will be based on gas-phase processes for applications to advanced submicron-featured ICs with high aspect ratio structures, such as 1-Gbit DRAMs with 0.1 micron feature sizes and dielectric films of less than 10 nm thickness. The entire cleaning process will be incorporated in an integrated single-wafer vacuum system and be part of a process sequence.

The processes and techniques most likely to be used have been outlined in Ch. 1 and discussed further in Chs. 5, 6, 7, and 8; they are not, therefore, addressed here except for some general remarks.

We can expect to see refinements in well established processes, including UV/ozone for removal of organic contaminants, HF-vapor etching for particle-free removal of oxide films, and plasma etching (including ECR) and reactive ion etching methods for the removal of a variety of surface impurities. Remote reactions with hydrogen plasma have been demonstrated capable of producing atomically clean and hydrogen-passivated silicon (100)-surfaces, and may be adapted for full-scale chip manufacturing after development of suitable production equipment.

A great deal of research and development effort is required to come up with chemical gas-phase processes for reacting metallic trace contaminants and removing them by volatilization at sufficiently low temperature to prevent their diffusion into the semiconductor and thus adversely affecting device functionality. Feasibility for elimination, or at least for reduction of the concentration, of a few specific metals has been demonstrated by use of such reactions, which are usually based on the formation of volatile organometallic complexes.

Photochemically activated halide reactions are especially promising for the removal of trace metallic impurities. Incorporation of HF vapor in the reaction mixture (e.g., Cl_2) would make it possible to simultaneously etch oxide films and expose metals trapped in oxides. The main problem in this approach is to find ways to maintain a low temperature and to avoid etching of the semiconductor substrate.

5.0 FUTURE NEEDS OF PROCESSING CHEMICALS

Several key points concerning the most frequently used semiconductor liquid processing chemicals are summarized in Table 3. Gases are generally less of a problem and can be obtained in ultrahigh purity grades. Moisture in trace amounts is sometimes a critical impurity that must by carefully checked.

More widespread use of chelating reagents and surfactants in wet-chemical cleaning is expected, and certain aqueous cleaning solutions may be employed at higher levels of dilution than in the past. Photoresist chemicals are the dirtiest materials used in wafer processing and require rigorous purification; metallic impurities will have to be reduced to 10 ppb each by 1995, and particle concentrations to not more than 1 per ml for particles of $\geq$0.1 µm in size. The purity of DI water is constantly being improved in terms of contaminant concentration and particle count. Colloidal silica in DI water is an insidious impurity that is hard to detect and control, and has to be eliminated more effectively in future cleaning and rinsing applications prior to critical process steps.

Table 3. Future Needs of the 1990s: Processing Chemicals

- ▲ Further reduction of total metal content in liquid chemicals to 1 ppb, particles to < 10 per mL
- ▲ General reduction of liquid chemicals for processing and ecology reasons
- ▲ Shift from liquid to gaseous reactants
- ▲ Development of gaseous analogs to conventional liquid reactants in wafer cleaning operations

Sources of secondary contamination in the transferring and storing of ultrapure liquid process chemicals, listed in Table 4, must be controlled much more effectively than in the past to maintain the ultrahigh purity required for advanced wafer cleaning. Similar considerations hold for gaseous reactants.

Table 4. Secondary Contamination Sources of Ultrapure Liquid Chemicals

▲ Containers	▲ Valves
▲ Packaging	▲ Pumps
▲ Handling	▲ Seals
▲ Transfering	▲ Filter media
▲ Piping	▲ Dispensing

Guidelines for maximum allowable impurity levels of metals, anions, and particles in several high-purity liquid chemicals up to 1993 are listed in Table 5. Note that the minimum particle size specified in the density count is continuously decreasing with time. The most critical and detrimental types of wafer impurities that must be controlled vigorously are metals. Figure 1 depicts the trends of metallic total impurities in inorganic liquid chemicals as decreasing exponentially from a concentration of well over 10 ppm in 1985 to a projected average range of 0.1 ppb in 1994, a decrease of over 10^5. The concentration of trace *organic* contaminants in chemicals has been somewhat neglected in the past and should also be reduced substantially to avoid possibly deleterious effects.

The preparation of ultrapure, particle-free aqueous chemical solutions by in situ gas injection will become widespread. Reactant gases will include O_3, HCl, NH_3, HF, and others.

Finally, reprocessing of chemicals will become mandatory, since waste disposal will be not longer acceptable for ecological reasons. The recycling of chemicals will become economically more attractive because a high degree of ultrapurification can be achieved by reprocessing. The general trend in industry is a reduction in the use of liquid chemicals and their substitution with gaseous reactants.

Table 5. Maximum Impurity Guidelines for High-Purity Chemicals. *(Source: SEMATECH Working Group, Forecast to 1993.)*

Impurity Type (ppb)	*1988*	*1990*	*1993*
▲ Individual metals	<25	<1-3	<1
▲ Total metals	<200	<50	<25
▲ Individual anions	<10	<5	<2
▲ Total anions	<50	<25	<12
▲ Particles	<1000	<5600	<2700
(per L)	≥ 0.2μm	≥ 0.05μm	≥ 0.03μm

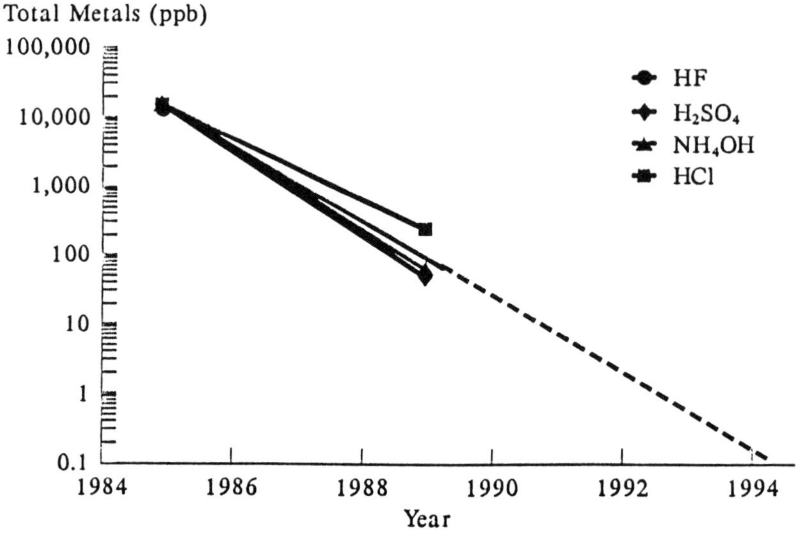

Figure 1. Trends of metallic impurities in inorganic liquid chemicals. *(Source: Adapted from V. Memon, Microcontamination Conf., Oct. 1990.)*

6.0 WAFER CLEANING EQUIPMENT

Several aspects that must be considered for refining existing equipment and for developing new types of cleaning systems for the future are listed in Table 6. The general trend is for designing fully automated ultrapure cleaning systems capable of performing a sequential series of liquid-phase or gas-phase cleaning operations. Flexibility and adaptability to different processes, as well as low cost of ownership, are also important considerations for the production engineer.

Further improvements and novel designs in sonic equipment for wet-chemical batch processing, and for liquid-displacement processors for single-wafer cleaning, are likely to be made in the next several years. Contamination-free wafer drying equipment technology is sure to see extensive development for full-scale manufacturing applications.

It has been pointed out in previous chapters that exposure of semiconductor wafers to room air during transference between individual processing systems can lead to recontamination and loss of controllability. For this reason, stringent measures must be imposed on certain critical processing steps, such as polysilicon gate and emitter formation, gate thermal oxidation, silicide formation, and the processing of device wafers with contact openings. Dry cleaning technology lends itself ideally for robotic and integrated processing where wafer cleaning can be combined with other process steps without exposing the wafers to room air, making it possible to establish and maintain ultraclean process ambients. The entire cleaning sequence can be conducted in situ by applying a sequence of gas or vapor-phase reactions at low or reduced pressure or in vacuum at elevated temperature in a multichamber or cluster tool that can be integrated with other cluster modules for film deposition and dry etching. Major innovations in this area of dry wafer cleaning equipment development can be expected to emerge in the next few years as a result of the intensive efforts that are now underway. It is most likely that these multichamber gas-phase cleaning systems and/or multiprocess cluster tools will be designed for single-wafer processing of 200-mm diameter and larger silicon wafers.

Table 6. Future Needs of the 1990s: Cleaning Equipment

- ▲ Fully automated system needed for gas-phase cleaning of one or several wafers
- ▲ Entire cleaning sequence to be conducted in this system
- ▲ System to be integrated or clustered with processing equipment
- ▲ Also, cleaner fab equipment required in general to reduce particle density

7.0 CONTROL OF MICROCONTAMINATION

The importance of controlling and preventing microcontamination in semiconductor wafer processing has been emphasized in several chapters of this book. Strictly speaking, contamination control is a different subject than wafer cleaning; nevertheless, it is an integral part of wafer cleaning technology. Future efforts should concentrate on contamination *prevention* technology, since it is less difficult to prevent contamination than to remove deposited impurities by cleaning. Therefore, wide-ranging measures must be taken to detect, identify, and reduce contaminants in wafer processing equipment, chemicals, and fab environment throughout the hundreds of processing steps needed in the manufacture of advanced integrated circuits. Future needs of the 1990s in this crucial area of technology are described in the summary in Table 7. As noted, the primary objective in controlling contamination is its prevention.

Table 7. Future Needs of the 1990s: Contamination Control and Analysis

- ▲ Objective: avoid wafer contamination by stringent control
- ▲ Reduce particles in equipment by scheduled maintenance and electrostatic charge removal
- ▲ Use high-purity, low-particulate processing chemicals, gases, and DI water for critical process steps
- ▲ Dispense chemicals by bulk-distributed delivery subsystems with point-of-use ultra filtration
- ▲ Optimize processes and gas-flow systems to eliminate or reduce particle generation
- ▲ Transfer and store wafers in a controlled microenvironment, such as SMIF (Standard Mechanical Interface)
- ▲ Monitor impurities on wafers in-line (e.g., particle counting)
- ▲ Measure ultratrace levels of contaminants on wafers by advanced instrumental analysis methods

In addition to the specifications listed in Table 7, consideration should also be given to airborne particles; they pose an ever present threat to the fab environment and must be removed by HEPA filtration of the air in chemical processing areas. The use of controlled, superpure isolation micro-environments, such as SMIF micro-enclosures, will become more prevalent for transferring and storing semiconductor wafers so as to effectively prevent their contamination from the ambient. Stringent control of electrostatics in and on equipment and processing surfaces etc. is a critical area that must not be overlooked to realize contamination-free superpure wafer processing. Careful attention must be given to the chemical cleaning of wafer cassettes on a scheduled basis to eliminate contaminants.

Finally, analytical methods will have to be improved significantly in the years to come to allow measurement of ultratrace levels of chemical contaminants on wafer surfaces and in processing liquids, gases, and materials. The present detection limit of many ultratrace impurities for the most advanced analytical systems, for example, is close to the level now achieved on ultraclean semiconductor surfaces, which is approaching 10^8 atoms/cm^2. The most sensitive surface analytical methods currently available for metallic ultratrace contaminants are SIMS, TXRF, and VPD/AAS, as described in Ch. 12. Future advances in the VPD (Vapor-Phase Decomposition) technique in combination with TXRF, SIMS, or ICP/MS analysis may offer the best chances for achieving increased sensitivity limits.

Nondestructive diagnostic electro-optical test methods, such as the successfully introduced surface-charge analysis (SCA, Ch. 11) will be applied more extensively to monitor surface contamination, the efficiency of cleaning processes, and process-induced damage on the production line. The combination of advanced electronic test methods with the instrumental ultratrace analytical methods described above will be an extremely powerful tool for detecting and measuring chemical contaminants.

Particle metrology requires major innovations to allow measurement of microparticles of sizes down to at least 0.05 µm; the limit at present is about 0.1 µm on planar surfaces and 0.01 µm in water. High-speed and low-cost in-line monitoring is required in submicron wafer processing for in situ detection and measurement of ultrafine particles. Reliable measurements on nonplanar, patterned product device wafers is highly desirable as well, preferably on an automated monitoring basis during actual processing. It is likely that new concepts need to be invented to develop practical instruments capable of detection, measurement, and inspection of wafers for submicron particles in semiconductor chip manufacturing.

8.0 SUMMARY AND CONCLUSIONS

The probable directions of semiconductor wafer cleaning and associated technologies have been projected up to the year 2000. The requirements for ultrapure silicon wafers have been summarized, and the future of liquid-phase cleaning and gas-phase cleaning processes for silicon wafers was discussed. Future developments for process chemicals and wafer cleaning equipment were then addressed. Contamination control was examined in a separate section in which we considered the future needs for the 1990s in ultraclean processing and the prevention of microcontamination. The future requirements for analytical ultramicro-methods for the detection and measurements of trace contaminants, the monitoring of surface contamination by use of electro-optical test methods, and the future of ultrafine particle measurements were then discussed.

We can conclude that for attaining the ultimate goal of ultraclean silicon wafers for IC chip manufacturing in the future the following key items need to be addressed:

1. Microcontamination must be controlled and prevented by use of superpure chemicals, equipment, and fab facilities to practically obviate wafer cleaning.
2. Innovative analytical methods must be developed for detecting, measuring, and monitoring ultratrace surface contaminants and microparticles on semiconductor wafers, devices, and in-process materials.
3. Semiconductor wafer cleaning science and technology must be improved substantially to meet the purity goals by the year 2000. New processes, new or improved equipment, and novel approaches are required.
4. Wet-chemical wafer cleaning will continue to be used for less critical devices. Modified and improved hydrogen peroxide-based cleaning solutions and dilute acids will be used in conjunction with megasonic treatments.
5. Gas-phase or dry cleaning methods will be needed for advanced device manufacturing. Established processes will be further refined and new processes need to be developed for the removal of trace metallic contaminants and particles from wafer surfaces. This area of wafer cleaning will continue to be in an especially evolutionary phase for years to come, since it involves new chemical processes, physical treatments, and implementation by integrated cluster tool equipment.

REFERENCES (July through December 1992)

1. Thirty-five papers presented at the *Fourth Symposium on Particles on Surfaces: Detection, Adhesion, and Removal,* The 23rd Annual Meeting of the Fine Particle Soc., Las Vegas, NV (July 13 - 17, 1992)
2. Hsieh, T. Y., Jung, K. H., Kwong, D. L., Koschmieder, T. H., and Thompson, J. C., Study of Rapid Thermal Precleaning for Si Epitaxial Growth, *J. Electrochem. Soc.,* 139(7):1971 - 1978 (July 1992)
3. Series of Short Papers and Comments on Cleaning Technologies, special issue of *Microcontamination,* 10(7):11-96 (July/August 1992)
4. Ma, Y., Yasuda, T., Habermehl, S., and Lucovsky, G., Si/SiO_2 Interfaces Formed by Remote Plasma-enhanced Chemical Vapor Deposition of SiO_2 on Plasma-Processed Si Substrates, *J. Vac. Sci. Technol.,* A10(4), Part I, 781-787 (July/Aug 1992)
5. Yapsir, A. S., Dry Surface Cleaning in Integrated Vacuum Reactive Ion Etching Processes. *ibid.*, pp. 792-794
6. Kasi, S. R., and Liehr, M., Preoxidation Si Cleaning and its Impact on Metal Oxide Semiconductor Characteristics. ibid., pp. 795-801
7. Haring, R. A. and Liehr, M., Reactivity of a Fluorine Passivated Silicon Surface. *ibid.*, pp. 802-805
8. Helms, C. R. and Deal, B. E., Mechanism of the HF/H_2O Vapor Phase Etching of SiO_2, *ibid.*, pp. 806-811
9. Huang, L. J. and Lau, M., Surface Electrical Properties of HF-Treated Si(100), *ibid.*, pp. 812-816
10. Thomas, R. E., Mantini, M. J., Rudder, R. A., Malta, D. P., and Hattangady, S. V., Carbon and Oxygen Removal from Silicon (100) Surfaces by Remote Plasma Cleaning Techniques, *ibid.*, pp. 817-822
11. Watanabe, J. K., and Gibson, U. J., Excimer Laser Cleaning and Processing of Si(100) Substrates in Ultrahigh Vacuum and Reactive Gases, *ibid.,* pp. 823-829
12. Ingrey, S., III-V Surface Processing, *ibid.,* pp. 829-836
13. Yota, J. and Borrows, V. A., Surface Chemistry of GaAs Treated with Buffered HF and NH_4F Solutions: Slow Reactions of Process Residuals, *ibid.,* pp. 837-842
14. Selwyn, G. S. and Patterson, E. F., Plasma Particulate Contamination Control. II. Self-Cleaning Tool Design, *ibid.,* pp. 1053-1059

15. O'Hanlon, J. F., and Parks, H. G., Impact of Vacuum Equipment Contamination on Semiconductor Yield, *J. Vac. Sci. Technol.,* A 10(4), Part II, 1863-1868 (July/Aug. 1992)

16. Sullivan, J., Schaffer, S., Manos, D., and Dylla, H. F., Contamination Control in the Design and Manufacture of Gas Flow Components, *ibid.,* pp. 1869-1874

17. Logan, J. S. and McGill, J. J., Study of Particle Emission in Vacuum from Film Deposits, *ibid.,* pp. 1875-1879

18. Seelmann-Eggebert, M., Carey, G., Krishnamurthy, V., and Helms, C. R., Effects of Cleanings on the Composition of HgCdTe Surfaces, *J. Vac. Sci. Technol.,* B 10(4):1297-1311 (July/Aug. 1992)

19. Kuiper, A. E. T., and Lathouwers, E. G. C., Room-Temperature HF Vapor-Phase Cleaning for Low-Pressure Chemical Vapor Deposition of Epitaxial Si and SiGe Layers, *J. Electrochem. Soc.,* 139(9):2594-2599 (Sept. 1992)

20. Baechle, T., Marvell, G, and Lynch, M., Evaluating the Capabilities of Minienvironments using Polished Silicon Monitor Wafers, *Microcontamination,* 10(8):35-39 (Sept. 1992)

21. Camenzind, M. J. and Balazs, M. K., TOC Levels in Incoming Chemicals, *Semiconductor International,* 15(10):89-92 (Sept. 1992)

22. Numerous Papers presented at the First International Symposium on Ultra Clean Processing of Silicon Substrates, *UCPSS '92 Extended Abstracts,* Leuven, Belgium (Sept. 17-19, 1992)

23. Kondo, I., Yoneyama, Kondo, K., and Takenaka, O., Effects of Different Pretreatments on the Surface Structure of Silicon and the Adhesion of Metal Films, *J. Vac. Sci. Technol.,* A10(5):3166-3170 (Sept./Oct. 1992)

24. Niwano, M., Suemitsu, M., Ishibashi, Y., Takeda, Y., Miyamoto, N., and Honma, K., Ultraviolet Ozone Oxidation of Si Surface Studied by Photoemission and Surface Infrared Spectroscopy, *ibid.,* pp. 3171-3175

25. Philipossian, A., The Activity of HF/H_2O Treated Silicon Surfaces in Ambient Air Before and After Gate Oxidation, *J. Electrochem. Soc.,* 139(10):2956-2961 (Oct. 1992)

26. Burggraaf, P., What's the Status of 200 mm Wafers? *Semiconductor International,* 15(11):48-50 (Oct. 1992)

27. Periasamy, R., Donovan, R. P., Clayton, A. C., and Ensor, D. S., Using Electrical Fields to Control Particle Deposition on Wafers in Vacuum Chambers, *Microcontamination,* 10(9):39-44 (Oct. 1992)

28. Steinman, A., Dealing with Electrostatic Charge: A Primer on the Invisible Contaminant and Air Ionization, *ibid.,* pp. 46-51

29. Batchelder, J. S., and Taubenblatt, M. A., Real-time Single Particle Composition Detection in Liquids, *Solid State Technology,* 35(10)S1-S9 (Oct. 1992)

30. Numerous papers presented at the *Microcontamination Conference '92,* Santa Clara, CA (Oct. 27-30, 1993), according to the following sessions:

    103: Particle Measurement, pp. 1-63

    104: Advanced Semiconductor Specialty Gas Technology, pp, 64-142

    108: Contamination Reduction in Equipment and Clean Rooms, pp. 143-180

    109: Disk Drives, pp. 181-223

    110: Limits of Detection of Analytical Methods, pp. 224-315

    201: The Practical Application of Minienvironments to High-Yield Semiconductor Manufacturing, pp. 316-363

    202: APIMS and Its Applications-I, pp. 364-400

    204: Process Improvements through Wafer Chemisphere Analysis, pp. 401-442

    205: APIMS and Its Applications-II, pp. 443-486

    301: Corrosive Gases: Analysis and Control of Purity Levels, pp. 487-542

    302: Developments in Wafer Cleaning Technology, pp. 543-613

    305: Ultrapure Process Chemicals, pp. 614-680

    306: Surface Analysis and Reduction of Contamination, pp. 681-728

    403: Ultrapure Water, pp. 729-789

    404: Process Optimization through Metrology, pp. 790-849

31. Ohmi, T., Imaoka, T., Sugiyama, I., and Kezuka, T., Metallic Impurities Segregation at the Interface between Si Wafer and Liquid during Wet Cleaning, *J. Electrochem. Soc.,* 139(11):3317-3335 (Nov. 1992)

32. Ohmi, T., Isagawa, T., Imaoka, T., and Sugiyama, I., Ozone Decomposition in Ultrapure Water and Continuous Ozone Sterilization for a Semiconductor Ultrapure Water System, *ibid.,* pp. 3336-3345

33. Peters, L., 20 Good Reasons to use In Situ Particle Monitors, *Semiconductor International,* 15(12):52-57 (Nov. 1992)

34. Hattori, T., Systems Approach Key to Attaining Cleanliness Goals, *Microcontamination,* 10(10):18-21 (Nov. 1992)

35. Papers related to Semiconductor Cleaning Processes presented at the *39th National Symposium of the American Vacuum Society*, Chicago, IL, (Nov. 9-13, 1992); Abstracts in Final Program Book, pp. 86,150-152,157,280,324,325,and 373-376

36. Surface-Chemical Papers, E5.11, E5.12,E5.26, E7.2, E7.7, E7.8, E7.9, E7.10, and E7.11 at the *1992 Fall Meeting of the Materials Research Society*, Boston, MA, (Nov. 30 - Dec. 4, 1992)

37. Hsu, E., Parks, H. G., Craigin, R., Tomooka, S, Ramberg, J. S., and Lowry, R. K., Deposition Characteristics of Metal Contaminants from HF-Based Process Solutions onto Silicon Wafer Surfaces, *J. Electrochem. Soc.,* 139(12):3659-3664 (Dec. 1992)

38. Jastrzebski, L., Henley, W., and Nuese, C. J., Surface Photovoltage Monitoring of Heavy Metal Contamination in IC Manufacturing, *Solid State Technology,* 35(2):27-35 (Dec. 1992)

39. Singer, P. H., Trends in Wafer Cleaning, *Semiconductor International,* 15(13):36-39 (Dec. 1992)

40. Ohmi, T. and Shibata, T., Scientific ULSI Manufacturing in the 21st Century, The Electrochemical Society *Interphase,* 1(1):32-37 (Winter 1992)

Index